Encyclopaedia of Mathematical Sciences
Volume 100

Mathematical Physics I

Subseries Editors:
J. Fröhlich S. P. Novikov D. Ruelle

Springer
Berlin
Heidelberg
New York
Barcelona
Hong Kong
London
Milan
Paris
Singapore
Tokyo

L. A. Bunimovich S. G. Dani R. L. Dobrushin
M. V. Jakobson I. P. Kornfeld N. B. Maslova
Ya. B. Pesin Ya. G. Sinai
J. Smillie Yu. M. Sukhov A. M. Vershik

Dynamical Systems, Ergodic Theory and Applications

Edited by
Ya. G. Sinai

Second, Expanded and Revised Edition
With 25 Figures

Springer

The first edition of this book was published as
Dynamical Systems II, Volume 2 of the Encyclopaedia of Mathematical Sciences

ISSN 0938-0396

ISBN 3-540-66316-9 Springer-Verlag Berlin Heidelberg New York

Printed in Germany
Typesetting: Camera ready by authors
Production: PRO EDIT GmbH, 69126 Heidelberg, Germany
Cover Design: E. Kirchner, Heidelberg, Germany
Printed on acid-free paper SPIN: 10654592 46/3143 hs 5 4 3 2 1 0

Publisher's Note

While work on this new expanded edition was progressing, Springer-Verlag implemented a new concept for the Encyclopaedia of Mathematical Sciences. Part of this is the new subseries *Mathematical Physics*. A consensus between the editor of this volume, the editors of this new subseries, and Springer-Verlag was quickly established that this volume should become part of the *Mathematical Physics* subseries.

December 1999

Preface to the Second Edition

The first edition of this Encyclopaedia volume was published as Encyclopaedia of Mathematical Sciences Volume 2, "Dynamical Systems II". For this second edition, published as the first volume of the "Mathematical Physics" subseries, two new parts have been added, comprising the contributions by S.G. Dani and J. Smillie.

R.L. Dobrushin and N.B. Maslova, who played a very essential role in the first edition, passed away during the last few years. Their contributions have been left unchanged. The parts by L.A. Bunimovich, M.V. Jakobson and Ya.B. Pesin were essentially revised, updated and extended. In the other contributions of the previous volume some additional references have been added and some stylistic changes have been carried out.

The authors would like to thank J. Mattingly for his critical reading of the manuscript.

December 1999 Ya.G. Sinai

Preface

Each author who took part in the creation of this issue intended, according to the idea of the whole edition, to present his understanding and impressions of the corresponding part of ergodic theory or its applications. Therefore the reader has an opportunity to get both concrete information concerning this quickly developing branch of mathematics and an impression about the variety of styles and tastes of workers in this field.

Ya.G. Sinai

List of Editor, Authors and Translators

Consulting Editor

Ya.G. Sinai, Princeton University, Dept. of Mathematics, 708 Fine Hall, Washington Road, Princeton, NJ 08544-1000, USA;
e-mail: sinai@math.princeton.edu;
Russian Academy of Sciences, L.D. Landau Institute for Theoretical Physics, ul. Kosygina 2, 117940 Moscow V-334, GSP-1, Russia

Authors

L.A. Bunimovich, Georgia Institute of Technology, School of Mathematics, Atlanta, GA 30332-016, USA; e-mail: bunimov@math.gatech.edu

S.G. Dani, School of Mathematics, Tata Institute of Fundamental Research, Homi Bhabha Road, Colaba, Bombay 400005, India;
e-mail: dani@math.tifr.res.in

R.L. Dobrushin †

M.V. Jakobson, Department of Mathematics, University of Maryland, College Park, MD 20742, USA; e-mail: mvy@math.umd.edu

I.P. Kornfeld, North Dakota State University, College of Science and Mathematics, Minard Hall 302C, Fargo, ND 58105, USA;
e-mail: kornfeld@plains.nodak.edu

N.B. Maslova †

Ya.B. Pesin, Department of Mathematics, Pennsylvania State University, University Park, PA 16802, USA; e-mail: pesin@math.psu.edu

Ya.G. Sinai, Princeton University, Dept. of Mathematics, 708 Fine Hall, Washington Road, Princeton, NJ 08544-1000, USA;
e-mail: sinai@math.princeton.edu;
Russian Academy of Sciences, L.D. Landau Institute for Theoretical Physics, ul. Kosygina 2, 117940 Moscow V-334, GSP-1, Russia

J. Smillie, Department of Mathematics, Cornell University, Malott Hall, Ithaca, NY 14853, USA; e-mail: smillie@math.cornell.edu

Y.M. Sukhov, Statistical Laboratory, DPMMS, University of Cambridge, 16 Mill Lane, Cambridge CB2 1SB, UK;
e-mail: y.m.suhov@statslab.cam.ac.uk

A.M. Vershik, Institute of Mathematics of the Russian Academy of Sciences, Fontanka 27, 191011 St. Petersburg, Russia; e-mail: vershik@pdmi.ras.ru

Translators

L.A. Bunimovich, Georgia Institute of Technology, School of Mathematics, Atlanta, GA 30332-016, USA; e-mail: bunimov@math.gatech.edu

M.V. Jakobson, Department of Mathematics, University of Maryland, College Park, MD 20742, USA; e-mail: mvy@math.umd.edu

I.P. Kornfeld, North Dakota State University, College of Science and Mathematics, Minard Hall 302C, Fargo, ND 58105, USA; e-mail: kornfeld@plains.nodak.edu

Y.M. Sukhov, Statistical Laboratory, DPMMS, University of Cambridge, 16 Mill Lane, Cambridge CB2 1SB, UK; e-mail: y.m.suhov@statslab.cam.ac.uk

Contents

I. General Ergodic Theory of Groups of Measure Preserving Transformations

Contents

Chapter 1
Basic Notions of Ergodic Theory and Examples of Dynamical Systems

I.P. Kornfeld, Ya.G. Sinai

§ 1. Dynamical Systems with Invariant Measures

Abstract ergodic theory deals with the measurable actions of groups and semigroups of transformations. This means, from the point of view of applications, that the functions defining such transformations need not satisfy any smoothness conditions and should be only measurable.

A pair $(M, \mathscr{M})$ where M is an abstract set and $\mathscr{M}$ is some σ-algebra of subsets of M, is called a measurable space. In the sequel M will be the phase space of a dynamical system. The choice of $\mathcal{M}$ will always be clear from the context. We shall make use of the notions of the direct product of measurable spaces and of $\mathcal{M}$-measurable functions.

Definition 1.1. A transformation $T: M \to M$ is measurable if $T^{-1}C \in \mathcal{M}$ for any $C \in \mathcal{M}$.

A measurable transformation T is also called *an endomorphism of the measurable space* $(M, \mathcal{M})$. Any endomorphism generates a cyclic semigroup $\{T^n\}$ of endomorphisms ($n = 0, 1, 2, \dots$).

If T is invertible and T^{-1} (as well as T) is measurable, then T is said to be *an automorphism of the measurable space* $(M, \mathcal{M})$. Any automorphism generates the cyclic group $\{T^n\}$ of automorphisms, $-\infty < n < \infty$.

A natural generalization of the above notions can be achieved by considering an arbitrary countable group or semigroup G and by fixing for each $g \in G$ a measurable transformation T_g such that $T_{g_1} \cdot T_{g_2} = T_{g_1 g_2}$ for all $g_1, g_2 \in G$, $T_e = id$.

Definition 1.2. The family $\{T_g\}$, $g \in G$, is said to be a *measurable action* of the countable group (semigroup) G.

The simplest example is as follows. Suppose that $(X, \mathcal{X})$ is a measurable space and M is the space of all X-valued functions on G, i.e. any $x \in M$ is a sequence $\{x_g\}$, $x_g \in X$, $g \in G$. For any $g_0 \in G$ define the transformation $T_{g_0} : M \to M$ by the formula $T_{g_0}x = x'$, where $x'_g = x_{g_0 g}$. In this case $\{T_g\}$ is called a group (semigroup) of shifts. In particular,

1) if G is the semigroup $\mathbb{Z}^1_+ = \{n: n \geqslant 0, n \text{ is an integer}\}$, then M is the space of all 1-sided X-valued sequences, i.e. the points $x \in M$ are of the form $x = \{x_n\}$, $x_n \in X$, $n \geqslant 0$, and $T_m x = \{x_{n+m}\}$, $m \in \mathbb{Z}^1_+$. T_1 is called a 1-sided shift.

2) if G is the group $\mathbb{Z}^1 = \{n: -\infty < n < \infty, n \text{ is an integer}\}$, then M is the space of all 2-sided sequences $x = \{x_n\}$, $x_n \in X$, $-\infty < n < \infty$, and $T_m x = \{x_{n+m}\}$, $m \in \mathbb{Z}^1$. T_1 is called a 2-sided shift, or, simply, a shift.

3) if $G = \mathbb{Z}^d = \{(n_1, n_2, \dots, n_d): n_i \in \mathbb{Z}^1, 1 \leqslant i \leqslant d\}$, $d \geqslant 1$, then M is the space of all sequences x of the form $x = \{x_n\} = \{x_{n_1,\dots,n_d}\}$, while $T^m x = \{x_{n+m}\}$, $m = \{m_1, \dots, m_d\} \in \mathbb{Z}^d$.

The above examples arise naturally in probability theory, where the role of M is played by the space of all realizations of d-dimensional random field.

Now suppose G is an arbitrary group or semigroup endowed with the structure of measurable space $(G, \mathcal{G})$ compatible with its group structure, i.e. all transformations $T_{g_0}: g \mapsto g_0 g$ $(g, g_0 \in G)$ are measurable.

Definition 1.3. The family $\{T_g : M \to M\}$, $g \in G$, where G is a measurable group, is called a *measurable action of the group* G (or a G-*flow*) if

1) $T_{g_1} \cdot T_{g_2} = T_{g_1 g_2}$ for all $g_1, g_2 \in G$;

2) for any $\mathcal{M}$-measurable function $f: M \to \mathbb{R}^1$ the function $f(T_g x)$ considered as a function on the direct product $(M, \mathcal{M}) \times (G, \mathcal{G})$ is also measurable.

Our main example is $G = \mathbb{R}^1$ with the Borel σ-algebra of subsets of $\mathbb{R}^1$ as $\mathcal{G}$. There also exist natural examples with $G = \mathbb{R}^d$, $d > 1$ (cf Chap. 12).

Let now $G = \mathbb{R}^1$. If T^t is the transformation in $\mathbb{R}^1$-flow corresponding to a $t \in \mathbb{R}^1$, then we have T^{t_1}. $T^{t_2} = T^{t_1+t_2}$. We will describe a natural situation in which the actions of $\mathbb{R}^1$ arise.

Suppose M is a smooth compact manifold and α is a smooth vector field on M. Consider the transformation T^t sending each point $x \in M$ to the point $T^t x$ which can be obtained from x by moving x along the trajectory of α for the period of time t (T^t is well defined because of compactness of M). Then $T^{t_1+t_2} = T^{t_1} \cdot T^{t_2}$ and T^t is a measurable action of $\mathbb{R}^1$.

Measurable actions of $\mathbb{R}^1$ are usually called *flows*, and those of $\mathbb{R}^1_+$—*semiflows*. The cyclic groups and semigroups of measurable transformations are also known as *dynamical systems with discrete time*, while flows and semiflows are known as *dynamical systems with continuous time.*

Now, let $(M, \mathcal{M}, \mu)$ be a measure space (probability space), i.e. $(M, \mathcal{M})$ is a measurable space and μ is a nonnegative normalized ($\mu(M) = 1$) measure on $\mathcal{M}$. Consider a measure ν on $\mathcal{M}$ given by $\nu(C) = \mu(T^{-1}C)$, $C \in \mathcal{M}$. This measure is said to be the image of the measure μ under T (notation: $\nu = T\mu$).

Definition 1.4. A measure μ is *invariant* under a measurable transformation $T: M \to M$ if $T\mu = \mu$.

If μ is invariant under T, then T is called an *endomorphism of the measure space* $(M, \mathcal{M}, \mu)$. If, in addition, T is invertible, it is called an *automorphism* of $(M, \mathcal{M}, \mu)$. If $\{T^t\}$ is a measurable action of $\mathbb{R}^1$ and each T^t, $-\infty < t < \infty$, preserves the measure μ, then $\{T^t\}$ is called a *flow* on the measure space $(M, \mathcal{M}, \mu)$.

Now consider the general case.

Definition 1.5. Let $\{T_g\}$ be a measurable action of a measurable group $(G, \mathcal{G})$ on the space $(M, \mathcal{M})$. A measure μ on $\mathcal{M}$ is called *invariant* under this action if, for any $g \in G$, μ is invariant under T_g.

We now introduce the general notion of metric isomorphism of dynamical systems which allows us to identify systems having similar metric properties.

Definition 1.6. Suppose $(G, \mathcal{G})$ is a measurable group and $\{T_g^{(1)}\}$, $\{T_g^{(2)}\}$ are two G-flows acting on $(M_1, \mathcal{M}_1)$, $(M_2, \mathcal{M}_2)$ respectively and having invariant measures μ_1, μ_2. Such flows are said to be *metrically isomorphic* if there exist G-invariant subsets $M_1' \subset M_1$, $M_2' \subset M_2$, $\mu_1(M_1') = \mu_2(M_2') = 1$, as well as an isomorphism $\varphi: (M_1', \mathcal{M}_1, \mu_1) \to (M_2', \mathcal{M}_2, \mu_2)$ of measure spaces M_1', M_2' such that $T_g^{(2)} \varphi x^{(1)} = \varphi T_g^{(1)} x^{(1)}$ for all $g \in G$, $x^{(1)} \in M_1'$.

Ergodic theory also studies measurable actions of groups on the space $(M, \mathcal{M}, \mu)$ which are not necessarily measure-preserving.

Definition 1.7. Suppose $\{T_g\}$ is a measurable action of a measurable group $(G, \mathcal{G})$ on $(M, \mathcal{M})$. The measure μ on $\mathcal{M}$ is said to be *quasi-invariant* under this action if for any $g \in G$ the measure $\mu_g \overset{\text{def}}{=} T_g \mu$, i.e. the image of μ under T_g, is equivalent to μ. In other words, μ and $T_g \mu$ have the same sets of zero measure.

In the case of quasi-invariant measure the notions of endomorphisms, automorphisms, G-flows can also be introduced in a natural way. The metric isomorphism of groups of transformations with quasi-invariant measure is naturally defined "up to a transformation with quasi-invariant measure", i.e. our requirement imposed on the transformation φ in Definition 1.6 is that the measure $\varphi\mu_1$ should be equivalent but not necessarily equal to μ_2.

The basic properties of transformations with invariant measure will be discussed in Section 2, and now we turn to the problem of existence of such measures.

Suppose α is a smooth vector field on a m-dimensional manifold M, $\{T^t\}$ is the corresponding group of shifts along the trajectories of α, and μ is an absolutely continuous measure, i.e. in any system of local coordinates $(x_1, \ldots, x_m)$ μ is given by its density: $d\mu = \rho(x_1, \ldots, x_m)\, dx_1 \ldots dx_m$.

The well-known Liouville's theorem says that μ is invariant under $\{T^t\}$ if ρ satisfies the Liouville equation: $\operatorname{div}(\rho\alpha) = 0$. Such a measure μ is known as a Liouville measure, or integral invariant of the dynamical system $\{T^t\}$. We will give now some applications of Liouville's theorem.

1. Let M be a $2m$-dimensional symplectic manifold and let the vector field α be given by a Hamilton function H not depending on time. In a local system of coordinates $(q^1, \ldots, q^m, p_1, \ldots, p_m)$ such that the symplectic form ω can be written as $\omega = \sum_{i=1}^m dq^1 \wedge dp_i$, the vector field α is given by the Hamilton system of equations

$$\frac{dq^i}{dt} = \frac{\partial H}{\partial p_i}, \qquad \frac{dp_i}{dt} = -\frac{\partial H}{\partial q^i} \tag{1.1}$$

The measure μ with density $\rho(q, p) = 1$ is invariant.

Dynamical systems given by the equations (1.1) are called *Hamiltonian dynamical systems*. The class of such systems includes, in particular, geodesic flows which frequently appear in applications. They can be introduced by the following construction.

Let Q be a smooth compact m-dimensional Riemannian manifold. Denote by $\mathscr{T}_q(\mathscr{T}_q^*)$ the tangent (cotangent) space to Q at a point $q \in Q$ and consider the unit tangent bundle $M = \{(q, v): q \in Q, v \in \mathscr{T}_q, \|v\| = 1\}$ whose points are called linear elements on Q.

The geodesic flow on Q is a group $\{T^t\}$ of transformations of M such that a specific transformation T^t consists in moving a linear element $x = (q, v)$ along the geodesic line which it determines, by a distance t. The measure μ on M with $d\mu = d\sigma(q)\, d\omega_q$, where $d\sigma(q)$ is the element of the Riemann volume, and ω_q is the Lebesgue measure on the unit sphere S^{m-1} in $\mathscr{T}_q$, is invariant under $\{T^t\}$.

Another way to introduce the geodesic flow is as follows:

The tangent bundle $\mathscr{T}_q = \{(q, v): q \in Q,\ v \in \mathscr{T}_q\}$ may be naturally identified with the cotangent bundle $\mathscr{T}^*Q = \{(q, p): q \in Q, p \in \mathscr{T}_q^*\}$. Let $(q^1, \ldots, q^m)$ be the local coordinates at the point $q \in Q$. Each point $p \in \mathscr{T}_q^*$ is uniquely determined by its components $(p_1, \ldots, p_m)$. The nondegenerate differential 2-form $\omega = \sum_{i=1}^m dq^i \wedge dp_i$ induces the symplectic structure on $\mathcal{T}^*Q$, and the geodesic flow $\{T^t\}$ which we have just introduced is naturally isomorphic to the restriction to the unit tangent bundle of the Hamiltonian dynamical system with Hamiltonian $H(q, p) = \frac{1}{2}\|p\|^2$. Ergodic properties of the geodesic flow are uniquely determined by the Riemannian structure on Q.

2. Let $M = \mathbb{R}^m \times \mathbb{R}^m$ be the tangent bundle over $\mathbb{R}^m$, $m \geqslant 1$, and the vector field $\alpha = (\alpha_1, \ldots, \alpha_m)$ be given by

$$\frac{dx_i}{dt} = y_i, \qquad \frac{dy_i}{dt} = \alpha_i(x_1, \ldots, x_m), \qquad 1 \leqslant i \leqslant m, \tag{1.2}$$

or

$$\frac{d^2 x_i}{dt^2} = \alpha_i(x_1, \ldots, x_m).$$

The measure μ with $d\mu = dx_1 \ldots dx_m \, d\dot{x}_1 \ldots d\dot{x}_m$ is invariant under the group $\{T^t\}$ of translations along the solutions of the system (1.2).

3. $M = \mathbb{R}^m$, $m \geqslant 1$, and the vector field α is given by

$$\frac{dx_i}{dt} = \alpha_i(x_1, \ldots, x_m), \qquad 1 \leqslant i \leqslant m,$$

and $\operatorname{div} \alpha = \sum_{i=1}^m \frac{\partial \alpha_i}{\partial x_i} = 0$. Then the measure μ with $d\mu = dx_1 \ldots dx_m$ will be invariant under the action of the group $\{T^t\}$ corresponding to the vector field α. The important case is $m = 3$. The trajectories of the vector field are in this case the magnetic flux of the stationary magnetic field with tension α. The equality $\operatorname{div} \alpha = 0$ is one of the Maxwell equations.

Remark. It was mentioned above that the invariant measure given by Liouville's theorem is usually infinite. If the system under consideration has a prime integral $H(x)$ with compact "level surfaces" $M_c = \{x \in M : H(x) = c\}$, $c \in \mathbb{R}^1$, a finite invariant measure can be constructed out of this infinite measure. In this case any trajectory lies on a single manifold M_c. If $\rho(x)$ is the density of the Liouville measure, the measure μ_c concentrated on M_c with $d\mu_c = \frac{1}{|\nabla H|} \rho \, d\sigma$, where $d\sigma$ is the element of Riemann volume, will also be invariant. This measure may turn out to be finite and, therefore, can be normalized. The main example: for a Hamiltonian system (cf Example 1) the Hamilton function H itself is a prime integral. The measure μ_c induced by the Liouville measure on the surface $H = c$ is called a microcanonical distribution.

4. *Billiards*. Suppose Q_0 is a closed m-dimensional Riemannian manifold of class C^∞, and Q is a subset of Q_0 given by the system of inequalities of the form $f_i(q) \geqslant 0$, $q \in Q_0$, $f_i \in C^\infty(Q_0)$, $1 \leqslant i \leqslant r < \infty$. The phase space of

the billiard in Q is the set M whose points are the linear elements $x = (q, v)$, $x \in \operatorname{Int} Q$, $v \in S^{m-1}$, as well as those $x = (q, v)$ for which $x \in \partial Q$, $v \in S^{m-1}$ and v is directed inside Q. The motion of a point $x = (q, v)$ under the billiard flow is the motion with unit speed along the trajectory of the geodesic flow until the boundary ∂Q is reached. At such moments the point reflects from the boundary according to the "incidence angle equals reflection angle" rule and then continues its motion. The measure μ on M such that $d\mu = d\rho(q) d\omega_q$, where $d\rho(q)$ is the element of Riemann volume, ω_q is the Lebesgue measure on S^{m-1}, is invariant under $\{T^t\}$. A more detailed construction of billiard flow see in [CFS].

5. M is a commutative compact group, μ is the Haar measure on M. If T is a group translation on M, i.e. the transformation of the form $Tx = x + g$, $(x, g \in M)$ then μ is invariant under T. This fact follows immediately from the definition of Haar measure.

Now let T be a group automorphism of the group M, i.e. T is a continuous and one-to-one mapping of M onto itself such that $T(x_1 + x_2) = Tx_1 + Tx_2$ for all $x_1, x_2 \in M$. The invariance of μ under T follows in this case from the uniqueness of the Haar measure.

6. Suppose $(X, \mathscr{X}, \lambda)$ is a probability space, M is the space of all sequences of the form $x = \{x_n\}$ where $x_n \in X$, $n \in \mathbb{Z}^1$ or $n \in \mathbb{Z}^1_+$, T is the shift on M, i.e. $Tx = x'$, $x'_n = x_{n+1}$. Let a measure μ on M be the product-measure of measure λ. In other words, the random variables x_n are mutually independent and have the distribution λ. The shift T is called in this case *a Bernoulli automorphism* (*endomorphism*). It gives us one of the most important examples of automorphisms (endomorphisms) in ergodic theory. The space $(X, \mathscr{X}, \lambda)$ is called *the state space* of Bernoulli automorphism (endomorphism).

7. The above example can be generalized as follows. Suppose $(Y, \mathscr{Y}, \lambda)$ is the probability space and we are given the transition operator $P_y(\cdot)$. This means that for any $y \in Y$ there is a probability measure P_y on $\mathscr{Y}$ and the family of all P_y is measurable in the sense that for any $\mathscr{Y}$-measurable function f the integral $\int_Y f(z)\, d P_y(z)$ is also a $\mathscr{Y}$-measurable function on Y. Assume further that λ is an invariant measure for $P_y(\cdot)$, i.e.

$$\lambda(C) = \int_Y P_y(C)\, d\lambda(y). \tag{1.3}$$

The space M of all sequences $x = \{y_n\}$, $y_n \in Y$, $n \in \mathbb{Z}^1$ or $n \in \mathbb{Z}^1_+$, and the shift T on M are just as in the previous example, but the construction of the measure μ differs from it. The measure is now given by

$$\mu(\{y\colon y_i \in C_0, y_{i+1} \in C_1, \dots, y_{i+k} \in C_k\})$$
$$= \int_{C_0 \times C_1 \times \dots C_{k-1}} d\lambda(y_i)\, dP_{y_i}(y_{i+1}) \cdot \ldots \cdot dP_{y_{i+k-2}}(y_{i+k-1}) \cdot P_{y_{i+k-1}}(C_k),$$

where $C_0, C_1, \dots, C_k \in \mathscr{Y}$, $1 \leqslant k < \infty$.

It follows from (1.3) that this measure is invariant under T. The transformation T is called in this case *a Markov automorphism* (*endomorphism*).

There are many groups and semigroups of transformations for which the existence of at least one invariant measure is not self-evident. We will present now the general approach to this problem due to N.N. Bogolyubov and N.M. Krylov.

Suppose M is a compact metric space, $\mathcal{M}$ is its Borel σ-algebra and $T: M \to M$ is a continuous mapping.

Theorem 1.1 (N.N. Bogolyubov, N.M. Krylov [BK]). *There exists at least one normalized Borel measure invariant under T.*

Indeed, let μ be an arbitrary normalized Borel measure on $\mathcal{M}$. For $n = 1, 2, \ldots$ consider the measures μ_n given by $\mu_n(C) = \mu_0(T^{-n}C)$, $C \in \mathcal{M}$, as well as $\mu^{(n)} = \frac{1}{n}\sum_{k=0}^{n-1}\mu_k$. The compactness of M implies that the space of all normalized Borel measures on M is weakly compact. Therefore the sequence $\{n_s\}$ of integers, $n_s \to \infty$ as $s \to \infty$, exists such that $\mu^{(n_s)}$ weakly converges as $s \to \infty$ to some measure μ. This limit measure μ will be invariant: for any continuous function f on M we have

$$\begin{aligned}\int_M f(Tx)\,d\mu &= \lim_{s\to\infty}\int_M f(Tx)\,d\mu^{(n_s)}\\ &= \lim_{s\to\infty}\frac{1}{n_s}\sum_{k=0}^{n_s-1}\int_M f(Tx)\,d\mu_k = \lim_{s\to\infty}\frac{1}{n_s}\sum_{k=0}^{n_s-1}\int_M f(x)\,d\mu_{k+1}\\ &= \lim_{s\to\infty}\frac{1}{n_s}\sum_{k=1}^{n_s}\int_M f(x)\,d\mu_k = \lim_{s\to\infty}\frac{1}{n_s}\sum_{k=0}^{n_s-1}\int_M f(x)\,d\mu_k\\ &= \int_M f(x)\,d\mu,\end{aligned}$$

which obviously signifies the invariance of μ.

If $\{T^t\}$ is a continuous one-parameter group of homeomorphisms of a compact metric space, one can use the same argument to prove that $\{T^t\}$ has at least one invariant measure.

Definition 1.8. A homeomorphism T of a compact metric space M is said to be *uniquely ergodic* if it has precisely one normalized Borel invariant measure. A homeomorphism T is said to be minimal if the trajectory $\{T^n x\colon -\infty < n < \infty\}$ of any point $x \in M$ is dense in M. A homeomorphism T is said to be topologically transitive if the trajectory of some point $x \in M$ is dense in M.

The notions just introduced characterize in different senses the property of "topological nondecomposability" of T. These notions are sometimes being applied not only to homeomorphisms but also to more general Borel transformations of topological spaces. In Section 3 for the dynamical systems on general measure spaces we shall introduce the notion of ergodicity playing the central role in ergodic theory and characterizing the "nondecomposability" in measure theoretical sense.

The properties of unique ergodicity and minimality are close to each other in the sense that there is a great number of natural examples for which both of them are satisfied or not satisfied simultaneously. However, in the general case neither of them implies the other. The minimality of T means that T has no nontrivial invariant closed set. The following theorem makes clear the meaning of the notion of unique ergodicity.

Theorem 1.2 (J.C. Oxtoby [O]). *Suppose T is a homeomorphism of the compact metric space M and μ is a normalized Borel measure invariant under T. The following statements are equivalent:*

1) *T is uniquely ergodic;*
2) *for any continuous function f on M and any $x \in M$ one has*

$$\lim_{n\to\infty} \frac{1}{n}\sum_{k=0}^{n-1} f(T^k x) = \int_M f\,d\mu;$$

3) *for any continuous function f on M the convergence $\frac{1}{n}\sum_{k=0}^{n-1} f(T^k x) \to \int_M f\,d\mu$ is uniform on M.*

In situations where the Krylov-Bogolyubov theory is applicable it may occur that many invariant measures exist. It seems useful to have a criterion for indicating the most important invariant measures. A new approach to this problem appeared recently in connection with progress in the theory of hyperbolic dynamical systems (cf Part II).

Suppose M is a smooth manifold; T: $M \to M$ is a diffeomorphism, ($\{T^t\}$ is a 1-parameter group of translations along the trajectories of some smooth vector field on M). Consider any absolutely continuous measure μ on M and its translations μ_n, $\mu_n(C) = \mu_0(T^{-n}C)$ (in the case of discrete time), μ_t, $\mu_t(C) = \mu_0(T^{-t}C)$ (in the case of continuous time). It may occur that the translated measures $\mu_n(\mu_t)$ converge to some limit measure μ as $n \to \infty$ (respectively, $t \to \infty$) and μ does not depend on the choice of the initial measure μ_0. This limit measure μ will necessarily be invariant, and it may be considered as the most important invariant measure for the dynamical system under consideration.

The dissipative systems may also possess such measures. Suppose $\{T^t\}$ is a 1-parameter group of translations along the solutions of the system of differential equations

$$\frac{dx_i}{dt} = f_i(x_1,\dots,x_m), \qquad 1 \leqslant i \leqslant m < \infty. \tag{1.4}$$

The system (1.4) is said to be dissipative if $\operatorname{div} f = \sum_{i=1}^{m} \frac{\partial f_i}{\partial x_i} < 0$. Liouville's theorem mentioned above implies that for any $C_0 \subset \mathbb{R}^m$ the m-dimensional volume of the set $C_t = T^t C_0$ decreases in time and tends to zero. This fact, however, by no means signifies that the initial measures become degenerate in some sense under the action of dynamics. On the contrary, they may converge to some nontrivial measures concentrated on the invariant sets whose m-dimensional volume vanish. The situation may be interpreted by saying that the dynamics itself produces a natural invariant measure as a result of the evolution of smooth measures. Such a situation arises frequently in many problems concerning so called strange attractors (see Part II Chap. 7). The question of the existence of such measures was raised repeatedly in connection with mathematical approaches to the analysis of turbulence.

Remark on Lebesgue spaces. A number of results presented below necessitate the assumption that the phase space of the dynamical system is the so-called *Lebesgue space.* We shall not give the full definition of the Lebesgue space (cf [Ro2]). Notice only that among all nonatomic measure spaces the Lebesgue spaces are precisely the ones which are isomorphic to the closed interval [0,1] with the Lebesgue measure. The assumption that the measure space is Lebesgue is not restrictive: the spaces arising in applications are, as a rule, automatically Lebesgue. In particular, any separable metric space with a measure defined on the Borel σ-algebra is Lebesgue. On the other hand, under this (Lebesgue) assumption one can apply the theory of measurable partitions which does not exist in the general case.

The partition of the space $(M, \mathcal{M}, \mu)$ is a system $\xi = \{C\}$ of nonempty and pairwise nonintersecting measurable sets such that $\bigcup_{C \in \xi} C = M$. The fact that ξ is measurable means that one can define the measures μ_C on its elements $C \in \xi$ in such a way that they play the role of conditional probabilities. The formal definition is as follows:

Definition 1.9. By a *canonical system of conditional measures* belonging to the partition ξ we mean a system of measures $\{\mu_C\}$, $C \in \xi$, possessing the following properties:

1) μ_C is defined on some σ-algebra $\mathcal{M}_C$ of subsets of C;

2) the space $(C, \mathcal{M}_C, \mu_C)$ is Lebesgue;

3) for any $A \in \mathcal{M}$ the set $A \cap C$ belongs to $\mathcal{M}_C$ for almost all $C \in \xi$, the function $f(x) = \mu_{C_\xi(x)}(A \cap C_\xi(x))$, where $C_\xi(x)$ is the element of ξ containing $x \in M$, is measurable and

$$\mu(A) = \int \mu_{C_\xi(x)}(A \cap C_\xi(x))\, d\mu.$$

A partition ξ is said to be *measurable* if it possesses a canonical system of conditional measures. If $\{\mu_C\}$, $\{\mu'_C\}$ are two canonical systems for a partition ξ, then $\mu_C = \mu'_C$ for almost all $C \in \xi$; in this sense the canonical system for ξ is unique.

§ 2. First Corollaries of the Existence of Invariant Measures. Ergodic Theorems[1]

The following theorem due to H. Poincaré gives the basic information about the behavior of trajectories of transformations with invariant measures.

Theorem 2.1 (Poincaré Recurrence Theorem) (cf [CFS]). *Suppose* $(M, \mathcal{M}, \mu)$ *is a measure space,* $T\colon M \to M$ *is its endomorphism. Then for any* $C \in \mathcal{M}$, $\mu(C) > 0$, *almost all points* $x \in C$ *return to* C *infinitely many times. In other words, there exists an infinite sequence* $\{n_i\}$ *of integers,* $n_i \to \infty$ *as* $i \to \infty$, *such that* $T^{n_i}x \in C$. *Therefore, any set* $A \in \mathcal{M}$ *for which* $T^n A \cap A = \varnothing$ *for all* n *sufficiently large, has zero* μ*-measure.*

The so-called "Zermelo paradox" in statistical mechanics is closely related to the Poincaré recurrence theorem. Consider the closed box containing N pointlike masses (molecules), that move under the action of interaction forces and reflect elastically from the boundaries. The differential equations describing the dynamics of such a system are Hamiltonian, so the one-parameter group of translations along the trajectories of this system preserves the Liouville measure. The manifolds of constant energy for the system under consideration are compact, and the Liouville measure induces the finite invariant measures concentrated on them. Therefore the Poincaré recurrence theorem may be applied. Suppose now that the set $C \subset M$ consists of such points of the phase space for which all molecules are located in one half of the box at the moment $t = 0$. The Poincaré recurrence theorem says that at some moments $t > 0$ all the moving molecules will be again in the same half of the box. At first glance we have a contradiction, since nobody has ever seen a gas not occupying entirely the volume available.

However, this phenomenon may be explained quite simply. The probability of the event C is estimated as $\exp(-\text{const} \cdot N)$, where N is the total number of molecules, and const depends on temperature, density and so on. In real conditions we have $N \sim 10^{23}$ molecules/cm^3, so $\mu(C)$ is extremely small. It will be shown later (see Sect. 4) that the time intervals between two consecutive realizations of the event C are estimated as $[\mu(C)]^{-1}$, so in our case they are extremely large. On the other hand, if N is sufficiently small, for example, if $N \sim 10$, it is quite possible that at some moment $t > 0$ all the molecules will be again in one half of the box. One can realize such an experiment with the help of a computer.

The important consequence of the existence of an invariant measure for a given dynamical system is the possibility of averaging over time (for the functions on the phase space). The following theorem is one of the cornerstones of ergodic theory.

[1] This section was written in collaboration with Ya.B. Pesin.

Theorem 2.2 (Birkhoff-Khinchin Ergodic Theorem, (cf [CFS]). *Suppose $(M, \mathcal{M}, \mu)$ is a space with normalized measure and $f \in L^1(M, \mathcal{M}, \mu)$. Then for almost every $x \in M$ the following limits exist:*

1) in the case of an endomorphism T

$$\lim_{n\to\infty} \frac{1}{n} \sum_{k=0}^{n-1} f(T^k x) \stackrel{\text{def}}{=} \bar{f}(x);$$

2) in the case of an automorphism T

$$\lim_{n\to\infty} \frac{1}{n} \sum_{k=0}^{n-1} f(T^k x) = \lim_{n\to\infty} \frac{1}{n} \sum_{k=0}^{n-1} f(T^{-k} x) \stackrel{\text{def}}{=} \bar{f}(x);$$

3) in the case of a flow $\{T^t\}$

$$\lim_{T\to\infty} \frac{1}{T} \int_0^T f(T^t x)\, dt = \lim_{T\to\infty} \frac{1}{T} \int_0^T f(T^{-t} x)\, dt \stackrel{\text{def}}{=} \bar{f}(x);$$

4) in the case of a semiflow $\{T^t\}$

$$\lim_{T\to\infty} \frac{1}{T} \int_0^T f(T^t x)\, dt \stackrel{\text{def}}{=} \bar{f}(x).$$

Moreover, $\bar{f} \in L^1(M, \mathcal{M}, \mu)$ and $\int_M \bar{f}\, d\mu = \int_M f\, d\mu$. The function $\bar{f}$ is invariant, i.e. $\bar{f}(T^n x) = \bar{f}(x)$ for $n \geqslant 0$ in the case of an endomorphism and for $-\infty < n < \infty$ in the case of an automorphism; $\bar{f}(T^t x) = \bar{f}(x)$ for $-\infty < t < \infty$ in the case of a flow and for $t \geqslant 0$ in the case of a semiflow.

Consider now the case when $f = \chi_C$, i.e. f is the indicator of some set $C \in \mathcal{M}$. The time mean $\frac{1}{n} \sum_{k=0}^{n-1} f(T^k x)$ is the relative frequency of visits of the points $T^k x$, $0 \leqslant k < n$, to the measurable set C. By the Birkhoff-Khinchin ergodic theorem, the limit value of such frequencies as $n \to \infty$ exists. This theorem is therefore analogous to the strong law of large numbers of probability theory.

The Birkhoff-Khinchin ergodic theorem was preceded by another important result due to von Neumann and also related to the convergence of the means $\frac{1}{n} \sum_{k=0}^{n-1} f(T^k x)$, $\frac{1}{T} \int_0^T f(T^t x)\, dt$ (for $f \in L^2(M, \mathcal{M}, \mu)$), but instead of almost everywhere convergence, the convergence in the metric of L^2 was studied. The general form of this result deals with the isometric linear operators and groups (semigroups) of such operators in Hilbert space. We shall formulate it only in the case of a single operator.

Theorem 2.3 (The von Neumann Ergodic Theorem, cf [CFS]). *Suppose U is an isometric operator in complex Hilbert space H; H_U is the subspace of vectors $f \in H$ invariant with respect to U, i.e. $H_U = \{f \in H : Uf = f\}$; P_U is the (operator of the) orthogonal projection to H_U. Then*

$$\lim_{n\to\infty}\left\|\frac{1}{n}\sum_{k=0}^{n-1}U^k f - P_U f\right\|_H = 0 \quad \textit{for any } f \in H.$$

To derive from this theorem the convergence in L^2 of the means $\frac{1}{n}\sum_{k=0}^{n-1} f(T^k x)$ for an endomorphism $T : (M, \mathcal{M}, \mu) \to (M, \mathcal{M}, \mu)$ it suffices to apply it to the operator U given by $U(f(x)) = f(Tx)$, $f \in L^2(M, \mathcal{M}, \mu)$. The invariance of the measure μ under T implies that U is an isometric operator. There are various generalizations of the von Neumann and Birkhoff-Khinchin Ergodic theorems. They are related to the measurable transformations without invariant measure or with the infinite invariant measure, to general groups of transformations, to the functions taking values in Banach spaces and so on. We will not discuss these results here (cf [Kr] and Sect. 3 of Chap. 4). A comprehensive treatment of various ergodic theorems is presented in [Kr1]. Measure preserving transformations in the spaces with infinite measures and related topics are discussed in [Aa]. Note only that in the simplest case (for a single transformation with finite invariant measure, flow or semiflow and a function $f \in L^2$) these general statements are, as a rule, equivalent to the Birkhoff-Khinchin or von Neumann theorems.

There is, however, a recent deep result due to J.F.C. Kingman, giving additional information even in this case.

Suppose T is an automorphism or an endomorphism of the measure space $(M, \mathcal{M}, \mu)$ and $f \in L^1(M, \mathcal{M}, \mu)$. The Birkhoff-Khinchin ergodic theorem deals with the almost everywhere convergence of the sequences of functions

$$\frac{1}{n}g_n(x) = \frac{1}{n}g_n(x; f) \stackrel{\text{def}}{=} \frac{1}{n}\sum_{k=0}^{n-1} f(T^k x).$$

All sequences $\{g_n(x)\}$ obviously satisfy the relation

$$g_{m+n}(x) = g_m(x) + g_n(T^m x); \quad x \in M; \quad m, n \in \mathbb{Z}^1_+, \tag{1.5}$$

and it is easily seen that (1.5) can be considered as an intrinsic characterization of such sequences. Indeed, if (1.5) holds for some $\{g_n(x)\}$, we have $g_n(x) = g_n(x; f)$ for $f = g_1(x)$.

Now consider the sequence $\{g_n(x)\}$ of the real-valued measurable functions on M satisfying, for almost all $x \in M$, the inequality

$$g_{m+n}(x) \leqslant g_m(x) + g_n(T^m x); \quad m, n \in \mathbb{Z}^1_+. \tag{1.6}$$

instead of the equality (1.5). It turns out that even under this weaker condition $\lim_{n\to\infty}\frac{1}{n}g_n(x)$ exists.

Theorem 2.4 (Subadditive ergodic theorem, J.F.C. Kingman [Kin]). *Suppose T is an endomorphism of the measure space $(M, \mathcal{M}, \mu)$, and $\{g_n(x)\}$, $n > 0$, is a* *sequence of measurable functions, g_n: $M \to \mathbb{R}^1 \cup \{-\infty\}$, $g_1^+ \in L^1(M, \mathcal{M}, \mu)$[2] satisfying the condition* (1.6). *Then there exists a function $\bar{g}$: $M \to \mathbb{R}^1 \cup \{-\infty\}$, $g \in L^1$, invariant under T and such that*

$$\lim_{n\to\infty} \frac{1}{n} g_n(x) = \bar{g}(x) \quad \text{for almost every } x \in M,$$

$$\lim_{n\to\infty} \frac{1}{n} \int_M g_n(x)\, d\mu = \inf_n \frac{1}{n} \int_M g_n(x)\, d\mu = \int_M \bar{g}(x)\, d\mu.$$

The following important result is an immediate consequence of Theorem 2.4.

Theorem 2.5 (The Furstenberg-Kesten theorem on the products of random matrices, H. Furstenberg, H. Kesten [FK]). *Suppose we are given an endomorphism T of a measure space $(M, \mathcal{M}, \mu)$ and a measurable function $G(x)$ on M taking values in the space of $m \times m$ real matrices ($m \geqslant 1$). Set $G_x^{(n)} = G(x) \cdot G(Tx) \cdot \ldots \cdot G(T^{n-1}x)$. If $\log^+ \|G(\cdot)\| \in L^1(M, \mathcal{M}, \mu)$, then the limit*

$$\lambda(x) = \lim_{n\to\infty} \frac{1}{n} \log \|G_x^{(n)}\|$$

exists for almost all $x \in M$, and $\lambda(x)$ is invariant under T. Moreover, $\lambda \in L^1(M, \mathcal{M}, \mu)$ and

$$\int_M \lambda\, d\mu = \lim_{n\to\infty} \frac{1}{n} \int_M \log \|G_x^{(n)}\|\, d\mu = \inf_n \frac{1}{n} \int_M \log \|G_x^{(n)}\|\, d\mu.$$

The next statement considerably strengthens the Furstenberg-Kesten theorem and is, in turn, a special case of the multiplicative ergodic theorem that will be formulated below.

Theorem 2.6 (V.I. Oseledets [Os2]). *Suppose T is an endomorphism of a measure space $(M, \mathcal{M}, \mu)$ and $G(x)$, $x \in M$, is a measurable function on M taking values in the space of $m \times m$ real matrices ($m \geqslant 1$), $G_x^{(n)} \stackrel{\text{def}}{=} G(x) \cdot G(Tx) \cdot \ldots \cdot G(T^{n-1}x)$. Suppose, further, that $\log^+ \|G(\cdot)\| \in L^1(M, \mathcal{M}, \mu)$. Then*

1) *there exists an invariant set $\Gamma \in \mathcal{M}$, $\mu(\Gamma) = 1$, such that the limit*

$$\Lambda_x = \lim_{n\to\infty} [G_x^{(n)*} G_x^{(n)}]^{1/2n}$$

exists for all $x \in \Gamma$ and Λ_x is a symmetric non-negative definite $m \times m$ matrix (we take the non-negative definite root);

2) *if $\exp \lambda_x^{(1)} < \exp \lambda_x^{(2)} < \cdots < \exp \lambda_x^{(s)}$ ($x \in \Gamma$) is the ordered set of all different eigenvalues of Λ_x (we have $s = s(x) \leqslant m$; the case $\lambda_x^{(1)} = -\infty$ is not excluded); $E_x^{(1)}$, $E_x^{(2)}, \ldots, E_x^{(s)}$ is the corresponding set of eigenspaces, $\dim E_x^{(r)} \stackrel{\text{def}}{=} m_x^{(r)}$, $1 \leqslant r \leqslant s$, then the functions $x \mapsto s(x)$, $x \mapsto \lambda_x^{(r)}$, $x \mapsto m_x^{(r)}$ ($1 \leqslant r \leqslant s$) are measurable and invariant with respect to T;*

3) *for any $x \in \Gamma$ and any $u \in \mathbb{R}^m$, $u \neq 0$, the limit*

[2] For any function $f(x)$ we set $f^+(x) = \max(0, f(x))$.

$$\lim_{n\to\infty} \frac{1}{n} \log \|G_x^{(n)} u\|,$$

exists and this limit equals $\lambda_x^{(r)}$, *where* r, $1 \leqslant r \leqslant s$, *is uniquely determined by the relations*

$$u \in E_x^{(1)} \oplus E_x^{(2)} \oplus \cdots \oplus E_x^{(r)},$$

$$u \notin E_x^{(1)} \oplus E_x^{(2)} \oplus \cdots \oplus E_x^{(r-1)}.$$

Some important notions will be necessary for the formulation of the multiplicative ergodic theorem, and we are going to introduce them.

Definition 2.1. By *a linear measurable bundle* we mean the triple (N, M, π), where N, M are measurable spaces, $\pi: N \to M$ is a measurable map, and there exists an isomorphism $\psi: N \to M \times \mathbb{R}^m$ such that

1) the partition of N, whose elements are the sets $\pi^{-1}(x)$, $x \in M$, goes under ψ to the partition of $M \times \mathbb{R}^m$ with the elements of the form $\{x\} \times \mathbb{R}^m$, $x \in M$.

2) the map $\pi \circ \psi \circ \pi^{-1}$ is the identity map of M onto itself.

In other words, a linear measurable bundle is the image of the direct product $M \times \mathbb{R}^m$ under some measurable map. The following terminology will be used: N is the space of the bundle, M is the base, π is the projection, $\pi^{-1}(x)$ is the fiber over x. The map ψ induces a structure of normed vector space in each fiber $\pi^{-1}(x)$.

Any continuous subbundle of the tangent bundle of a smooth manifold (in particular, the tangent bundle itself) gives us an example of a linear measurable bundle.

Definition 2.2. *A characteristic exponent* is a measurable function $\chi: N \to \mathbb{R}^1$ such that for almost every $x \in M$ and any $v, v_1, v_2 \in \pi^{-1}(x)$ one has

1) $-\infty < \chi(x, v) < \infty$ if $v \neq 0$; $\chi(x, 0) = -\infty$;

2) $\chi(x, \alpha v) = \chi(x, v)$, $\alpha \in \mathbb{R}^1$, $\alpha \neq 0$;

3) $\chi(x, v_1 + v_2) \leqslant \max\{\chi(x, v_1), \chi(x, v_2)\}$.

It may be shown that for any $x \in M$ the restriction of χ to $\pi^{-1}(x)$ takes at most m values different from $-\infty$. Denote these values by $\chi_i(x)$, $1 \leqslant i \leqslant s(x) \leqslant m$, and assume that

$$\chi_1(x) < \chi_2(x) < \cdots < \chi_{s(x)}(x). \tag{1.7}$$

Let $L_i(x)$ be the subspace $\{v \in \pi^{-1}(x): \chi(x, v) \leqslant \chi_i(x)\}$. The subspaces $L_i(x)$, $1 \leqslant i \leqslant s(x)$ determine the filtering of $\pi^{-1}(x)$, i.e.

$$\{0\} = L_0(x) \subset L_1(x) \subset \cdots \subset L_{s(\alpha)}(x) = \pi^{-1}(x). \tag{1.8}$$

Let $k_i(x) = \dim L_i(x)$, $k_0(x) = 0$. The integer-valued functions $s(x)$, $k_1(x), \ldots, k_{s(x)}(x)$, as well as the families of subspaces $L_i(x)$, $1 \leqslant i \leqslant s(x)$ depend measurably upon x. Conversely, suppose we are given the integer-valued measurable function $s(x) \leqslant m$, the measurable functions $\chi_1(x), \ldots, \chi_{s(x)}(x)$ satisfying (1.7), and the filtering (1.8) depending measurably upon x with $\dim L_i(x) = k_i(x)$. Then the function $\chi(x, v)$ given by $\chi(x, v) = \chi_i(x)$ for $x \in M$, $v \in L_i(x) \setminus L_{i-1}(x)$, is measurable and defines the characteristic exponent on N.

Let T be an endomorphism of M preserving a normalized measure μ.

Definition 2.3. *A measurable multiplicative cocycle* with respect to T is a measurable function $a(n, x)$, $x \in M$, taking values in the space of $m \times m$ matrices and satisfying the relation $a(n+k, x) = a(n, T^k x) \cdot a(k, x)$ (a more general definition of a cocycle will be given in Section 4).

The function $a(n, x) = G(x) \cdot \ldots \cdot G(T^{n-1}x)$, where $G(x)$ is a measurable function taking values in the space of $m \times m$ matrices, is an example of a measurable multiplicative cocycle.

Let $a(n, x)$ be a measurable multiplicative cocycle with respect to T. Consider the function

$$\chi^+(x, v) = \overline{\lim_{n\to\infty}} \frac{1}{n} \log \|a(n, x)v\|, \quad x \in M, \quad v \in \pi^{-1}(x). \tag{1.9}$$

It may be shown that χ^+ is measurable and satisfies the conditions of Definition 2.2. Thus it defines some characteristic exponent which will be called the Lyapunov characteristic exponent corresponding to T and to the cocycle $a(n, x)$. It is easily seen that the functions $s(x)$, $\chi_i(x)$, $k_i(x)$ and the subspaces $L_i(x)$ corresponding to χ^+ are invariant with respect to T.

Fix $x \in M$ and consider the filtering (1.8) at the point x. A normalized basis $\bar{e}(x) = \{e_i(x)\}$ of the space $\pi^{-1}(x)$ is said to be regular if the vectors $e_i(x)$, $1 \leqslant i \leqslant k_1(x)$ belong to $L_1(x)$, the vectors $e_i(x)$, $k_1(x) + 1 \leqslant i \leqslant k_2(x)$ belong to $L_2(x) \setminus L_1(x)$ and so on. If $\{e_i(x)\}$ is a regular basis of $\pi^{-1}(x)$, then $\{e_i^{(n)}(x)\}$, where $e_i^{(n)}(x) = a(n, x)e_i(x)/\|a(n, x)e_i(x)\|$ is a regular basis of $\pi^{-1}(T^n x)$.

Definition 2.4. A point $x \in M$ is said to be *forward regular* if for some regular basis $\bar{e}(x) = \{e_i(x)\}$ we have

$$\sum_{i=1}^{m} \chi^+(x, e_i(x)) = \underline{\lim_{n\to\infty}} \log|\det a(n, x)|.$$

In a somewhat different but equivalent form the definition of forward regularity may be found in [BVGN], [Os2].

If in the right hand side of (1.9) we take the upper limit as $n \to -\infty$ rather than as $n \to +\infty$, we obtain the definition of the characteristic exponent χ^- on N. A point $x \in M$ is called backward regular if it is forward regular with respect to χ^-.

The notions of forward and backward regularity are classical ones and go back to A.M. Lyapunov and O. Perron who studied the stability properties of the solutions of linear ordinary differential equations with nonconstant coefficients (for the most part, the one-sided solutions—for $t > 0$ and for $t < 0$—were studied). In order to investigate the stability properties of the solutions of the Jacobi equations along the two-sided trajectories of the invertible dynamical systems, it is necessary to consider the points which are not only forward and backward regular but also have the characteristic exponents χ^+ and χ^-, as well as corresponding filterings, compatible with each other. This signifies that there exist subspaces $E_i(x)$, $i = 1, \ldots, s(x)$, such that

a) $L_i^+(x) = \bigoplus_{j=1}^{k_i(x)} E_j(x)$, $L_i^-(x) = \bigoplus_{j=k_i(x)+1}^{s(x)} E_j(x)$, where $\{L_i^+(x)\}$, $\{L_i^-(x)\}$ are the filterings related to χ^+, χ^- respectively;

b) $\lim_{n\to\pm\infty} \frac{1}{|n|} \log\|a(n,x)v\| = \pm\chi_j(x)$ uniformly over $v \in E_j(x)$.

c) $\chi^-(\Gamma_j(x)) = \chi^+(\Gamma_j(x)) = (k_j(x) - k_{j-1}(x))\chi_j(x)$, where $\Gamma_j(x)$ is the volume of the parallelepiped in the space $E_j(x)$ and

$$\chi^{\pm}(\Gamma_j(x)) \stackrel{\text{def}}{=} \lim_{n\to\pm\infty} \frac{1}{|n|} \log|\Gamma_j(T^n x)|.$$

The points which are both forward and backward regular and satisfy the conditions a), b), c) are called *Lyapunov regular*, or *biregular* (see [Mi1]).

It may be shown that if x is Lyapunov regular, then so are all points of the form $T^n x$, $n \in \mathbb{Z}$ (so it is convenient to speak of Lyapunov regular trajectories), and the subspaces $E_i(T^n x)$ at the point $T^n x$ satisfy the relations $E_i(T^n x) = a(n,x)E_i(x)$.

Denote by M^+, M^-, M_0 the sets of forward, backward and Lyapunov regular points in M respectively. These sets are invariant with respect to T and, by Theorem 2.6, $\mu(M^+) = \mu(M^-) = 1$. We also have $M_0 \subset M^+ \cap M^-$, where the inclusion may in general be strict. Nevertheless, it may be shown that the set M_0 is of full measure. More precisely, the following is true.

Theorem 2.7 (The multiplicative ergodic theorem, V.I. Oseledets [Os2]; in a somewhat different form this theorem was proved by V.M. Millionshchikov [Mi2]). Let $a(n,x)$ be the multiplicative cocycle on the linear measurable bundle (N, M, π). Assume that $\int_M \log\|a(1,x)\|\, d\mu < \infty$ (such a cocycle is called Lyapunov). Then μ-almost every point $x \in M$ is Lyapunov regular.

Note that, unlike the theorem on forward and backward regularity, which is an immediate consequence of the subadditive ergodic theorem, the above statement requires some additional ideas for its proof.

In the special case of irrational rotations on the circle the Birkhoff-Khinchin ergodic theorem can be considered as a generalization (in "almost everywhere sense") of the fact that, for every irrational α, the sequence of fractional parts of $n\alpha$, $n = 1, 2, \ldots$ is uniformly distributed. It was discovered by J. Bourgain in the late 1980's that much more delicate results on uniform distribution, such as Weyl's theorem on the uniform distribution of fractional parts of polynomials, or Vinogradov's theorem concerning the sequence of fractional parts of prime numbers, admit similar generalizations in the form of pointwise ergodic theorem along subsequences. It was shown in [Bo1], [Bo2], [Bo3] that for every automorphism T of a probability space $(M, \mathcal{M}, \mu)$, and any $f \in L^p(M,\mu)$, $p > 1$, the sequence $\frac{1}{n}\sum_{k=0}^{n-1} f(T^{p(n)}x)$ converges μ-almost everywhere if $p(n)$ is given by any integer-valued polynomial. In [Bo3] and in Wierdl [W] the same result has been obtained for the sequence $\{p(n)\}$ of all primes. These facts are very deep, and their proofs combine methods of harmonic analysis with number-theoretical estimates of exponential sums.

In [Bo3] the following "return-time" ergodic theorem by J. Bourgain, H. Furstenberg, Y. Katznelson and D. Ornstein is presented. Given any automorphism T of a probability space $(M, \mathcal{M}, \mu)$, and $f \in L^2(\mu)$, there is a subset M_0 of M of full μ-measure such that for any $x_0 \in M_0$, any second automorphism S of some probability space $(Y, \mathcal{B}, \nu)$, and any $g \in L^2(\nu)$, there is a subset X_0 of full ν-measure so that for any $y_0 \in Y_0$, the sequence

$$\frac{1}{n}\sum_{k=0}^{n-1} g(S^k y_0) f(T^k x_0)$$

converges. The "return-time" interpretation of this theorem can be easily understood if one takes for g an indicator function of some set in Y.

A presentation of Bourgain's results and his method can be found in [Th1]. The relationship between ergodic theory and harmonic analysis is discussed in [BJR]. A survey of recent work on pointwise ergodic theorems and related topics is given in [RW].

§ 3. Ergodicity. Decomposition into Ergodic Components. Various Mixing Conditions

The Birkhoff-Khinchin ergodic theorem shows that the only fact of existence of an invariant measure for a given dynamical system guarantees the possibility of averaging along its trajectories almost everywhere. For $f \in L^1(M, \mathcal{M}, \mu)$ denote by $\bar{f}$ the time mean appearing in the Birkhoff-Khinchin theorem. Let T be an endomorphism of the space $(M, \mathcal{M}, \mu)$.

Definition 3.1. A set $A \in \mathcal{M}$ is said to be *invariant* mod 0 *with respect to* T if $\mu(A \triangle T^{-1}A) = 0$.

The invariance mod 0 of A with respect to T implies its invariance mod 0 with respect to all T^n.

Let $\{T^t\}$ be a flow or a semiflow.

Definition 3.2. A set $A \in \mathcal{M}$ is said to be *invariant* mod 0 *with respect to* $\{T^t\}$ if $\mu(A \triangle T^t A) = 0$ for all t.

All invariant mod 0 sets form a σ-algebra which will be denoted by $\mathcal{M}^{\text{inv}}$. The function $\bar{f}$ in the Birkhoff-Khinchin ergodic theorem is the conditional expectation of f with respect to $\mathcal{M}^{\text{inv}}$.

The next definition plays a fundamental role in ergodic theory.

Definition 3.3. A dynamical system is *ergodic* (with respect to an invariant measure μ) if $\mathcal{M}^{\text{inv}} = \mathcal{N}$ where $\mathcal{N}$ is the trivial σ-algebra consisting of the sets of measure 0 or 1.

In various fields of mathematics it is often useful to single out a class of elementary, in some sense indecomposable objects and then represent the objects of general form as some combinations of these elementary ones (prime numbers in number theory, irreducible representations, extreme points of convex sets etc.). The ergodic systems may be thought of as such elementary objects in ergodic theory. If T is ergodic, for any $f \in L^1(M, \mathcal{M}, \mu)$ we have $\bar{f} = \int_M f \, d\mu$, where $\bar{f}$ is the time mean for f, so the statement of the ergodic theorem coincides in this case with the statement of the strong law of large numbers.

Let T be an ergodic automorphism and $f = \chi_C$ be the indicator of a set $C \in \mathcal{M}$. Then $\frac{1}{n}\sum_{k=0}^{n-1} f(T^k x)$ is the relative frequency of visits to C by the trajectory of the point x in the time interval $[0, n-1]$, and $\lim_{n\to\infty} \frac{1}{n}\sum_{k=0}^{n-1} f(T^k x) = \mu(C)$. In this case therefore, the Birkhoff-Khinchin ergodic theorem may be formulated as follows: the time mean equals the space mean almost everywhere.

If T is a uniquely ergodic homeomorphism of a compact metric space M preserving a normalized Borel measure μ, then T, if considered as an automorphism of the space (M, μ), is ergodic. The convergence in the Birkhoff-Khinchin theorem holds in this case at every point $x \in M$, not only almost everywhere.

The following result concerning the uniquely ergodic realizations of automorphisms was first obtained by R. Jewett under some restrictions and then by W. Krieger in the general case.

Theorem 3.1 (W. Krieger [Kri2], R. Jewett [J]). *For any ergodic automorphic T of the Lebesgue space M there exists a uniquely ergodic homeomorphism T_1 of some compact metric space M_1 such that T_1 as an automorphism of M_1 with its invariant measure is metrically isomorphic to T.*

For an arbitrary, not necessarily ergodic, dynamical system $\{T^t\}$ on the Lebesgue space $(M, \mathcal{M}, \mu)$ introduce the measurable hull ξ of the partition ζ of M into separate trajectories of $\{T^t\}$, i.e. the most refined of those partitions of M whose elements consist of entire trajectories of $\{T^t\}$. Let $\{\mu_C\colon C \in \xi\}$ be the canonical system of conditional measures for ξ.

Theorem 3.2 (on the decomposition into ergodic components, von Neumann [N], V.A. Rokhlin [Ro2]). *For almost every $C \in \xi$ the dynamical system $\{T^t\}$ induces the dynamical system $\{T_C^t\}$ on (C, μ_C) which is ergodic with respect to μ_C.*

The elements of the partition ξ are sometimes called *ergodic components*.

Examples. 1. Let M be the m-dimensional torus with the Haar measure, T be a group translation on M. Then T is of the form $Tx = (x_1 + \alpha_1, \ldots, x_m + \alpha_m)$ for $x = (x_1, \ldots, x_m)$ (we write the group operation additively). T is ergodic if the numbers $1, \alpha_1, \ldots, \alpha_m$ are linearly independent over the field of rational numbers (rationally independent).

2. More generally, suppose M is a commutative compact group, μ is the Haar measure on M, T is a group translation, i.e. $Tx = x + g$ $(x, g \in M)$. Denote by χ_n the characters of the group $M, (n = 0, 1, \dots)$, $\chi_0 \equiv 1$. Then T is ergodic if $\chi_n(g) \neq 1$ for any $n \neq 0$. This condition guarantees also the unique ergodicity and minimality of T.

3. As in Example 1, M is the m-dimensional torus with the Haar measure, $\{T^t\}$ is a group translation on M, i.e. $T^t x = (x_1 + \alpha_1 t, \dots, x_m + \alpha_m t)$. This flow is sometimes referred to as a conditionally periodic flow, or a conditionally periodic winding of the torus. The flow $\{T^t\}$ is ergodic if $\alpha_1, \dots, \alpha_m$ are rationally independent.

4. Integrable systems of classical mechanics. A Hamiltonian system in $2m$-dimensional phase space is said to be integrable (cf [Ar1]) if it has m prime integrals in involution. The Liouville theorem says that if all trajectories of the system are concentrated in a bounded part of the space, there exist locally m prime integrals $I_1, \dots, I_m$, such that the m-dimensional manifolds $I_1 = \text{const}, \dots, I_m = \text{const}$ are m-dimensional tori, and the motions on them are conditionally periodic. The Liouville theorem shows, therefore, that the system considered is not ergodic and its ergodic components are m-dimensional tori. In particular, geodesic flows on the surfaces of rotation having the additional prime integral which is known as the Clairaut integral, are not ergodic, and their ergodic components are two-dimensional tori. It follows from Jacobi's results that the same statement holds for geodesic flows on ellipsoids.

5. Group automorphisms. Suppose M is a commutative compact group, μ is the normalized Haar measure, T is a group automorphism of M. Denote by T^* the adjoint automorphism for T acting in the group M^* of characters of M by $(T^*\chi)(x) = \chi(Tx)$. Then T is ergodic if and only if the equality $(T^*)^n \chi = \chi, n \neq 0$, is impossible unless $\chi \equiv 1$. In particular, if M is the m-dimensional torus and T is given by an integer matrix $\|a_{ij}\|$, the above condition means that there are no roots of unity among the eigenvalues of $\|a_{ij}\|$.

6. Suppose M is the space of sequences $x = (\dots, x_{-1}, x_0, x_{1,\dots})$, $x_i \in X$, T is the shift in M, μ is an invariant measure. Denote by $\mathscr{A}^{(k)}$, $-\infty < k < \infty$, the σ-algebra generated by the random variables x_i, $-\infty < i \leqslant k$. It is easily seen that $\mathscr{M}^{\text{inv}} \subseteq \bigcap_k \mathscr{A}^{(k)}$. A random process $(M, \mathscr{M}, \mu)$ is said to satisfy the zero-one law due to Kolmogorov if $\bigcap_k \mathscr{A}^{(k)} = \mathscr{N}$. The shift T corresponding to such a process is necessarily ergodic. In particular, each Bernoulli shift is ergodic.

If T is an ergodic automorphism, for any pair of functions $f, g \in L^2(M, \mathscr{M}, \mu)$ we have, by the Birkhoff-Khinchin theorem,

$$\lim_{n \to \infty} \frac{1}{n} \sum_{k=0}^{n-1} \int_M f(T^k x) g(x) \, d\mu = \int_M f \, d\mu \cdot \int_M g \, d\mu$$

almost everywhere.

The statistical properties of ergodic systems in the general case are fairly poor. We shall introduce now some notions characterizing the systems having more or less developed statistical properties. For the sake of simplicity we restrict ourselves to the case of automorphisms.

Definition 3.4. An automorphism T is *weak mixing* if for any $f, g \in L^2(M, \mathcal{M}, \mu)$ we have

$$\lim_{n\to\infty} \frac{1}{n} \sum_{k=0}^{n-1} \left[\int_M f(T^k x) g(x)\, d\mu - \int_M f\, d\mu \cdot \int_M g\, d\mu \right]^2 = 0.$$

Definition 3.5. An automorphism T is *mixing* if for any $f, g \in L^2(M, \mathcal{M}, \mu)$

$$\lim_{n\to\infty} \int_M f(T^n x) g(x)\, d\mu = \int_M f\, d\mu \cdot \int_M g\, d\mu.$$

Definition 3.6. An automorphism T is *r-fold mixing* ($r \geqslant 1$) if for any $f_1, \ldots, f_r$, $g \in L^{r+1}(M, \mathcal{M}, \mu)$

$$\lim_{n_1\to\infty, \ldots, n_r\to\infty} \int_M f_1(T^{n_1}x) f_2(T^{n_1+n_2}x) \cdots f_r(T^{n_1+\cdots+n_r}x) g(x)\, d\mu$$

$$= \prod_{i=1}^{r} \int_M f_i\, d\mu \cdot \int_M g\, d\mu.$$

Definition 3.7. An automorphism T is said to be a *K-automorphism* (*Kolmogorov automorphism*) if for any $A_0, A_1, \ldots, A_r \in \mathcal{M}$, $1 \leqslant r < \infty$,

$$\lim_{n\to\infty} \sup |\mu(A_0 \cap B^{(n)}) - \mu(A_0)\mu(B^{(n)})| = 0,$$

where the supremum is taken over all sets $B^{(n)}$ in the σ-algebra generated by the sets of the form $T^k A_i$, $k \geqslant n$, $1 \leqslant i \leqslant r$.

A flow $\{T^t\}$ is said to be a K-flow if there exists a t_0 such that T^{t_0} is K-automorphism.

The mixing property implies weak mixing, and weak mixing implies ergodicity.

Suppose μ_0 is a measure on $\mathcal{M}$ absolutely continuous with respect to μ and $d\mu_0/d\mu = f_0(x)$. From the physical point of view it is natural to call μ_0 a non-equilibrium state. Denote by μ_n the measure defined by $\mu_n(C) = \mu_0(T^{-n}C)$, $C \in \mathcal{M}$. Then

$$\mu_n(C) = \int_{T^{-n}C} f_0(x)\, d\mu = \int_C f_0(T^n x)\, d\mu.$$

In other words, μ_n is absolutely continuous with respect to μ and $d\mu_n/d\mu = f_0(T^n x)$. In the case of mixing the measure μ_n converge to μ in the following sense: for any $g \in L^2(M, \mathcal{M}, \mu)$ we have $\int g\, d\mu_n \to \int g\, d\mu$. So, under the action of dynamics, any non-equilibrium state converges to some limit, which may be naturally called the equilibrium state.

K-systems are r-fold mixing for any $r \geqslant 1$. They play an important role in the entropy theory of dynamical systems (cf Chap. 3). The definition of K-automorphism may be also formulated in a somewhat different form, as follows: suppose that $\mathscr{A}_0$ is the σ-algebra generated by the sets A_i, $0 \leqslant i \leqslant r$, and $\mathscr{A}_k = T^k \mathscr{A}_0$; let, further, $\mathscr{A}_l^\infty$ be the minimal σ-algebra containing all $\mathscr{A}_k$, $k \geqslant l$. An automorphism T is K-automorphism if and only if $\bigcap_l \mathscr{A}_l^\infty = \mathscr{N}$ for any initial σ-algebra $\mathscr{A}_0$.

If $\{T^t\}$ is a flow and T^{t_0} is a K-automorphism for some $t_0 \in \mathbb{R}^1$, then so are T^t for all $t \in \mathbb{R}^1$, $t \neq 0$, (cf Chap. 3).

It is not known whether or not the mixing (1-fold mixing) property implies r-fold mixing for $r > 1$. We will now state a deep result, due to H. Furstenberg, on the relationship between weak mixing and a kind of r-fold weak mixing.

Theorem 3.3 (H. Furstenberg [FKO]). *Let T be a weak mixing automorphism of a Lebesgue space $(M, \mathscr{M}, \mu)$ and $r \geqslant 1$ is an integer. For any $f_0, f_1, \ldots, f_r \in L^\infty(M, \mathscr{M}, \mu)$ one has*

$$\lim_{n\to\infty} \frac{1}{n} \sum_{k=1}^{n} \left[\int_M \prod_{l=0}^{r} T^{kl} f_l \, d\mu - \prod_{l=0}^{r} \int_M f_l \, d\mu \right]^2 = 0.$$

There is a very interesting application of the above theorem to some number-theoretical problems. We begin by the formulation of a general statement based on Theorem 3.2.

Theorem 3.4 (H. Furstenberg [FKO]). *Let T be an automorphism of the Lebesgue space $(M, \mathscr{M}, \mu)$. For any set $A \in \mathscr{M}$, $\mu(A) > 0$, and any natural number r there exists a natural number k such that*

$$\mu\left(\bigcap_{l=0}^{r} T^{-lk} A\right) > 0.$$

This statement is known as the Furstenberg ergodic theorem. It can be easily verified that in the case of weak mixing theorem, 3.4 is an immediate consequence of Theorem 3.3. On the other hand, there is yet another special case, namely that of ergodic group translations of the commutative compact groups, for which the proof is not difficult and can be obtained by standard tools. Turning to the general case, we may assume the automorphism T to be ergodic: otherwise, the decomposition into ergodic components can be used to complete the proof, and for T ergodic the statement can be obtained from two special cases mentioned above.

We are going to formulate now a remarkable result giving an answer to a well-known number-theoretical question which was open for a long time. It will be seen a little later how ergodic theory, namely the Furstenberg ergodic theorem, might work to obtain an independent and very elegant proof of it.

Theorem 3.5 (E. Szemeredi, cf [FKO]). *Suppose $\Lambda \subset \mathbb{Z}^1$ is a set of integers with positive upper density*[3]. *Then for any natural number r it contains some arithmetic progression of length r.*

To derive the Szemeredi theorem from Theorem 3.3 it suffices to apply it to the automorphism T and the set A, where T is the restriction of the shift in the space X of all sequences $x = \{x_i\}$, $x_i = 0$ or 1, to the closure M of the trajectory of the point $x^{(\Lambda)}$ with $x_i^{(\Lambda)} = 1$ if $i \in \Lambda$, $x_i^{(\Lambda)} = 0$ if $i \notin \Lambda$[4], and $A = \{x \in M: x_0 = 1\}$.

The positiveness of the upper density of Λ implies that there exists an invariant measure μ for T such that $\mu(A) > 0$.

Theorem 3.3 of H. Furstenberg initiated the study of the so called "non-conventional ergodic theorems" establishing, for a given automorphism T of a probability space $(M, \mathcal{M}, \mu)$, the norm convergence (usually $L^2(\mu)$-convergence) of the expressions of the form

$$\frac{1}{n}\sum_{k=0}^{n-1} f_1(T^{p_1(k)}x) \cdot f_2(T^{p_2(k)}x) \ldots f_s(T^{p_s(k)}x)$$

where the $p_i(n)$ are integer sequences converging to infinity, and $f_1, f_2, \ldots f_s \in L^\infty(M, \mu)$. These theorems are closely related to questions of recurrence and multiple recurrence (see [F]). V. Bergelson [Ber] proved the polynomial ergodic theorem where L^2-convergence of the expressions of the above form is established if T is weak mixing (see §3), and $p_1, p_2, \ldots, p_s$ are nonconstant polynomials with all differences $p_i - p_j$, $i \neq j$, also nonconstant. Some related results appear in the work by J.-P. Conze and E. Lesigne [CL], [L].

§4. General Constructions

In the section the descriptions of the most important general constructions of ergodic theory are collected together. Using these constructions one can obtain new examples of dynamical systems as some combinations of the known ones.

4.1. Direct Products of Dynamical Systems. For the sake of simplicity we restrict ourselves to the case of automorphisms. Suppose we are given the automorphisms T_1, T_2 with phase spaces $(M_1, \mathcal{M}_1, \mu_1)$, $(M_2, \mathcal{M}_2, \mu_2)$ respectively. The automorphism T of the product-space $M = M_1 \times M_2$ given by $Tx = (T_1x_1, T_2x_2)$ for $x = (x_1, x_2)$ is called the *direct product* $T_1 \times T_2$ of the automorphisms T_1, T_2. The direct product of several automorphisms is defined in a similar way.

---

[3] By the upper density of Λ we mean the number $\bar{\rho}(\Lambda) = \overline{\lim}_{n-m\to\infty}$ card $(\Lambda \cap [m, n])$.

[4] We consider X as the direct product, $X = \prod_{-\infty}^{\infty} \{0, 1\}$ with the Tikhonov topology.

Theorem 4.1. 1) *If T_1 is ergodic and T_2 is weak mixing, then $T_1 \times T_2$ is ergodic.*
2) *If T_1, T_2 are weak mixing, so is $T_1 \times T_2$.*
3) *If T_1, T_2 are mixing, so is $T_1 \times T_2$.*
4) *If T_1, T_2 are K-automorphisms, so is $T_1 \times T_2$.*

The direct product of two ergodic automorphisms may be non-ergodic. Example: $M_1 = M_2$ is the unit circle S^1 with the Haar measure and $T_1 = T_2$ is an irrational rotation. Then $T_1 \times T_2$ is not ergodic.

The direct products of endomorphisms and flows can also be defined in an obvious way. The statements of Theorem 4.1 remain valid in the case of flows.

4.2. Skew Products of Dynamical Systems. Suppose $(M, \mathscr{M}, \mu)$ is the direct product of the measure spaces $(M_1, \mathscr{M}_1, \mu_1)$, $(M_2, \mathscr{M}_2, \mu_2)$. Consider an automorphism T_1 of $(M_1, \mathscr{M}_1, \mu_1)$ and a family $\{T_2(x_1)\}$ of automorphisms of M_2 depending measurably on $x_1 \in M_1$. The measurability means here that for any measurable function $f(x_1, x_2)$ on $M_1 \times M_2$ the functions $f_n(x_1, x_2) = f(T_1^n x_1, T_2^n(x_1)x_2)$ are also measurable for all n. It may be easily checked that T preserves the measure μ. The automorphism T is called the *skew product* of the automorphism T_1 and the family $\{T_2(x_1)\}$.

Examples. 1. Suppose M_2 is a commutative compact group, μ_2 is the Haar measure on M_2 and the family $\{T_2(x_1)\}$ consists of group translations, i.e. $T_2(x_1)x_2 = x_2 + \varphi(x_1)$, where φ is a measurable map from M_1 into M_2. Sometimes $T = T_1 \times \{T_2(x_1)\}$ is called a group extension of T_1. It is easy to formulate in this case the criterion of ergodicity for T: T is ergodic if and only if:

1) T_1 is ergodic;
2) for any nontrivial character χ of the group M_2 the equation $c(x_1) = c(T_1 x_1) \cdot \chi(\varphi(x_1))$ has only the trivial solution $c = 0$.

If, in particular, $M_1 = M_2 = S^1$, and T_1 is a translation: $T_1 x_1 = x_1 + \alpha$, $\alpha \in S^1$, then the skew product $T(x_1, x_2) = (x_1 + \alpha, x_2 + \varphi(x_1)$ is known as a skew translation of the torus.

By iterating this construction we get the so called compound skew translations of the m-dimensional torus ($m \geqslant 2$):

$$T(x_1, \ldots, x_m) = (x_1 + \alpha, x_2 + \varphi_1(x_1), x_3 + \varphi_2(x_1, x_2), \ldots, x_m + \varphi_{m-1}(x_1, \ldots, x_{m-1})).$$

It was proved by H. Furstenberg that the ergodicity of a compound skew translation implies its unique ergodicity as a homeomorphism of the torus (cf [F]).

2. T_1 is a Bernoulli automorphism in the space M_1 of sequences $\{x_n^{(1)}\}_{-\infty}^{\infty}$ of 0's and 1's, $p(0) = p(1) = 1/2$, $M_2 = S^1$ with the Haar measure. For an irrational number $\alpha \in S^1$ consider the family $\{T_2(x_1)\}$ given by

$$T_2(x_1)x_2 = \begin{cases} x_2, & \text{if } x_0^{(1)} = 0 \\ x_2 + \alpha, & x_0^{(1)} = 1, \end{cases}$$

where $x_1 = \{x_n^{(1)}\}_{-\infty}^{\infty}$, $x_2 \in M_2$.

The corresponding skew product is a K-automorphism.

One of the most important notions in the theory of skew products is that of a cocycle. Suppose T_1 is an automorphism of the space $(M_1, \mathcal{M}_1, \mu_1)$ and G is a measurable group, i.e. the set endowed with the structures of both a group and a measurable space compatible with each other. The measurable map $\phi : M_1 \times \mathbb{Z}^1 \to G$ such that $\phi(x_1, m+n) = \phi(x, m) \cdot \phi(T_1^m x_1, n)$, $x_1 \in M_1, m, n \in \mathbb{Z}^1$, is called a cocycle for T_1 with values in G. The cocycles ϕ_1 and ϕ_2 are *cohomologous* if there exists a measurable map $\psi : M_1 \to G$ such that $\phi_1(x_1, n) = [\psi(T_1^n x_1)]^{-1} \cdot \phi_2(x_1, n) \cdot \psi(x_1)$. To each skew product $T = T_1 \times \{T_2(x_1)\}$ one can associate the cocycle ϕ taking values in the group $\mathfrak{M}$ of automorphisms of the space M_2:

$$\phi(x_1, n) = T_2(x_1) \cdot T_2(T_1 x_1) \cdot \ldots \cdot T_2(T_1^{n-1} x_1).$$

If two such cocycles are cohomologous, the associated skew products are metrically isomorphic.

4.3. Factor-Systems. As before we shall deal with automorphisms only, the other cases being similar. Suppose the automorphisms T, T_1 of the measure spaces $(M, \mathcal{M}, \mu)$, $(M_1, \mathcal{M}_1, \mu_1)$ respectively are given. If there exists a homomorphism $\varphi: M \to M_1$ such that $\varphi(Tx) = T_1 \varphi(x)$ for all $x \in M$, then T_1 is called a *factor-automorphism* of T. Example: if $T = T_1 \times T_2$ is the direct product of two automorphisms, then both T_1 and T_2 are factor-automorphisms of T. Let T_1 be a factor-automorphism of T and $\varphi: M \to M_1$ be the corresponding homomorphism. This homomorphism induces naturally the partition ξ of M into the preimages of the points $x_1 \in M_1$ under φ. If the spaces M, M_1 are Lebesgue, then ξ is measurable, and we may speak of the conditional measures on the elements of ξ. Suppose now that for almost every point $x_1 \in M_1$ the corresponding element of ξ is finite and consists of, say, N points, where N does not depend on x_1. If T is ergodic, the conditional measure on such an element equals $1/N$ at any of its points. In this case the automorphism T can be represented as a skew product over T_1, such that M_2 consists of N points and $\{T_2(x_1)\}$ is a family of permutations of M_2.

Now consider the general case, when the elements of ξ are not necessarily finite. As above, we have the canonical systems of the conditional measures $\{\mu_C: C \in \xi\}$. Since T is ergodic, for almost all $C \in \xi$ the Lebesgue spaces (C, μ_C) are metrically isomorphic. They are either the spaces with non-atomic measure, or the finite measure spaces consisting of $N < \infty$ points. The automorphism T induces the automorphisms of the spaces (C, μ_C). The family of such automorphisms may be identified in a natural way with the measurable family $\{T_2(x_1)\}$, so T is represented as a skew product $T_1 \times \{T_2(x_1)\}$.

4.4. Integral and Induced Automorphisms. Let T be an automorphism of the measure space $(M, \mathcal{M}, \mu)$ and $E \in \mathcal{M}$, $\mu(E) > 0$. Introduce the measure μ_E on E by $\mu_E(A) = \mu(A)[\mu(E)]^{-1}$, $A \in \mathcal{M}$, $A \subset E$. Then E may be viewed as a space with normalized measure. Consider the integer valued function k_E on E such that

$k_E(x) = \min\{n \geqslant 1: T^n x \in E\}$. It follows from the Poincaré recurrence theorem that k_E is well defined for almost all $x \in E$. The function k_E is known as the *return time function* into E. For T ergodic the following Kac formula is true: $\int_E k_E(x)\,d\mu_E(x) = [\mu(E)]^{-1}$. This formula signifies that the mean return time into E is equal to $[\mu(E)]^{-1}$. For almost every $x \in E$ define the transformation T_E by $T_E x = T^{k_E(x)}x$, $x \in E$. It may be easily verfied that T_E is an automorphism of the measure space $(E, \mathscr{M}_E, \mu_E)$, where $\mathscr{M}_E$ is the σ-algebra of the sets A, $A \in \mathscr{M}$, $A \subset E$.

Definition 4.1. The automorphism T_E is called the *induced automorphism* constructed from the automorphism T and the set E.

Now consider the "dual" construction. Suppose T_1 is an automorphism of the space $(M_1, \mathscr{M}_1, \mu_1)$ and $f \in L^1(M_1, \mathscr{M}_1, \mu_1)$ is a positive integer valued function. Introduce the measure space M whose points are of the form (x_1, i), where $x_1 \in M_1$, $0 \leqslant i < f(x_1)$, i is an integer. Define the measure μ on M by

$$\mu(A_i) = \mu(A)\cdot\left(\int_{M_1} f\,d\mu_1\right)^{-1},$$

where $A_i \subset M$ is the set of the form (A, i), $A \in \mathscr{M}_1$. The transformation T^f of M given by

$$T^f(x_1, i) = \begin{cases} (x_1, i+1) & \text{if } i+1 < f(x_1), \\ (T_1 x_1, 0) & \text{if } i+1 = f(x_1) \end{cases}$$

is an automorphism of M.

Definition 4.2. T^f is called the *integral automorphism* corresponding to T_1 and f.

The space M_1 may be naturally identified with the subset of M consisting of the points $(x_1, 0)$. Under this identification one may consider T_1 as the induced automorphism of T corresponding to the set M_1. If an automorphism T is ergodic, so are its integral and induced automorphisms. The situation is much more complicated with mixing properties. For example, any ergodic automorphism has mixing induced automorphisms.

Example. Let T be a Markov automorphism in the space of 2-sided infinite sequences $x = \{x_n\}_{-\infty}^{\infty}$, $x_n \in Y$, where Y is a finite set. Suppose T is ergodic and define $E_y = \{x: x_0 = y\}$, $y \in Y$. Then the induced automorphism of T corresponding to E_y is Bernoulli. The well known Doeblin method in the theory of Markov processes is based on the transition from T to T_{E_y}.

4.5. Special Flows and Special Representations of Flows. Let T_1 be an automorphism of the measure space $(M_1, \mathscr{M}_1, \mu_1)$ and $f \in L^1(M_1, \mathscr{M}_1, \mu_1)$, $f > 0$. Consider the space M whose points are of the form (x_1, s), $x_1 \in M_1$, $0 \leqslant s < f(x_1)$. The measure μ on M is defined as a restriction to M of the direct product $\mu_1 \times \lambda$, where λ is the Lebesgue measure on $\mathbb{R}^1$. We may suppose μ to be normalized.

Introduce the flow $\{T^t\}$ on M under which any point $(x_1, s) \in M$ moves vertically upward with unit speed until its intersection with the graph of f, then jumps instantly to the point $(T_1 x_1, 0)$ and continues its vertical motion. Formally $\{T^t\}$ may be defined for $t > 0$ by

$$T^t(x_1, s) = \left(T_1^n x_1, s + t - \sum_{k=0}^{n-1} f(T_1^k x_1) \right),$$

where n is uniquely determined from the unequality

$$\sum_{k=0}^{n-1} f(T_1^k x_1) \leqslant s + t < \sum_{k=0}^{n} f(T_1^k x_1).$$

For $t < 0$ the flow $\{T^t\}$ may be defined analogously, or else by the relation $T^{-t} = (T^t)^{-1}$.

Definition 4.3. The flow $\{T^t\}$ is called the *special flow* corresponding to the automorphism T_1 and the function f.

Theorem 4.2 (W. Ambrose, S. Kakutani, cf [AK]). *Any flow $\{T^t\}$ on a Lebesgue space, for which the set of fixed points is of zero measure, is metrically isomorphic to some special flow.*

The metric isomorphism in Theorem 4.2 is sometimes referred to as a *special representation* of the flow $\{T^t\}$. The idea of special representations goes back to Poincaré who used the so called "first return map" on the transversals to vector fields in his study of the topological behavior of the solutions of ordinary differential equations.

Examples of special representations. 1. Suppose we are given the system of differential equations

$$\frac{du}{dt} = A(u, v), \qquad \frac{dv}{dt} = B(u, v) \tag{1.10}$$

on the 2-dimensional torus with cyclic coordinates (u, v), and the functions A, $B \in C^r$, $r \geqslant 2$, $A^2 + B^2 > 0$. Suppose further that the 1-parameter group $\{T^t\}$ of translations along the solutions of this system has an absolutely continuous invariant measure μ, $d\mu = P(u, v)\,du\,dv$, $P(u, v) > 0$. It is known that there exists a smooth non-self-intersecting curve Γ on the torus transversal to the vector field (1.10) and having the property that for any point $p \in \Gamma$ the trajectory starting at p will intersect Γ again at some moment $t = f(p)$ (the so-called Siegel curve). Denote by $q = q(p) \in \Gamma$ the point of this intersection. One may choose the parameter on Γ in such a way that the transformation $p \mapsto q$ becomes a rotation of the circle by a certain angle. This implies that the flow $\{T^t\}$ has a representation as a special flow corresponding to a rotation of the circle and the function f.

2. Suppose $Q \subset \mathbb{R}^d$, $d \geqslant 2$, is a compact domain with piecewise smooth boundary, M is the unit tangent bundle over Q, $\{T^t\}$ is the billiard in Q (cf Sect. 1). Denote by M_1 the set of the unit tangent vectors having supports in

∂Q and directed into Q. Introduce the following transformation $T_1 : M_1 \to M_1$ sending any point $x \in M_1$ to some point $y = T_1 x$: x moves along its billiard trajectory until the intersection with the boundary ∂Q and then reflects from the boundary according to billiards law; the point y is just the result of this reflection. T_1 has an invariant measure μ_1 of the form $d\mu_1 = d\sigma(q) d\omega_q \cdot |(n(q), x)|$, where $d\sigma$ is the element of the volume of ∂Q, $d\omega$ is the element of the volume of S^{d-1}, $n(q)$ is the unit normal vector to ∂Q at a point $q \in \partial Q$. Denoting by $f(x)$ the time interval between two consecutive reflections (at x and at y), we obtain the representation of $\{T^t\}$ as a special flow corresponding to T_1 and f.

The following result strengthens considerably Theorem 4.2. In a slightly weaker form it was given in D. Rudolph [Ru1]; the formulation below is taken from U. Krengel [Kr2].

Theorem 4.3 (D. Rudolph [Ru1], [Kr2]). *Suppose $\{T^t\}$ is an ergodic flow on the Lebesgue space $(M, \mathcal{M}, \mu)$, and we are given the positive numbers p, q, ρ; p/q is irrational. There exists a special representation of $\{T^t\}$ such that the function f appearing in this representation takes only two values, p and q, and*

$$\mu_1(\{x_1 \in M_1 : f(x_1) = p\}) = \rho \cdot \mu_1(\{x_1 \in M_1 : f(x_1) = q\}).$$

The explicit construction of such a representation for a given flow may often be very non-trivial.

4.6. Natural Extensions of Endomorphisms. Suppose T_0 is an endomorphism of the space $(M_0, \mathcal{M}_0, \mu_0)$. Define a new space M whose points are the infinite sequences of the form $x = (x_1^{(0)}, x_2^{(0)}, x_3^{(0)}, \ldots)$, where $x_i^{(0)} \in M_0$ and $T_0 x_{i+1}^{(0)} = x_i^{(0)}$ for any $i > 0$. Denote by $\mathcal{M}$ the σ-algebra generated by the subsets $A_{i,C}$ of M of the form $A_{i,C} = \{x \in M : x_i^{(0)} \in C\}$, where $i > 0$, $C \in \mathcal{M}_0$. Consider the measure μ on $\mathcal{M}$ given by $\mu(A_{i,C}) = \mu_0(C)$, and define the transformation T of $(M, \mathcal{M}, \mu)$ by $T(x_1^{(0)}, x_2^{(0)}, x_3^{(0)}, \ldots) = (T_0 x_1^{(0)}, T_0 x_2^{(0)}, T_0 x_3^{(0)}, \ldots)$. This transformation is invertible and its inverse T^{-1} is given by $T^{-1}(x_1^{(0)}, x_2^{(0)}, x_3^{(0)}, \ldots) = (x_2^{(0)}, x_3^{(0)}, \ldots)$. The measure μ is invariant under T.

Definition 4.4. The automorphism T is the *natural extension* of the endomorphism T_0.

Theorem 4.4. *T is ergodic (mixing, weak mixing) if and only if T_0 is also ergodic.*

Given an endomorphism T_0, one may construct the decreasing sequence of sub-σ-algebras $\mathcal{M}_k$, where $\mathcal{M}_k$ consists of the sets of the form $T_0^{-k} C$, $C \in \mathcal{M}_0$.

Definition 4.5. An endomorphism T_0 is said to be *exact* if $\bigcap_{k \geqslant 0} \mathcal{M}_k = \mathcal{N}$.

If an endomorphism T_0 is exact, its natural extension T is a K-automorphism. The notion of exact endomorphism plays an important role in the theory of one-dimensional mappings (cf Part II, Chap. 9).

Example. Suppose M_0 is the space of 1-sided sequences $(x_1^{(0)}, x_2^{(0)}, \dots)$ of 0's and 1's; T_0 is the shift in M_0, μ_0 is an invariant measure for T_0. The phase space M of the natural extension of T_0 is the space of 2-sided sequences of 0's and 1's; the extension itself is the shift in M, and for any cylinder $C \subset M$ the invariant measure (for T) $\mu(C)$ equals $\mu_0(C_0)$, where C_0 is the corresponding cylinder in M_0.

Joinings. Let $r \geqslant 2$ be an integer, and T_i, $1 \leqslant i \leqslant r$, be an automorphism of a probability space $(M_i, \mathcal{M}_i, \mu_i)$. A *joining* of $T_1, \dots T_r$ is a probability measure μ on the product σ-algebra $\prod_{i=1}^r \mathcal{M}_i$ which is invariant under the direct product $\prod_{i=1}^r T_i$ and whose projection on each M_i (that is, the marginal measure on $(\prod_{j \neq i} \mathcal{M}_j) \times M_i)$ coincides with μ_i, for any i, $1 \leqslant i \leqslant r$. If $r = 2$, and the measure $\mu_1 \times \mu_2$, which is obviously a joining, is the only joining for T_1, T_2, then T_1 and T_2 are called *disjoint*.

The notion of disjointness was introduced by H. Furstenberg [F4]. Disjoint automorphisms can be visualized, to some extent, as being "relatively prime": they cannot have nontrivial common factors. This analogy with number theory, however, has its limitations; in particular, the absence of nontrivial common factors does not guarantee disjointness [Ru2].

It turned out that joinings, which appeared as a natural generalization of the idea of disjointness, provide a powerful tool for constructing transformations with highly nontrivial properties. D. Rudolph [Ru2] studied the self-joinings, i.e., the joinings of an automorphism T with itself, and introduced the class of automorphisms with *minimal self-joinings* (m.s.j.) having in a sense as few self-joinings as possible. Any m.s.j.-automorphism has trivial commutator, i.e., it commutes with its powers only (the first example with this property was constructed by D. Ornstein). One of the numerous corollaries of the construction in [Ru2] is an example of two metrically non-isomorphic automorphisms which are weakly isomorphic (each of them is isomorphic to a factor of the other).

A class of the so called *simple* automorphisms which are generalizations of m.s.j. automorphisms was studied in [dJR]. This class plays (again, with some limitations) the role of prime numbers in number theory; the unique factorization property holds for weakly mixing simple transformations (see [Th1] where a survey of main results on joinings is given).

M. Ratner [Ra2] studied the self-joinings of the horocycle flows and proved that these flows give examples of simple and m.s.j. dynamical systems coming from smooth (actually, algebraic) actions. These results imply the rigidity of the horocycle flows [Ra1].

B. Host [Ho] studied the pairwise independent joinings of $T_1, \dots T_r$, that is, the joinings whose projections on each $M_i \times M_j$, $i \neq j$, coincide with $\mu_i \times \mu_j$. He showed that every pairwise independent joining of $r \geqslant 3$ weakly mixing automorphisms with singular spectrum is actually independent. In other words, a pairwise independent joining of such systems coincides with $\prod_{i=1}^r \mu_i$. As a corollary of this result, he showed that a mixing automorphism with singular maximal spectral type (see Ch. 2, §1) is r-fold mixing, for every

r. Some other applications of joinings to the question on the relationship between mixing and r-fold mixing can be found in [Ry1], [Ry2], [Ry3].

Several facts that have been originally obtained by different methods admit "joining proofs": Bourgain's return time theorem [Ru4], the result of B. Marcus on r-fold mixing of horocycle flows [Ry1], the von Neumann theorem on metric isomorphism of ergodic automorphisms with the same pure point spectrum (M. Lemanczyk, see [Th1]).

Chapter 2
Spectral Theory of Dynamical Systems

I.P. Kornfeld, Ya.G. Sinai

§ 1. Groups of Unitary Operators and Semigroups of Isometric Operators Adjoint to Dynamical Systems

We will introduce now the notions of correlation functions and their Fourier transforms which play an important role in various applications of the theory of dynamical systems. Suppose $\{T^t\}$ is a flow on a measure space $(M, \mathcal{M}, \mu)$ and $f \in L^2(M, \mathcal{M}, \mu)$. By the correlation function corresponding to f we mean the function $b_f(t) = \int_M f(T^t x) f(x)\, d\mu$. A number of statistical properties of the dynamical system may be characterized by the limit behavior of the differences $[b_f(t) - (\int_M f\, d\mu)^2]$. In the case of mixing, these expressions tend to zero as $t \to \infty$. If the convergence is fast enough one may write the above expressions in the form

$$b_f(t) - \left(\int_M f\, d\mu\right)^2 = \int_{-\infty}^{\infty} \exp(i\lambda t) \cdot \rho_f(\lambda)\, d\lambda.$$

The function $\rho_f(\lambda)$ in this representation is called the spectral density of f. The set of those λ for which $\rho_f(\lambda)$ is essentially non-zero for typical f, characterizes in some sense the frequencies playing the crucial role in the dynamics of the system under consideration. It is sometimes said that the system "produces a noise" on this set. The investigation of the behavior of the functions $b_f(t)$, as well as of their analogs $b_f(n)$, $n \in \mathbb{Z}$, in the case of discrete time, is very important both for theory and for applications.

In this chapter the basic information concerning the properties of the functions b_f will be given.

Let $\{T^t\}$ (respectively, $\{T^n\}$) be a 1-parameter (respectively, cyclic) group of automorphisms or a semigroup of endomorphisms. It induces the adjoint group or semigroup of operators in $L^2(M, \mathcal{M}, \mu)$ which acts according to the formula

$U^t f(x) = f(T^t x)$ (respectively, $U^n f(x) = f(T^n x)$), $f \in L^2$. In the case of automorphisms (endomorphisms) these operators are unitary (isometric). If $\{T^t\}$ is a flow and $L^2(M, \mathscr{M}, \mu)$ is separable, the group $\{U^t\}$ is continuous. The function $b_f(t)$ ($b_f(n)$) may be expressed in terms of $U^t(U^n)$:

$$b_f(t) = (U^t f, f) \quad (t \in \mathbb{R}^1 \text{ or } \mathbb{R}^1_+)$$

$$b_f(n) = (U^n f, f) \quad (n \in \mathbb{Z}^1 \text{ or } \mathbb{Z}^1_+).$$

Definition 1.1. Suppose we are given two dynamical systems in the spaces $(M_1, \mathscr{M}_1, \mu_1)$, $(M_2, \mathscr{M}_2, \mu_2)$ respectively. These systems are said to be *spectrally equivalent* if there exists an isomorphism of the Hilbert spaces $L^2(M_1, \mathscr{M}_1, \mu_1)$, $L^2(M_2, \mathscr{M}_2, \mu_2)$ intertwining the actions of groups (semigroups) $\{U_1^t\}$, $\{U_2^t\}$ (or else $\{U_1^n\}$, $\{U_n^2\}$).

The metric isomorphism of two dynamical systems implies their spectral equivalence. The converse is, generally, false.

The properties of a dynamical system which can be expressed in terms of the spectral properties of the operators $U^t(U^n)$ are said to be its spectral properties. In any case, there exists the 1-dimensional subspace of L^2 consisting of the eigenfunctions of $U^t(U^n)$ which are constant almost everywhere (the corresponding eigenvalue is 1).

Theorem 1.1. 1) *A dynamical system is ergodic if and only if the space of eigenfunctions with eigenvalue 1 is one-dimensional*;

2) *A dynamical system is weak mixing if and only if any eigenfunction of the adjoint group of unitary operators (or the semigroup of isometric operators) is a constant.*

3) *A dynamical system is mixing if and only if for any $f \in L^2(M, \mathscr{M}, \mu)$ one has*

$$\lim_{t \to \infty} b_f(t) = \left(\int f \, d\mu \right)^2$$

in the case of continuous time,

$$\lim_{n \to \infty} b_f(n) = \left(\int f \, d\mu \right)^2$$

in the case of discrete time.

Therefore, ergodicity, mixing and weak mixing are spectral properties.

In what follows we shall consider the invertible case only, i.e. the automorphisms (the cyclic groups of automorphisms) and the flows on a Lebesgue space $(M, \mathscr{M}, \mu)$, as well as corresponding adjoint groups of unitary operators. We begin by recalling the main results of the theory of spectral equivalence of unitary operators.

1. *The case of automorphisms.* There exists a finite Borel measure σ on the circle S^1 such that for any $f \in L^2(M, \mathscr{M}, \mu)$ the function $b_f(n)$ may be expressed in the form

$$b_f(n) = \int_0^1 \exp(2\pi i\lambda n) \cdot p_f(\lambda) d\sigma(\lambda), \tag{2.1}$$

and for some $f_0 \in L^2(M, \mathcal{M}, \mu)$ we have $p_{f_0}(\lambda) \equiv 1$. The measure σ is called the measure of maximal spectral type. There exists a partition of the circle S^1 into countably many measurable subsets $A_1, A_2, \ldots, A_k, \ldots, A_\infty$, and a decomposition of $L^2(M, \mathcal{M}, \mu)$:

$$L^2(M, \mathcal{M}, \mu) = \bigoplus_{k=1}^{\infty} H_k \oplus H_\infty,$$

such that the subspaces H_k are pairwise orthogonal, $\dim H_k = k$, and in any H_k one can choose a basis $\{f_{k,i}\}$, $i = 1, \ldots, k$, for which $p_{f_{k,i}}(\lambda) = 1$ for $\lambda \in A_k$, $p_{f_{k,i}}(\lambda) = 0$ for $\lambda \notin A_k$.

The function $m(\lambda)$ such that $m(\lambda) = k$, $\lambda \in A_k$, is called the spectral multiplicity function. If $\sigma(A_k) > 0$ for infinitely many k, the operator U is said to have the spectrum of unbounded multiplicity. Otherwise the spectrum of U is of bounded multiplicity. If there is only one k with $\sigma(A_k) > 0$, the operator U has the homogeneous spectrum. In particular, if $k = \infty$ and σ is the Lebesgue measure, the operator U has the countable Lebesgue spectrum.

2. *The case of flows.* Instead of (2.1), we have the representation

$$b_f(t) = \int_{-\infty}^{\infty} \exp(2\pi i\lambda t) p_f(\lambda)\, d\sigma(\lambda),$$

where σ is a finite Borel measure on $\mathbb{R}^1$. The sets A_k in this case become the subsets of $\mathbb{R}^1$, and in the definition of the Lebesgue spectrum the Lebesgue measure on S^1 should be replaced by the Lebesgue measure on $\mathbb{R}^1$.

Suppose T is an ergodic automorphism, $\{U^n\}$ is the adjoint group of operators and $\Lambda_d(T)$ is the set of eigenvalues of the operator $U_T = U^1$. U_T being unitary, we have $\Lambda_d(T) \subset S^1$. Similarly, for an ergodic flow $\{T^t\}$, denote by $\Lambda_d(\{T^t\}) \subset \mathbb{R}^1$ the set of eigenvalues of the adjoint group of operators $\{U^t\}$.

Theorem 1.2 (cf [CFS]). *$\Lambda_d(T)$ is a subgroup of S^1, any eigenvalue $\lambda \in \Lambda_d(T)$ is of multiplicity one and the absolute value of any eigenfunction is constant almost everywhere. For a flow $\{T^t\}$ the similar assertions are true, but $\Lambda_d(\{T^t\})$ in this case is a subgroup of $\mathbb{R}^1$.*

Theorem 1.3 (cf [CFS]). *Suppose T is a K-automorphism or $\{T^t\}$ is a K-flow. Then the adjoint group of operators has the countable Lebesgue spectrum on the invariant subspace of functions with zero mean.*

The proofs of these spectral theorems are based on the following important property of the operators adjoint to automorphisms: if f, g and $f \cdot g \in L^2(M, \mathcal{M}, \mu)$, then $Uf \cdot Ug = U(f \cdot g)$, i.e. these operators preserve the additional structure in L^2 related to the existence of the partial multiplication in L^2—the structure of the so-called unitary ring.

§ 2. The Structure of the Dynamical Systems with Pure Point and Quasidiscrete Spectra

Definition 2.1. An ergodic automorphism T (or a flow $\{T^t\}$) is said to be an *automorphism* (a *flow*) *with pure point spectrum* if the cyclic (1-parameter) group $\{U^n\}$ ($\{U^t\}$) has a system of the orthogonal eigenfunctions that is complete in $L^2(M, \mathcal{M}, \mu)$.

Unlike the general case, the unitary equivalence of the groups of operators adjoint to the dynamical systems with pure point spectrum implies the metric isomorphism of the systems themselves. This fact enables us to obtain the complete metric classification of such systems.

Theorem 2.1 (J. von Neumann [N]). *Suppose T_1, T_2 are the ergodic automorphisms with the pure point spectrum of the Lebesgue spaces $(M_1, \mathcal{M}_1, \mu_1)$, $(M_2, \mathcal{M}_2, \mu_2)$. They are metrically isomorphic if and only if $\Lambda_d(T_1) = \Lambda_d(T_2)$, where $\Lambda_d(T)$ is the countable subgroup of the circle S^1 consisting of the eigenvalues of the operator U_T adjoint to T.*

Using the Pontryagin duality theory one can easily construct for arbitrary countable subgroup $\Lambda \subset S^1$ the automorphism T with pure point spectrum such that $\Lambda_d(T) = \Lambda$. This automorphism T is a group translation on the character group M of the group Λ with the Haar measure μ. The group M is compact in the case considered and T is given by $Tg = g \cdot g_0$, $(g, g_0 \in M)$ where $g_0(\lambda) = \lambda$, $\lambda \in \Lambda$. Combining this assertion and Theorem 2.1, we get the following fact: any ergodic automorphism with pure point spectrum is metrically isomorphic to a certain group translation on the character group of its spectrum. In the case of continuous time the similar statement holds.

Theorem 2.2. *For two ergodic flow $\{T_1^t\}$, $\{T_2^t\}$ on the Lebesgue spaces $(M_1, \mathcal{M}_1, \mu_1)$, $(M_2, \mathcal{M}_2, \mu_2)$ with pure point spectrum to be metrically isomorphic, it is necessary and sufficient that $\Lambda_d(\{T_1^t\}) = \Lambda_d(\{T_2^t\})$, where $\Lambda_d(\{T^t\})$ is a countable subgroup of $\mathbb{R}^1$ consisting of the eigenvalues of the 1-parameter group $\{U^t\}$.*

Just as in the case of automorphisms, for any countable subgroup Λ of $\mathbb{R}^1$, one can construct the flow $\{T^t\}$ with pure point spectrum $\Lambda_d(\{T^t\}) = \Lambda$ such that any T^t is a group translation on the character group of Λ. Hence, any ergodic flow with pure point spectrum is metrically isomorphic to a flow generated by the group translations along a certain 1-parameter subgroup of the character group of the spectrum.

Examples. 1. T is an ergodic group translation on the m-dimensional torus: $Tx = ((x_1 + \alpha_1) \bmod 1, \ldots, (x_m + \alpha_m) \bmod 1)$ for $x = (x_1, \ldots, x_m)$. The functions $\chi_{n_1, \ldots, n_m}(x) = \exp 2\pi i \sum_{k=1}^m n_k x_k$ form a complete system of characters of the torus. The automorphism T has the pure point spectrum consisting of numbers $\exp 2\pi i \sum_{k=1}^m n_k \alpha_k$; $(n_1, \ldots, n_m \in \mathbb{Z}^1)$.

2. Suppose M is the additive group of integer 2-adic numbers with the normalized Haar measure μ. Associating to each point $x = \sum_{k=0}^{\infty} a_k 2^k \in M$ ($a_k = 0$ or 1) the infinite sequence $(a_0, a_1, \dots)$ of 0's and 1's, one may identify M with the space of all such sequences. Fix $x_0 = 1 \cdot 2^0 + 0 \cdot 2^1 + 0 \cdot 2^2 + \cdots \in M$. The transformation $T : M \to M$ given by $Tx = x \oplus x_0$, ($\oplus$ stands for the addition of 2-adic numbers) is a group translation and, on the other hand, is an automorphism of the space (M, μ). Let us evaluate the spectrum of T.

For any n-tuple $i^{(n)} = (i_0, i_1, \dots, i_{n-1})$ of 0's and 1's define the cylinder set $C_{i^{(n)}}^{(n)} = \{x = \sum_{k=0}^{\infty} a_k 2^k \in M\colon a_k = i_k \text{ for } 0 \leqslant k \leqslant n-1\}$. For each n there are 2^n sets $C_{i^{(n)}}^{(n)}$, and the transformation T permutes them cyclically. We can enumerate these sets $C_0^{(n)}, C_1^{(n)}, \dots, C_{2^n-1}^{(n)}$, so that we have $TC_p^{(n)} = C_{p+1}^{(n)}$, $0 \leqslant p < 2^n$. The functions $f_{r/2^n}(x) = \exp \dfrac{2\pi i r p}{2^n}$, $x \in C_p^{(n)}$ $(n = 0, 1, \dots; 0 \leqslant r < 2^n)$ form a complete system of characters of M, and $U_T \chi_{r/2^n} = \exp \dfrac{2\pi i r}{2^n} \cdot \chi_{r/2^n}$. Hence, T is an automorphism with pure point spectrum consisting of numbers of the form $\exp \dfrac{2\pi i r}{2^n}$.

There is a generalization of the theory of pure point spectrum to a wider class of dynamical systems.

Suppose T is an automorphism of a Lebesgue space $(M, \mathcal{M}, \mu)$; $\Lambda_0 \stackrel{\text{def}}{=} \Lambda_d(T)$ is the group of eigenvalues of the unitary operator U_T; Φ_0 is the (multiplicative) group of normalized in $L^2(M)$ eigenfunctions of U_T. The groups Λ_0 and Φ_0 may be considered as the subsets of $L^2(M)$, and the inclusion $\Lambda_0 \subset \Phi_0$ (in $L^2(M)$) is obviously true.

For any $n \geqslant 1$ set $\Phi_n = \{f \in L^2(M)\colon \|f\| = 1,\ U_T f = \lambda f \text{ for some } \lambda \in \Phi_{n-1}\}$, $\Lambda_n = \{\lambda \in L^2(M)\colon \|\lambda\| = 1,\ U_T f = \lambda f \text{ for some } f \in \Phi_n\}$.

By induction over n, the groups Φ_n, Λ_n are defined for all $n \geqslant 0$. Their elements are called quasi-eigenfunctions and quasi-eigenvalues of rank n respectively. Since $\Phi_n \subset \Phi_{n+1}$, $\Lambda_n \subset \Lambda_{n+1}$, one may define the groups $\Phi = \bigcup_{n=0}^{\infty} \Phi_n$, $\Lambda = \bigcup_{n=0}^{\infty} \Lambda_n$. Any eigenvalue of rank $n \geqslant 1$ may be considered as an eigenfunction of rank $n-1$, so the corresponding eigenvalue of rank $n-1$ exists. This argument shows that a certain homomorphism $\theta\colon \Lambda \to \Lambda$ arises in a natural way. It is given by $\theta f = \lambda$ for $f \in \Lambda$, if $U_T f = \lambda f$, $\lambda \in \Lambda$. Restricting θ to Λ_n, one obtains the homeomorphisms $\theta_n\colon \Lambda_n \to \Lambda_n$.

Definition 2.2. An ergodic automorphism T such that Φ is a complete system of functions in $L^2(M)$ is called an *automorphism with quasi-discrete spectrum.*

Any automorphism with pure point spectrum has, obviously, quasi-discrete spectrum. The ergodic skew translation on the 2-dimensional torus: $T(x, y) = (x + \alpha, y + x)$, $x, y \in S^1$, α is irrational, is an example of the automorphism with quasi-discrete but not pure point spectrum.

We restrict ourselves to the case of the so called totally ergodic automorphisms, i.e. such automorphisms T that all T^n, $n \neq 0$, are ergodic.

Theorem 2.3 (L.M. Abramov [A]). *Suppose T_1, T_2 are totally ergodic automorphisms of the Lebesgue spaces $(M_1, \mathcal{M}_1, \mu_1)$, $(M_2, \mathcal{M}_2, \mu_2)$ with quasi-discrete spectrum. They are metrically isomorphic if and only if the corresponding groups $\Lambda^{(1)} = \bigcup_{n=0}^{\infty} \Lambda_n^{(1)}$, $\Lambda^{(2)} = \bigcup_{n=0}^{\infty} \Lambda_n^{(2)}$ of quasi-eigenvalues and the homeomorphisms $\theta^{(1)}: \Lambda^{(1)} \to \Lambda^{(1)}$, $\theta^{(2)}: \Lambda^{(2)} \to \Lambda^{(2)}$ satisfy the conditions:*

1) $\Lambda_0^{(1)} = \Lambda_0^{(2)} \overset{\text{def}}{=} \Lambda_0$;
2) *there exists an isomorphism V between the groups $\Lambda^{(1)}$, $\Lambda^{(2)}$ for which*
a) $V\lambda = \lambda$ for $\lambda \in \Lambda_0$;
b) $V\Lambda_n^{(1)} = \Lambda_n^{(2)}$, $n = 0, 1, \ldots$;
c) $\theta^{(2)} = V\theta^{(1)}V^{-1}$.

There is a representation theorem for the automorphisms with quasi-discrete spectrum similar to that for pure point spectrum. It says that for any sequence of countable commutative groups $\Lambda_0 \subseteq \Lambda_1 \subseteq \ldots$, $\bigcup_{n=0}^{\infty} \Lambda_n = \Lambda$, where Λ_0 is a subgroup of S^1 which does not contain roots of unity (except 1) there exists a totally ergodic automorphism T with quasi-discrete spectrum acting on the group M of characters of Λ, such that for any n the group of its quasi-eigenvalues of rank n is isomorphic to Λ_n.

§ 3. Examples of Spectral Analysis of Dynamical Systems

There are many examples of dynamical systems for which the spectra of the adjoint groups of operators may be completely evaluated.

Theorem 1.3 says that K-systems have the countable Lebesgue spectrum. Beyond the class of K-systems, the dynamical systems having in $L_2^{(0)}$ the countable Lebesgue spectrum also exist, for example, the horocycle flows on compact surfaces of constant negative curvature.

For a wide class of ergodic compound skew products on tori, i.e. the transformations of the form

$$T(x_1, \ldots, x_m) = (x_1 + \alpha, x_2 + f(x_1), \ldots, x_m + f_{m-1}(x_1, \ldots, x_{m-1})),$$

where $x_1, \ldots, x_m \in S^1$, α is irrational, the spectrum may be calculated. If the functions $f_k(x_1, \ldots, x_k)$ are smooth and their derivatives satisfy certain inequalities, the spectrum of U_T in the invariant subspace $H_1 \subset L^2$ consisting of function depending on x_1 only, is pure point, while in the orthogonal complement $H_1^{\perp}$ it is countable Lebesgue (A.G. Kushnirenko, cf [CFS]).

For a long time the study of spectra of various classes of dynamical systems gave ground to hope that the theory of pure point spectrum might be extended in some sense to the dynamical systems of general form. However, it is now clear that the situation is much more complicated. The examples of systems with very unexpected spectral properties have been constructed, most of them with the help of theory of periodic approximation. Some of these examples will be described now (cf [CFS]).

1. Let M be the measure space, $M = S^1 \times \mathbb{Z}_2$, where S^1 is the unit circle with the Lebesgue measure, $\mathbb{Z}_2 = \{1, -1\}$ with the measure $p(\{1\}) = p(\{-1\}) = 1/2$. Consider the automorphism T of M which is a skew product over a rotation of the circle: $T(x, z) = (x + \alpha, g(x)z)$, $x \in S^1$, $z \in \mathbb{Z}_2$. In the subspace $H \subset L^2(M)$ consisting of functions depending on x only, the unitary operator U_T has a pure point spectrum, while in the orthogonal complement $H^\perp$ the spectrum of T is continuous. Since any function $f \in H^\perp$ satisfies $f(x, -z) = -f(x, z)$, the product of any two functions from $H^\perp$ lies in H.

2. For any automorphism T with pure point spectrum, the maximal spectral type ρ (i.e. the type of discrete measure concentrated on the group $\Lambda_d(T)$) obviously dominates the convolution $\rho * \rho$. This property is known as the group property of spectrum. It is also satisfied by many dynamical systems with continuous and mixed spectra which arise naturally in applications. Generally, however, this is not true. A counterexample may be constructed, as before, in the phase space $M = S^1 \times \mathbb{Z}_2$. Namely, there are certain irrational numbers α, $\beta \in [0, 1]$ for which the skew product T, $T(x, z) = ((x + \alpha) \bmod 1, w(x)z)$, $x \in S^1$, $z \in \mathbb{Z}_2$, where $w(x) = -1$ for $x \in [0, \beta)$, $w(x) = 1$ for $x \in [\beta, 1)$, lacks the group property of spectrum.

3. The spectral multiplicity function has been evaluated for many classes of dynamical systems, and in all "natural" cases this function is either unbounded, or equals 1 on the set of full measure of maximal spectral type. But again this observation cannot be extended to the general case. There exist automorphisms (constructed also as skew products over the rotations of the circle) with non-simple continuous spectrum of finite multiplicity.

The problem of finding the exact conditions to be satisfied by the spectrum of a group of unitary operators for this group to be the adjoint group of operators of some dynamical system, is extremely difficult.

In particular, it is not known whether the dynamical systems with simple Lebesgue spectrum, or even absolutely continuous spectrum of finite multiplicity, exist.

§4. Spectral Analysis of Gauss Dynamical Systems

An important class of dynamical systems for which complete spectral analysis has been carried out, is related to Gauss distributions and stationary Gauss processes of probability theory.

Consider the space M of sequences of real numbers, infinite in both directions, $x(s)$, where s is an integer and $-\infty < s < \infty$. Suppose $\mathcal{M}$ is the σ-algebra generated by all finite dimensional cylinders, i.e. sets of the form $A = \{x(s) \in M : x(s_1) \in C_1, \dots, x(s_r) \in C_r\}$, where $C_1, \dots, C_r$ are Borel subsets of $\mathbb{R}^1$. The measure μ on $\mathcal{M}$ is said to be a Gauss measure if the joint distribution of any family of random variables $\{x(s_1), \dots, x(s_r)\}$ is an r-dimensional Gauss distribution. If the mean value $m = \mathbb{E}x(s)$ does not depend on s and the correlation function $b(s_1, s_2) = \int x(s_1)x(s_2)d\mu(x)$ depends on $s_1 - s_2$ only, the Gauss measure is stationary. Without loss of generality we may assume that $m = 0$.

For a stationary measure μ the sequence $b(s) \stackrel{\text{def}}{=} b(s, 0)$ is positive definite and thus may be represented in the form $b(s) = {}_0^1\!\int \exp(2\pi i s\lambda)\, d\sigma(\lambda)$, where σ is a finite measure on the circle S^1. The measure σ is known as the spectral measure of the Gauss measure μ.

Definition 4.1. The cyclic group $\{T^n\}$ of shift transformations in the space M provided with a stationary Gauss measure is said to be *a Gauss dynamical system*. The generator T^1 of this group is said to be *a Gauss automorphism*.

One may consider a space M of all real valued functions $x(s)$, $s \in \mathbb{R}^1$, rather than sequences $x(s)$, s is an integer. In this case the σ-algebra $\mathscr{M}$ and Gauss stationary measures on it may be defined in the same way.

Definition 4.2. The one-parameter group $\{T^t\}$ of shift transformations in the space M provided with a stationary Gauss measure is said to be *a Gauss dynamical system with continuous time* (*Gauss flow*).

We shall formulate the results for the case of discrete time only. Describe first the unitary operator U_T adjoint to the Gauss automorphism.

The complex Hilbert space $L^2(M, \mathscr{M}, \mu)$ can be decomposed into the countable orthogonal direct sum of subspaces, $L^2(M, \mathscr{M}, \mu) = \bigoplus_{m=0}^{\infty} H_m$, such that all H_m are invariant under U_T. Any H_m is of the form $H_m = H_m^{(r)} + iH_m^{(r)}$, where $H_m^{(r)}$ is a real subspace of the real Hilbert space $L^2_{(r)}(M, \mathscr{M}, \mu)$. $H_0^{(r)}$ (respectively, H_0) is the subspace of real (respectively, complex) constants. $H_1^{(r)}$ is the subspace spanned by the vectors y of the form $y = \sum a_k x(s_k)$, a_k are real. All random variables $y \in H_1^{(r)}$ have Gauss distributions. The space $H_m^{(r)}$, $m > 1$, is spanned by all possible Hermite-Ito polynomials $:y_1 \dots y_m:$ of Gauss random variables $y_1, \dots, y_m \in H_1^{(r)}$.[1] Describe now the action of the operator U_T in the subspaces H_m, $m \geqslant 1$. First we introduce the real Hilbert space $Q_m^{(r)}$ consisting of complex-valued functions $\varphi(\lambda_1, \dots, \lambda_m)$ defined for $\lambda_1, \dots, \lambda_m \in S^1$ (i.e. for $\lambda = (\lambda_1, \dots, \lambda_m)$ belonging to the torus $Tor^{\,m}$), symmetric with respect to their variables, satisfying the relation $\varphi(-\lambda_1, \dots, -\lambda_m) = \overline{\varphi(\lambda_1, \dots, \lambda_m)}$ and having finite norms $\|\varphi\|$, where $\|\varphi\| \stackrel{\text{def}}{=} [\int_{Tor^m} |\varphi(\lambda_1, \dots, \lambda_m)|^2\, d\sigma(\lambda_1) \dots d\sigma(\lambda_m)]^{1/2} < \infty$. Next, introduce the complex Hilbert space Q_m by $Q_m = Q_m^{(r)} \oplus iQ_m^{(r)}$.

Theorem 4.1 (cf [CFS]). *For any $m \geqslant 1$ there exists an isometric map θ_m: $Q_m \to H_m$ such that*

1) $\theta_m Q_m = H_m$;

2) *under the isomorphism θ_m^{-1}: $H_m \to Q_m$ the operator U_T is mapped into the operator of multiplication by the function* $\exp 2\pi i(\lambda_1 + \dots + \lambda_m)$.

---

[1] The Hermite-Ito polynomial $:y_1 \cdot y_2 \cdot \ldots \cdot y_m:$ $(y_i \in H_1^{(r)})$ is the perpendicular lowered from the extremity of vector $y_1 \cdot y_2 \cdot \ldots \cdot y_m$ to the subspace generated by all possible products $y_1' \cdot y_2' \cdot \ldots \cdot y_p'$, where $p < m$, $y_i' \in H_1^{(r)}$.

Such a complete description of the structure of the operator U_T enables us to connect a number of ergodic and spectral properties of the automorphism T to the properties of the spectral measure σ.

The necessary and sufficient condition for a Gauss automorphism T to be ergodic is that the measure σ be continuous. This condition is also necessary and sufficient for T to be weak mixing. For a Gauss automorphism T to be mixing, as well as mixing of all orders, it is necessary and sufficient that the Fourier coefficients $b_s = \int \exp(2\pi i s\lambda)\, d\sigma(\lambda)$ of the measure σ tend to zero as $|s| \to \infty$. The maximal spectral type of the operator U_T is the type of the measure $e^{\sigma} \overset{\text{def}}{=} \sum_{k=0}^{\infty} \frac{\sigma^{(k)}}{k!}$ where, for $k \geqslant 1$, the measure $\sigma^{(k)}$ is the k-fold convolution of σ with itself, and for $k = 0$, $\sigma^{(0)}$ is the normalized measure supported at the point $\lambda = 1$.

The description of the structure of the operator U_T also enables us to construct the examples of Gauss dynamical systems with non-trivial spectral properties. In particular, there exist Gauss automorphisms with simple continuous spectrum (cf [CFS]).

Chapter 3
Entropy Theory of Dynamical Systems

I.P. Kornfeld, Ya.G. Sinai

The notion of entropy was introduced in the XIX century in the works of founders of statistical mechanics, R. Clausius, J.C. Maxwell, L. Boltzmann and others, in connection with the analysis of irreversibility phenomena. Later, entropy appeared and became the fundamental concept in the information theory created by C. Shannon in the 1940's which was concerned with the problems of the transmission of information in the presence of noise. Though the formal expression for entropy was the same in both cases, there were some differences in its meaning. A.N. Kolmogorov in his work [Kol1] applied the ideas of information theory and the notion of entropy to the analysis of some problems of ergodic theory. This work gave rise to a new branch of ergodic theory with numerous results and applications – the so-called entropy theory of dynamical systems. At the present time one may consider the developing of this theory to be mostly completed. This chapter is devoted to its exposition.

§ 1. Entropy and Conditional Entropy of a Partition

Entropy theory is based on rather elementary concepts of entropy and conditional entropy of a finite or countable partition of a measure space. Let ξ be such a partition of a Lebesgue space $(M, \mathcal{M}, \mu)$. Denote the elements of ξ by C_i, $i = 1, 2, \dots$.

Definition 1.1. The *entropy of the partition* ξ is the number

$$H(\xi) = -\sum_i \mu(C_i) \log \mu(C_i).$$

If $\mu(C_i) = 0$, we adopt the convention that $\mu(C_i) \log \mu(C_i) = 0$. Further, all logarithms are to the base e. It is clear that $0 \leqslant H(\xi) \leqslant \infty$. If the partition ξ is uncountable, we set by definition $H(\xi) = \infty$. Thus, $H(\xi)$ is defined for all partitions ξ. The expression for $H(\xi)$ may be rewritten in a somewhat different form. Namely, if $C_\xi(x)$ is the element of ξ containing the point $x \in M$, we have

$$H(\xi) = -\int_M \log \mu(C_\xi(x))\, d\mu(x).$$

The properties of measurable partitions will be used below. For reader's convenience we collect them here.

1. The partition ξ_1 is not finer than ξ_2 (we write $\xi_1 \preccurlyeq \xi_2$) if ξ_2 is a subpartition of ξ_1 (mod 0). We need to explain the notation (mod 0) that we have just used. It means that one can throw out from the space M a certain set of zero measure in such a way that on the complement to this set each element $C_{\xi_1} \in \xi_1$ is the union of the entire elements $C_{\xi_2} \in \xi_2$. The inequality $\preccurlyeq$ introduces the partial order in the set of all partitions. The maximal element of this set is the partition ε of M into separate points, while the minimal one is the partition ν whose unique element is M itself. For any family of measurable partitions $\{\xi_a\}$ the partitions $\sup_a \xi_a$ (notation $\vee_a \xi_a$) and $\inf_a \xi_a$ (notation $\wedge_a \xi_a$) are well defined and are called the measurable product and measurable intersection of $\{\xi_a\}$ respectively. For any measurable partition ξ denote by $\mathcal{M}(\xi)$ the sub-σ-algebra of the σ-algebra $\mathcal{M}$ whose elements are the subsets of M consisting (mod 0) of the entire elements of ξ. Then $\mathcal{M}(\wedge_a \xi_a) = \cap_a \mathcal{M}(\xi_a)$, and $\mathcal{M}(\vee_a \xi_a)$ is the smallest σ-algebra containing all $\mathcal{M}(\xi_a)$.

Suppose η is a measurable partition. Almost every element D_η of η may itself be considered as a Lebesgue space with the measure $\mu(\cdot|D_\eta)$. Any measurable partition ξ induces the measurable partition on almost each D_η. The entropy of this partition is denoted by $H(\xi|D_\eta)$ and is called the conditional entropy of ξ under the condition D_η.

Definition 1.2. By the *conditional entropy* of the partition ξ with respect to the partition η we mean the number

$$H(\xi|\eta) = \int_{M/\eta} H(\xi|D_\eta)\, d\mu,$$

where M/η is the factor-space of M corresponding to the partition η, and the measure on M/η is induced by μ. If $\mu(C_\xi(x)|\eta)$ is the conditional measure of $C_\xi(x)$ under the condition $D_\eta(x)$, then $H(\xi|\eta) = -\int_M \log \mu(C_\xi(x)|\eta)\, d\mu$. It is also clear that $H(\xi|\nu) = H(\xi)$.

The partitions ξ, η are said to be independent if for any $A \in \mathcal{M}(\xi)$, $B \in \mathcal{M}(\eta)$ we have $\mu(A \cap B) = \mu(A) \cdot \mu(B)$. If ξ, η are independent, we have $H(\xi|\eta) = H(\xi)$.

We outline now the properties of conditional entropy. It follows from what was said above that the properties of (unconditional) entropy of a partition are the special cases of those for conditional entropy.

1) $H(\xi|\eta) \geqslant 0$; the equality holds only if $\xi \preccurlyeq \eta$;

2) $H(\xi|\eta) \leqslant H(\xi)$; if $H(\xi) < \infty$, the equality holds only if ξ, η are independent;

3) if $\xi_1 \preccurlyeq \xi_2$ then $H(\xi_1|D_\eta) \leqslant H(\xi_2|D_\eta)$ on almost every D_η; hence, $H(\xi_1|\eta) \leqslant H(\xi_2|\eta)$; if $\xi_1 \preccurlyeq \xi_2$ and $H(\xi_1|\eta) = H(\xi_2|\eta)$, then $\xi_1 = \xi_2 (\mathrm{mod}\, 0)$;

4) $H(\xi_1 \vee \xi_2|\eta) = H(\xi_1|\eta) + H(\xi_2|\xi_1 \vee \eta)$;

5) if $\eta_1 \succcurlyeq \eta_2 (\mathrm{mod}\, 0)$, then $H(\xi|\eta_1) \leqslant H(\xi|\eta_2)$;

It follows immediately from 4) and 5) that

$$H(\xi_1 \wedge \xi_2|\eta) \leqslant H(\xi_1|\eta) + H(\xi_2|\eta);$$

if $H(\xi_1|\eta)$, $H(\xi_2|\eta) < \infty$, the equality holds if and only if ξ_1 and ξ_2 are independent on almost every D_η;

6) if $\xi_1 \preccurlyeq \xi_2 \preccurlyeq \ldots, \xi = \bigvee_n \xi_n$, then $H(\xi|\eta) = \lim_{n\to\infty} H(\xi_n|\eta)$;

7) if $\xi_1 \succcurlyeq \xi_2 \succcurlyeq \ldots, \xi = \bigwedge_n \xi_n$ and $H(\xi_n|\eta) < \infty$ for at least one value of n, then $H(\xi|\eta) = \lim_{n\to\infty} H(\xi_n|\eta)$;

8) if $\eta_1 \preccurlyeq \eta_2 \preccurlyeq \ldots, \eta = \bigvee_n \eta_n$ and $H(\xi|\eta_n) < \infty$ for at least one value of n, then $H(\xi|\eta) = \lim_{n\to\infty} H(\xi|\eta_n)$;

9) if $\eta_1 \succcurlyeq \eta_2 \succcurlyeq \ldots$, and $\eta = \bigwedge_n \eta_n$, then $H(\xi|\eta) = \lim_{n\to\infty} H(\xi|\eta_n)$;

10) if T is an endomorphism of a Lebesgue space, then

$$H(T^{-1}\xi|T^{-1}\eta) = H(\xi|\eta).$$

Denote by Z the space of partitions ξ with $H(\xi) < \infty$ and define the metric ρ on Z by $\rho(\xi, \eta) = H(\xi|\eta) + H(\eta|\xi)$. Then (Z, ρ) is a complete metric space.

§ 2. Entropy of a Dynamical System

In this section we shall formulate the definition of entropy of a dynamical system and calculate the entropy of some dynamical systems. The meaning of the notion of entropy will become more clear step by step in the subsequent sections and in part II. We begin with the case of discrete time.

Suppose T is an endomorphism of the Lebesgue space $(M, \mathcal{M}, \mu)$ and ξ is a measurable partition. Set $\xi_T^- = \bigvee_{n=0}^\infty T^{-n}\xi$. The partition $T^{-1}\xi_T^-$ is sometimes referred to as "the past" of the partition ξ with respect to T.

Definition 2.1. By the *entropy per unit time* of the partition ξ with respect to the endomorphism T we mean the number $h(T, \xi) = H(\xi | T^{-1} \xi_T^-)$.

The properties of $h(T, \xi)$ (cf [Ro3]):

1) $h(T, \xi) \leqslant h(\xi)$;

2) $h(T, \xi_1 \vee \xi_2) \leqslant h(T, \xi_1) + h(T, \xi_2)$; if $(\xi_1^-)_T$ and $(\xi_2^-)_T$ are independent, the equality is reached;

3) if $H(\xi) < \infty$, then

$$h(T, \xi) = \lim_{n \to \infty} \frac{1}{n} H(\xi \vee T^{-1} \xi \vee \cdots \vee T^{-n+1} \xi);$$

4) $h(T^n, \bigvee_{k=0}^{n-1} T^{-k} \xi) = n \cdot h(T, \xi)$, $n = 1, 2, \ldots$;

5) if T is an automorphism and $\xi \in Z$, then $h(T, \xi) = h(T^{-1}, \xi)$;

6) $h(T, \xi)$ as a function in ξ is continuous on Z;

7) if $\xi_1, \xi_2 \in Z$ and $\xi_1 \preccurlyeq \xi_2$, then $h(T, \xi_1) \leqslant h(T, \xi_2)$;

8) if $H(\xi_1 \vee \xi_2 | T^{-1} \xi_1^-) < \infty$, then

$$h(T, \xi_1) = \lim_{n \to \infty} H(\xi_1 | T^{-1} \xi_1^- \vee T^{-n} \xi_2^-).$$

The next definition plays a central role in the whole entropy theory. Let $\{T^n\}$ be a cyclic semigroup of endomorphisms or a cyclic group of automorphisms of the Lebesgue space with the generator $T = T^1$.

Definition 2.2. The *entropy of* T (of the dynamical system $\{T^n\}$) is the number

$$h(T) = \sup h(T, \xi),$$

where supremum is taken over all measurable partitions ξ of M.

The entropy $h(T)$ is obviously a metric invariant of T, in the sense that $h(T_1) = h(T_2)$ if T_1 and T_2 are metrically isomorphic. $h(T)$ is also referred to as metric entropy, Kolmogorov entropy, Kolmogorov-Sinai entropy.

The properties of $h(T)$ (cf [Ro3]):

1) $h(T) = \sup h(T, \xi)$, where supremum is taken only over finite partitions ξ;

2) if $\xi_1 \preccurlyeq \xi_2 \preccurlyeq \ldots, \xi_n \in Z$ for all $n \geqslant 1$ and $\bigvee_n \xi_n = \varepsilon$, then $\lim_{n \to \infty} h(T, \xi_n)$ exists and equals $h(T)$;

3) $h(T^n) = n \cdot h(T)$ for all $n \geqslant 0$; if T is an automorphism, then $h(T^{-1}) = h(T)$ and $h(T^n) = |n| \cdot h(T)$ for all integers n, $-\infty < n < \infty$;

4) if T_1 is a factor-endomorphism of an endomorphism T, (cf Chap. 1, Sect. 4), then $h(T_1) \leqslant h(T)$;

5) $h(T_1 \times T_2) = h(T_1) + h(T_2)$;

6) if T is an ergodic automorphism, a set $E \in \mathcal{M}$, $\mu(E) > 0$, and T_E is the induced automorphism, then $h(T_E) = [\mu(E)]^{-1} \cdot h(T)$. This relation is known as Abramov's formula;

7) if T^f is the integral automorphism corresponding to an ergodic automorphism T and an integer valued function $f > 0$ (cf Chap. 1, Sect. 4), then $h(T^f) = (\int_M f \, d\mu)^{-1} \cdot h(T)$;

8) the entropy of an endomorphism equals the entropy of its natural extension;

9) if η is the partition of M into ergodic components of T, then $h(T) = \int_{M/\eta} h(T|C_\eta)\,d\mu$.

Define now the entropy of a flow.

Definition 2.3. The *entropy of a flow* $\{T^t\}$ is the number $h(T^1)$.

This definition is motivated by the following theorem.

Theorem 2.1 (L.M. Abramov, cf [CSF]). *If* $\{T^t\}$ *is a flow*, $h(T^t) = |t| \cdot h(T^1)$ *for any* $t \in \mathbb{R}^1$.

If $\{T^t\}$ *is the special flow corresponding to an ergodic automorphism* T_1 *and a function* f (cf Chap. 1 Sect. 4), *then*

$$h(\{T^t\}) = \left(\int f\,d\mu_1\right)^{-1} \cdot h(T_1).$$

Definition 2.4. A measurable partition ξ is called a (1-sided) *generating partition* for an endomorphism T if $\xi_T^- = \varepsilon$. A measurable partition ξ is called a 2-sided generating partition or, simply, a generating partition for an automorphism T if $\xi_T \overset{\text{def}}{=} \bigvee_{n=-\infty}^{\infty} T^n\xi = \varepsilon$.

In many cases the evaluation of entropy is based on the following theorem.

Theorem 2.2 (A.N. Kolmogorov [Kol1], Ya.G. Sinai [Si1]). *If* $\xi \in Z$ *and* ξ *is a* 1-*sided generating partition for an endomorphism* T *or a* 2-*sided generating partition for an automorphism* T, *then* $h(T) = h(T, \xi)$.

Examples of entropy computation

1. Suppose T is a periodic automorphism, i.e. there exists a natural number m such that $T^m x = x$ for almost all x. Then $h(T) = 0$. This equality follows from the fact that for any partition ξ of $r < \infty$ elements and any $n > 0$, the partition $\bigvee_{k=0}^{n-1} T^{-k}\xi$ has at most r^n elements. Thus

$$h(T, \xi) = \lim_{n\to\infty} \frac{1}{n} H\left(\bigvee_{k=0}^{n-1} T^{-k}\xi\right) \leqslant \lim_{n\to\infty} \frac{m}{n} \log r = 0.$$

2. If T is the rotation of the unit circle S^1 by the angle α, then $h(T) = 0$. For α rational this is a special case of Example 1, while for α irrational the partition $\xi = \{[0, \frac{1}{2}), [\frac{1}{2}, 1)\}$ is generating (for T), and $\bigvee_{k=0}^{n-1} T^{-k}\xi$ consists of $2n$ intervals. Thus,

$$h(T) = h(T, \xi) = \lim_{n\to\infty} \frac{1}{n} H\left(\bigvee_{k=0}^{n-1} T^{-k}\xi\right) \leqslant \lim_{n\to\infty} \frac{1}{n} \log 2n = 0.$$

3. Let T be a Bernoulli automorphism in the space $(M, \mathscr{M}, \mu)$ of sequences $x = (\ldots, x_{-1}, x_0, x_1, \ldots)$, $x_i \in (Y, \mathscr{Y}, \nu)$, $Y = \{a_1, \ldots, a_r\}$, $\nu(\{a_k\}) = p_k$, $1 \leqslant k \leqslant r$. The partition $\xi = (C_1, \ldots, C_r)$, where $C_k = \{x \in M\colon x_0 = a_k\}$, is generating (for T). Therefore, $h(T) = h(T, \xi) = H(\xi) = -\sum_{k=1}^{r} p_k \log p_k$. The immediate consequence of this fact is that for any $h > 0$ there are Bernoulli automorphisms T

with $h(T) = h$ and, therefore, there exist continuum of pairwise non-isomorphic Bernoulli automorphisms. All metric invariants of dynamical systems that were known before entropy gave no possibility to distinguish between different Bernoulli automorphisms, and it was not known whether non-isomorphic Bernoulli automorphisms existed.

4. Suppose M is the same space as in Example 3 and T is a Markov automorphism with transition matrix $\prod = \|p_{ij}\|$, $1 \leqslant i, j \leqslant r$, and stationary probabilities $(p_1, \ldots, p_r)$. The partition ξ of Example 3 is still generating, and

$$h(T) = h(T, \xi) = -\sum_{i,j=1}^{r} p_i p_{ij} \log p_{ij}.$$

5. Suppose T is an ergodic algebraic automorphism of the m-dimensional torus with the invariant Haar measure. T is given by an integer $m \times m$ matrix A with $\det A = 1$. Let $\lambda_1, \ldots, \lambda_m$ be the eigenvalues of A. Then

$$h(T) = \log \prod_{i:|\lambda_i|>1} |\lambda_i|.$$

6. Suppose M is $[0, 1]$ and T is the endomorphism of M given by a function f defined on $[0, 1]$, i.e. $Tx = f(x)$. Suppose, further, that f has finitely many discontinuities and its derivative f' exists and satisfies the inequality $|f'| > 1$ on the intervals between any two discontinuities. Let μ be an absolutely continuous Borel measure on $[0, 1]$ invariant under T. Then $h(T) = {}_0^1\int \log|f'(x)|\rho(x)dx$, where $\rho(x)$ is the density of the measure μ.

For the endomorphism T with $f(x) = \{1/x\}$ (it has a countable number of discontinuities, but the above formula is still valid) we have

$$h(T) = \frac{2}{\log 2} \int_0^1 \frac{|\log x|}{1+x} dx - \frac{\pi^2}{6 \log 2}.$$

This endomorphism is closely related to the decomposition of real numbers into continuous fractions.

7. For an integrable Hamiltonian system $\{T^t\}$ $h(\{T^t\}) = 0$.

8. The entropy of a billard in any polygon or polyhedron is equal to zero.

9. For a geodesic flow $\{T^t\}$ on a surface of negative constant curvature-K we have $h(\{T^t\}) = \sqrt{K}$.

§ 3. The Structure of Dynamical Systems of Positive Entropy

The following theorem gives important information about the meaning of the concept of entropy.

Theorem 3.1 (The Shannon-McMillan-Breiman theorem, cf [Bi]). *Suppose T is an ergodic automorphism of a Lebesgue space $(M, \mathcal{M}, \mu)$, ξ is a finite partition of M, $C_n(x)$ is the element of the partition $\xi \vee T\xi \vee \cdots \vee T^{n-1}\xi$ containing $x \in M$.*

Then

$$\lim_{n\to\infty}\left[-\frac{1}{n}\log\mu(C_n(x))\right] = h(T,\xi)$$

for almost all $x \in M$.

Another important fact about entropy is the Krieger theorem on generating partitions.

Theorem 3.2 (W. Krieger [Kri1]. *Suppose T is an ergodic automorphism, $h(T) < \infty$. Then for any $\varepsilon > 0$ there exists a finite generating partition for T such that $H(\xi) \leqslant h(T) + \varepsilon$. If $h(T) < \log k$ for some integer $k > 1$, there exists a generating partition of k elements.*

This theorem shows that any ergodic automorphism of finite entropy can be represented as a stationary random process with discrete time and a finite number of states.

Turning back to arbitrary endomorphisms consider the partitions ξ of the phase space with $h(T,\xi) = 0$. For such partitions we have $\xi_T^- = T^{-1}\xi_T^- = T^{-2}\xi_T^- = \dots$, i.e. $\xi \preccurlyeq T^{-1}\xi_T^-$. From the probabilistic point of view the equality $h(T,\xi) = 0$ means that we deal with such a random process that the "infinitely remote past" entirely determines its "future".

For an ergodic endomorphism T, M.S. Pinsker (cf [Ro3]) defined the partition $\pi(T)$ which is the least upper bound of all partitions ξ with $h(T,\xi) = 0$. If $\{T^t\}$ is a flow, the partition $\pi(T^t)$ does not depend on t. It is denoted by $\pi(\{T^t\})$. The endomorphisms T with $\pi(T) = v$ are called the automorphisms of completely positive entropy.

Theorem 3.3 (cf [Ro3]). *Suppose T is an ergodic automorphism, $h(T) > 0$. Denote by $\mathscr{H}^+$ the orthogonal complement to the subspace of $L^2(M, \mathscr{M}, \mu)$ consisting of functions constant* (mod 0) *on the elements of $\pi(T)$. Then $U_T\mathscr{H}^+ = \mathscr{H}^+$, and the operator U_T has the countable Lebesgue spectrum on $\mathscr{H}^+$.*

This theorem implies, in particular, that any automorphism with pure point spectrum, singular spectrum, as well as spectrum of finite multiplicity, is of zero entropy.

Theorem 3.4 (cf [Ro3]). *If T is an endomorphism of completely positive entropy, then T is exact. If T is an automorphism of completely positive entropy, then T is a K-automorphism.*

Definition 3.1. A partition ζ is said to be *exhaustive* with respect to an automorphism T, if $T\zeta \succcurlyeq \zeta$, $\bigvee_{k=0}^{\infty} T^k\zeta = \varepsilon$.

From the probabilistic point of view the σ-algebra $\mathscr{M}(\zeta)$ is the analog of the σ-algebra corresponding to "the past" of a random process.

Theorem 3.5 (cf [Ro3]). *If ζ is an exhaustive partition for T, then $\bigwedge_{n=0}^{\infty} T^{-n}\zeta \succcurlyeq \pi(T)$.*

It follows from this theorem that $\pi(T) = \nu$ for any K-automorphism T. Therefore, the equality $\pi(T) = \nu$ is equivalent to the fact that T is a K-automorphism.

This implies, in particular, that any factor-automorphism and any power of a K-automorphism also is a K-automorphism.

Definition 3.2. An exhaustive partition ζ is said to be *extremal* if $\bigwedge_{n=0}^{\infty} T^{-n}\zeta = \pi(T)$. An extremal partition is said to be *perfect* if $h(T, \zeta) = h(T)$.

Theorem 3.6 (V.A. Rokhlin, Ya.G. Sinai (cf [Ro3]). *Any automorphism possesses perfect partitions.*

The above definitions can be easily extended to the case of continuous time.

Definition 3.3. A measurable partition ζ is said to be *extremal* with respect to a flow $\{T^t\}$ if 1) $T^t\zeta \succcurlyeq \zeta$ for $t > 0$; 2) $\bigvee_{t>0} T^t\zeta = \varepsilon$; 3) $\bigwedge_t T^t\zeta = \pi(\{T^t\})$. An extremal partition is said to be *perfect* if $H(T^s\zeta|\zeta) = s \cdot h(\{T^t\})$ for any $s > 0$.

Theorem 3.7 (P. Blanchard, B.M. Gurevich, D. Rudolph, [Bl], [Gu], [Ru1]). *Any ergodic flow possesses perfect partitions.*

This theorem implies that if there exists $t_0 \in \mathbb{R}^1$ such that T^{t_0} is a K-automorphism, then $\{T^t\}$ is a K-flow, and, therefore, all T^t, $t \neq 0$, are K-automorphisms.

For the automorphisms of zero entropy we always have $\pi(T) = \varepsilon$; in this sense the properties of such automorphisms are opposite to the properties of K-automorphisms, because $\pi(T) = \nu$ for them.

Theorem 3.8 (cf [Ro3]. *For an automorphism T to be of zero entropy it is necessary and sufficient that any of the following properties be satisfied*:

i) *the only exhaustive partition for T is ε;*

ii) *any 2-sided generating partition for T is also a 1-sided generating partition;*

iii) *there exists a 1-sided generating partition ξ for T with $H(\xi) < \infty$.*

§4. The Isomorphy Problem for Bernoulli Automorphisms and K-Systems

The problem that will be discussed in this section is the so-called isomorphy problem, i.e. the problem of classifying the dynamical systems up to their metric isomorphism. As for the general setting of this problem, i.e. the one related to the class of all dynamical systems, it is, unfortunately, clear now that there is no complete solution to it in reasonable terms: there are too many examples showing that the metric properties of dynamical systems can be very complicated and, sometimes, very unexpected. When the notion of entropy was introduced and the existence of continuum pairwise non-isomorphic Bernoulli automorphisms had been proved with its help, attention was attracted to a certain special case of the isomorphy problem, namely, to the problem of finding the exact conditions under which two Bernoulli automorphisms and, more generally, two K-

automorphisms are metrically isomorphic. The question may be formulated as a specific coding problem.

Suppose T_1, T_2 are two Bernoulli automorphisms with distinct state spaces (distinct alphabets). In order to prove that they are metrically isomorphic one must find a coding procedure translating the sequences written in one alphabet into the sequences written in the other one and having the property that the shifted sequences are coded by the shifted ones.

Suppose the state spaces of T_1, T_2 consist of m_1, m_2 elements ($m_1, m_2 < \infty$) and the measures in them are given by the vectors $(p_1, \ldots, p_{m_1})$, $(q_1, \ldots, q_{m_2})$ respectively. Then T_1, T_2 are obviously non-isomorphic if $h(T_1) \neq h(T_2)$, i.e. $-\sum p_i \log p_i \neq -\sum q_i \log q_i$. The following example due to L.D. Meshalkin shows that the Bernoulli automorphisms T_1, T_2 with $h(T_1) = h(T_2)$ may be metrically isomorphic even if their state spaces are non-isomorphic.

Suppose $Y_1 = \{a_i\}_{i=1}^4$, $Y_2 = \{b_j\}_{j=0}^4$, and the measures $\sigma^{(1)}$, $\sigma^{(2)}$ on Y_1, Y_2 are given by the vectors $(\frac{1}{4}, \frac{1}{4}, \frac{1}{4}, \frac{1}{4})$ and $(\frac{1}{2}, \frac{1}{8}, \frac{1}{8}, \frac{1}{8}, \frac{1}{8})$ respectively. Let T_l ($l = 1, 2$) be the Bernoulli automorphism acting in $M_l = \prod_{n=-\infty}^{\infty} Y_n^{(l)}$, $Y_n^{(l)} \equiv Y_l$, with the invariant measure $\mu_l = \prod_{n=-\infty}^{\infty} \sigma_n^{(l)}$, $\sigma_n^{(l)} \equiv \sigma^{(l)}$. Consider the map $\phi: M_1 \to M_2$, which sends $x^{(1)} = (\ldots, a_{i_{-1}}, a_{i_0}, a_{i_1}, \ldots)$ to $x^{(2)} = (\ldots, b_{j_{-1}}, b_{j_0}, b_{j_1}, \ldots)$ according to the following rule: if $i_n = 1$ or 2, we set $j_n = 0$; if $i_n = 3$ (respectively 4), we find first the maximal $n_1 < n$ for which $\operatorname{card}(\{k\colon n_1 \leqslant k \leqslant n, i_k = 1 \text{ or } 2\}) = \operatorname{card}(\{k\colon n_1 \leqslant k \leqslant n, i_k = 3 \text{ or } 4\})$. Such an n_1 exists with probability 1. Moreover, it is clear that $i_{n_1} = 1$ or 2, and we set in this case $j_n = 1$ (respectively 3), if $i_{n_1} = 1$; $j_n = 2$ (respectively 4) if $i_{n_1} = 2$. It may be easily verified that ϕ defines the metric isomorphism between T_1 and T_2. There are some generalizations of the above construction, but even in its generalized form the method can be applied only to some special sub-classes of Bernoulli automorphisms. The first general result concerning the isomorphism problem of Bernoulli automorphisms was stated in terms of weak isomorphism.

Definition 4.1. Two dynamical systems are said to be *weakly isomorphic* if each of them is metrically isomorphic to a certain factor-system of the other one.

Theorem 4.1 (Ya.G. Sinai [Si3]). *Any two Bernoulli automorphisms with the same entropy are weakly isomorphic.*

This theorem is an immediate consequence of the following more general result.

Theorem 4.2 (Ya.G. Sinai [Si3]). *If T_1 is an ergodic automorphism of a Lebesgue space, T_2 is a Bernoulli automorphism of finite entropy and $h(T_2) \leqslant h(T_1)$, then T_2 is metrically isomorphic to some factor automorphism of T_1.*

The weak isomorphism of two dynamical systems does not generally imply its metric isomorphism, but corresponding examples are rather complicated. (cf D. Rudolph [Ru2]).

The final solution to the isomorphism problem of the Bernoulli automorphisms was given by the following theorem due to D. Ornstein.

Theorem 4.3 (D. Ornstein, cf [Or]). *Any two Bernoulli automorphisms with the same entropy are metrically isomorphic.*

The state spaces of T_1, T_2 in this theorem are not assumed to be finite or countable, the case $h(T_1) = h(T_2) = \infty$ is not excluded. Therefore, entropy is a complete metric invariant among the class of Bernoulli automorphisms.

We will outline the ideas involved in the proof of this very important theorem and will then present a series of further results concerning the isomorphy problem for K-automorphisms and K-flows.

Definition 4.2. A measurable partition ξ of a Lebesgue space $(M, \mathcal{M}, \mu)$ is said to be *Bernoulli with respect to an automorphism* T: $M \to M$ if all its shifts, i.e. the partitions $T^n\xi$, $-\infty < n < \infty$, are independent. An automorphism T is called *B-automorphism* if there are Bernoulli generating partitions for T.

The class of B-automorphisms obviously coincides with the class of automorphisms which are metrically isomorphic to some Bernoulli automorphisms. The above definition is simply the "coordinates-free" version of the standard definition of Bernoulli automorphisms which is convenient in studying the questions related to the isomorphy problem.

If an automorphism T has a finite Bernoulli generating partition, then the state space of the corresponding Bernoulli automorphism is finite.

We shall often deal in this section with pairs (T, ξ), where T is an ergodic automorphism of $(M, \mathcal{M}, \mu)$, and ξ is a partition of M. It will be assumed without any special mentioning that $\xi = (C_1, \ldots, C_r)$ is a finite measurable partition with the fixed order of its elements, i.e. ξ is an ordered partition, while M is a Lebesgue space with continuous measure. To each such pair (T, ξ) one can naturally associate a random process with r states, or, in other words, a shift-invariant measure in the space $M^{(r)}$ of 2-sided infinite sequences of r symbols. Two pairs (T_1, ξ_1), (T_2, ξ_2) will be called equivalent (notation: $(T_1, \xi_1) \sim (T_2, \xi_2)$) if they correspond to the same measure on $M^{(r)}$. A pair (T, ξ) is sometimes identified with the corresponding random process and, so, (T, ξ) itself is called a process.

The Ornstein distance between the pairs $(T_1, \xi_1), (T_2, \xi_2)$.

We start with the definition of a certain distance ρ between two ordered partitions $\xi_1 = (C_1, \ldots, C_r)$, $\xi_2 = (D_1, \ldots, D_r)$ of the space $(M, \mathcal{M}, \mu)$ by the formula

$$\rho(\xi_1, \xi_2) = \frac{1}{2} \sum_{\substack{1 \leqslant i, j \leqslant r \\ i \neq j}} \mu(C_i \cap D_j).$$

Next suppose $\{\xi_1^{(k)}\}_{k=0}^{n-1}$ (respectively $\{\xi_2^{(k)}\}_{k=0}^{n-1}$) is a sequence of $n \geqslant 1$ partitions of the space $(M_1, \mathcal{M}_1, \mu_1)$ (respectively $(M_2, \mathcal{M}_2, \mu_2)$), each partition being of r elements, $r < \infty$. Consider another measure space $(M, \mathcal{M}, \mu)$ and the maps ϕ_i:

$M_l \to M$ $(l = 1, 2)$ defining the isomorphisms between M_l and M. The map ϕ_l sends the partition $\xi_l^{(k)}$ to a certain partition, $\phi_l(\xi_l^{(k)})$, of $M(l = 1, 2; k = 0, 1, \ldots, n-1)$. Set

$$\bar{d}_{\phi_1, \phi_2}(\{\xi_1^{(k)}\}_0^{n-1}, \{\xi_2^{(k)}\}_0^{n-1}) = \frac{1}{n} \sum_{k=0}^{n-1} \rho(\phi_1(\xi_1^{(k)}), \phi_2(\xi_2^{(k)})),$$

$$\bar{d}(\{\xi_1^{(k)}\}_0^{n-1}, \{\xi_2^{(k)}\}_0^{n-1}) = \inf_{\phi_1, \phi_2} \bar{d}_{\phi_1, \phi_2}(\{\xi_1^{(k)}\}_0^{n-1}, \{\xi_2^{(k)}\}_0^{n-1}),$$

where inf is taken over all possible maps ϕ_1, ϕ_2 defining the isomorphisms of M_1, M_2 and M. Clearly, $\bar{d}$ does not depend on the choice of M and the inequality $0 \leqslant \bar{d} \leqslant 1$ is always true. If T_l is an automorphism of M_l and ξ_l is a partition of $M_l (l = 1, 2)$, then for any $n \geqslant 1$ we set

$$\bar{d}_n((T_1, \xi_1), (T_2, \xi_2)) = \bar{d}(\{T_1^k \xi_1\}_0^{n-1}, \{T_2^k \xi_2\}_0^{n-1}).$$

Definition 4.3. The *Ornstein distance* between the pairs (T_1, ξ_1), (T_2, ξ_2) is the number

$$\bar{d}((T_1, \xi_1), (T_2, \xi_2)) = \overline{\lim_{n \to \infty}}\, \bar{d}_n((T_1, \xi_1), (T_2, \xi_2)). \tag{3.1}$$

The right-hand side of the latter equality may be rewritten with $\lim_n$ instead of upper limit, since such limit always exists. This remark shows that the triangle inequality hold for $\bar{d}$. To gain familiarity with the Ornstein metric it is useful to associate with the above sequences $\{\xi_1^{(k)}\}_0^{n-1}$, $\{\xi_2^{(k)}\}_0^{n-1}$ the measures $\nu_n^{(1)}$, $\nu_n^{(2)}$ on the space $M_n^{(r)}$ of all sequences $\mathscr{I}$ of the form $\mathscr{I} = (i_0, \ldots, i_{n-1})$, ($i_k$ are integers), given by

$$\nu_n^{(l)}(\mathscr{I}) = \mu_l(C_{i_0, 0}^{(l)} \cap C_{i_1, 1}^{(l)} \cap \cdots \cap C_{i_{n-1}, n-1}^{(l)}),$$

where $C_{1,k}^{(l)}, \ldots, C_{r,k}^{(l)}$ are the elements of the partition $\xi_l^{(k)}$, $k = 0, 1, \ldots, n-1$; $l = 1, 2$. If $\{\xi_1^{(k)}\}$, $\{\xi_2^{(k)}\}$ are of the form $\{T_1^k \xi_1\}_0^{n-1}$, $\{T_2^k \xi_2\}_0^{n-1}$, then the measures $\nu_n^{(1)}$, $\nu_n^{(2)}$ are precisely the finite dimensional distributions of the random processes corresponding to the pairs $(T_1, \xi_1), (T_2, \xi_2)$. Hence, $\bar{d}_n((T_1, \xi_1), (T_2, \xi_2))$ may actually be regarded as a distance between $\nu_n^{(1)}$ and $\nu_n^{(2)}$. The equality $\bar{d}((T_1, \xi_1), (T_2, \xi_2)) = 0$ means that $(T_1, \xi_1) \sim (T_2, \xi_2)$, that is, the factor-automorphisms $T_1 | (\xi_1)_{T_1}$, $T_2 | (\xi_2)_{T_2}$ are metrically isomorphic (ξ_T stands for $\bigvee_{-\infty}^{\infty} T^n \xi$). Therefore, $\bar{d}$ is the metric on the space of equivalence classes of the pairs (T, ξ). The metric $\bar{d}(\{\xi_1^{(k)}\}_0^{n-1}, \{\xi_2^{(k)}\}_0^{n-1})$ may be obtained from the following general construction due to L.V. Kantorovich and G.S. Rubinstein ([Or], Russian edition). Suppose X is a metrizable topological space, Mes(X) is the space of all normalized Borel measures on X. Any metric ρ on X may be associated with a metric d_p on Mes(X):

$$d_\rho(\nu_1, \nu_2) = \inf_\lambda \int_{X \times X} \rho(x_1, x_2)\, d\lambda(x_1, x_2),$$

where inf is taken over all normalized Borel measures λ on $X \times X$ such that their marginal measures (projections to the factors of $X \times X$) equal ν_1, ν_2 respectively (i.e. $\lambda(A \times X) = \nu_1(A)$, $\lambda(X \times B) = \nu_2(B)$ for all measurable $A, B \subset X$). Now take the space $M_n^{(r)}$ as X, and the Hamming metric χ on X as ρ, i.e. $\chi((i_0^{(1)}, \ldots, i_{n-1}^{(1)}), (i_0^{(2)}, \ldots, i_{n-1}^{(2)}))$ is equal to the relative fraction of those k for which $i_k^{(1)} \neq i_k^{(2)}$. Then the Kantorovich-Rubinstein construction leads to a certain metric d_ρ which is equivalent to $\bar{d}$ (i.e. the corresponding topologies are the same).

Finitely determined partitions. Suppose T is an automorphism of a space $(M, \mathcal{M}, \mu)$, $\xi = (C_1, \ldots, C_r)$ is a partition of M.

Definition 4.4. ξ is said to be *finitely determined* with respect to T if for any $\varepsilon > 0$ there exist $\delta > 0$ and an integer $n \geqslant 1$ such that for any pair $(\bar{T}, \bar{\xi})$, where $\bar{T}$ is an automorphism of a Lebesgue space $(\bar{M}, \bar{\mathcal{M}}, \bar{\mu})$, $\bar{\xi} = (\bar{C}_1, \ldots, \bar{C}_r)$ is a partition of $\bar{M}$, satisfying the conditions:

1) $|h(\bar{T}, \bar{\xi}) - h(T, \xi)| < \delta$,
2) $\sum_{1 \leqslant i_0, i_1, \ldots, i_{n-1} \leqslant r} |\bar{\mu}(\bigcap_{k=0}^{n-1} \bar{T}^k \bar{C}_{i_k}) - \mu(\bigcap_{k=0}^{n-1} T^k C_{i_k})| < \delta$,

one has $\bar{d}((T, \xi), (\bar{T}, \bar{\xi})) < \varepsilon$.

Sometimes the pair (T, ξ) itself is called finitely determined if ξ is finitely determined with respect to T.

The fact that (T, ξ) is finitely determined means (in terms of the corresponding random process) that if any pair (process) $(\bar{T}, \bar{\xi})$ has a sufficiently large collection of finite dimensional distributions which are close to the corresponding distributions of (T, ξ) and, moreover, $h(\bar{T}, \bar{\xi})$ and $h(T, \xi)$ are close to each other, then (T, ξ) and $(\bar{T}, \bar{\xi})$ are $\bar{d}$-close.

Theorem 4.4 (cf [Or]). *If ξ is a finite Bernoulli partition for T, then ξ is finitely determined with respect to T.*

We will outline now the main ideas involved in the proof of Theorem 4.3. Only the case of Bernoulli automorphism with finite state space will be considered. In this case Theorem 4.3 is the consequence of the following general statement.

Theorem 4.5. *If T_1 and T_2 have finitely determined generating partitions and $h(T_1) = h(T_2)$, then T_1 and T_2 are metrically isomorphic.*

To derive Theorem 4.3 from this statement, it suffices to observe that, according to Theorem 4.4, any Bernoulli generating partition is necessarily finitely determined. The proof of Theorem 4.5 splits up into 2 steps.

A. Suppose the pairs (T_1, ξ_1), (T_2, ξ_2) are such that:
1) $h(T_2, \xi_2) \leqslant h(T_1, \xi_1)$;
2) ξ_2 is finitely determined with respect to T_2. Then there exists a partition $\tilde{\xi}_1$ of the space M_1 such that the factor-automorphisms $T_1|(\tilde{\xi}_1)_{T_1}$, $T_2|(\xi_2)_{T_2}$ are metrically isomorphic, and $\rho(\xi_1, \tilde{\xi}_1) \leqslant \text{const} \cdot \mathcal{D}^{1/2}$, where $\mathcal{D} = \bar{d}((T_1, \xi_1), (T_2, \xi_2))$.

The assertion A in the case when T_2 is a Bernoulli automorphism and ξ_2 is its Bernoulli generating partition, is a strengthened form of Theorem 4.2. The partition $\tilde{\xi}_1$ appearing in A is obtained as the limit of a certain sequence $\{\xi_1^{(n)}\}_{n=0}^{\infty}$, $\xi_1^{(0)} = \xi_1$, such that $\bar{d}((T_1, \xi_1^{(n)}), (T_2, \xi_2)) \to 0$. For the limit partition $\tilde{\xi}_1$ we have $\bar{d}((T_1, \tilde{\xi}_1), (T_2, \xi_2)) = 0$, that is $T_1|(\tilde{\xi}_1)_{T_1}$ and $T_2|(\xi_2)_{T_2}$ are isomorphic, The inductive choice of $\xi_1^{(n)}$ is based on the following lemma playing the central role in the proof of A.

Lemma. *Suppose* (T_1, ξ_1), (T_2, ξ_2) *and* $\mathscr{D}$ *are as in the statement of A. Then for any* $\delta > 0$ *there exists a partition* $\xi^{(\delta)}$ *of* M_1 *such that* $\bar{d}((T_1, \xi^{(\delta)}), (T_2, \xi_2)) < \delta$ *and* $\rho(\xi_1, \xi^{(\delta)}) \leqslant \mathrm{const} \cdot \mathscr{D}^{1/2}$.

Applying the statement A to the pairs (T_1, ξ_1), (T_2, ξ_2) appearing in Theorem 4.5, we obtain two partitions $\tilde{\xi}_1$ and $\tilde{\xi}_2$ of the space M_1, and to complete the proof it suffices to replace $\tilde{\xi}_1$ by a generating partition η (of the space M_1) having the same properties as $\tilde{\xi}_1$. The possibility of this replacement is assured by the following statement.

B. Suppose T is an automorphism of $(M, \mathscr{M}, \mu)$, and the partitions ξ, $\tilde{\xi}$ are finitely determined with respect to T. Suppose, further, that ξ is a generating partition. Then for any $\varepsilon > 0$ there is a partition $\eta^{(\varepsilon)}$ such that the factor-automorphisms $T|\tilde{\xi}_T$, $T|\eta_T^{(\varepsilon)}$ are metrically isomorphic and $\rho(\tilde{\xi}, \eta^{(\varepsilon)}) < \varepsilon$, $\eta_T^{(\varepsilon)} \overset{\varepsilon}{\succ} \xi$.[1]

Using the assertion B with T_1 in the role of T, one can construct the partition η as the limit of a sequence $\{\eta_n\}_{n=0}^{\infty}$, $\eta_0 = \tilde{\xi}$, where $\eta_{n+1} = \eta_n^{(\varepsilon_n)}$, $\varepsilon_n \to 0$. As n goes to infinity, the partitions η_n become "more and more generating", and the limit partition η gives us the needed metric isomorphism.

The structure of B-automorphisms. It follows from Theorems 4.4, 4.5 that any automorphism having a finitely determined generating partition is a B-automorphism. The next result shows that the property of being finitely determined, if held by at least one generating partition, is necessarily inherited by all finite partitions of the phase space.

Theorem 4.6. *If* T *is a B-automorphism of* $(M, \mathscr{M}, \mu)$, $h(T) < \infty$, *then any finite partition* ξ *of* M *is finitely determined with respect to* T.

Corollary. *Any factor-automorphism* T_1 *of a B-automorphism* T *is also a B-automorphism.*

To prove this statement, it suffices to apply Theorem 4.6 to a finite generating partition ξ for T_1 which we consider as a partition of the phase space of T.

The assertion of the corollary is also valid in the case $h(T) = \infty$.

It may be easily deduced from the Ornstein isomorphism theorem that any B-automorphism has the roots of all degrees (the root of n-th degree from T is an automorphism Q_n such that $(Q_n)^n$ is metrically isomorphic to T). One may take as Q_n any Bernoulli automorphism of entropy $\frac{1}{n}h(T)$. Any other root of n-th degree from T is metrically isomorphic to this one.

In order to prove that a given automorphism is B, one must, according to the Ornstein theorem, find a finitely determined generating partition for it. In many cases, however, it is useful to replace the "finitely determined" property by

[1] We write $\xi \overset{\varepsilon}{\succ} \eta$ if there exists a partition η' for which $\rho(\eta, \eta') < \varepsilon$ and $\xi \succ \eta'$.

an equivalent one which is often easier to be verified. We shall give now the corresponding definition.

Given an ordered partition $\xi = (C_1, \dots, C_r)$ of a space $(M, \mathcal{M}, \mu)$ and a set $D \in \mathcal{M}$, $\mu(D) > 0$, denote by $\xi | D$ the ordered partition $(C_1 \cap D, \dots, C_r \cap D)$ of the space $(D, \mathcal{M}|D, [\mu(D)]^{-1} \cdot \mu)$, where $\mathcal{M}|D = \{A \in \mathcal{M}: A \subseteq D\}$.

Definition 4.5. A partition ξ of a space $(M, \mathcal{M}, \mu)$ is said to be *very weak Bernoulli* (*vwB*) with respect to an automorphism $T: M \to M$, if for any $\varepsilon > 0$ there is an integer $N = N(\varepsilon) > 0$ such that for any $m \geqslant 0$, $n \geqslant N$ there exists a set $A \in \mathcal{M}$ consisting of the entire elements of the partition $\eta_m = \bigvee_{k=-m}^{0} T^k \xi$ and satisfying the conditions:

1) $\mu(A) > 1 - \varepsilon$;

2) $\bar{d}(\{T^k \xi | D_1\}_0^{n-1}, \{T^k \xi | D_2\}_0^{n-1}) < \varepsilon$ for any elements D_1, D_2 of η_m such that $D_1, D_2 \subseteq A$.

The condition 2) may be obviously replaced by

2′) $\bar{d}(\{T^k \xi\}_0^{n-1}, \{T^k \xi | D\}_0^{n-1}) < \varepsilon$

for any $D \in \eta_m$, $D \subseteq A$.

Theorem 4.7. *For any automorphism T the class of vwB partitions coincides with that of finitely determined partitions.*

Corollary. *If ξ is a vwB partition for T, then the factor-automorphism $T|\xi_T$ is a B-automorphism.*

We shall describe now the so-called weak Bernoulli property of a partition which is stronger than *vwB*.

Definition 4.6. A partition ξ is said to be *weak Bernoulli* with respect to T, if for any $\varepsilon > 0$ there exists an integer $n > 0$ such that the partitions $\bigvee_{k=-m}^{0} T^k \xi$ are ε-independent from $\bigvee_{k=n}^{n+m} T^k \xi$ for all $m \geqslant 0$.[2]

Any weak Bernoulli partition is *vwB*. Though the converse is generally false (cf [Sm]), in many important cases the *vwB* property can be deduced from this, simpler condition.

Let T be a mixing Markov automorphism acting in the space M of 2-sided sequences $x = (\dots, y_{-1}, y_0, y_1, \dots)$, $y_i \in Y$, where $Y = \{a_1, \dots, a_r\}$ is a finite set. The generating partition $\xi = (C_1, \dots, C_r)$, $C_i = \{x \in M: y_0 = a_i\}$, $1 \leqslant i \leqslant r$, is weak Bernoulli. This implies that any mixing Markov automorphism is B (B-automorphism).

There are examples of finite partitions for B-automorphisms which are not weak Bernoulli (cf Smorodinsky [Sm]). On the other hand, Theorems 4.6 and 4.7 show that such partitions are necessarily *vwB*.

The next theorem enables us to establish the B-property of a given automorphism without finding a *vwB* generating partition for it.

[2] The partitions $\xi = \{C_i\}$ and $\eta = \{D_j\}$ are ε-independent if $\sum_{i,j} |\mu(C_i \cap D_j) - \mu(C_i)\mu(D_j)| < \varepsilon$.

Theorem 4.8. *Suppose $\xi_1 \preccurlyeq \xi_2 \preccurlyeq \ldots, \xi_n \to \varepsilon$, is an increasing sequence of finite partitions of a space $(M, \mathscr{M}, \mu)$ which are vwB with respect to an automorphism T acting on M. Then T is a B-automorphism.*

B-flows

Definition 4.7. A flow $\{T^t\}$ on a Lebesgue space $(M, \mathscr{M}, \mu)$ is said to be a *B-flow* if there exists $t_0 \in \mathbf{R}^1$ such that T^{t_0} is a B-automorphism.

By now examples of B-flows of various origins have been constructed (cf part II). It should be noted, however, that even the fact of existence of at least one B-flow is far from being trivial. Only the Ornstein theory gave tools for imbedding Bernoulli automorphisms of finite entropy into flows.

Example. Suppose M_1 is the space of 2-sided sequences $x = (\ldots, x_{-1}, x_0, x_1, \ldots)$ of 0's and 1's, and the Bernoulli measure μ_1 on M_1 is given by the vector $(\frac{1}{2}, \frac{1}{2})$. Consider the special flow $\{T^t\}$ corresponding to the shift automorphism T_1 and the function $f\colon M_1 \to \mathbb{R}^1$ such that $f(x) = \alpha$ if $x_0 = 0$, $f(x) = \beta$ if $x_0 = 1$, and α/β is irrational. The phase space of $\{T^t\}$ is $M = \{(x, t)\colon x \in M_1, 0 \leqslant t < f(x)\}$. The partition $\xi = (C_1, C_2)$ of M with $C_1 = \{(x, t) \in M\colon x_0 = 0\}$, $C_2 = \{(x, t) \in M\colon x_0 = 1\}$ is generating for T^{t_0} if $|t_0|$ is small enough. It may be shown that ξ is *vwB*, so $\{T^t\}$ is a B-flow.

The Ornstein isomorphism theorem was extended to the case of B-flows.

Theorem 4.9 (D. Ornstein [Or]). *If $\{T_1^t\}$, $\{T_2^t\}$ are B-flows and $h(\{T_1^t\}) = h(\{T_2^t\})$, then $\{T_1^t\}$, $\{T_2^t\}$ are metrically isomorphic.*

The proof of this statement involves the same ideas that the proof of the corresponding theorem about automorphisms. The important tool used in the proof is the continuous-time version of the $\bar{d}$-metric (i.e. the metric in the space of pairs $((\{T_1^t\}, \xi_1), (\{T_2^t\}, \xi_2))$).

The isomorphy problem for K-systems. Unlike the case of Bernoulli automorphisms where a complete metric classification was given by the Ornstein theorem, the situation in the class of K-systems is much more complicated. What we have in this case is a collection of counterexamples. D. Ornstein [Or] produced an example of a K-automorphism which is not isomorphic to any Bernoulli automorphism. Using some modification of the construction of this example, he also constructed continuum of pairwise non-isomorphic K-automorphisms with the same entropy.

M.S. Pinsker conjectured that any ergodic automorphism of positive entropy is metrically isomorphic to a direct product $T_1 \times T_2$, where T is a K-automorphism, while $h(T_2) = 0$. If this conjecture were true, it would signify that the general isomorphism problem might be reduced to the special cases concerned with automorphisms of zero entropy and K-automorphisms. However, the counterexample to the Pinsker hypothesis also due to D. Ornstein [Or] showed that such a reduction is impossible.

Finitary isomorphism. Suppose M_l, $l = 1, 2$, is the space of all sequences $x^{(l)} = (\ldots y_{-1}^{(l)}, y_0^{(l)}, y_1^{(l)}, \ldots)$, $y_i^{(l)} \in Y_l = \{a_1, \ldots, a_{r_l}\}$, $r_1, r_2 < \infty$; $T_l\colon M_l \to M_l$ is the shift automorphism, i.e. $T_l x^{(l)} = \tilde{x}^{(l)} = (\ldots, \tilde{y}_{-1}^{(l)}, \tilde{y}_0^{(l)}, \tilde{y}_1^{(l)}, \ldots)$, $\tilde{y}_i^{(l)} = y_{i+1}^{(l)}$, with an invariant measure μ_l defined on the Borel σ-algebra $\mathcal{M}_l$ of subsets of M_l.[3] Let $\pi_0\colon M_l \to Y$ be the projection: $\pi_0 x^{(l)} = y_0^{(l)}$ if $x^{(l)} = (\ldots y_{-1}^{(l)}, y_0^{(l)}, y_1^{(l)} \ldots) \in M_l$. Given $x^{(l)} \in M_l$ and a natural number n, denote by $C_n(x^{(l)})$ the cylinder subset of M_l of the form $\{\bar{x}^{(l)} \in M_l\colon \pi_0(T^k \bar{x}^{(l)}) = \pi_0(T^k x^{(l)}) \text{ if } |k| \leqslant n\}$.

Definition 4.8. The shifts T_1, T_2 are said to be *finitarily isomorphic* if they are metrically isomorphic and the maps $\phi_1\colon M_1 \to M_2$, $\phi_2\colon M_2 \to M_1$, $\phi_2 = \phi_1^{-1}$, intertwining T_1 and T_2, can be chosen in such a way that there are subsets $A_l \subset M_l$ $(l = 1; 2)$, $\mu_l(A_l) = 1$, such that for any point $x^{(l)} \in A_l$ $(l = 1; 2)$ one can find a natural $n = n(x^{(l)})$ for which the inclusion $\bar{x}^{(l)} \in C_n(x^{(l)})$ implies $\pi_0(\phi_l \bar{x}^{(l)}) = \pi_0(\phi_l x^{(l)})$.

In other words, the finitarity of an isomorphism means that for almost every point x of the space $M^{(l)}$ any coordinate of its image under the isomorphism map can be uniquely determined by a finite number of coordinates of x. For example, the map ϕ in the Meshalkin's example (see page 46) defines a finitary isomorphism.

The general Ornstein construction in his proof of isomorphism theorem for Bernoulli automorphisms with the same entropy leads to non-finitary isomorphism. Examples of shift automorphisms which are metrically but not finitarily isomorphic are known. Nevertheless, the following assertion, sharpening the Ornstein isomorphism theorem, is true.

Theorem 4.10 (Keane-Smorodinsky theorem on finitary isomorphism [KeS]). *Any two Bernoulli automorphisms T_1, T_2 with finite state spaces and $h(T_1) = h(T_2)$ are finitarily isomorphic.*

This theorem was also extended to the class of mixing Markov automorphisms (M. Keane, M. Smorodinsky).

§ 5. Equivalence of Dynamical Systems in the Sense of Kakutani

Since the isomorphism problem of dynamical systems is extremely difficult, various attempts were made to weaken the metric isomorphism condition in order to obtain a more compact picture.

One of the most interesting attempts was based on the notion of equivalence of dynamical systems due to S. Kakutani.

Definition 5.1. Ergodic automorphisms T_1 and T_2 of the Lebesgue spaces $(M_1, \mathcal{M}_1, \mu_1)$, $(M_2, \mathcal{M}_2, \mu_2)$ are said to be *equivalent in the sense of Kakutani* or, simply, equivalent if one of the following two conditions is satisfied:

1) there exist subsets $E_1 \in \mathcal{M}_1$, $E_2 \in \mathcal{M}_2$, $\mu_1(E_1) > 0$, $\mu_2(E_2) > 0$ such that the induced automorphisms $(T_1)_{E_1}$, $(T_2)_{E_2}$ are metrically isomorphic.

---

[3] M_l are provided with the direct product topology.

2) the natural valued functions $f_1 \in L^1(M_1, \mathcal{M}_1, \mu_1)$, $f_2 \in L^1(M_2, \mathcal{M}_2, \mu_2)$ exist such that the integral automorphisms $(T_1)^{f_1}$, $(T_2)^{f_2}$ are metrically isomorphic.

In fact, it may be easily proved that the above conditions 1), 2) are equivalent to each other, and it is for reasons of symmetry only that both of them appear in Definition 5.1. The above relation is transitive, so it is really an equivalence relation.

S. Kakutani introduced this relation in connection with his theorem on special representations of flows. The purpose was to describe the class of all possible special representations for a given flow.

Theorem 5.1 (S. Kakutani, cf [ORW]). *The ergodic automorphisms T_1, T_2 can be considered as base automorphisms in two special representations of the same flow if and only if they are equivalent.*

According to the Abramov theorem (the properties 6, 7 of entropy of an automorphism), the properties of an automorphism to have zero, positive or infinite entropy are invariant under Kakutani equivalence. For a long time only these 3 entropy classes of non-equivalent systems were known.

The examples of non-equivalent systems belonging to the same entropy class were given in the 1970's by J. Feldman who used the methods influenced by the Ornstein isomorphism theory. We shall give a brief exposition of these methods and results.

The distance $\bar{f}$ between the pairs (T_1, ξ_1), (T_2, ξ_2).

It was mentioned in Section 4 that $\bar{d}$-distance between two sequences of partitions $\{\xi_1^k\}_0^{n-1}$, $\{\xi_2^{(k)}\}_0^{n-1}$, where each $\xi_1^{(k)}$, $\xi_2^{(k)}$ contains $r < \infty$ elements, may be obtained from the Hamming metric χ in the space $M_n^{(r)}$ by using the Kantorovich-Rubinstein construction. Now we shall define another metric, χ' in $M_n^{(r)}$. For $\mathcal{I}^{(1)}, \mathcal{I}^{(2)} \in M_n^{(r)}$, $\mathcal{I}^{(1)} = (i_0^{(1)}, \dots, i_{n-1}^{(1)})$, $\mathcal{I}^{(2)} = (i_0^{(2)}, \dots, i_{n-1}^{(2)})$, we set $\chi'(\mathcal{I}^{(1)}, \mathcal{I}^{(2)}) = 1 - s/n$, where s is the maximal integer for which one can find two sequences $k_1 < k_2 < \cdots < k_s$, $m_1 < m_2 < \cdots < m_s$ with $i_{k_p}^{(1)} = i_{m_p}^{(2)}$, $1 \leqslant p \leqslant s$. It is clear that

$$\chi'(\mathcal{I}^{(1)}, \mathcal{I}^{(2)}) \leqslant \chi(\mathcal{I}^{(1)}, \mathcal{I}^{(2)}) \tag{3.2}$$

for any $\mathcal{I}^{(1)}, \mathcal{I}^{(2)} \in M_n^{(r)}$.

The Kantorovich-Rubinstein construction being applied to χ' instead of χ leads to a new metric, $\bar{f}$, measuring the distance between the measures ν_1, ν_2 on $M_n^{(r)}$, or else, between the corresponding sequences of partitions $\{\xi_1^{(k)}\}_0^{n-1}$, $\{\xi_2^{(k)}\}_0^{n-1}$:

$$\bar{f}(\{\xi_1^{(k)}\}_0^{n-1}, \{\xi_2^{(k)}\}_0^{n-1}) = \inf_\lambda \int_{M_n^{(r)} \times M_n^{(r)}} \chi'(\mathcal{I}^{(1)}, \mathcal{I}^{(2)})\, d\lambda,$$

where inf is taken over all normalized measures λ on $M_n^{(r)} \times M_n^{(r)}$ with $\lambda(A \times M_n^{(r)}) = \nu_1(A)$, $\lambda(M_n^{(r)} \times B) = \nu_2(B)$ for all $A, B \subset M_n^{(r)}$. It follows from (3.2) that $\bar{f}(\{\xi_1^{(k)}\}, \{\xi_2^{(k)}\}) \leqslant \bar{d}(\{\xi_1^{(k)}\}, \{\xi_2^{(k)}\})$.

Definition 5.2. $\bar{f}$-distance between the pairs (T_1, ξ_1), (T_2, ξ_2) is given by the formula

$$\bar{f}((T_1, \xi_1), (T_2, \xi_2)) = \overline{\lim_n}\, \bar{f}_n((T_1, \xi_1), (T_2, \xi_2)),$$

where

$$\bar{f}_n((T_1, \xi_1), (T_2, \xi_2)) = \bar{f}(\{T_1^k \xi_1\}_0^{n-1}, \{T_2^k \xi_2\}_0^{n-1}).$$

It may be thought that the $\bar{f}$-metric is related to the notion of Kakutani equivalence just in the same way as $\bar{d}$-metric is related to the notion of metric isomorphism. A considerable part of the Ornstein isomorphism theory may be translated from "$\bar{d}$-language" into "$\bar{f}$-language", and this leads to a new "equivalence theory" with many striking results. We begin with the translation of the notion of vwB partition.

LB-partitions and LB-automorphisms

Definition 5.3. A finite partition ξ of a Lebesgue space $(M, \mathcal{M}, \mu)$ is said to be *Loosely Bernoulli* (LB) with respect to an ergodic automorphism T of M, if for any $\varepsilon > 0$ there exists an integer $N = N(\varepsilon) > 0$ such that for any $m \geqslant 0$ and any $n \geqslant N$ one can find a set A consisting of the entire elements of the partition $\eta_m = \bigvee_{k=-m}^{0} T^k \xi$ and satisfying

1) $\mu(A) > 1 - \varepsilon$;

2) $\bar{f}(\{T^k \xi\}_0^{n-1}, \{T^k \xi | D\}_0^{n-1}) < \varepsilon$ for every element $D \in \eta_m$ such that $D \subseteq A$.

An automorphism T is said to be Loosely Bernoulli (LB) if it possesses a generating LB-partition.

The inequality connecting $\bar{f}$- and $\bar{d}$-metric yields the fact that the class of LB-automorphisms contains all B-automorphisms of finite entropy. Actually, this class is much wider. Many automorphisms of zero entropy, in particular, ergodic translations on commutative compact groups, ergodic interval exchange transformations (cf Sect. 2, Chap. 4) are LB. The LB-property was introduced by J. Feldman who used it in his construction of new examples of non-Bernoulli K-automorphisms.

Theorem 5.2 (J. Feldman, cf [ORW]). *If T is an LB-automorphism, then*

1) *any induced automorphism of T is LB;*

2) *any integral automorphism of T is LB;*

3) *any factor-automorphism of T is LB.*

It follows from 1) and 2) that the LB-property is stable under Kakutani equivalence.

J. Feldman has constructed an example of an ergodic automorphism T_0 of zero entropy which is not LB, i.e. in particular, is non-equivalent to any group translation. Using this example together with Theorem 5.2, one can construct automorphisms of positive entropy and even K-automorphisms without LB-property.

Consider the Bernoulli automorphism T_1 with two states having probabilities $\frac{1}{2}$, $\frac{1}{2}$ and acting in the space $(M_1, \mathcal{M}_1, \mu_1)$. Let $\xi = (C_1, C_2)$ be the Bernoulli generating partition for T_1, $\mu_1(C_1) = \mu_1(C_2) = \frac{1}{2}$. For an arbitrary ergodic automorphism T_2 of a space $(M_2, \mathcal{M}_2, \mu_2)$ consider the family of automorphisms $\{T_2(x_1)\}, x_1 \in M_1$, of the space M_2: $T_2(x_1) = T_2$ if $x_1 \in C_1$; $T_2(x_1) = Id$ if $x_1 \in C_2$. Let T be the corresponding skew product on $M = M_1 \times M_2$: $T(x_1, x_2) = (T_1 x_1, T_2 x_2)$ if $x_1 \in C_1$; $T(x_1, x_2) = (T_1 x_1, x_2)$ if $x_1 \in C_2$. I. Meilijson (cf [ORW]) proved that all such automorphisms are K-automorphisms. The induced automorphism $T_{C_1 \times M_2}$ is obviously isomorphic to the direct product $(T_1)_{C_1} \times T_2$ and thus T_2 is a factor-automorphism of $T_{C_1 \times M_2}$. In view of Theorem 5.2, the necessary condition for T to be LB is that T_2 be LB. If we take the above mentioned Feldman automorphism T_0 in the role of T_2, the corresponding skew product T will be a K-automorphism which is not only non-isomorphic to any B-automorphism, but even is non-equivalent to it in the sense of Kakutani.

S. Kalikow [K] showed that the same properties are satisfied by the following, much simpler, example. Let T_1 be, as before, the Bernoulli automorphism of the space $(M_1, \mathcal{M}_1, \mu_1)$ with two states and the probability vector $(\frac{1}{2}, \frac{1}{2})$, and $\xi = (C_1, C_2)$ be its Bernoulli generating partition, $\mu_1(C_1) = \mu_1(C_2) = \frac{1}{2}$. Let T_2 be the same automorphism acting on the space $(M_2, \mathcal{M}_2, \mu_2)$. The automorphism T of the space $M_1 \times M_2$ defined by

$$T(x_1, x_2) = \begin{cases} (T_1 x_1,\ T_2 x_2) & \text{if } x_1 \in C_1, \\ (T_1 x_1,\ T_2^{-1} x_2) & \text{if } x_1 \in C_2, \end{cases}$$

is also a K-automorphism and it lacks the LB-property.

FF-partitions. A very important tool in Ornstein's proof of the isomorphism theorem is the notion of a finitely-determined partition. The similar role in the "equivalence theory" is played by its translation into "$\bar{f}$-language".

Definition 5.4. A partition $\xi = (C_1, \ldots, C_r)$ of a Lebesgue space $(M, \mathcal{M}, \mu)$ is said to be *finitely-fixed* (FF) with respect to an ergodic automorphism T of M, if for any $\varepsilon > 0$ there exist $\delta > 0$ and $n \geqslant 1$ such that any pair $(\bar{T}, \bar{\xi})$, where $\bar{T}$ is an ergodic automorphism of a Lebesgue space $(\bar{M}, \bar{\mathcal{M}}, \bar{\mu})$, $\bar{\xi} = (\bar{C}_1, \ldots, \bar{C}_r)$ is a partition of $\bar{M}$ that satisfies

1) $|h(\bar{T}, \bar{\xi}) - h(T, \xi)| < \delta$,

2) $\sum_{0 \leqslant i_0, i_1, \ldots, i_{n-1} \leqslant r} |\bar{\mu}(\bigcap_{k=0}^{n-1} \bar{T}^k \bar{C}_{i_k}) - \mu(\bigcap_{k=0}^{n-1} T^k C_{i_k})| < \delta$,

also satisfies $\bar{f}((T, \xi), (\bar{T}, \bar{\xi})) < \varepsilon$.

An automorphism T is said to be finitely fixed if it possesses a finitely fixed generating partition.

Theorem 5.3. *For any ergodic automorphism T the classes of FF- and LB-partition coincide.*

This assertion together with Theorem 5.2 shows that FF-property is stable under Kakutani equivalence.

Theorems 5.4 and 5.5 below are the main positive results in the theory of equivalence in the sense of Kakutani. They may be viewed as "$\bar{f}$-translations" of Theorems 4.2 and 4.5.

Theorem 5.4 (cf [ORW]). *Suppose that T_1 is an ergodic automorphism of a Lebesgue space, T_2 is a FF-automorphism, and $h(T_2) \leqslant h(T_1)$. Then there exists an automorphism T_1', Kakutani equivalent to T_1 and having a factor-automorphism which is metrically isomorphic to T_2.*

Theorem 5.5 (cf [ORW]). *If T_1, T_2 are FF-automorphisms and both are either of zero entropy, or of positive entropy, then they are Kakutani equivalent.*

The analogy between the "f-theory" which was sketched above and the Ornstein "$\bar{d}$-theory" turns out to be so complete that, unfortunately, beyond the class of LB-automorphisms (just as in the Ornstein theory—beyond the class of B-automorphisms) the results are mostly negative.

There exists an example of an automorphism T which is non-equivalent in the sense of Kakutani to T^{-1} (cf [ORW]). For any h, $0 \leqslant h \leqslant \infty$, there exists uncountably many pairwise non-equivalent automorphisms of entropy h (cf [ORW]). The examples of LB-automorphisms T whose cartesian squares $T \times T$ are not LB have also been constructed (cf [ORW]).

§ 6. Shifts in the Spaces of Sequences and Gibbs Measures

Suppose M is the space of all sequences $x = \{x_n\}_{-\infty}^{\infty}$, where x_n take values in the finite space $C = \{C_1, \ldots, C_r\}$. Consider the shift T in M: $Tx = x'$, where $x_n' = x_{n+1}$. As was mentioned above, this transformation arises naturally in various problems of ergodic theory. We are now going to describe a wide class of invariant measures for T having strong mixing properties. This class includes Bernoulli and many Markov measures. The thermodynamic formalism (cf [Rue], [Si4]) is used in its construction.

Take an arbitrary function $U = U(x_0; x_1, x_{-1}, x_2, x_{-2}, \ldots)$. It will be called below a potential of interaction of the variable x_0 with all other variables x_n, $n \neq 0$ (the terminology in this section was borrowed from the statistical mechanics). It is natural to require that the main contribution to this interaction should be made by the variables x_n with n sufficiently small. Fix a sequence of non-negative numbers $\alpha = \{\alpha_n\}_{n=1}^{\infty}$, $\alpha_n \to 0$ as $n \to \infty$.

Definition 6.1. A potential U belongs to the class $\mathfrak{U}(\alpha)$ if there exist a number $C = C(U) > 0$ and a sequence of functions $U_n = U_n(x_0; x_1, x_{-1}, \ldots, x_{-n})$, such that

$$\sup_{x_m : |m| \geqslant n+1} |U(x_0; x_1, x_{-1}, \ldots, x_n, x_{-n}, x_{n+1}, x_{-n-1}, \ldots) - U_n(x_0, x_1, x_{-1}, \ldots, x_n, x_{-n})| \leqslant C\alpha_n.$$

The latter relation means that the value of the function U alters by no more than $2C\alpha_n$ under the arbitrary variations of all x_m, $|m| \geqslant n+1$. We shall consider the classes $\mathfrak{U}(\alpha)$ with $\sum \alpha_n < \infty$. Let us explain the meaning of this condition. Take a sequence $x = \{x_n\}$ and a segment $[a,b]$. Set $x' = \{x'_n\}$, where $x'_n = x_n, n \notin [a,b]$; $x'_n = C = \text{const}, n \in [a,b]$. Introduce the sum

$$H(x_k\}_a^b|\{x_k\}, k \notin [a,b]) = \sum_{k=-\infty}^{\infty} [U(x_k; x_{k+1}, x_{k-1}, \dots) - U(x'_k; x'_{k+1}, x'_{k-1}, \dots)].$$

If $\sum \alpha_n < \infty$, this sum is finite. We shall see later (cf Definition 6.2) that the exact value of $x'_n, n \in [a,b]$ does not matter for our purpose; one may define x' by fixing another sequence in $[a,b]$. The value $H(\{x_k\}_a^b|\{x_k\}, k \notin [a,b])$ will be called the energy of the configuration $\{x_k\}_a^b$ under the boundary condition $\{x_k\}, k \notin [a,b]$.

Definition 6.2. By the *conditional Gibbs state* in the segment $[a,b]$ with the boundary condition $\{x_k\}$, $k \notin [a,b]$, corresponding to a potential U, we mean the probability distribution on the space of finite sequences $\{x_k\}, a \leqslant k \leqslant b$, such that

$$P(\{x_k\}_a^b|\{x_k\}, k \notin [a,b]) = \frac{\exp(-H(\{x_k\}_a^b|\{x_k\}, k \notin [a,b])}{\Xi(\{x_k\}, k \notin [a,b])}$$

Here $\Xi(\{x_k\}, k \notin [a,b])$ is a normalizing factor which is called a statistical sum,

$$\Xi(\{x_k\}, k \notin [a,b]) = \sum_{\{x_k\}_a^b} \exp(-H(\{x_k\}_a^b|\{x_k\}, k \notin [a,b])).$$

Now let μ be an arbitrary probability measure on M. For any segment $[a,b]$ consider the measurable partition $\xi_{[a,b]}$ whose elements are obtained by fixing all x_n, $n \notin [a,b]$.

Definition 6.3. A measure μ in the space M is called *a Gibbs measure* with respect to a potential U, if for any segment $[a,b]$ its conditional measure on μ-a.e. $C_{\xi_{[a,b]}}$ is a conditional Gibbs state.

For another way to introduce the Gibbs measures see Part V, Chapter 12.

Theorem 6.1. *If $\sum \alpha_n < \infty$, then at least one Gibbs measure with respect to the potential U exists.*

A Gibbs measure which is not necessarily invariant under T may be constructed rather simply. For any increasing sequence of segments $[a_i, b_i]$, $\bigcup_i [a_i, b_i] = \mathbb{Z}^1$, and any self-consistent sequence of boundary conditions $\{x_k\}$, $k \notin [a_i, b_i]$, one may construct, using Definition 6.2, the sequence of conditional Gibbs states corresponding to the configurations in the segments $[a_i, b_i]$. It can be easily verified that any weak limit point of this sequence is a Gibbs measure. The translations of this measure, i.e. its images under T^n, $n \in \mathbb{Z}^1$, and the arithmetical means of these translations will also be Gibbs. In order to obtain a shift invariant Gibbs measure, we need only take a weak limit point of the sequence of these arithmetical means.

There are some examples of the potentials of classes $\mathfrak{U}(\alpha)$, where α is a slowly decreasing sequence, for which a Gibbs measure is not unique. The problem of construction and analysis of such measures is closely related to the theory of phase transitions in statistical physics, and we shall not discuss it. Instead, we shall be interested in the opposite case when α_n decrease rapidly.

Theorem 6.2 (cf [Rue]). *If $\alpha = \{\alpha_n\}$ satisfies $\sum n\alpha_n < \infty$, then the Gibbs measure μ_0, corresponding to a potential $U \in \mathfrak{U}(\alpha)$, is unique. The dynamical system (M, μ_0, T) is Bernoulli.*

Explain the meaning of the condition $\sum n\alpha_n < \infty$. Fix a large segment $[a,b]$ and consider the energy $H(\{x_k\}_a^b | \{x_k\}, k \notin [a,b])$. Then, if $\sum n\alpha_n < \infty$, the difference $H(\{x_k\}_a^b | \{\bar{x}_k\}, k \notin [a,b]) - H(\{x_k\}_a^b | \{\bar{\bar{x}}_k\}, k \notin [a,b])$, corresponding to boundary values $\{\bar{x}_k\}$, $\{\bar{\bar{x}}_k\}$, is bounded from above by a constant not depending on $[a,b]$, i.e. the conditional Gibbs states under the distinct boundary conditions are equivalent to each other, and the density is uniformly bounded from above and from below. The uniqueness of the Gibbs measure is an easy consequence of this fact.

The variational principle for Gibbs measures. For an arbitrary invariant under T measure μ, consider the expression

$$P(\mu) = -\int U(x)\, d\mu + h_\mu(T),$$

where $h_\mu(T)$ is the entropy of T with respect to μ. The Gibbs measure μ_0 is uniquely determined by the property

$$P(\mu_0) = \max_{\mu} P(\mu),$$

where max is taken over all measures invariant under T. There is a similar variational principle in the theory of smooth dynamical systems (cf Ch. 7, Section 3 (part II)).

The convenience of dealing with the Gibbs measures is related to the fact that they are determined by a simple functional parameter—the potential U.

Consider a sequence α with $\sum n\alpha_n < \infty$ and U_1, $U_2 \in \mathfrak{U}(\alpha)$. Suppose that the corresponding Gibbs measures are equal to each other. Then U_1, U_2 satisfy the so-called homological equation

$$U_1(x) = U_2(x) + V(Tx) - V(x),$$

where $V \in \mathfrak{U}(\alpha')$, $\alpha'_n = n\alpha_n$ (cf [Si4]). For the Gibbs measures the rate of decay of correlations has been studied in sufficient detail. The simplest case is the exponential one: $\alpha_n = \rho^n$, $0 < \rho < 1$.

Theorem 6.3 (cf [Rue]). *Suppose $U \in \mathfrak{U}(\{\rho^n\})$, and μ is the corresponding Gibbs measure. If for some function $f(x)$, $x \in M$, there exist a sequence of functions $f_n(x) = f_n(x_{-n}, \ldots, x_n)$ and positive numbers $C < \infty$, $\rho_1 < 1$ satisfying*

$$\sup_x |f(x) - f_n(x)| \leqslant C\rho_1^n,$$

then

$$\left|\int f(T^n x) f(x)\, d\mu - \left(\int f\, d\mu\right)^2\right| \leqslant K\lambda^n$$

for some positive K *and* $\lambda < 1$.

In applications, the Gibbs measures arise in a more general framework. Let $\Pi = \|\pi_{ij}\|$ be a square $r \times r$-matrix of 0's and 1's. We shall consider the so-called transitive case when for some $m > 0$ all entries of the matrix Π^m are strictly positive. Introduce the space M_Π consisting of such sequences $x = (\ldots x_{-1}, x_0, x_1, \ldots) \in M$ that $\pi_{x_n, x_{n+1}} = 1$, $-\infty < n < \infty$. In other words, the matrix Π dictates which symbols may occur as neighbours in the sequences $x \in M_\Pi$. Our initial situation may now be considered as a special case, when $\pi_{ij} \equiv 1$.

The shift T on M_Π is called a topological Markov chain. By a measure of maximal entropy for the shift T we mean an invariant measure μ_0 such that $h_{\mu_0}(T) = \max_\mu h_\mu(T)$, where max is taken over all invariant measures μ. In the case considered there is an explicit way to construct such a measure μ_0. Suppose $e = \{e_i\}_1^r$ is the eigenvector for Π with positive coordinates corresponding to a positive eigenvalue λ, i.e. $\Pi e = \lambda e$, and $e^* = \{e_i^*\}_1^r$ is a similar vector for Π^*, i.e. $\Pi^* e^* = \lambda e^*$. The existence of e, e^* follows from the well-known Perron-Frobenius theorem. Since all π_{ij} are integers, we have $\lambda > 1$. The measure μ_0 is the Markov measure with the transition probabilities $p_{ij} = \pi_{ij} e_j / \lambda e_i$ and the stationary distribution $\{e_i e_i^*\}_1^r$, where e, e^* are normalized in order to have $\sum_i e_i e_i^* = 1$.

Theorem 6.4. *Let* $\mathcal{O}_n (n = 1, 2, \ldots)$ *be the set of all points of period* n *for* T *(i.e.* $T^n x = x$, $x \in \mathcal{O}_n$*). Then*

1) $\lim\limits_{n\to\infty} \frac{1}{n} \log \operatorname{card}(\mathcal{O}_n) = \log \lambda$,
2) *for any* $f \in C(M_\Pi)$ *one has*

$$\lim_{n\to\infty} \frac{1}{\operatorname{card}(\mathcal{O}_n)} \sum_{x \in \mathcal{O}_n} f(x) = \int f\, d\mu_0.$$

In other words, the periodic points for T are uniformly distributed in the space M_Π with respect to μ_0.

A considerable generalization of this theorem will be given in Part II, Chapter 7.

The definitions of conditional Gibbs states and Gibbs measures may be carried over to the case of shifts in M_Π without any changement. Theorems 6.1 and 6.2 are still valid in this situation as well as the variational principle. In particular, the measure of maximal entropy may be defined by means of the variational principle with $U \equiv 0$.

Chapter 4
Periodic Approximations and Their Applications. Ergodic Theorems, Spectral and Entropy Theory for the General Group Actions[1]

I.P. Kornfeld, A.M. Vershik

§1. Approximation Theory of Dynamical Systems by Periodic Ones. Flows on the Two-Dimensional Torus

Dynamical systems whose trajectories all have the same period are usually called periodic. It seems natural to consider these systems, i.e. the ones with the simplest possible behavior of the trajectories, as an appropriate tool for approximating the systems of general form. In this regard their role is similar to that of polynomials and rational functions in the constructive theory of functions, where the functions of general form are approximated by them.

The starting point in the study of the periodic approximations of dynamical systems is the following fact which may be considered as "the existence theorem" for such approximations.

Definition 1.1. An automorphism T is said to be *aperiodic* if the set of its periodic points is of zero measure.

Theorem 1.1 (The Rokhlin-Halmos Lemma, cf P. Halmos [H]). *If T is an aperiodic automorphism of a Lebesgue space $(M, \mathcal{M}, \mu)$, then for any $\varepsilon > 0$ and any natural number n there is a set $E \in \mathcal{M}$ such that*

1) $T^iE \cap T^jE = \varnothing, 0 \leqslant i \neq j \leqslant n-1$;
2) $\mu(\bigcup_{k=0}^{n-1} T^iE) > 1 - \varepsilon$.

An immediate consequence of the Rokhlin-Halmos lemma is the following assertion.

Corollary. *The set of periodic automorphisms is dense in the space of all automorphisms of the Lebesgue space $(M, \mathcal{M}, \mu)$ provided with uniform topology i.e. the one defined by the metric $d(T_1, T_2) = \sup_{E \in \mathcal{M}} \mu(T_1 E \triangle T_2 E)$.*

Various properties of the dynamical systems are connected to the rapidity of their approximation by the periodic ones. To obtain concrete results of this kind, it is necessary to specify the notion of speed of approximation.

Definition 1.2. Suppose $f(n) \searrow 0$. An automorphism T of the space $(M, \mathcal{M}, \mu)$ admits an approximation of the first type by periodic transformations (aptI) with

---

[1] Sections 1 and 2 were written in collaboration with E.A. Sataev.

speed $f(n)$, if one can find a sequence of partitions $\{\xi_n\}$, $\xi_n \to \varepsilon$, and a sequence of automorphisms $\{T_n\}$, such that

1) T_n preserves ξ_n, i.e. it sends each element of ξ_n to an element of the same partition;

2) $\sum_{i=1}^{q_n} \mu(TC_i^{(n)} \triangle T_n C_i^{(n)}) < f(q_n)$, $n = 1, 2, \dots$, where $\{C_i^{(n)}\}$, $1 \leqslant i \leqslant q_n$, is the collection of all elements of ξ_n.

If T satisfies 1), 2) and

3) T_n cyclically permutes the elements of ξ_n,

then T is said to admit a cyclic approximation with speed $f(n)$.

If for the sequences of partitions $\{\xi_n\}$, $\xi_n = \{C_i^{(n)}\}_{i=1}^{q_n}$, and periodic automorphisms T_n we have

1′) T_n preserves ξ_n;

2′) $\sum_{i=1}^{q_n} \mu(TC_i^{(n)} \triangle T_n C_i^{(n)}) < f(p_n)$, where $p_n = \min\{p \geqslant 1\colon T_n^p = Id\}$;

3′) $U_{T_n} \to U_T$ in strong operator topology in $L^2(M, \mathscr{M}, \mu)$, where U_{T_n}, U_T are the unitary operators adjoint to T_n, T respectively,

then T is said to admit an approximation of the second type (aptII) with speed $f(n)$.

A similar definition of various types of approximation can be given in the case of continuous time.

Definition 1.3. Suppose $g(u) \searrow 0$. A flow $\{T^t\}$ on a Lebesgue space $(M, \mathscr{M}, \mu)$ is said to admit an approximation of the first type by periodic transformations (aptI), if one can indicate sequences of real numbers t_n, of partitions ξ_n of the space M into q_n sets $C_i^{(n)} \in \mathscr{M}$, and of automorphisms S_n of the space M such that

1) $\xi_n \to \varepsilon$;

2) S_n preserves ξ_n;

3) $\sum_{i=1}^{q_n} \mu(T^{t_n} C_i^{(n)} \triangle S_n C_i^{(n)}) < g(q_n)$;

4) $p_n t_n \to \infty$, where p_n is the order of S_n as a permutation of the sets $C_i^{(n)}$ i.e. $p_n = \min\{p \geqslant 1\colon S_n^p C_i^{(n)} = C_i^{(n)}, 1 \leqslant i \leqslant q_n\}$.

If we have, in addition,

5) S_n cyclically permutes the sets $C_i^{(n)}$,

then the flow $\{T^t\}$ is said to admit a cyclic approximation with speed $g(u)$.

If the sequences of real numbers t_n, of partitions ξ_n and of periodic automorphisms S_n satisfy:

1′) $\xi_n \to \varepsilon$;

2′) S_n preserves ξ_n;

3′) $p_n t_n \to \infty$, where p_n is the order of S_n as a permutation of the elements $C_i^{(n)}$ of ξ_n;

4′) $\sum_{i=1}^{q_n} \mu(T^{t_n} C_i^{(n)} \triangle S_n C_i^{(n)}) < g(p_n)$;

5′) for any element $f \in L^2(M, \mathscr{M}, \mu)$ we have

$$\lim_{n \to \infty} \|U_{T^{t_n}} f - U_{S_n} f\| = 0,$$

then the flow $\{T^t\}$ is said to admit approximation of the second type by periodic transformations (aptII) with speed $g(u)$.

Example. Suppose $T = T_\alpha$ is a rotation of the circle S^1 by an irrational angle α, i.e. $Tx = (x + \alpha)(\mathrm{mod}\ 1)$, $x \in [0, 1)$, and $\alpha_n = p_n/q_n$ is a sequence of irreducible fractions such that $\lim_{n\to\infty} \alpha_n = \alpha$. Suppose, further, that for some function $f(n)$ satisfying $n \cdot f(n) \searrow 0$, we have

$$|\alpha - p_n/q_n| < f(q_n), \qquad n = 1, 2, \ldots. \tag{4.1}$$

Taking $\xi_n = \{C_i^{(n)}\}_1^{q_n}$, $C_i^{(n)} = \left[\dfrac{i-1}{q_n}, \dfrac{i}{q_n}\right)$, and a sequence of the rotations of the circle $T_n = T_{\alpha_n}$ by the angles α_n, we can prove that T_α admits a cyclic approximation with speed $2n \cdot f(n)$. It is known (from the theory of continuous fractions) that for every α there exists a sequence $\{p_n/q_n\}_1^\infty$ such that (4.1) holds with $f(n) = \dfrac{1}{\sqrt{5} \cdot n^2}$. Any rotation T_α therefore admits a cyclic approximation with speed $f(n) = \dfrac{2}{\sqrt{5} \cdot n}$.

It is natural to suggest that the faster a dynamical system can be approximated by periodic transformations, the worse its statistical properties are.

We will formulate a series of rigorous results confirming this suggestion and concerning the relationship of mixing, spectral and entropy properties with the speed of approximation.

Theorem 1.2 (cf [KS]). *If an automorphism T (a flow $\{T^t\}$) admits an aptII with speed $f(n) = \theta/n$, where $\theta < 2$, then $T(\{T^t\})$ is not mixing.*

Theorem 1.3 (cf [KS]). *If an automorphism T (a flow $\{T^t\}$) admits a cyclic approximation with speed $f(n) = \theta/n$, $\theta < \frac{1}{2}$, then the unitary operator U_T (the group of unitary operators $\{U_{T^t}\}$) has a simple spectrum.*

Theorem 1.4 (cf [KS]). *If an automorphism T (a flow $\{T^t\}$) admits an aptII with speed $f(n) = \theta/n$, $\theta < \frac{1}{2}$, then the maximal spectral type of the unitary operator U_T (of the group of unitary operators $\{U_{T^t}\}$) is singular with respect to the Lebesgue measure.*

Theorem 1.5 (cf [KS]). *If T is an ergodic automorphism of entropy $h(T)$, then $h(T) = \frac{1}{2}c(T)$, where $c(T)$ is the infimum of the set of positive numbers θ, for which T admits aptI with speed $f(n) = \theta/\log n$.*

The situation becomes quite different when we are interested in relations between ergodicity and the speed of (cyclic) approximation: a sufficiently fast cyclic approximation guarantees the ergodicity of an automorphism.

Theorem 1.6 (cf [KS]). *If an automorphism T admits cyclic approximation with speed $f(n) = \theta/n$, $\theta < 4$, then T is ergodic.*

There is an estimate from below for the speed of approximation (of aptI) which is valid for all automorphisms. This estimate may be considered as a sharpened version of Rokhlin-Halmos lemma.

Theorem 1.7 (cf [KS]). *Any automorphism T admits aptI with speed $f(n) = a_n/\log n$, where a_n is an arbitrary monotonic sequence of real numbers tending to infinity.*

As for the estimates from above, no such estimate valid for all automorphisms can be obtained. Moreover, there exist automorphisms such that for any $f(n) \searrow 0$, they admit cyclic approximation with speed $f(n)$. This property is satisfied, for example, by the automorphism T with pure point spectrum consisting of the numbers of the form $\exp 2\pi ip/2^q$ (see Chap. 2, Sect. 2, p. 34).

To a great extent the significance of the approximation method in ergodic theory is due to the fact that it enables us to construct the concrete examples of dynamical systems having various non-trivial metric and spectral properties.

With the help of the approximation theory, the automorphism T such that the maximal spectral type σ of the operator U_T does not dominate its convolutional square $\sigma * \sigma$ (i.e. the automorphism with a spectrum lacking the group property) was constructed. We now explain the meaning of this example. For an ergodic automorphism T with pure point spectrum, to say that the maximal spectral type σ of U_T dominates the type $\sigma * \sigma$ is simply to reformulate the fact that the set of eigenvalues of U_T is a subgroup of S^1. The above example showed that in this problem, like in many others, the situation with continuous and mixed spectra is more complicated. Another application of the approximation method is the example of the automorphism with continuous spectrum which has no square roots (cf [KS]).

The complete investigation of spectral properties of smooth dynamical systems on the 2-dimensional torus was also carried out with the help of the approximation theory.

Suppose $M = \mathbb{R}^2/\mathbb{Z}^2$ is the two-dimensional torus with cyclic coordinates u, v and normalized Lebesgue measure $du\,dv$. Consider the system of differentional equations

$$\frac{du}{dt} = A(u,v), \qquad \frac{dv}{dt} = B(u,v) \tag{4.2}$$

on it with right-hand sides of class C^r, $r \geqslant 2$. Define a flow $\{T^t\}$ on M as the one-parameter group of translations along the solutions of (4.2). We will assume that T preserves a measure μ that is absolutely continuous with respect to $du\,dv$ with density $P(u,v)$ of class C^5 and that $A^2 + B^2 > 0$, i.e. that system (4.2) has no fixed points. The number $\lambda = \lambda_1/\lambda_2$, where $\lambda_1 = \iint_M PA\,du\,dv$, $\lambda_2 = \iint_M PB\,du\,dv$, is called a rotation number of the system (4.2). Spectral properties of the flow $\{T^t\}$ are connected with the speed of approximation of λ by rational numbers.

If λ is rational or if at least one of the numbers λ_1, λ_2 is equal to zero, then the flow $\{T^t\}$ cannot be ergodic. If λ is irrational, the study of the properties of the flow $\{T^t\}$ is based on a special representation of this flow.

Theorem 1.8 (A.N. Kolmogorov [Kol2]). *If λ is irrational, then the flow $\{T^t\}$ is metrically isomorphic to the special flow constructed from the automorphism T_1*

of rotation of the circle S^1 by a certain irrational angle α, where α is of the form

$$\alpha = \frac{m\lambda + n}{p\lambda + q}; \qquad m, n, p, q \in \mathbb{Z}^1, \qquad \det \|{}^{m}_{p}\,{}^{n}_{q}\| = 1,$$

and a function $F\colon S^1 \to \mathbb{R}^1$ of class C^5.

The proof of Theorem 1.8 is based on the fact that there is a smooth closed non-self-intersecting curve Γ on the torus, which is transversal to the trajectories of the flow $\{T^t\}$ at all its points, and such that for any trajectory of $\{T^t\}$ there are infinitely many moments $t > 0$ and infinitely many moments $t < 0$, when it intersects the curve Γ. Such a Γ is known as the Siegel curve for $\{T^t\}$. The transformation of Γ which sends any point $x \in \Gamma$ to a point $T^{t_0}x$, where $t_0 = \min\{t > 0\colon T^t x \subset \Gamma\}$, is conjugate to a certain rotation of S^1, which is the base automorphism of the special representation of $\{T^t\}$. If the number λ and, therefore, all numbers α of the form $\alpha = \dfrac{m\lambda + n}{p\lambda + q}$, are poorly approximable by rational numbers, then the special flow $\{T^t\}$, appearing in the statement of Theorem 1.8, is metrically isomorphic to the special flow constructed from the same base automorphism and a constant function. With such an argument, we can obtain the following result.

Theorem 1.9 (A.N. Kolmogorov [Kol2]). *If λ satisfies*

$$|\lambda - p/q| \geqslant \mathrm{const}\cdot q^{-4}, \qquad \mathrm{const} > 0,$$

for all integers $p, q, q \neq 0$, then the group of unitary operators adjoint to the flow $\{T^t\}$ has pure point spectrum consisting of numbers of the form const$(k + l\lambda)$, $-\infty < k, l < \infty$, where k, l are integers.

A sufficiently fast approximation of the number λ by rational numbers guarantees a sufficiently fast cyclic approximation of the special flow from Theorem 1.8.

Theorem 1.10 (A. Katok [Kat1]). *If for the number λ there is a sequence $\{p_n/q_n\}$ of irreducible fractions such that*

$$q_n^4 |\lambda - p_n/q_n| \to 0$$

when $n \to \infty$, then the flow $\{T^t\}$ admits a cyclic approximation with speed $g(u) = o(u^{-2})$.

Theorems 1.9 and 1.10, together with general theorems about approximations (Theorems 1.2, 1.3, 1.4) show that for any irrational λ the flow $\{T^t\}$ is not mixing, the spectrum of the adjoint group $\{U^t\}$ of unitary operators is simple and the maximal spectral type is singular with respect to the Lebesgue measure.

It was indicated by A.N. Kolmogorov [Kol2] that a smooth flow $\{T^t\}$ given by the equations (4.2) and having no fixed points may be weak mixing, i.e. the spectrum of the adjoint group $\{U^t\}$ may be continuous in the orthogonal complement to the subspace of constant functions. The explicit construction of such examples was developed by M.D. Shklover [Sh]. There is a modification of this construction which enables one to obtain more general examples (D.V. Anosov [An]).

Up to this point it was assumed that the flow $\{T^t\}$ has no fixed points. If we omit this assumption, the situation becomes quite different. Such flows may be mixing [Ko2]. If the right-hand sides of (4.2) are continuous but not necessarily smooth, then the spectrum of the adjoint group $\{U^t\}$ may have both continuous and discrete components (even under the "no fixed point" assumption [Kry].

§ 2. Flows on the Surfaces of Genus $p \geqslant 1$ and Interval Exchange Transformations

The approximation method turned out to be an appropriate tool in the study of ergodic properties of smooth flows not only on the 2-dimensional torus but also on general orientable surfaces of genus $p \geqslant 1$. For such a flow under some weak conditions, a transversal closed curve (similar to the Siegel curve for the flows on torus) can also be constructed, and this curve, in turn, enables one to construct the special representation of the flow. However, the base automorphisms in this case are of more general form than the rotations of the circle—they are the so-called interval exchange transformations. The study of such transformations is also of intrinsic interest.

Suppose the space M is the semi-interval $[0, 1)$, $\xi = (\Delta_1, \ldots, \Delta_r)$ is a partition of M into r, $2 \leqslant r < \infty$, disjoint semi-intervals numbered from left to right, $\pi = (\pi_1, \ldots, \pi_r)$ is a permutation of the integers $1, 2, \ldots, r$.

Definition 2.1. Suppose the transformation $T : M \to M$ is a translation on each of the semi-intervals Δ_i, $1 \leqslant i \leqslant r$, and it "exchanges" these semi-intervals according to the permutation π, i.e. the semi-intervals $\Delta'_i = T\Delta_i$ adhere to each other in the order $\Delta'_{\pi_1}, \ldots, \Delta'_{\pi_r}$. Then T is said to be the *interval exchange transformation* corresponding to the partition ξ and the permutation π.

Any interval exchange transformation is an invertible transformation of M with finitely many discontinuities, preserving the Lebesgue measure. If T has at least one periodic point, it cannot be ergodic with respect to the Lebesgue measure, since there exists in this case a non-trivial union of finite number of intervals which is invariant with respect to T.

Theorem 2.1 (M. Keane [Ke1]). *The following properties of T are equivalent:*

i) *T is aperiodic* (*i.e. T has no periodic points*)

ii) *T is topologically transitive* (*i.e. T has an everywhere dense trajectory*);

iii) *T is minimal* (*i.e. all its trajectories are dense in M*);

iv) $\max_{1 \leqslant i_0, i_1, \ldots, i_n \leqslant r} \operatorname{diam}(\Delta_{i_0} \cap T\Delta_{i_1} \cap \ldots \cap T^n \Delta_{i_n}) \to 0$ *when* $n \to \infty$;

v) *the trajectories of the discontinuity points of T are infinite and distinct.*

If $r = 2$ (recall that r stands for the number of the interval exchanged), the interval exchange transformation is isomorphic to some rotation of the circle; for $r = 3$ it is isomorphic to the induced transformation constructed from a rotation automorphism and some interval $\Delta \subset [0, 1)$. This implies that for $r = 2, 3$ any aperiodic (minimal) interval exchange transformation is ergodic with respect to the Lebesgue measure, and, moreover, is uniquely ergodic. In other words, the Lebesgue measure is the unique invariant normalized Borel measure for it. For any $r \geqslant 4$ this assertion is not true (Keane [Ke2]).

Even the ergodicity with respect to the Lebesgue measure does not guarantee the unique ergodicity of the interval exchange transformation [Ke2]. However, the number of pairwise distinct ergodic normalized invariant measures for an arbitrary interval exchange transformation is always finite. The following estimate holds.

Theorem 2.2 (W. Veech [V1]). *There are at most $[r/2]$ ergodic normalized invariant measures for any exchange transformation of r intervals.*

The cases when the interval exchange transformations have more than one invariant normalized measure (i.e. they are not uniquely ergodic) may be considered in some sense as exceptional ones. In order to explain the exact meaning of this assertion, note that any interval exchange transformation is entirely determined by the pair (λ, π), where $\lambda = (\lambda_1, \ldots, \lambda_r)$, $\lambda_i \geqslant 0$, $\sum_{i=1}^{r} \lambda_i = 1$, is the vector of the lengths of the exchanged intervals, and π is the corresponding permutation.

Denote by $T_{\lambda,\pi}$ the transformation corresponding to the pair (λ, π). Then $T_{\lambda,\pi}$ is obviously non-ergodic if the permutation π is reducible, that is $\pi(\{1, 2, \ldots, j\}) = \{1, 2, \ldots, j\}$ for some j, $1 \leqslant j < r$.

Theorem 2.3 (M. Keane, G. Rauzy, V.A. Chulajevsky [Ch]). *Let an integer $r \geqslant 2$ and an irreducible permutation π of the set $\{1, \ldots, r\}$ be fixed. Then the set of all $\lambda \in \mathbb{R}^r$ for which $T_{\lambda,\pi}$ is non-uniquely ergodic is of first category in the sense of Baire.*

M. Keane conjectured that the measure-theoretic version of Theorem 2.3 is also true, i.e. that "almost all" interval exchange transformations are uniquely ergodic. This conjecture was settled by H. Measur and W. Veech independently, and both proofs use very deep methods.

Theorem 2.4 (H. Masur [Mas], W. Veech [V2]). *If $r \geqslant 2$ and π is an irreducible permutation of the set $\{1, 2, \ldots, r\}$, then the set of all $\lambda \in \mathbb{R}^r$ for which $T_{\lambda,\pi}$ is not uniquely ergodic, is of zero Lebesgue measure.*

At the present time several independent proofs of this theorem are known. A rather "elementary" proof was proposed recently by M. Boshernitzan. In this proof a certain "property $\mathscr{P}$" of an interval exchange transformation was defined in such a way that a) $\mathscr{P}$ is satisfied "almost everywhere", b) minimality of an interval exchange transformation together with property $\mathscr{P}$ imply its unique ergodicity. We shall now give the exact definition of $\mathscr{P}$.

A subset A of positive integers will be called essential if for any integer $l \geqslant 2$ there is a real number $c > 1$ for which the system of inequalities

$$\begin{cases} n_{i+1} > 2n_i, & 1 \leqslant i \leqslant l-1, \\ n_l < cn_1 \end{cases}$$

has infinitely many solutions $(n_1, \ldots, n_l)$, all $n_i \in A$.

Let T be an interval exchange transformation. Denote by $\lambda^{(n)} = (\lambda_1^{(n)}, \ldots, \lambda_{r_n}^{(n)})$ the vector of the lengths of intervals exchanged by T^n, $n = 1, 2, \ldots$. Set $m_n(T) = \min_{1 \leqslant i \leqslant r_n} \lambda_i^{(n)}$.

Definition 2.2. *An interval exchange transformation has Property $\mathscr{P}$ if for some $\varepsilon > 0$ the set $A = A(T, \varepsilon) = \{n \in N : m_n(T) \geqslant \varepsilon/n\}$ is essential.*

Theorem 2.5 (M. Boshernitzan). *Let T be a minimal interval exchange transformation which satisfies Property $\mathscr{P}$. Then T is uniquely ergodic. For any $r \geqslant 2$ and any permutation π of r symbols, the set of those $\lambda \in \mathbb{R}^r$,* for which *$T_{\lambda,\pi}$ does not satisfy Property $\mathscr{P}$, is of zero Lebesgue measure.*

Further results on the ergodic and spectral properties of "typical" interval exchange transformations were obtained by W. Veech [V3]. He proved that almost all such transformations (in the same sense as in Theorem 2.4) are totally ergodic (i.e. all powers T^n are ergodic with respect to the Lebesgue measure) and have simple spectrum [V3]. On the other hand, there is the following negative result concerning the strong mixing property.

Theorem 2.6 (A.B. Katok, cf [CFS]). *Suppose T is an interval exchange transformation, μ is an arbitrary invariant Borel measure for T. Then T is not mixing with respect to μ.*

Some properties of the flows on surfaces of genus $p \geqslant 1$ were established by methods similar to those of the theory of interval exchange transformations or else were deduced from the corresponding results about interval exchanges using the above mentioned special representation of such flows.

Suppose $\{T^t\}$ is a topologically transitive flow of class C^1 on the 2-dimensional compact oriented manifold of genus $p \geqslant 1$ (cf Vol. 1, Part II) with a finite number of fixed points all of them being non-degenerate saddles. Assume that $\{T^t\}$ has no wandering points (i.e. points x such that for some neighborhood $U \ni x$ and some t_0 one has $U \cap T^t U = \varnothing$ for $|t| \geqslant t_0$). It was proved in [Kat2] that the number of non-trivial normalized ergodic measures for such flows (i.e. measures such that any trajectory of the flow is of zero measure) does not exceed p. This estimate is exact: for any pair of natural numbers $p, k, p \geqslant k$, there is a topologically transitive flow of class C^∞ on a surface M_p of genus p having k non-trivial ergodic normalized measures and $2p - 2$ fixed points which are non-degenerate saddles (E.A. Sataev [S]). In [Blo] the examples of uniquely ergodic flows have been constructed on all surfaces except for the sphere, the projective plane and the Klein bottle, where the non-existence of such flows was already known. In

[Ko2] the examples of mixing flows of class C^∞ for which the invariant measure has the density of class C^∞, have been constructed on all surfaces, again except for the three mentioned above.

§ 3. General Group Actions

3.1. Introduction. Ergodic theory of general group actons (theory of dynamical systems with "general time") deals with arbitrary groups of transformations with invariant and quasi-invariant measures. The classical cases are those of groups $\mathbb{Z}^1$ and $\mathbb{R}^1$. The investigation of other group actions was initially motivated mainly by its applications to the study of classical ones. One of the earliest and most interesting examples of this kind is the geodesic flow on a closed surface of constant negative curvature. It was shown by I.M. Gelfand and S.V. Fomin [GF] that this flow can be represented as an action of a certain 1-parameter (hyperbolic) subgroup of the group $SL(2, \mathbb{R})$ on the homogeneous space $M = SL(2, \mathbb{R})/\Gamma$, where Γ is a discrete group. The information about the unitrary representations of $SL(2, \mathbb{R})$ in $L^2(M)$ enabled them to prove easily that the spectrum of a geodesic flow is countable Lebesgue, a fact that was first obtained in the 1940's by E. Hopf and G. Hedlund with the use of rather difficult methods (cf [Mau]). The same idea was employed later in some other problems (the horocycle flow on the surface of constant negative curvature), and it stimulated the systematic study of the flows on homogeneous spaces in connection with the actions of Lie groups on them (cf [AGH]).

Another example is related to the approximation theory. It is useful for the study of approximations to consider the actions of locally finite groups, such as $\Sigma \mathbb{Z}_2$, and of the quasicyclic group. The general theory of actions of such groups (the isomorphy problem, spectral theory, the construction of metric invariants) is no simpler that the corresponding theory for $\mathbb{Z}$ and in many points is parallel to it; on the other hand, there are many questions about approximations which are much simpler in these cases than in the case of $\mathbb{Z}$. Therefore, $\Sigma \mathbb{Z}_2$ and the quasicyclic group may be taken as natural model examples in the approximation theory.

In the 1970's the advantages of systematic study of general group actions became quite evident, and the corresponding theory, closely related to the group representations theory, the theory of Lie groups and differential geometry, was intensely developed. The ergodic methods, in turn, were applied to some problems of Lie groups (i.e. the Mostow-Margulis theory of arithmetic subgroups) and group representatons theory. It should be noted that some metric problems concerning the actions of $\mathbb{Z}^n$, $\mathbb{R}^n$ found their applications to mathematical physics. In recent years, much effort has been devoted to the study of infinite dimensional ("large") groups, such as the groups of diffeomorphisms and of currents. There is a very important difference between the properties of actions of locally-compact and of non-locally compact groups: namely, in non-locally compact case it may occur that even a quasi-invariant measure does not exist and, so, we may have no correct definitions of such notions as the decomposition into ergodic components, the trajectory partition. For locally compact groups these questions may be settled by the same methods as for $\mathbb{Z}^1$ and $\mathbb{R}^1$. We will restrict ourselves to the case of locally compact groups and consider some general questions: the definition of group actions, ergodic theorems, the characterization of discrete spectrum.

3.2. General Definition of the Actions of Locally Compact Groups on Lebesgue Spaces. Let G be an arbitrary locally compact separable group and $(M, \mathcal{M}, \mu)$ be a Lebesgue space. There are two natural ways to define a measure preserving action of G on $(M, \mathcal{M}, \mu)$. These two definitions obviously coincide in the case of discrete groups, while in the case of continuous groups (including the case $G = \mathbb{R}^1$) they differ considerably.

1) *Definition of a measurable action* (cf Chap. 1, Sect. 1). Consider a map T: $G \times M \to M$, $T(g, x) = T_g x$, satisfying the following conditions: a) measurability, i.e. T is measurable as a map from $(G \times M, m \times \mu)$ into (M, μ), where m is a Haar measure on G (we need not specify it, since it is the measurability itself, and not the numerical value of the measure, that counts); b) invariance: $\mu(T_g^{-1} A) = \mu(A)$, $A \in \mathcal{M}$; c) the group property: $T_{g_1 g_2} x = T_{g_1} T_{g_2} x$. $T_e x = x$. In these equalities x runs through some set of full measure which does not depend on g_1, g_2.

2) *Definition of a continuous action.* Consider a map $g \mapsto \tilde{T}_g$, associating to any $g \in G$ some element of the group of coinciding (mod 0) automorphisms of the space $(M, \mathcal{M}, \mu)$ provided with weak topology.

Assume that the following conditions are satisfied: a) the function $g \mapsto \mu(\tilde{T}_g A \cap B)$ is continuous on G for any A, $B \in \mathcal{M}$; b) $\mu(\tilde{T}_g A) = \mu(A)$; c) $\tilde{T}_{g_1 g_2} = \tilde{T}_{g_1} \tilde{T}_{g_2}$, where $\tilde{T}_g$ is a class of coinciding (mod 0) automorphisms.

It follows from general facts that to define a continuous action is just the same as to define the continuous homomorphism $g \mapsto U_g$ of G into the group of real unitary operators of Hilbert space preserving the structure of "partial multiplication" in L^2: $(U_g f)(x) = f(T_g^{-1} x)$. Hence, a continuous action determines in a standard way a certain weakly continuous representation $g \mapsto U_g$ of the group G. It is known as the Koopman representation.

It is easily seen that any measurable action is continuous. To prove this, consider the operators U_g and use the fact that if for all pairs f_1, f_2 of elements of the Hilbert space H the functions $F(g) = (U_g f_1, f_2)$ are measurable on G, then they are necessarily continuous.

The converse of our statement is also true, but it is more difficult. This problem is known as the group lifting problem. One has to prove that if the relation

$$T_{g_1 g_2} x = T_{g_1} T_{g_2} x$$

holds for almost every x for any pair (g_1, g_2), but the set of such x may depend on g_1, g_2, then there exists a measurable group $g \mapsto T'_g$, for which $T_g = T'_g \pmod 0$ for all $g \in G$.

In the case of countable groups the proof of this statement is rather simple (J. von Neumann, P. Halmos, cf [Ro2]), since we have only a countable number of $g \in G$, and for each g we can alter the definition of T_g on a set of zero measure. In the case of continuous groups more delicate methods for constructing the measurable realization are needed, since there may be continuum sets of zero measure, where the group condition is not satisfied, and their union may coincide with the entire space. For $G = \mathbb{R}^1$ and the pure point spectrum the question was settled positively by V.A. Rohlin in [Ro2]. In the general case of locally compact groups, the positive solutions (in somewhat different terms) were given by G. Mackey [M2] and A.M. Vershik [Ve1]; in the case $G = \mathbb{R}^1$—by C. Maruyama [Ma]. The uniqueness was proved in [Ve1]. A simpler proof was given in [Ve6].

The final statement is as follows:

Theorem 3.1. *Any continuous action of a separable locally compact group G with an invariant measure on the Lebesgue space $(M, \mathcal{M}, \mu)$ possesses a unique* (mod 0) *measurable realization.*

Therefore, one need not distinguish between continuous and measurable actions and may deal with the kind of action which is more convenient in a given situaton. So, in entropy theory it seems natural to deal with continuous actons, while in trajectory theory (cf Chap. 5) the measurable actions are usually considered. The statement of Theorem 3.1 is also valid for the action with quasi-invariant measures. However, it is not true without the assumption of local compactness of G: the existence of Haar measure is essential for its proof.

3.3. Ergodic Theorems. First investigations of the actions of general groups on measure spaces were concerned, in particular, with the generalizations of ergodic theorems (see, for example [C]). From the probabilistic point of view these theorems may be considered as the laws of large numbers stationary in narrow sense random processes, while from the physical point of view they justify the interchangeability of the "time" means (the integrals over G) and the space means.

Suppose that G is a group acting on $(M, \mathcal{M}, \mu)$ in the sense of definitions given in 3.2. Fix a family $\{G_n\}$, $n = 1, 2, \ldots$, of compact subsets of G. We say that the individual ergodic theorem is satisfied for the action $g \mapsto T_g$ with respect to $\{G_n\}$ if for any $f \in L^1(M)$ the limit

$$\lim_{n\to\infty} \frac{1}{\operatorname{card}(G_n)} \sum_{g \in G_n} f(T_g x) = \bar{f}(x)$$

(in the case of discrete groups), or else

$$\lim_{n\to\infty} \frac{1}{m(G_n)} \int_{G_n} f(T_g x)\, dm(g) = \bar{f}(x),$$

where m is the Haar measure (in the case of continuous groups) exists almost everywhere. The function $\bar{f}$ is the projection of f to the subspace of G-invariant functions; if the action is ergodic, the $\bar{f} = \text{const} = \int_M f\,d\mu$.

Recall that for $\mathbb{Z}^1$ and $\mathbb{R}^1$ the Birkhoff-Khinchin ergodic theorem claims that:

$$\lim_{n\to\infty} \frac{1}{n} \sum_{k=0}^{n-1} f(T^k x) = \bar{f}(x) \quad \text{almost everywhere,}$$

$$\lim_{T\to\infty} \frac{1}{T} \int_0^T f(T^t x)\,dt = \bar{f}(x) \quad \text{almost everywhere.}$$

A new and very simple proof of this theorem for the group $\mathbb{Z}^1$ was given recently by Y. Katznelson and B. Weiss [KW2]. Their proof uses parts of the Kamae arguments which he used to obtain a proof of Birkhoff-Khinchin theorem based on of the non-standard analysis.

A sequence $\{G_n\}$, $G_n \subset G$, is called a universal averaging sequence for G if for all measure preserving actions of G the individual ergodic theorem is satisfied with respect to $\{G_n\}$. There are universal sequences for the groups $\mathbb{Z}^m$, $\mathbb{R}^m$: one can take as $\{G_n\}$ the sequence of cubes whose sides increase to infinity. The existence of such sequences is also known for solvable groups, but their construction is more difficult. A general method of proving such theorems (based on the properties of the martingales) was proposed by A.M. Vershik.

Unfortunately, at the present time there is no final solution to this problem for arbitrary locally compact groups.

A natural hypothesis that in the case of the amenable groups, the so-called Følner sequences, i.e. the sequences $\{G_n\}$, $G_n \subset G$, such that for any $h_1, \ldots, h_k \in G$

$$\lim_{n\to\infty} \frac{m\left(\bigcap_{i=1}^{k} h_i G_n\right)}{m(G_n)} = 1,$$

(m is a Haar measure), should be universal, turned out to be false.

A series of works by A.A. Tempelman (cf. [T1], [T2]) contains, in particular, a proof of the individual ergodic theorem for the groups with polynomial growth of number of words; some sufficient conditions for a Følner sequence to be universal have been given in his theorems. These results seem to be the most general known "universal" (i.e. not depending on a specific action) individual ergodic theorems for the locally compact groups. However, more general methods are needed for arbitrary amenable groups.[2]

There are various generalizations of the individual ergodic theorem dealing with the so-called weighted averages. We will formulate a result of this kind.

[2] Because by well-known Gromov's theorem the polynomial growth of the set of words is equivalent to almost nilpotency of the discrete groups the individual ergodic theorems from [T2] are valid only for those groups. So, up to now (1997) we have no individual ergodic theorems for general action of even exponential solvable groups (say, for the group "$ax + b$", $a, b \in \mathbb{Q}$).

Theorem 3.2 (V.I. Oseledets [Os1]). *Suppose* $\{T_g\}$ *is a measure preserving action of a countable group* G *on a Lebesgue space* $(M, \mathcal{M}, \mu)$, and ν *is a measure on* G. *Denote by* Γ *the support of* ν *and assume that*

1) $\mathcal{U}_\Gamma = \mathcal{U}_{\Gamma \cdot \Gamma^{-1}}$, *where* $\mathcal{U}_\Gamma$ *(respectively,* $\mathcal{U}_{\Gamma\cdot\Gamma^{-1}}$*) is the* σ*-algebra of the subsets* $A \in \mathcal{M}$ *invariant under all* T_g, $g \in \Gamma$ *(respectively,* $g \in \Gamma \cdot \Gamma^{-1}$*);*
2) ν *is symmetric, i.e.* $\nu(E) = \bar{\nu}(E^{-1})$, $E \subset G$. *Then for any* $f \in L^1(M, \mathcal{M}, \mu)$ *the limit*

$$\lim_{n\to\infty} \sum_{g\in G} f(T_g x)\nu^{(n)}(g)$$

exists, where $\nu^{(n)}$ *is the n-fold convolution of* ν.[3]

Mention also the multiplicative ergodic theorem [Os2] related to the study of random products of the elements of a group.

A strong generalization of the Birkhoff-Khinchin theorem in the case of a single operator is the so-called ratio ergodic theorem.

Theorem 3.3 (D. Ornstein, R. Chacon [CO]). *Let* $(M, \mathcal{M}, \mu)$ *be a space with* σ*-finite measure and* T *be a positive linear operator in* $L^1(M, \mathcal{M}, \mu)$ *with* $\|T\| \leqslant 1$. *For any pair* $f, g \in L^1(M, \mathcal{M}, \mu)$, $g \geqslant 0$, *the limit*

$$\lim_{n\to\infty} \frac{\sum_{k=0}^{n-1} T^k f(x)}{\sum_{k=0}^{n-1} T^k g(x)}$$

exists almost everywhere on the set $\{x \in M\colon \sup_k T^k g(x) > 0\}$.

Numerous papers were devoted to modifications and generalizations of this statement (cf [Kr]). In particular, its continuous-time analog has been proved (M. Akcoglu, J. Consolc).

General ergodic theorems for the amenable semigroups have been established in [T1].

The von Neumann statistical ergodic theorem on the strong convergence in $L^2(M)$ of the operators $\frac{1}{n}\sum_{k=0}^{n-1} U^k$, where U is the operator adjoint to an automorphism T, was also generalized by many authors. These generalizations can in a natural way be considered as part of operator theory rather than of ergodic theory. They are usually formulated for the operators in Banach spaces. The most general results are due to A.A. Tempelman [T1]. He proved, in particular, that for all connected simple groups with finite center the statistical ergodic theorem is satisfied for any sequence of averaging sets whose measures tend to infinity [T1].

---

[3] In the recent survey of A.G. Kachurovskii "Rates of convergence in ergodic theorems" (Russian Math. Surveys *51* (1996), 653–703) some new results concerning the speed of convergence in classical ergodic theorems for the group $\mathbb{Z}$ can be found.

The problem of finding the exact conditions for the sequence of subsets of a group to be an averaging sequence is very delicate. Some interesting counterexamples exist: if W_2 is the free group with two generators, G_n is the set of words of length $\leqslant n$, then there exists an action for which neither individual nor statistical ergodic theorem is satisfied with respect to $\{G_n\}$.

3.4. Spectral Theory. The spectral theory of measure preserving group actions is part of representation theory. Suppose that we are given an action $\{T_g\}$ of a group G on $(M, \mathcal{M}, \mu)$ with invariant or quasi-invariant measure. Associate to any $g \in G$ an operator U_g in $L^2(M)$ by the formula

$$(U_g f)(x) = f(T_g^{-1}x)\sqrt{p_g(x)},$$

where $p_g(x) = \dfrac{d\mu_g(x)}{d\mu(x)}$ is the density of μ_g with respect to μ. Note that p_g is a multiplicative 1-cocycle f the group G taking values in L^1, and it is cohomological to zero if and only if there is a finite invariant measure equivalent to μ; if μ itself is invariant, then $(U_g f)(x) = f(T_g^{-1}x)$. The correspondence $g \mapsto U_g$ is a unitary representation of G.

The principal questions are as follows:

a) to describe the decomposition of the above representation into irreducible ones;

b) to describe the representations which may appear in this situation.

In the case $G = \mathbb{Z}^1$ question a) is equivalent to the calculation of the spectral measure and the multiplicity function, while question b) requires the description of all possible spectra of the dynamical systems. Both problems are very delicate, although important information has been obtained in the case $G = \mathbb{Z}^1$ (cf [CFS]). In the general locally compact case the situation is, of course, much more complicated.

One of the first works devoted to the spectral theory of general actions was the paper of G.W. Mackey [M2], where the exact generalization of the von Neumann theorem on the pure point spectrum has been obtained. We will formulate this result.

A unitary representation of a group G has, by definition, a pure point (or a discrete) spectrum, if it is a direct sum of finite-dimensional irreducible representations. For $G = \mathbb{Z}^1$ this is equivalent to the standard definition of the pure point spectrum (we need not mention the "finite-dimensional" property). The von Neumann theorem says that for any ergodic action of $\mathbb{Z}^1$ the set of eigenvalues (the spectrum) is a countable subgroup of S^1; moreover, the action itself can be reconstructed from its spectrum up to metric isomorphism and can be represented as a translation on a commutative compact group, more precisely, on the group of characters of the spectrum. Hence, any countable subgroup of S^1 is the spectrum of some dynamical system (cf Chap. 2, Sect. 2). The Mackey generalization is as follows.

Theorem 3.4. *Suppose G is a separable locally compact group which acts ergodically on a Lebesgue space $(M, \mathcal{M}, \mu)$ with invariant measure, and the spectrum of G is discrete. Then there exist a compact subgroup K, a homomorphism $\varphi: G \to K$ onto a dense subgroup in K, as well as a closed subgroup $H \subset K$ such that the action of G on $(M, \mathcal{M}, \mu)$ is metrically isomorphic to the action of G on the homogeneous space K/H by the translations by the elements $\varphi(g)$, $g \in G$, the invariant measure being the image of the Haar measure on K.*

This theorem gives a complete solution to the problem of description of the dynamical systems with "time" G having a discrete spectrum. Unlike the commutative case, the representation does not uniquely determine the dynamical system because there is an example of a group (even a finite group) with two of its non-conjugate subgroups H_1, H_2 for which the representations of K in $L^2(K/H_1)$ and in $L^2(K/H_2)$ are equivalent. However, the group K can be uniquely reconstructed from the representation: it is the closure of the group $\{U_g: g \in G\}$ in the group of unitary operators in $L^2(M)$ with weak topology.

Such a complete analogy cannot be extended from the case of discrete spectrum to the general case. It is not clear even, what are the representations that can appear in the spectra of actions. The interesting special case is $G = \mathrm{SL}(2, \mathbb{R})$ and, more generally, the groups having additional series (which are not contained in the regular representation), i.e. the non-amenable groups.

The role of the countable Lebesgue spectrum is played by the so-called countable Plancherel spectrum.

Let us give an example. Suppose G is a discrete group and $M = [0, 1]^G$ is the space of all sequences $\{x_g\}$, $g \in G$ such that $x_g \in [0, 1]$. Take some measure μ_0 on $[0, 1]$ and consider the corresponding product-measure μ on M. Then the action of G on itself by the left translations brings about an action of G on M preserving the measure μ. It is called a Bernoulli action, by analogy with the Bernoulli shifts in the case $G = \mathbb{Z}^1$.

Theorem 3.5. *The spectrum of a Bernoulli action is countable Plancherel, i.e. $L^2([0, 1]^G, \mu)$ can be decomposed into a direct sum of subspaces in such a way that on each of them, the left regular representation of G acts.*

The spectral theory of Gauss dynamical systems (cf [CFS]) can be easily generalized to general locally compact groups. For the groups of type II[4] the spectral theory is closely related to the theory of factors [Kir].

[4] The groups of type II are the groups having such representations that the algebra spanned by the operators of the representation is of type II, i.e. in its central decomposition the factors of type II appear on a set of positive measure.

§ 4. Entropy Theory for the Actions of General Groups

The definition of entropy-type invariants for the group actions generalizing in a natural way the notion of entropy of an automorphism, was given in [Kir] (cf also [Con], [KW1], [OW]). Like the situation in general ergodic theorems, a rather developed theory exists for the class of amenable groups. We consider only the groups $\mathbb{Z}^m$, $m \geqslant 2$, for which the most complete results have been obtained. For notational reasons we set $m = 2$ and begin with the definition of entropy in this case.

Let (T_1, T_2) be a pair of commuting automorphisms of a Lebesgue space $(M, \mathcal{M}, \mu)$ with continuous measure. They define the action $\{T_g\}$ of the group $G = \mathbb{Z}^2$ on M. Denote by T_g the automorphism $T_1^{n_1} T_2^{n_2}$ corresponding to the element $g = (n_1, n_2) \in \mathbb{Z}^2$.

The first step is the definition of entropy of a measurable partition ξ with respect to $\{T_g\}$. It is convenient to take as a starting point property 3) of $h(T, \xi)$ (cf page 41) rather than the definition of $h(T, \xi)$ in the classical case of actions of $\mathbb{Z}^1$ (Definition 2.1, Chap. 3). We retain the notations introduced in Section 2, Chapter 3.

For any $E \subset \mathbb{Z}^2$ denote by ξ_E the partition $\bigvee_{g \in E} T_g \xi$. The set $\Pi \subset \mathbb{Z}^2$ will be called a parallelogram if $\Pi = \tilde{\Pi} \cap \mathbb{Z}^2$, where $\tilde{\Pi}$ is a parallelogram on $\mathbb{R}^2 \supset \mathbb{Z}^2$. Let $m(\Pi)$ be the length of the minimal side of $\tilde{\Pi}$ and $|\Pi|$ be the cardinality of Π.

Theorem 4.1 (cf [Con]). *Suppose that $\xi \in Z$ and $\{\Pi_n\}$ is a sequence of parallelograms on $\mathbb{Z}^2$ such that $m(\Pi_n) \to \infty$. There exists the limit*

$$h(G, \xi) = \lim_{n \to \infty} \frac{1}{|\Pi_n|} H(\xi_{\Pi_n}),$$

not depending on the choice of $\{\Pi_n\}$.

Note that we often use in this section the notations like $h(G, \xi)$, where under G we mean the action $\{T_g)$ of G rather than the group G itself.

Definition 4.1. The number $h(G, \xi)$ is said to be the *entropy of the partition ξ with respect to the action $\{T_g\}$.*

In the case of general discrete amenable group G any Følner sequence $\{\Phi_n\}$ of subsets of G can be taken in the role of $\{\Pi_n\}$. The corresponding limit also exists and does not depend on the choice of $\{\Phi_n\}$.

The properties of $h(G, \xi)$ (cf J. Conze [Con]):

1) $h(G, \xi) = H(\xi | \xi_{T_1}^- \vee (\xi_{T_2})_{T_1}^-)$. This property shows that the partition $\xi_G^- \stackrel{\text{def}}{=} \xi_{T_1}^- \vee (\xi_{T_2})_{T_1}^-$ may be thought of as "the past" of the partition ξ with respect to G. "The past" depends not only on the action of G, but also on the ordered set (T_1, T_2) of generators of G. All natural possibilities for defining "the past" for the actions of countable amenable groups are described in [P];

2) $h(G, \xi) \leqslant H(\xi)$;

3) $h(G, \xi_1 \vee \xi_2) \leqslant h(G, \xi_1) + h(G, \xi_2)$;

4) suppose G_n is a subgroup of index n of the group $G(n \geqslant 1)$, Γ_n is the fundamental domain for G_n containing the unity of G. Then $h(G_n, \xi_{\Gamma_n}) = n \cdot h(G, \xi)$;

5) for any automorphism T_g in the group G

$$h(G, \xi) \leqslant h(T_g, \xi);$$

6) $h(G, \xi)$ as a function of ξ is continuous on Z;

7) if $\xi_1 \preccurlyeq \xi_2$, then $h(G, \xi_1) \leqslant h(G, \xi_2)$;

8) for any $\xi_1, \xi_2 \in Z$

$$h(G, \xi_1) = \lim_{n \to \infty} H(\xi_1 | (\xi_1)_G^- \vee T_1^{-n}(\xi_2)_G^-);$$

9) if ξ is a generating partition for G, i.e. $\xi_G \overset{\text{def}}{=} \xi_{\mathbb{Z}^2} = \varepsilon$, then for any $\eta \in Z$

$$h(G, \eta) \leqslant h(G, \xi).$$

Definition 4.2. By the *entropy of the action* $\{T_g\}$ we mean the number

$$h(G) = \sup_{\xi \in Z} h(G, \xi).$$

The entropy $h(G)$ is a metric invariant in the sense that if two actions G_1, G_2 of the same group are metrically isomorphic, then $h(G_1) = h(G_2)$.

The properties of entropy $h(G)$:

1) If $G = G_1 \times G_2$, i.e. the action G of a certain group is the direct product of the actions G_1 and G_2 of the same group, (the formal definition of a direct product is a simple generalization of the one for the actions of $\mathbb{Z}^1$), then

$$h(G) = h(G_1) + h(G_2);$$

2) if G_1 is a factor-action of an action G (the definition is still the same as in the case of $\mathbb{Z}^1$), then $h(G_1) \leqslant h(G)$;

3) for any subgroup G_n of index n

$$h(G_n) = n \cdot h(G);$$

4) for any automorphism T_g in G we have

$$h(T_g) \geqslant h(G);$$

5) if T_1, T_2 are the generators of an action of $\mathbb{Z}^2$ and either $h(T_1) < \infty$ or $h(T_2) < \infty$, then $h(G) = 0$.

Examples. 1) Suppose M is the space of all sequences $x = \{x_{n_1,n_2}\}, (n_1, n_2) \in \mathbb{Z}^2$, where each x_{n_1,n_2} is an element of a finite set X. The measure λ on X is given by a probability vector $(p_1, \dots, p_m)$, $p_k \geqslant 0$, $\sum_{k=1}^m p_k = 1$. Introduce the measure μ in M which is the product-measure of λ, and set $T_1 x = x'$, $T_2 x = x''$, where $x'_{n_1,n_2} = x_{n_1+1,n_2}$; $x''_{n_1,n_2} = x_{n_1,n_2+1}$. It is clear that T_1 and T_2 commute and that they define the action G of $\mathbb{Z}^2$ on M. This action is called a Bernoulli action. For its entropy we have the formula $h(G) = -\sum_{k=1}^m p_k \log p_k$. It can be easily checked that in this case $h(T_1) = h(T_2) = \infty$.

2) Suppose M is the space of all sequences $x = \{x_n\}$, $n \in \mathbb{Z}^1$ where each x_n is a point of a Lebesgue space $(X, \mathscr{X}, \lambda)$. Let S be an automorphism of $(X, \mathscr{X}, \lambda)$. The measure μ on M is, as before, the product-measure of measure λ. Define the automorphisms T_1, T_2 of M by the formulae $T_1(\{x_n\}) = \{Sx_n\}$, $T_2(\{x_n\}) = \{x_{n+1}\}$. The entropy of the action of $\mathbb{Z}^2$ generated by T_1 and T_2 is equal to $h(S)$.

Many fundamental facts of entropy theory for the actions of $\mathbb{Z}^1$ can be carried over to the case of general group actions. More precisely, for a free and ergodic action of a countable amenable group G with finite entropy, there exists a finite generating partition (cf [Šu]). (Note that an action $\{T_g\}$ is free if $\mu(\{x \in M: T_g x = x\}) = 0$ for any $g \in G \setminus \{e\}$). There are generalizations of the Ornstein isomorphism theorem for the Bernoulli actions with the same entropy (A.M. Stepin [St]). A group generalization of the Shannon-McMillan-Breiman theorem exists, as far as we know, only in the case of the action of the so-called quasi-cyclic group (the group of all dyadic rationals of the unit circle) (cf B.S. Pickel, A.M. Stepin [PS]). If we are interested in the L^1-convergence in this theorem rather than in almost everywhere convergence, the corresponding result (an analog of the McMillan theorem) has been proved for any discrete amenable group (cf J.C. Kieffer [Ki]).

In [Av] the notion of entropy of a random walk on a group was introduced. For the detailed study of this notion and its applications to the problem of the boundaries of random walks, see [KV].

The theory of dynamical systems with positive entropy has been extended to the actions of groups $\mathbb{Z}^m$, $m \geqslant 2$. For notational reasons we set again $m = 2$.

Just as in the case of the actions of $\mathbb{Z}^1$, the Pinsker partition π can be defined: $\pi(G) = \sup\{\xi\colon \xi \in Z, h(G, \xi) = 0\}$. The actions G with $\pi(G) = \nu$ are called the actions of completely positive entropy. The next definition is very important for entropy theory of the actions of $\mathbb{Z}^2$ and has no analogs in $\mathbb{Z}^1$-theory.

Definition 4.3. Suppose (T_1, T_2) is an ordered pair of commuting automorphisms of the Lebesgue space $(M, \mathscr{M}, \mu)$; ζ is a measurable partition. The partition ζ is said to be (T_1, T_2)-*strong invariant* if 1) $T_1\zeta \preccurlyeq \zeta$;
2) $\bigwedge_{n=0}^{\infty} T_1^{-n}\zeta = T_2^{-1}\zeta_{T_1}$.

The strong invariance of ζ implies that $T_1^{n_1}T_2^{n_2}\zeta \preccurlyeq \zeta$ if $(n_1, n_2) < (0, 0)$, the pairs (n_1, n_2) being ordered lexicographically. The partitions ζ with this last property are called (simply) (T_1, T_2)-invariant. Generally, they need not be strong invariant. However, any action G with $h(G) > 0$ has non-trivial strong invariant partitions. In particular, if $\xi \in Z$ is such that all $T_g\xi$, $g \in \mathbb{Z}^2$, are independent, then the partition $\zeta = \xi_G^-$ is strong invariant.

In the theory of actions of $\mathbb{Z}^2$ with positive entropy the strong invariant partitions play the role similar to that of invariant partitions in the classical case $G = \mathbb{Z}^1$.

Definition 4.4. A partition $\zeta \in Z$ is said to be (T_1, T_2)-*exhaustive* if ζ is (T_1, T_2)-strong invariant and $\bigvee_{n=0}^{\infty} T_2^n\zeta_{T_1} = \varepsilon$.

Theorem 4.2. (B. Kaminski [Ka2]). *If ζ is (T_1, T_2)-exhaustive, then*

$$\bigwedge_{n=0}^{\infty} T_2^{-n}\zeta_{T_1} \geqslant \pi(G).$$

This assertion becomes false if ζ is assumed only to be (T_1, T_2)-invariant rather than (T_1, T_2)-strong invariant.

Definition 4.5. A (T_1, T_2)-exhaustive partition ζ is said to be (T_1, T_2)-*extremal* if $\bigwedge_{n=0}^{\infty} T_2^{-n}\zeta_{T_1} = \pi(G)$. A (T_1, T_2)-extremal partition ζ is said to be (T_1, T_2)-*perfect* if $h(G) = h(G, \zeta) = H(\zeta | T_1^{-1}\zeta)$.

Theorem 4.3 (B. Kaminski [Ka2]). *For any ordered pair (T_1, T_2) of generators of the group G there are (T_1, T_2)- perfect partitions.*

Definition 4.6. An action G of the group $\mathbb{Z}^2$ is said to be a K-*action* (G is said to be a K-group) if for any pair (T_1, T_2) of generators of G and for any A_0, A_1, $\ldots$, $A_r \in \mathcal{M}$ $(1 \leqslant r < \infty)$ we have

$$\lim_{n\to\infty} \sup_{B^{(n)} \in \mathscr{B}^n} |\mu(A_0 \cap B^{(n)}) - \mu(A_0)\cdot \mu(B^{(n)})| = 0,$$

where $\mathscr{B}^{(n)}$ is the σ-algebra corresponding to the partition $T_1^{-n}\xi_{T_1}^{-} \vee T_2^{-n}(\xi_{T_1})_{T_2}^{-}$, and ξ is the partition generated by $A_1, \ldots, A_r$.

With the help of the theory of strong invariant partitions the following statement was proved.

Theorem 4.4. (B. Kaminski [Ka2]). *G is a K-group if and only if G is of completely positive entropy.*[5]

[5] Recently D. Rudolph and B. Weiss (in preparation, 1997) have proved that the action of any amenable group with completely positive entropy (= all quotient actions have positive entropy) has mixing of any order. It is interesting that the proof used the corresponding theorem due to Pinsker-Rokhlin-Sinai for the group $\mathbb{Z}$, in this case one having an alternative definition of complete positivity of entropy (K-property). Until now (1997) there is no such a definition for other groups.

Chapter 5
Trajectory Theory

A.M. Vershik

§1. Statements of Main Results

The starting point for trajectory theory of the dynamical systems is a natural question which goes back to H. Poincaré and is concerned with the classical (smooth) dynamical systems. Consider two topological dynamical systems, i.e. 1-parameter groups of continuous transformations of a compact space. We call them topologically (orbitally) equivalent if there is a homeomorphism of the phase space intertwining the orbits (the trajectories) of these systems and preserving the orientation (the order of the points) on the orbits. Such a rough notion of equivalence is useful for the study of phase portraits of the dynamical systems, i.e. the structure of the partitions of phase spaces into separate trajectories. Such properties as the existence (or non-existence) of the periodic trajectories, of invariant submanifolds and so on, turn out to be stable under the above equivalence. This notion is also very useful for the investigation of the so-called rough properties of the dynamical systems (cf [Ar2]).

In ergodic theory it is natural to consider the measure-theoretical orbital equivalence rather than the topological one.

The exact definition will be given first in the case of a single automorphism (for the group $\mathbb{Z}^1$), and then will be extended to the general case.

Let $(M, \mathcal{M}, \mu)$ be a Lebesgue space and T be its automorphism preserving the measure μ. Denote by $\tau(T)$ the partition of M into separate trajectories of T. In other words, the element of $\tau(T)$ containing a point $x \in M$ is a finite or countable set $\{y: y = T^n x\}$, $n \in \mathbb{Z}^1$. The partition $\tau(T)$ is well defined in the sense that for any pair of coinciding (mod 0) automorphisms T_1, T_2 the corresponding partitions $\tau(T_1)$, $\tau(T_2)$ also coincide (mod 0). It should be stressed that $\tau(T)$ measurable in general case, and therefore the factor-space $(M, \mathcal{M}, \mu)/\tau(T)$ is not necessarily Lebesgue; moreover, there might be no nontrivial measurable sets at all in this factor-space. By the same reason, there might be no measurable functions which are constant on the elements of $\tau(T)$. However, all these facts do not signify that $\tau(T)$ should be visualized as a kind of pathological object like non-measurable sets. On the contrary, they only signify that some special methods are needed in order to study the trajectory properties of dynamical systems. For example, there are no conditional measures on the elements of $\tau(T)$ (as it would be if $\tau(T)$ were measurable) but there are the so-called ratio set and the cocycle which play a similar role in many questions (cf. below).

Definition 1.1. Two automorphisms T_1, T_2 of the space $(M, \mathcal{M}, \mu)$ are said to be *orbitally equivalent* if their trajectory partitions $\tau(T_1)$, $\tau(T_2)$ are metrically isomorphic, i.e. if $S\tau(T_1) = \tau(T_2)$ for some automorphism S of $(M, \mathcal{M}, \mu)$.

To make this definition more clear we reformulate it in a somewhat different form. Assume first that the trajectory partitions for T_1 and T_2 coincide. This yields that if for $x, y \in M$ we have $y = T_2^n x$ for some n, then there is a number $m = m(n, x)$ for which $y = T_1^m x$. Thus, there are measurable (in x) functions $m_1 = m_1(n, x)$, $m_2 = m_2(n, x)$, $x \in M$, $n \in \mathbb{Z}^1$, taking value in $\mathbb{Z}^1$ and satisfying $T_1^n x = T_2^{m_1(n,x)} x$, $T_2^n x = T_1^{m_2(n,x)} x$; in particular, $T_1 x = T_2^{m_1(x)} x$, $T_2 x = T_1^{m_1(x)} x$, where $m_1(x) = m_1(1, x)$, $m_2(x) = m_2(1, x)$. In other words, T_1 (respectively, T_2) can be obtained from T_2 (respectively, T_1) by means of some measurable change of time. Therefore, two automorphisms are orbitally isomorphic if each of them is metrically isomorphic to some automorphism which can be obtained from the other one by means of a measurable change of time. It can be easily seen that $m_1(n, x)$ is uniquely determined by $m_1(x)$: $m_1(n, x) = \sum_{k=0}^{n-1} m_1(T_2^k x)$, $n > 0$; $m_1(-1, x) = -m_1(T_2^{-1} x)$. However, in order to obtain the generalizations to other groups, it is more convenient to deal with the function $m_1(\cdot, \cdot)\, \mathbb{Z}^1 \times M \to \mathbb{Z}^1$ (rather than with $m_1(\cdot)$) which will be called the function of the change of time.

Suppose now that we are given a measurable action of some locally compact group G on a Lebesgue space $(M, \mathcal{M}, \mu)$ by the automorphisms preserving the measure μ.

In order to define correctly the trajectory partition in this case, we use the theorem on the uniqueness (mod 0) of the measurable realization (cf [Ve1]). In view of this theorem, the trajectory partitions for two coinciding (mod 0) automorphisms may differ only on a set of zero measure. Denote by $\tau(T)$ the trajectory partition of the action T of group G. Identifying the action T with the group itself, we may write $\tau(G)$ instead of $\tau(T)$

Definition 1.2. The groups G_1, G_2 of automorphisms are said to be *orbitally isomorphic* if the partitions $\tau(G_1)$, $\tau(G_2)$ are metrically isomorphic.

Note that we do not assume in this definition that G_1 and G_2 are isomorphic as abstract groups.

Just as in the case of the group $\mathbb{Z}^1$, the equality $\tau(G_1) = \tau(G_2)$ implies the existence of such measurable functions $m_1\colon G_1 \times M \to G_2, m_2\colon G_2 \times M \to G_1$, that for $g_i \in G_i$, $i = 1, 2$, one has $T_1(g_1)x = T_2(m_1(g_1, x))x$, $T_2(g_2)x = T_1(m_2(g_2, x))x$. If only one of these equalities holds (for example, the first one) then we have $\tau(G_1) \succcurlyeq \tau(G_2)$, i.e. the trajectories of the group G_2 are the unions of some trajectories of the group G_1. The functions m_1, m_2 may be considered as the change-of-time functions. In particular, for the flows ($G_1 = G_2 = \mathbb{R}^1$) we have:

$$S_1^t x = S_2^{m_1(t,x)} x, \qquad S_2^t x = S_1^{m_2(t,x)} x;$$

$m_1, m_2\colon \mathbb{R}^1 \times M \to \mathbb{R}^1$ are measurable functions.

Note, further, that the above definition signifies (just as in the case $G = \mathbb{Z}^1$) that if two groups of automorphisms are orbitally equivalent, then each of them is metrically isomorphic to a group obtained from the other one by means of a change of time.

The orbital isomorphism of general dynamical systems is much weaker as an equivalent relation (i.e. the corresponding equivalence classes are larger) than topological orbital equivalence considered at the beginning of this section: the requirement of continuity of the intertwining map is replaced by that of its measurability and the invariance of the measure. Furthermore, this map need not preserve the order of points on the trajectories, and, finally, the sets of zero measure are ignored.

The central problem of measure-theoretical trajectory theory is the description of the equivalence classes (under the orbital isomorphism) of the measure preserving automorphisms groups G (the most important cases are, as usual, $G = \mathbb{Z}^1$ and $G = \mathbb{R}^1$). The groups of automorphisms which are not measure-preserving but have only a quasi-invariant measure can also be considered; in this case the interwining map S in the definition of the orbital isomorphism is also assumed to have only a quasi-invariant measure.

Since almost all non-trivial classification problems in the theory of dynamical systems are well known to have no solutions giving a sufficiently compact picture, the reader might be prepared to learn the same thing about the problem considered. This point of view was initially shared by a number of specialists. Actually, however, the situation is quite different, and in the cases of $\mathbb{Z}^1$, $\mathbb{R}^1$ and, more generally, of the amenable groups, there is an unexpectedly simple answer to our question.

However, in the general case (for the arbitrary non-amenable groups), there is no complete solution this problem and it is clear now that a simple classification cannot be obtained.

At the present time, trajectory theory is a vast area of the theory of dynamical systems. It is intimately related to the study of the so-called ergodic equivalence relations and measurable grouppoids generalizing the trajectory partitions. Other topics related to trajectory theory are the foliations on the manifolds, the classificational partitions and so on. The most detailed information has been obtained about the trajectory partitions.

It is difficult to say now who was the first to state the problem concerning the orbital classification for the actions of the group $\mathbb{Z}^1$. In any case, this question is contained implicitely in the paper by J. von Neumann and F. Murrey [NM] (1944) in connection with their study of the rings of operators. Other formulations were given by C.W. Mackey [M3] (in context of the theory of virtual subgroups) and by V.A. Rokhlin in the late 1950's.

H. Dye [D] in 1963 using some ideas of J. von Neumann gave the solution to this problem for $G = \mathbb{Z}^1$ in the case of measure preserving transformations. Another, purely measure-theoretical solution was obtained by A.M. Vershik [Ve2] and R.M. Belinskaya [B] in 1996.[1]

---

[1] See below in corollaries on lacunary isomorphism on page 84. For a detailed version of the theory of decreasing sequences of measurable partitions with applications to ergodic theory and new development see [A.M. Vershik "The theory of decreasing sequences of measurable partitions", St. Petersburg Math. J. *6* (1995), 705–761].

Following J. von Neumann, H. Dye connected the problem considered with the theory of II_1-factors, namely, the problem was reduced to the proof of the hyperfiniteness of the factor corresponding to a given automorphism.

We will formulate now the final result and give the sketch of the proof in the case $G = \mathbb{Z}^1$.

Theorem 1.1. *Any two measure preserving ergodic automorphisms of a Lebesgue space are orbitally isomorphic.*

Corollaries. 1) *For two measure preserving automorphisms of a Lebesgue space to be orbitally isomorphic, it is necessary and sufficient that their measurable partitions into ergodic components be isomorphic.*

This assertion can be proved in a standard way: we fix first the isomorphism of the partitions into ergodic components and then apply the above theorem to each component.

Theorem 1.1 shows that the notion of orbital isomorphism is in some sense degenerate in the case $G = \mathbb{Z}^1$: we have no dynamical invariants of an automorphism (such as its spectrum, entropy and so on) which are stable under orbital isomorphism; the type of the partition into ergodic components (this type may be regarded as a geometric invariant of an automorphisms) is the only invariant of the trajectory isomorphism.

2) *For any ergodic automorphism T there exists an automorphism S metrically isomorphic to T which can be constructed from the automorphism Q with pure point spectrum of the form $\{2\pi ip/2^r\}$, p, q are integers, $q \neq 0$, by means of some change of time:*

$$Sx = Q^{m(x)}x, \qquad S = VTV^{-1}.$$

The automorphism Q has been chosen in this corollary as the simplest one. Therefore, by rearrangements of the points within the trajectories of Q one can obtain, for example, a Bernoulli automorphism (in the role of S). The function $m(\cdot)$ in this case will certainly be very complicated and its distribution function will decrease very slowly.

All this enables us to study the invariants of the change-of-time function $m(\cdot)$ appearing in the construction of a given automorphism from Q, instead of the automorphism itself.

Now consider the groups of more general form. The following theorem which was obtained quite recently by A. Connes, D. Ornstein J. Feldman and B. Weiss [CFW] is a final result of the longterm investigations by these authors, as well as by A.M. Vershik and W. Krieger, devoted to the generalizations of Theorem 1.1.

Theorem 1.2. *Suppose we are given an ergodic, free and measure preserving action of a countable discrete group G on the space $(M, \mathcal{M}, \mu)$ (the property of being free for an action means that $\mu(\{x: gx = x\}) = 0$ for all $g \in G \setminus \{e\}$). For this action to be orbitally isomorphic to some action of the group $\mathbb{Z}^1$ it is necessary and sufficient that the group be amenable (i.e. G admits an invariant mean).*

We deduce immediately from this fact that any two ergodic and free actions of countable amenable groups are orbitally isomorphic; on the other hand, the orbital class of any such action of a non-amenable group is not similar to that for amenable groups.

This remarkable theorem (together with many other facts) shows that the class of the amenable groups should be viewed as a natural class of groups to which the most general theorems on the actions of $\mathbb{Z}^1$ may be extended.

A similar result concerning the actions of continuous groups is also true.

Theorem 1.3. *Suppose G_1 and G_2 are two non-discrete locally compact amenable (i.e. possessing a topological invariant mean) groups having a countable base of open sets. Any two ergodic and free measure preserving actions of these groups are orbitally isomorphic. In particular, any of such actions is orbitally isomorphic to the action of $\mathbb{R}^1$ by group translations on the two-dimensional torus.*

This statement is a rather easy consequence of the deep Theorem 1.2.

Theorems 1.1–1.3 give a solution to the problem of study of orbital isomorphism of the measure preserving actions of locally compact groups. Beyond this class of groups the situation is much more complicated. We shall formulate at a later stage some results in this direction, and we now proceed to the sketch of the proof of Theorem 1.2. Some new important notions will be needed; our exposition follows the papers [Ve2], [B] (cf also [FM], [Ve3], [Kri3], [Mo], [HO], [FV]).[2]

§2. Sketch of the Proof. Tame Partitions

Definition 2.1. A partition τ of a Lebesgue space $(M, \mathcal{M}, \mu)$ is said to be *tame* if it can be represented as a measure-theoretical intersection of some decreasing sequence of measurable partitions.

Recall that a partition ξ is smaller that η (ξ is not finer than η) if each element of ξ is the union of some elements of η[3]. Thus, the sequence of the partitions $\{\xi_n\}$ is decreasing if the sequence of the elements $C_n(x)$ containing a point $x \in M$ is increasing. The measure-theoretical intersection $\bigcap_n \xi_n$ is the partition with the elements $C(x) = \bigcup_n C_n(x)$, where $C_n(x)$ is the element of ξ containing $x \in M$. The partition $\bigcap_n \xi_n$ is not necessarily measurable.

Recall further that the measurable intersection $\bigwedge_n \xi_n$ is the maximal measurable partition, which is smaller than all ξ_n. Such a partition always exists, and in the case considered, it coincides with the measurable hull of the partition $\bigcap_n \xi_n$. A tame partition is said to be ergodic if its measurable hull is trivial, i.e. is equal to the partition ν having only one nonempty element.

---

[2] A version of the proof of Theorem 1.2 using new criteria of amenability was recently (1997) given by V. Kaimanovich who also had given an example of nonamenable equivalent relations relations with amenable leaves (in preparation).

[3] Another ordering of the partitions is usually employed in combinatorics.

Any measurable partition is obviously tame, but the example below playing an important role in the theory shows that the converse is not true.

Example. Suppose $M = [0, 1]$, $\xi_n^{(0)}$ is the partition of M such that the points $x, y \in M$ belong to the same element of $\xi_n^{(0)}$ iff $x - y$ is of the form $k/2^n$, $k = 0, 1, \ldots, 2^n - 1$. It is clear that $\xi_1^{(0)} > \xi_2^{(0)} > \cdots$. The partition $\tau^{(0)} = \bigcap_n \xi_n^{(0)}$ can be defined by postulating that $x, y \in M$ belong to the same element iff $x - y$ is of the form $k/2^n$, k, n are integers. It may be easily seen that $\tau^{(0)}$ is ergodic. The sequence $\{\xi_n^{(0)}\}$ (as well as any sequence isomorphic to it) is called the standard dyadic sequence of partitions.

The canonical system of conditional measures cannot be correctly defined for the tame partitions which are not measurable; its role is played by the so-called projective system. In order to define it, observe first that the conditional measures satisfy the following transitivity property: if $\xi < \zeta$ and $x, y \in C \in \xi$, $x, y \in D \in \zeta$, then $\mu_\xi^C(x)/\mu_\xi^C(y) = \mu_\zeta^D(x)/\mu_\zeta^D(y)$ almost everywhere; here $\mu_\eta^E(x)$ is the conditional measure of the point x in the element E of a partition η. Thus, for $\tau = \bigcap_n \xi_n$ the number $\mu^C(x)/\mu^C(y) \equiv \mu_{\xi_n}^{C_n}(x)/\mu_{\xi_n}^{C_n}(y)$ (n is sufficiently large) is well defined. Moreover, this number does not depend on the specific representation of the partition τ as the intersection of a decreasing sequence of measurable partitions (due to the same transitivity property of the conditional measures). Denote $\beta(x, y) = \mu^C(x)/\mu^C(y)$ for $x, y \in C \in \tau = \bigcap_n \xi_n$. The function β satisfies the following properties:

1) β is defined on the pairs of points belonging to the same elements of the tame partition; $\beta(x, y) \in \mathbb{R}_+$;

2) $\beta(x, x) = 1$, $\beta(x, y) = \beta(y, x)^{-1}$;

3) $\beta(x, y) \cdot \beta(y, z) = \beta(x, z)$.

Any function satisfying 1)–3) will be called a $\mathbb{R}_+$-valued 1-cocycle. The above mentioned cocycle is known as the Radon-Nicodym cocycle of the tame partition. If the sum $\sum_{x \in C(y)} \beta(x, y) = [\mu^C(y)]^{-1}$ were finite, then $\mu^C(\cdot)$ might be considered as a conditional measure. Hence, the fact of its finiteness is equivalent to the measurability of τ. If τ is not measurable, the above sum is infinite, and the conditional measure cannot be defined. However, the cocycle β can substitute this measure in various questions. It turns out that the full information concerning the metric type of a tame partition is contained in the Radon-Nicodym cocycle similarly to the fact that the system of conditional measures contains the full information about the metric type of a measurable partition. If $\beta \equiv 1$, the corresponding partition is called homogenous.[4]

We shall now formulate the results on the relationship between measurable and tame partitions.

---

[4] Many results concerning the theory of very general cocycles with arbitrary values and for actions of general groups one can find in the papers by K. Schmidt, V. Golodets and his school, for instance, A. Fedorov (see [Ve6F]).

Proposition 1. *The trajectory partition for an arbitrary automorphism T of a Lebesgue space with quasi-invariant measure is tame. This tame partition is homogeneous if and only if T is measure-preserving. Ergodicity of T is equivalent to ergodicity of its trajectory partition.*

Theorem 2.1. *There is only one (up to isomorphism) ergodic homogeneous tame partition.*

The last two facts imply Theorem 1.1. There are many ways to prove proposition 1; the most clear one is based on the iterated applications of the periodic approximations (analog of the Rokhlin-Halmos Lemma).

Since the trajectory partitions for periodic automorphisms (or for the periodic parts of non-aperoidic automorphisms) are measurable and, thus, are tame, T may be assumed to be aperiodic. Let μ be a quasi-invariant measure for T. Choose a set A_1, $\mu(A_1) < 1/2$, such that $\mu(\bigcup_k T^k A_1) = 1$, and denote by ξ_1 the partition with the elements of the form $\{x, Tx, \ldots, T^{k(x)}x\}$, where $x \in A_1$, $T^{k(x)+1}x \in A_1$, $T^i x \notin A_1$, $1 \leqslant i \leqslant k(x)$. We may assume that $k(x)$ is finite almost everywhere; if not, the set A_1 should be slightly icreased to meet this condition.

Now consider the induced automorphism T_{A_1} on A_1, choose a set $A_2 \subset A_1$, $\mu(A_2) < \frac{1}{4}$, and repeate the above procedure with A_2 in the role of A_1 and T_{A_1} in the role of T.

Let ξ_2' be the corresponding partition of A_1 (the analog of ξ_1). The projection $\pi\colon M \to A_1$ (along ξ_1) enables us to lift this partition to M. Set $\xi_2 = \pi^{-1}\xi_2'$. It is clear that $\xi_1 \succcurlyeq \xi_2$. Continuing this procedure, we get the decreasing sequence $\xi_1 \succcurlyeq \xi_2 \succcurlyeq \cdots$, $\bigcap_n \xi_n = \tau(T)$. This proves that $\tau(T)$ is tame. It may be checked that $\tau(T)$ is homogeneous whenever T is measure preserving.

The proof of Theorem 2.1 is more difficult. Consider first the representation of an arbitrary tame partition τ with countable elements in the form $\tau = \bigcap_n \xi_n$ and modify this representation: $\tau = \bigcap_n \tilde{\xi}_n$, so that almost each element of $\tilde{\xi}_n$ contains 2^n points. The sequence $\tilde{\xi}_1 \succ \tilde{\xi}_2 \succ \cdots$ is dyadic. Theorem 2.1 will then follow from the statement below on dyadic sequences.

Theorem 2.2 ([Ve2]). *Let $\{\xi_n\}_1^\infty$ be a homogeneous ergodic (i.e. $\bigwedge_n \xi_n = \nu$) dyadic sequence of measurable partitions. There is a sequence $\{n_k\}$, $n_k \to \infty$, of natural numbers such that the subsequence $\{\xi_{n_k}\}_{k=1}^\infty$ is metrically isomorphic to $\{\xi_{n_k}^{(0)}\}_{k=1}^\infty$, where $\{\xi_n^{(0)}\}$ is the standard dyadic sequence (cf above).*

Corollaries. 1) (on lacunary isomorphism)

Any two homogeneous ergodic dyadic sequences $\{\xi_n\}$, $\{\xi_n'\}$ are lacunarily isomorphic, i.e. there exists a sequence $\{n_k\}$, $n_k \to \infty$, for which $\{\xi_{n_k}\}$ and $\{\xi_{n_k}'\}$ are metrically isomorphic.

2) *The partitions $\bigcap_n \xi_n$ and $\bigcap_n \xi_n'$* (we retain the notations used in corollary 1) *are metrically isomorphic* (since $\bigcap_n \xi_n = \bigcap_k \xi_{n_k}$ if $n_k \to \infty$).

The statement of Theorem 2.1 follows from Corollary 2 and Proposition 1. The proof of Theorem 2.2 is more technical. Other proofs of Theorem 2.1 were given by H. Dye [D], S. Kakutani et al [H1K]. Note that the study of problems of the trajectory theory with the use of decreasing sequences of partitions initiated in [Ve2] turned out to be useful and productive in its own right; some new invariants of automorphisms have been found in this way.

Proposition 1 shows that trajectory theory for the automorphisms with quasi-invariant measure may be reduced to the classification problem of the tame partitions with countable elements. The following results have been obtained in this direction: a class of partitions for which the solution of the classification problem is as simple as in the homogeneous case have been determined; on the other hand, the classification in general case turned out to be equivalent to the metric classification of flows and, thus, it cannot be obtained in reasonable terms.

We proceed now to a more detailed exposition of the results (essentially due to W. Krieger) concerning the above mentioned reduction. These results are important on their own right and the methods employed are very elegant. We start off with an example.

Consider a space $M = \prod_{n=1}^{\infty} X$, $\operatorname{card}(X) = k$, and a probability vector $p = (p_1, \ldots, p_k)$. Let μ be the Bernoulli measure on M which is the product-measure of the measure on X defined by p. Define the so-called tail partition κ of M. Its elements are the equivalence classes under the following relation: $x = \{x_n\}$ and $y = \{y_n\}$ are equivalent if $x_n = y_n$ for n sufficiently large. It may be shown that κ is tame and ergodic with respect to μ. We will call κ the Bernoulli partition. If $p_i = \frac{1}{k}$, then κ is homogeneous.

In the statements below, under "isomorphism" we mean an isomorphism with quasi-invariant measure (i.e. sending a measure to an equivalent one).

Theorem 2.3. 1) *An ergodic tame partition has a homogeneous ergodic subpartition if and only if it is Bernoulli.*

2) *Any Bernoulli partition is uniquely determined* (*up to isomorphism*) *by the subgroup F of the multiplicative group $\mathbb{R}_+$, which is the closure of the group generated by the numbers p_i/p_j, $1 \leqslant i, j \leqslant k$.*

If $p_i = \frac{1}{k}$, then $F = \{1\}$, so we get the statement of Theorem 2.1.

The second part of Theorem 2.3 was obtained independently by W. Krieger and A.M. Vershik.

Since any closed subgroup of $\mathbb{R}_+$ is of one of the forms: $\{1\}$, $\mathbb{R}_+$ and $\{\lambda^n\}_{n \in \mathbb{Z}}$, $0 < \lambda < 1$, we have an explicite classification of tame partitions possessing homogeneous subpartitions. The above mentioned subgroups of $\mathbb{R}_+$ are sometimes called the ratio sets for trajectory partitions.

Note that a statement similar to Theorem 2.1 is true for Lebesgue spaces with σ-finite measure (i.e. the spaces which can be represented as countable unions of spaces with finite measure). In this case there also exists only one (up to isomorphism) ergodic homogeneous tame partition.

Now consider the general case. We begin with the definition of the Poincaré flow introduced by G. Mackey and applied to the problem considered by W. Krieger. Set $\Delta(x, y) = \log \beta(x, y)$, where $\beta(x, y)$ is the Radon-Nicodym cocycle. Δ is sometimes called a modular function. If S is an ergodic automorphism of a measure space (M, μ) with quasi-invariant measure, and $y = Sx$, then $\Delta(x, y) = \log \frac{d\mu(Sx)}{d\mu(x)}$. Let $Y' = M \times \mathbb{R}^1$ and $\mu' = \mu \times m$, where m is the Lebesgue measure on $\mathbb{R}^1$. Define a flow $\{T_t'\}$ on Y' by $T_t'(x, a) = (x, a + t)$, and an automorphism $\tilde{S}$ by $\tilde{S}(x, a) = (Sx, a + \Delta(x, Sx))$. It is clear that T_t' and $\tilde{S}$ commute. Let $Y = Y'/\tau(\tilde{S})$ be the factor-space of Y' with respect to the action $\tilde{S}$, i.e. with respect to the measurable hull of the partition into separate trajectories of $\tilde{S}$. Then Y, as a measure space with the factor-measure $\mu'/\tau(\tilde{S})$, is a Lebesgue space, (maybe with σ-finite measure) and the flow $\{T^t\} = \{T_t'/\tau(\tilde{S})\}$ acts on it. This flow is know as the Poincaré flow, or the flow associated with the trajectory partition of the automorphism S.

We will now state the main property of the above construction.

Theorem 2.4. *Two ergodic automorphisms S_1, S_2 with quasi-invariant measures are orbitally isomorphic (in the "quasi-invariant" sense) if and only if their Poincaré flows are metrically isomorphic as flows with quasi-invariant measures.*

This theorem reduces the classification problem of tame partitions (or else of trajectory partitions for the automorphisms with quasi-invariant measures) to the problem of metric isomorphism of flows with quasi-invariant measure. Unfortunately, the last problem has no complete solution in reasonable terms, but the reduction that we have described, enables one to obtain some nontrivial invariants of the trajectory partitions with the aid of the invariants of flows.

In the case of homogeneous tame partition ($\beta \equiv 1$, $\Delta \equiv 0$) the corresponding Poincaré flow is of the form $T^t a = a + t$, $a \in \mathbb{R}^1$; the Bernoulli partition with the group $\{\lambda^n\}$ corresponds to the periodic flow of period $(-\log \lambda)$, the Bernoulli partition with the group $\mathbb{R}^1$ to the trivial flow on the single-point space. Other tame partitions are associated with aperiodic flows (cf [Kri3], [Mo], [HO]).

These results may be considered as the answers to the main questions concerning the general problems of trajectory theory for the group $\mathbb{Z}^1$. It is interesting to note that a number of constructions and even statements appeared first in the theory of operator algebras which is closely related to trajectory theory. Although our exposition is purely measure-theoretic, the operator analogs of such notions as main cocycle, modular function etc, exist, and some of them actually appeared in ergodic theory as measure-theoretical translations from the algebraic language. The paper of H. Dye was already mentioned in this connection; there are also the papers of H. Araki and E. Woods [AW], A. Connes [Co1]. These works seem to be the first examples of applications of operator algebras to ergodic theory; earlier, conversely, the measure-theoretical constructions were being used in the theory of factors to solve some algebraic problems.

§ 3. Trajectory Theory for Amenable Groups

We shall now consider the groups of more general form, and begin with the continuous case.

Two partitions, τ_1, τ_2, of a Lebesgue space $(M, \mathcal{M}, \mu)$ are said to be *stably equivalent* if the partitions $(\tau_1 \times \varepsilon)$, $(\tau_2 \times \varepsilon)$ of the space $(M \times [0,1], \mu \times m)$, where m is the Lebesgue measure, ε is the partition of $[0,1]$ into separate points, are isomorphic.

Theorem 3.1 (A. Ramsey [R]). *The trajectory partition of an arbitrary locally compact non-discrete group G with a countable base, acting on a Lebesgue space with quasi-invariant measure, is stably equivalent to the trajectory partition of some discrete group.*

The simplest case is $G = \mathbb{R}^1$. The Ambrose-Kakutani theorem on special representation of flows yields in this case that the trajectory partition of a flow is stably equivalent to that of the base automorphism.

On the other hand, it may be shown that the stable equivalence of two tame partitions with countable elements is equivalent to their metric isomorphism (with quasi-invariant measure). Hence, the problem for flows and, more generally, for locally compact groups is reduced to the case of discrete groups. In particular, the above stated (Section 1) Theorem 1.3 follows from Theorems 1.2 and 3.1.

We will give a separate formulation for flows, i.e. one-parameter groups of measure preserving automorphisms.

Theorem 3.2 *Any two ergodic flows with invariant measures (either finite or σ-finite) are orbitally equivalent. Therefore, any ergodic flow with a finite invariant measure is metrically isomorphic to a flow which can be constructed from the conditionally periodic flow on the two-dimensional torus by means of some measurable change of time.*

Now we turn to the following question: what is the class of discrete groups, for which the trajectory partitions of the actions with invariant measure are tame? It is clear that the actions themselves in this case can be obtained from the actions of $\mathbb{Z}^1$ by means of changes of time. It was proved by H. Dye [D] that this class includes the countable groups for which the number of words of length $\leqslant n$ in any finitely generated subgroup grows polynomially in n. It was suggested at the same time that this class includes all amenable groups. The necessity of the amenability condition follows easily from Følner's criterion (cf [D]), but the converse is much more difficult. The converse statement was first proved in some special cases: for the solvable groups—by A. Connes and W. Krieger, for the completely hyperfinite groups—by A.M. Vershik.

A complete solution based on some generalization of Følner's criterion, was obtained recently by J. Feldman, A. Connes, D. Ornstein and B. Weiss (Theorem 1.2). They consider the problem as a special case of a more general question about the hyperfiniteness of a grouppoid. The notions of tame (= hyperfinite) and amenable grouppoids were proved to be equivalent. We shall not discuss here the details of the proof but note the following fact. The theorem shows that the amenable groups are just the class of groups having the property that the fundamental facts of ergodic theory for $\mathbb{Z}^1$ such as Rokhlin-Halmos lemma, Birkhoff-Khinchin theorem, the existence of entropy as a metric invariant, can be generalized to their free actions. It turned out that these metric facts depend only on the algebraic nature of the group but not on its specific action (this action is assumed only to be free and measure-preserving). In particular, all ergodic actions of such groups are orbitally equivalent to each other and orbitally isomorphic to the actions of $\mathbb{Z}^1$. Beyond the class of the amenable groups[5] the situation is much more complicated (cf Section 4).

To end the consideration of tame partitions we will formulate some other results. It is natural to ask whether or not the class of tame partitions coincides with that of the trajectory partitions of discrete and continuous locally compact groups. The answer is no.

Theorem 3.3 (A.M. Vershik [Ve4], V. G. Vinokurov, N. Ganikhodgaev [VG]). *There exists a unique, up to isomorphism, ergodic tame partition which is not stably isomorphic to any tame partition with countable elements.*

We shall call it the special tame partition. It can be constructed in a following way. Let $\{\xi_n\}_1^\infty$ be a decreasing sequence of measurable partitions such that almost all elements of the partitions ξ_n/ξ_{n-1} have no atoms. Then $\bigcap_n \xi_n$ is the required partition. For example, the tail partition of $M = \prod_1^\infty [0,1]$ with the product measure m^∞ is special. Recall that the elements of the tail partition are the classes of sequences which differ from each other by a finite number of coordinates only.

Here is an equivalent realization of this partition. Let $T^\infty = \prod_1^\infty S^1$ be the infinite-dimensional torus and $\sum_1^\infty S^1$ be the direct sum of the groups S^1 which acts on T^∞ in a natural way. The trajectory partition of this action is also the special tame partition. It can be represented as a trajectory partition for an inductive limit of compact groups but not for a locally compact group. At the present time there is no general trajectory therory of actions of non-locally-compact groups. It is unclear even what the fundamental concepts in this case are.

Our last remark is about the semigroups of endomorphisms of a Lebesgue space. Suppose G is a countable semigroup of endomorphisms of a Lebesgue

$$(M_\varphi f)(x,g) = \varphi(x)f(x,g), \qquad (U_h f)(x,g) = f(T_h x, hg),$$

[5] Note that all finite groups, all commutative groups are amenable, that amenability is stable under the transition to subgroups, factor-groups and extensions (hence, the solvable groups are amenable). Free groups with two or more generators as well as the lattices of semisimple groups are non-amenable. The examples of non-amenable group which do not contain the free group with 2 generators are known [O]. There also exists examples of amenable groups of subexponential but not polynomial growth of the number of words, which cannot be obtained by the standard constructions from the commutative and finite groups (R.I. Grigorchuk [G]).

space with quasi-invariant measure. Define the trajectory partition $\tau(G)$ by assuming that the points x, y are in the same element if there are $g_1, g_2 \in G$ such that $g_1 x = g_2 y$. In the case $G = \mathbb{Z}_+^1$ this definition is due to V.A. Rokhlin; if G is a group, this definition is equivalent to the standard one, since $y = g_2^{-1} g_1 x$. There is another useful partition which might be treated as tail partition: two points x, y are in the same element if there exists $g \in G$ for which $gx = gy$. There are no analogs of this notion for groups. The study of the trajectory partitions of semigroups is now in its very beginning; the general results have been obtained only for $\mathbb{Z}_+^1$, i.e. for the case of a single endomorphism.

Theorem 3.4 (R. Bowen [Bo2], A.M. Vershik [Ve5]). *The trajectory partition for any endomorphism is tame.*

§4. Trajectory Theory for Non-Amenable Groups. Rigidity

Except for the important works of R. Zimmer generalizing G.A. Margulis' results on the rigidity of arithmetic subgroups to the actions of semisimple groups, and a series of interesting examples, little is known about the trajectory partitions of non-amenable groups.

It is clear, however, that the situation differs considerably from the amenable case (cf Sects. 1–3). We shall consider mostly the measure-preserving actions. If the measure is assumed only to be quasi-invariant, then for any countable group (not necessarily amenable) one can construct actions for which the trajectory partitions are tame. (Recall that in measure-preserving case this partition is tame only if the group is amenable). To construct such examples, it is enough to observe that the action of any countable group G on itself by (left) group translations is tame. The phase space (G, m) (m stands for the Haar measure) in this example is discrete. In order to construct a non-discrete example, consider the space $M = G \times [0, 1]$ and an arbitrary free action of G on $[0, 1]$. The direct product of these two actions with a finite quasi-invariant measure equivalent to $m \times \rho$ (ρ is the Lebesgue measure on $[0, 1]$) is the required action. What is essential is the fact that the actions of non-amenable groups having tame trajectory partitions often arise in natural situations. Here is a non-trivial example.

Example 1 (R. Bowen [Bo2], A.M. Vershik [Ve5]). The action of $\mathrm{PSL}(2, \mathbb{Z})$ on $P_1\mathbb{R}$ by fractional-linear transformations.

Consider the action of the group of fractional-linear transformations with integer coefficients on $\mathbb{R}^1$: $x \to \dfrac{ax+b}{cx+d}$, $a, b, c, d \in \mathbb{Z}^1$, $ad - bc = 1$. Its trajectory partition is the partition into the equivalence classes in the sense of continuous fraction expansions (cf [Ca]). Although this action is free with respect to the Lebesgue measure and the group $\mathrm{PSL}(2, \mathbb{R})$ is non-amenable, the trajectory partition is tame (also with respect to the Lebesgue measure). This means, in particular, that there is a measurable transformation T such that any two points x, y are equivalent if and only if $T^n x = y$ for some $n \in \mathbb{Z}$.

This example is closely related to Theorem 3.4, since the above partition and the trajectory partition of the endomorphism are essentially the same. This fact was recently obtained by A. Connes, J. Feldman and B. Weiss (with the use of some Zimmer's results) as a corollary from the following general statement: if G is a locally compact group, P is its closed amenable subgroup and Γ is a discrete subgroup, then the partition $\tau(\Gamma)$ into the trajectories of the action of Γ on G/P is tame. In the above example we have $G = \mathrm{SL}(2, \mathbb{R})$, $P = \begin{pmatrix} * & * \\ 0 & * \end{pmatrix}$, $\Gamma = \mathrm{SL}(2, \mathbb{Z})$.

Example 2. A tame action of the free group (the boundary). Let W_k be the free group with k generators and $\widetilde{W}_k$ be the set of all infinite (in one direction) irreducible words, i.e. sequences $\{x_i\}_0^\infty$, where $x_i \in \{g_1, \ldots, g_k, g_1^{-1}, \ldots, g_k^{-1}\}$ and $x_i x_{i+1} \neq e$.

The group W_k acts on $\widetilde{W}_k$ (by concatenation with the subsequent elimination of pairs of the form xx^{-1}). Introduce the product-measure μ on $\widetilde{W}_k$ with uniform factors (x_i takes $2k$ values for $i = 0$ and $(2k - 1)$ values for $i > 0$). The partition $\tau(W_k)$ turns out to be a tame partition of $(\widetilde{W}_k, \mu)$. It is also related to the trajectory partition of the shift endomorphism. Note that $\widetilde{W}_k$ is the boundary of the random walk on W_k with a finite measure.

Unlike the amenable case, a non-amenable group may have many non-orbitally isomorphic measure-preserving actions.

Example 3. Suppose W_2 is the free group with two generators. Consider the following two actions of W_2: 1) by the rotations of the 2-dimensional sphere S^2 with the Lebesgue measure ($W_2 \subset \mathrm{SO}(3)$); 2) the Bernoulli action, i.e. action by left shifts in the space $\mathbb{Z}_2(W_2) = \mathbb{Z}_2^{W_2}$ of all functions on W_2 taking their values in $\mathbb{Z}_2$ and provided with the product-measure with factors $(1/2, 1/2)$. Both actions are free (mod 0), but they are not orbitally isomorphic. The last statement is a corollary from the following theorem.

Theorem 4.1. *Consider a measure-preserving action of a countable group G on the space $(M, \mathcal{M}, \mu)$. Let $g \mapsto U_g$ be the unitary representation of G given by $(U_g f)(x) = f(g^{-1}x)$. The following property of the representation $\{U_g\}$ is stable under the orbital equivalence: for any $\varepsilon > 0$ and any n-tuple $g_1, \ldots, g_n \in G$ there exists $f \in L^2(M)$, $f \perp 1$, such that* $\max_i \|U_{g_i} f - f\| < \varepsilon$ *(the almost invariant vector).*

The non-equivalence of the two actions from Example 3 follows from the fact that the first of them satisfies this last property, while the second one does not satisfy.

The proof is based on the decomposition of the corresponding unitary representations into irreducible ones. In view of Theorem 1.2, both trajectory partitions are not tame.

There is yet another and more geometrical way to describe the above property: it is equivalent to the existence of a non-trivial almost invariant set, i.e. for any

$\varepsilon > 0$ and any $g_1, \ldots, g_n \in G$ there must be a set $B \in \mathcal{M}$ for which $|\mu(B) - \frac{1}{2}| < \varepsilon$, $\max_i \mu(g_i B \triangle B) < \varepsilon$.

Some other trajectory properties, for example, the existence of almost invariant finite partitions and so on, may be formulated in similar terms. V. Ya. Golodets and S.I. Bezuglyi [BG] produced the example of countinuum groups and their measure preserving actions which are pairwise non-orbitally isomorphic. This example may be considered as an analog of D. McDuff's result concerning the existence of continuum of countable groups such that the factors generated by their regular representations are pairwise non-isomorphic.

It would be very interesting to find non-isomorphic trajectory partitions among the different actions of the same group. It seems natural that the Bernoulli actions of different entropies form such a family. Another open question is whether or not the Bernoulli actions of the groups W_k, W_s ($k \neq s$) are orbitally isomorphic.

We will now formulate some results due to R. Zimmer on the rigidity of the trajectory partitions. The main property of the actions of "large" groups discovered by R. Zimmer on the base of Margulis' and others' investigations on the arithematical subgroups of semisimple Lie grouups is that the orbital isomorphism and the metric isomorphism for such groups are essentially the same (unlike the amenable case).

Theorem 4.2 (R. Zimmer [Z]). *Suppose G_1, G_2 are two connected centerless semisimple groups of real rank > 1 without compact factor-groups. Suppose, further, that the restrictions of their free and measure preserving actions to any non-trivial normal subgroup are ergodic (this property is sometimes called irreducible ergodicity). These actions are orbitally isomorphic if and only if there exists an isomorphism between the groups G_1 and G_2 under which the actions are metrically isomorphic.*

Corollary. *Suppose we are given two free and ergodic actions T_1, T_2 of groups* $\mathrm{SL}(m, \mathbb{R})$, $\mathrm{SL}(n, \mathbb{R})$, $(m, n \geqslant 3)$ *respectively on a Lebesgue space $(M, \mathcal{M}, \mu)$. Then the orbital isomorphism between T_1 and T_2 yields that $m = n$ and $VT_1V^{-1} = T_2(r(g))$, where V is an automorphism of $(M, \mathcal{M}, \mu)$ and r is an automorphism of* $\mathrm{SL}(n, \mathbb{R})$.

This statement is not true for $n = 2$.

The following fact about the actions of discrete groups can also be obtained from Theorem 4.2.

Example 4. The actions of $\mathrm{SL}(n, \mathbb{R})$ on the tori $\mathbb{R}^n/\mathbb{Z}^n$ are orbitally non-isomorphic for distinct n. The actions of $\mathrm{SL}(n, \mathbb{R})$ on $P_{n-1}\mathbb{R}$ are also pairwise orbitally non-isomorphic. The corresponding rigidity theorem is as follows.

Theorem 4.3 (R. Zimmer [Z]). *Suppose G, G' are two centerless semisimple groups of real rank > 1 without compact factor groups, and S, S' are their discrete subgroups. Suppose further, that T, T' are irreducibly ergodic actions of S, S' respectively. If T and T' are orbitally isomorphic, then $G = G'$ and there exists an isomorphism between S and S' under which T and T' are metrically isomorphic.*

The above theorems show that the trajectory class of an action "almost uniquely" determines the action itself. This property of the action is called its rigidity. A similar property appeared earlier in Margulis-Mostow results saying that under certain conditions the fundamental group "almost uniquely" determines the Riemannian metric on the manifold.

The essential part of Zimmer's method is his reformulations of certain notions related to Lie groups such as amenability, rigidity, discrete spectrum and so on, in terms of group actions (the amenability of the group action etc.). It is interesting that such reformulations enabled him to simplify the proofs of some purely group results.[6]

§ 5. Concluding Remarks. Relationship Between Trajectory Theory and Operator Algebras

Although our exposition of trajectory theory is purely geometrical, there also exists its functional-algebraic version, and both methods were elaborated simultaneously. Recall that there is a natural correspondence between the measurable partitions and the rings of functions (from L^∞ or from L^2) which are constant on their elements. Therefore, the theory of measurable partitions or of sub-σ-algebras of the σ-algebra of measurable sets may be treated as a theory of subrings of the commutative ring L^∞. Since in most interesting cases the trajectory partitions are not measurable, they are not associated with subrings of L^∞. The idea is to consider non-commutative rings in this case.

Historically, this construction first appeared in [NM], where the cross-product corresponding to group actions were introduced, but it took a long period of time to understand that the above-mentioned cross-product depends only on the trajectory partitions of the action but not on the action itself.

This, now well-known construction, can be used for studying C^*- and W^*-algebras, factors, representations and for constructing the algebraic examples with the use of ergodic methods, as well as for studying the dynamical systems and their trajectory partitions, foliations and so on with the use of algebraic methods. Let us now describe this construction.

Suppose G is a discrete group and $T = \{T_g\}$ is its measure preserving action on a Lebesgue space $(M, \mathcal{M}, \mu)$. Consider the space $M \times G$ and the measure $\mu \times m$ on it, where m is a Haar measure on G. Let $W(G, M)$ be the weakly closed operator algebra in $L^2(M \times G)$ generated by the operators M_φ and U_h of the form

[6] Recently A. Furman, following a certain example due to R. Zimmer, gave an example of the smooth action of a discrete group such that there are no *free* measurable action of any discrete group with the same orbit partitions. This example solves an old problem.

where $\varphi \in L^\infty(M)$; $h, g \in G$; $f \in L^2(M \times G)$. This algebra is called the cross-product algebra of $L^\infty(M)$ and $l^1(G)$ with respect to the action T.

This construction is just the one contained in [NM], and it is orbitally invariant, i.e. if we consider, instead of the action T of G some orbitally equivalent action T' of G', this does not affect $W(G, M)$. Moreover, this algebra can be defined in the invariant terms, i.e. the specific action $\{T_g\}$ need not be mentioned [FM]. Hence, the algebraic invariants of $W(G, M)$ are the trajectory invariants of $\{T_g\}$. By the present time a widely developed theory generalizing the above construction to quasi-invariant measures, to foliations, to C^*-algebras, making use of cohomologies, exists (cf [FM], [Mo], [HO], [Co3]). The so called full group, or Dye group, consisting of all automorphism T for which the trajectory partition $\sigma(T)$ is finer than some given trajectory partition, say, $\tau(G)$, was intensely studied. It seems useful to keep in mind that the algebra $\mathfrak{M} = \{M_\varphi$: $\varphi \in L^\infty(M)$ is an analog of the Cartan algebra in $W(G, M)$, while the Dye group (denoted by [G]) is an analog of the Weil group in the theory of semisimple Lie algebras. This analogy plays a substantial role in the study of W^*-algebras ([Ve3], [FM]).

On the other hand, many results in trajectory theory (for example the Feldman-Connes-Ornstein-Weiss theorem, i.e. Theorem 1.2 of this chapter) were obtained only after their analogs in the theory of W^*-algebras had been proved (in our example, after the Connes theorem [Co2]).[7]

Bibliography

The relationship between ergodic theory on the one hand and classical mechanics and theory of ordinary differential equations on the other hand is discussed in [Ar1], [Ar2]. The book BVGN contains the exposition of the theory of Lyapunov characteristic exponents for the systems of ordinary differential equations. An important bibliography of works in ergodic theory up to 1975 may be found in the review article [KSS] Approximation theory of dynamical systems by periodic ones with some its applications is the topic of [KS].

Spectral theory, as well as numerous examples of dynamical systems are discussed in [CFS]. The review articles [Ro1], [Ro2], [Ro3] are devoted to the detailed exposition of measure theory and to the principal problems of general metric and entropy theory of dynamical systems. The book [Si5]

[7] The relationship between dynamical systems and C^*-, W^*-algebras was studied very intensively last decade. We can mention AF and locally semisimple algebras on one side and so-called adic transformations on the another (see A.M. Vershik Uniform algebraic approximation of shift and multiplication operator. Sov. Dokl. 259 (1981), 526–529 and [A.N. Livshits, A.M. Vershik, "Adic models of ergodic transformations, spectral theory and related topics", Adv. in Soviet Math. *9* (1992), 185–204], [R. Herman, I. Putnam, C. Skau "Ordered Bratteli diagrams, dimension groups and topological dynamics", Intern. J. Math. *3* (1992), 827–864] etc.). In particular, one may mention an application of K-theory of C^*-algebras to dynamical systems ("ergodic K-theory") and new constructions of algebras using endomorphisms and polymorphisms. (see A.M. Vershik Multivalued mappings with invariant measure (polymorphisms) and Markov operators. Journ. of Sov. Math. 23 (1983) 2243–2266.)

contains a brief and clear exposition of the elements of ergodic theory. Some chapters of [Bi] are also devoted to entropy theory. One of the first systematic expositions of ergodic theory is [H]. The book [Or] deals with the problem of metric isomorphism and contains the important results due mainly to its author, in particular, it contains the complete solution to the problem of metric isomorphism for Bernoulli automorphisms. The results of recent investigations concerning the problem of the equivalence of dynamical systems in the sense of Kakutani are presented in [ORW]. The relationship between ergodic theory and statistical mechanics of the lattice systems is considered in [Rue]. The main area of applications of thermodynamic formalism is the theory of hyperbolic dynamical systems.

For the convenience of the reader, references to reviews in Zentralblatt für Mathematik (Zbl.), complied using the MATH database, have, as far as possible, been included in this bibliography.

[A] Abramov, L.M.: Metric automorphisms with quasi-discrete spectra. Izv. Akad. Nauk SSSR, Ser. Mat. *26*, 513–530 (1962) [Russian]. Zbl. 132.359

[AK] Ambrose, W., Kakutani, S.: Structure and continuity of measurable flows. Duke Math. J. *9*, 25–42 (1942)

[An] Anosov, D.V.: On the additive homologic equation related to an ergodic rotation of the circle. Izv. Akad. Nauk SSSR. Ser. Mat. *37*, 1259–1274 (1973) [Russian]. Zbl. 298.28016. English transl.: Math. USSR, Izv. 7, 1257–1271 (1975)

[AW] Araki, H., Woods, E.: A classification of factors. Publ. Res. Inst. Math. Sci., Kyoto Univ., Ser. *A 4*, 51–130 (1968). Zbl. 206.129

[Ar1] Arnold, V.I.: Mathematical methods in classical mechanics. Nauka, Moscow 1974 [Russian]; English transl.: Springer-Verlag, New York–Berlin–Heidelberg 1978. 462 p. Zbl. 386.70001

[Ar2] Arnold, V.I.: Additional chapters of the theory of ordinary differential equations. Nauka, Moscow 1978 [Russian]

[AGH] Auslander, L., Green, L., Hahn, F.: Flows on homogeneous spaces. Princeton Univ. Press, Princeton N.J., 1963. 107 p. Zbl. 106.368

[Av] Avez, A.: Entropie des groupes de type fini. C.R. Acad. Sci., Paris, Ser. *A 275*, 1363–1366 (1972). Zbl. 252.94013

[B] Belinskaya, R.M.: Partitions of (the) Lebesgue space in(to) (the) trajectories of ergodic automorphisms. Funkts. Anal. Prilozh. *2*, No. 3, 4–16 (1968) [Russian]. Zbl. 176.447. English transl.: Funct. Anal. Appl. *2* (1968) 190–199 (1969)

[Be] Bewely, T.: Extensions of the Birkhoff and von Neumann ergodic theorems to semigroups actions. Ann. Inst. Heni Poincaré, Sect. *B* 7, 283–291 (1971). Zbl. 226.28009

[BG] Bezuglyi, S.I., Golodets, V.Ya.: Hyperfinite and II_1-actions for non-amenable groups. J. Funct. Anal. *40*, 30–44 (1981). Zbl. 496.22011

[Bi] Billingsley, P.: Ergodic theory and information. J. Wiley and Sons, New York–London–Sydney 1965. Zbl. 141.167

[Bl] Blanchard, F.: Partitions extrémales de flots d'entropie infinie. Z. Wahrscheinlichkeitstheor. Verw. Geb. *36*, 129–136 (1976). Zbl. 319.28012 (Zbl. 327.28012)

[Blo] Blokhin, A.A.: Smooth ergodic flows on surfaces. Tr. Mosk. Mat. O.-va *27*, 113–128 (1972) [Russian]. Zbl. 252.28008

[BK] Bogolyubov, N.N., Krylov, N.M.: La théorie générale de la mesure dans son application à l'étude de systèmes dynamiques de la mécanique non-linéaire. Ann. Math. II. Ser. *38*, 65–113 (1937). Zbl. 16.86

[Bo1] Bowen, R.: Equilibrium states and the ergodic theory of Anosov diffeomorphisms. Lect. Notes Math. 470 (1975). Zbl. 308.28010

[Bo2] Bowen, R.: Anosov foliations are hyperfinite. Ann. Math., II. Ser. *106*, 549–565 (1977). Zbl. 374.58008

[BVGN] Bylov, B.F., Vinograd, R.E., Grobman, D.M., Nemitsky V.V.: Theory of Lyapunov exponents and applications to stability problems. Nauka, Moscow, 1966. 576 p. [Russian]. Zbl. 144.107

[C] Calderon, A.: A general ergodic theorem. Ann. Math., II. Ser. *58*, 182–191 (1953). Zbl. 52.119

[Ca] Cassels, J.W.S.: Zn introduction to diophantine approximation. Cambridge Univ. Press, 1957. 166 p. Zbl. 77.48

[CO] Chacon, R.V., Ornstein, D.S.: A general ergodic theorem. Ill. J. Math. *4*, 153–160 (1960). Zbl. 134.121

[Ch] Chulaevsky, B.A.: Cyclic approximations of the interval exchange transformations. Usp. Mat. Nauk *34*, 215–216 (1979) [Russian]. Zbl. 415.28014

[Col] Connes, A.: Une classification des facteurs de type III. Ann. Sci. Ec. Norm. Supér. IV. Ser. *6*, 133–252 (1973). Zbl. 274.46050

[Co2] Connes, A.: Classification of injective factors of types II_1, II_∞, III_λ, $\lambda \neq 1$. Ann. Math., II. Ser. *104*, 73–115 (1976). Zbl. 343.46042

[Co3] Connes, A.: Classification des facteurs. Proc. Symp. Pure Math. *38*, 43–102 (1982). Zbl. 503.46043

[CFW] Connes, A., Feldman, J., Weiss, B.: An amenable equivalence relation is generated by single transformation. Ergodic Theory Dyn. Syst. *1*, 431–450 (1981). Zbl. 491.28018

[Con] Conze, J.P.: Entropie d'un groupe abélien de transformations. Z. Wahrscheinlichkeitstheor. Verw. Geb. *25*, 11–30 (1972). Zbl. 261.28015

[CFS] Cornfeld, I.P., Fomin, S.V., Sinai, Ya.G.: Ergodic Theory. Springer-Verlag, Berlin-Heidelberg—New York 1982. Zbl. 493.28007

[D] Dye, H.: On a group of measure preserving transformations. I, II. Am. J. Math. *81*, 119–159 (1959); *85*, 551–576 (1963). Zbl. 87.115 Zbl. 191.428

[FV] Fedorov, A.L., Vershik, A.M.: Trajectory theory. In: Modern problems of mathematics. New achievements, vol. 26, 121–212. VINITI, Moscow

[FM] Feldman, J., Moore, C.C.: Ergodic equivalence relations, cohomology and von Neumann algebras. I, II. Trans. Am. Math. Soc. *234*, 289–324 (1977). Zbl. 369.22009

[F] Furstenberg, H.: Strict ergodicity and transformations of the torus. Am. J. Math. *83*, 573–601 (1961). Zbl. 178.384

[FKO] Furstenberg, H., Katznelson, Y., Ornstein, D.: The theoretical proof of Szemeredi's theorem. Bull. Am. Math. Soc., New Ser. 7, 527–552 (1982). Zbl. 523.28017

[FK] Furstenberg, H., Kesten, H.: Products of random matrices. Ann. Math. Stat. *31*, 457–469 (1960). Zbl. 137.355

[GF] Gelfand, I.M., Fomin, S.V.: Geodesic flows on manifolds of constant negative curvature. Usp. Mat. Nauk 7, 118–137 (1952) [Russian]. Zbl. 48.92

[G] Grigorchuk, R.I.: On Milnor's problem concerning group growth. Dokl. Akad. Nauk SSSR *271*, 30–33 (1983). Zbl. 547.20025. English transl.: Sov. Math., Dokl. *28*, 23–26 (1983)

[Gu] Gurevich, B.M.: Perfect partitions for ergodic flows. Funkts. Anal. Prilozh. *11*, No. 3, 20–23 (1977) [Russian]. Zbl. 366.28006 English transl.: Funct. Anal. Appl. *11*, 179–182 (1978)

[HIK] Hajian, A., Ito, Y., Kakutani, S.: Full groups and a theorem of Dye. Adv. Math. *17*, 48–59 (1975). Zbl. 303.28017

[H] Halmos, P.R.: Lectures on ergodic theory. Math. Soc. Jap. Tokyo, 1956. 99p. Zbl. 73.93

[HO] Hamachi, T., Osikawa, M.: Ergodic groups of automorphisms and Krieger's theorem. Semin. Math. Sci., Keio University, *3*, 113 p. (1981). Zbl. 472.28015

[J] Jewett, R.: The prevalence of uniquely ergodic systems. J. Math. Mech. *19*, 717–729 (1970). Zbl. 192.406

[KV] Kaimanovich, V., Vershik, A.: Random walks on discrete groups boundary and entropy. Ann. Probab. *11*, 457–490 (1983)

[K] Kalikow, S.A.: T, T^{-1}-transformation is not loosely Bernoulli. Ann. Math., II. Ser. *115*, 393–409 (1982). Zbl. 523.28018

[Kal] Kaminski, B.: Mixing properties of two-dimensional dynamical systems with completely positive entropy. Bull. Acad. Pol. Sci., Ser. Sci. Math. *28*, 453–463 (1980). Zbl. 469.28013

[Ka2] Kaminski, B.: The theory of invariant partitions for actions. Bull. Acad. Pol. Sci., Ser. Sci. Math. *29*, 349–362 (1981). Zbl. 479.28016

[Kat1] Katok, A.B.: Spectral properties of dynamical systems with an integral invariant on the torus. Funkts. Anal. Prilozh. *1*, No. 4, 46–56 (1967) [Russian]. Zbl. 172.121. English transl.: Funct. Anal. Appl. *1*, 296–305 (1967)

[Kat2] Katok, A.B.: Invariant measures for the flows on oriented surfaces. Dokl. Akad. Nauk SSSR *211*, 775–778 (1973) [Russian]. Zbl. 298.28013. English transl.: Sov. Math., Dokl. *14*, 1104–1108 (1974)

[KSS] Katok, A.B., Sinai, Ya.G., Stepin, A.M.: Theory of dynamical systems and general transformation groups with invariant measure. Itogi Nauki Tekh., Ser. Mat. Anal. *13*, 129–262. Moscow, (1975) [Russian]. Zbl. 399.28011. English transl.: J. Sov. Math. *7*, 974–1065 (1977)

[KS] Katok, A.B., Stepin, A.M.: Approximations in ergodic theory. Usp. Mat. Nauk *22*, No. 5, 81–106 (1967) [Russian]. Zbl. 172.72. English transl.: Russ. Math. Surv. *22*, No. 5, 77–102 (1967)

[KW1] Katznelson, Y., Weiss, B.: Commuting measure-preserving transformations. Isr. J. Math. *12*, 161–173 (1972). Zbl. 239.28014

[KW2] Katznelson, Y., Weiss, B.: A simple proof of some ergodic theorems. Isr. J. Math. *42*, 291–296 (1982). Zbl. 546.28013

[Ke1] Keane, M.: Interval exchange transformations. Math. Z. *141*, 25–31 (1975). Zbl. 278.28010

[Ke2] Keane, M.: Nonergodic interval exchange transformations. Isr. J. Math. *26*, 188–196 (1977). Zbl. 351.28012

[KeS] Keane, M., Smorodinsky, M.: Bernoulli schemes of the same entropy are finitarily isomorphic. Ann. Math., II. Ser. *109*, 397–406 (1979). Zbl. 405.28017

[Ki] Kiefer, J.C.: A generalized Shannon-McMillan theorem for the actions of an amenable group on a probability space. Ann. Probab. *3*, 1031–1037 (1975). Zbl. 322.60032

[Kin] Kingman, J.F.C.: Subadditive ergodic theory. Ann. Probab. *1*, 883–909 (1973). Zbl. 311.60018

[Kir] Kirillov, A.A.: Dynamical systems, factors and group representations. Usp. Mat. Nauk. *22*, No. 5, 67–80 (1967) [Russian]. Zbl. 169.466. English transl.: Russ. Math. Surv. No. *5*, 63–75 (1967)

[Ko1] Kochergin, A.V.: On the absence of mixing for special flows over rotations of the circle and for flows on the two-dimensional torus. Dokl. Akad. Nauk SSSR, Ser. Mat. *205*, 515–518 (1972) [Russian]. Zbl. 262.28015; English transl.: Sov. Math., Dokl. *13*, 949–952 (1972)

[Ko2] Kochergin, A.V.: Change of time in flows and mixing. Izv. Akad. Nauk SSSR, Ser. Mat. *37*, 1275–1298 (1973). [Russian]; Zbl. 286.28013. English transl.: Math. USSR, Izv. *7*, 1273–1294 (1975)

[Kol1] Kolmogorov, A.N.: A new metric invariant of transient dynamical systems and automorphisms of Lebesgue spaces. Dokl. Akad. Nauk SSSR, Ser. Mat. *119*, 861–864 (1958) [Russian]. Zbl. 83.106

[Kol2] Kolmogorov, A.N.: On dynamical systems with an integral invariant on the torus. Dokl. Akad. Nauk SSSR, Ser. Mat. *93*, 763–766 (1953) [Russian]. Zbl. 52.319

[Kr] Krengel, U.: Recent progress in ergodic theorems. Astérisque *50*, 151–192 (1977). Zbl. 376.28016

[Kri1] Krieger, W.: On entropy and generators of measure preserving transformations. Trans. Am. Math. Soc. *149*, 453–464 (1970). Zbl. 204.79

[Kri2] Krieger, W.: On unique ergodicity. Proc. Sixth Berkeley Symp. Math. Stat. Prob. *2*, 337–346 (1970). Zbl. 262.28013

[Kri3] Krieger, W.: On ergodic flows and the isomorphism of factors. Math. Ann. *223*, 19–30 (1976). Zbl. 332.46045

[Kry] Krygin, A.B.: An example of continuous flow on the torus with mixed spectrum. Mat. Zametki *15*, 235–240 (1974) [Russian]. Zbl. 295.58011. English transl.: Math. Notes *15*, 133–136 (1974)

[M1] Mackey, G.W.: Point realizations of transformation groups. Ill. J. Math. *6*, 327–335 (1962). Zbl. 178.172

[M2] Mackey, G.W.: Ergodic transformations groups with a pure point spectrum. Ill. J. Math. *8*, 593–600 (1964). Zbl. 255.22014

[M3] Mackey, G.W.: Ergodic theory and virtual groups. Math. Ann. *166*, 187–207 (1966). Zbl. 178.388

[Ma] Maruyama, G.: Transformations of flows. J. Math. Soc. Japan *18*, 303–330 (1966). Zbl. 166.404

[Mas] Masur, H.: Interval exchange transformations and measured foliations. Ann. Math. II. Ser. *115*, 169–200 (1982). Zbl. 497.28012

[Mau] Mautner, F.: Geodesic flows on symmetric Riemann spaces. Ann. Math., II. Ser. *65*, 416–431 (1957). Zbl. 84.375

[Me] Meshalkin, L.D.: A case of isomorphism of Bernoulli schemes. Dokl. Akad. Nauk SSSR, Ser. Mat. *128*, 41–44 (1959) [Russian]. Zbl. 99.123

[Mil] Millionshchikov, V.M.: Metric theory of linear systems of differential equations. Mat. Sb., Nov. Ser. *77*, 163–173 (1968) [Russian]. Zbl. 176.46

[Mi2] Millionshchikov, V.M.: A criterion of stability of probability spectrum for linear systems of differential equations with reccurrent coefficients and criterion of quasi-reducibility for systems with quasi-periodic coefficients. Mat. Sb., Nov. Ser. *78*, 179–201 (1969) [Russian]. Zbl. 186.147

[Mo] Moore, C.C.: Ergodic theory and von Neumann algebras. Proc. Symp. Pure Math. *38*, 179–226 (1982). Zbl. 524.46045

[N] Neumann, J. von.: Zur Operatorenmethode in der klassischen Mechanik. Ann. Math., II. Ser. *33*, 587–642 (1932). Zbl. 5.122

[NM] Neumann J. von, Murrey, F.: On rings of operators IV. Ann. Math., II. Ser. *44*, 716–808 (1943). Zbl. 60.269

[O] Olshansky, A.Yu.: On the problem of existence of an invariant mean of a group. Usp. Mat. Nauk *35*, No. 4, 199–200 (1980) [Russian]. Zbl. 452.20032. English transl: Russ. Math. Surv. *35*, No. 4, 180–181 (1980)

[Or] Ornstein, D.S.: Ergodic theory, randomness and dynamical systems. Yale Univ. Press, New Haven and London, 141 p. 1974. Zbl. 296.28016

[OW] Ornstein, D.S., Weiss, B.: Entropy and isomorphism theorems for actions of amenable groups. Isr. J. Math. 1987

[ORW] Ornstein, D.S., Rudolph, D.J., Weiss, B.: Equivalence of measure preserving transformations. Mem. Am. Math. Soc. *262*, 116 p. (1982). Zbl. 504.28019

[Os1] Oseledets, V.I.: Markov chains, skew products and ergodic theorems for "general" dynamical systems. Teor. Veroyatn. Primen. *10*, 551–557 (1965) [Russian]. Zbl. 142.144. English transl.: Theor. Probab. Appl. *10*, 499–504 (1965)

[Os2] Oseledets, V.I.: A multiplicative ergodic theorem. Lyapunov characteristic exponents of dynamical systems. Tr. Mosk. Mat. O.-va *19*, 179–210 (1968) [Russian]. Zbl. 236.93034. English transl.: Trans. Mosc. Math. Soc. *19*, 197–231 (1969)

[P] Pitskel, B.S. On informational "futures" of amenable groups. Dokl. Akad. Nauk SSSR, Ser. Mat. *223*, 1067–1070 (1975) [Russian]. Zbl. 326.28027. English transl.: Sov. Math., Dokl. *16*, 1037–1041 (1976)

[PS] Pitskel, B.S., Stepin, A.M.: On the equidistribution property of entropy of commutative groups of automorphisms. Dokl. Akad. Nauk SSSR, Ser. Mat. *198*, 1021–1024 (1971) [Russian]. Zbl. 232.28017. English transl.: Sov. Math., Dokl. *12*, 938–942 (1971)

[R] Ramsay, A.: Virtual groups and group actions. Adv. Math. *6*, 253–322 (1971). Zbl. 216.149

[Re] Renault, J.: A groupoid approach to C*-algebras. Lect. Notes Math. *793* 160 p. (1980). Zbl. 433.46049

[Rol] Rokhlin, V.A.: On the principal notions of measure theory. Mat. Sb., Nov. Ser. *67*, 107–150 (1949) [Russian]. Zbl. 33.169

[Ro2] Rokhlin, V.A.: Selected topics of metric theory of dynamical systems. Usp. Mat. Nauk *4*, No. 2, 57–128 (1949) [Russian]. Zbl. 32.284

[Ro3] Rokhlin, V.A.: Lectures on the entropy theory of measure preserving transformations. Usp. Mat. Nauk *22*, No. 5, 3–56 (1967) [Russian]. Zbl. 174.455. English transl.: Russ. Math. Surv. *22*, No. 5, 1–52 (1967)

[RS] Rokhlin, V.A., Sinai, Ya.G.: Construction and properties of invariant measurable partitions. Dokl. Akad. Nauk SSSR, Ser. Mat. *141*, 1038–1041 (1961) [Russian]. Zbl. 161,343. English transl.: Sov. Math., Dokl. *2*, 1611–1614 (1961)

[Ru1] Rudolph, D.J.: A two-valued step-coding for ergodic flows. Math. Z. *150*, 201–220 (1976). Zbl. 318.28010 (Zbl. 325.28019).

[Ru2] Rudolph, D.J.: An example of a measure preserving map with minimal self-joinings and applications. J. Anal. Math. *35*, 97–122 (1979). Zbl. 446.28018

[Rue] Ruelle, D.: Thermodynamic formalism. Addison–Wesley, London–Amsterdam–Don Mills, Ontario–Sydney–Tokyo 1978. 183 p. Zbl. 401.28016

[S] Sataev, E.A.: On the number of invariant measures for the flows on oriented surfaces. Izv. Akad. Nauk SSSR, Ser. Mat. *39*, 860–878 (1975) [Russian]. Zbl. 323.28012. Math. USSR, Izv. 9, 813–830 (1976)

[Sh] Shklover, M.D.: On classical dynamical systems on the torus with continuous spectrum. Izv. Vyssh. Uchebn. Zaved. Mat. 1967, No. *10*, 113–124 (1967) [Russian]. Zbl. 153.126

[Si1] Sinai, Ya.G.: On the concept of entropy for a dynamical system. Dokl. Akad. Nauk SSSR, Ser. Mat. *124*, 768–771 (1959) [Russian]. Zbl. 86.101

[Si2] Sinai, Ya.G.: Dynamical systems with countable Lebesgue spectrum I. Izv. Akad. Nauk SSSR, Ser. Mat. *25*, 899–924 (1961) [Russian]. Zbl. 109.112

[Si3] Sinai, Ya.G.: On weak isomorphism on measure preserving transformations. Mat. Sb. *63*, 23–42 (1964) [Russian]. English transl.: Transl., II. Ser., Am. Math. Soc. *57*, 123–143 (1966).

[Si4] Sinai, Ya.G.: Gibbs measures in ergodic theory. [Usp., Nov. Ser. Mat. Nauk *27*, 21–64] (1972) [Russian]. Russ. Math. Surv. *27*, No. 4, 21–69 (1972). Zbl. 255.28016

[Si5] Sinai, Ya.G.: Introduction to ergodic theory. Erevan Univ. Press, Erevan 1973. English transl.: Math. Notes *18*, Princeton, Princeton Univ. Press (1976). Zbl. 375.28011

[Sm] Smorodinsky, M.: A partition on a Bernoulli shift which is not weak Bernoulli. Math. Syst. Theory. *5*, 201–203 (1971). Zbl. 226.60066

[St] Stepin, A.M.: Bernoulli shifts on groups. Dokl. Akad. Nauk SSSR, Ser. Mat. *223*, 300–302 (1975) [Russian]. Zbl. 326.28026. English transl.: Sov. Math., Dokl. *16*, 886–889 (1976)

[Su] Sujan, S.: Generators of an Abelian group of invertible measure preserving transformations. Monatsh. Math. *90*, 67–79 (1980). Zbl. 432.28016

[T1] Tempelman, A.A.: Ergodic theorems for general dynamical systems. Tr. Mosk. Mat. O.-va *26*, 95–132 (1972) [Russian]. Zbl. 249.28015. English transl.: Trans. Mosc. Math. Soc. *26*, 94–132 (1974)

[T2] Tempelman, A.A.: Ergodic theorems on groups. Vilnius 1986

[V1] Veech, W.A.: Interval exchange transformations. J. Anal. Math. *33*, 222–272 (1978). Zbl. 455.28006

[V2] Veech, W.A.: Gauss measures for transformations on the space of interval exchange maps. Ann. Math. II. Ser. *115*, 201–242 (1982). Zbl. 486.28014

[V3] Veech, W.A.: The metric theory of interval exchange transformations I. Generic spectral properties. Am. J. Math. 1983

[Ve1] Vershik, A.M.: A measurable realization of continuous groups of automorphisms of a unitary ring. Izv. Akad. Nauk SSSR, Ser. Mat. *29*, 127–136 (1965) [Russian]. Zbl. 194.163. English transl.: Transl., II. Ser. Am. Math. Soc. 84, 69–81 (1969)

[Ve2] Vershik, A.M.: On the lacunary isomorphism of sequences of measurable partitions. Funkts Anal. Prilozh. *2*, No. 3, 17–21 (1968) [Russian]. Zbl. 186.202. English transl.: Funct. Anal. Appl. 2 (1968), 200–203 (1969)

[Ve3] Vershik, A.M.: Non-measurable partitions, trajectory theory and operator algebras. Dokl. Akad. Nauk SSSR *199*, 1004–1007 (1971) [Russian]. Zbl. 228.28013. English transl.: Sov. Math., Dokl. *12*, 1218–1222 (1971)

[Ve4] Vershik, A.M.: Decreasing sequences of σ-algebras. Teor. Veroyatn. Primen. *19*, 657–658 (1974) [Russian].

[Ve5] Vershik, A.M.: The action of PSL $(2, \mathbb{R})$ on $P_1 R$ is approximable. Usp. Mat. Nauk *33*, No. 1, 209–210 (1978) [Russian]. Zbl. 391.28008. English transl.: Russ. Math. Surv. *33*, No. 1, 221–222 (1978)

[Ve6] Vershik, A.M.: Measurable realizations of groups of automorphisms and integral representations of positive operators. Siberian Math. J. *28*, N 1, 52–60

[Ve6F] Vershik, A.M., Fedorov, A.A.: Trajectory theory. Itogi Nauki Tekh., Ser. Sovrem. Probl. Mat., vol. *26*, 171–211. (1985) [Russian]. Zbl. 614.28018

[VG] Vinokurov, V.G., Ganikhodgaev, N.: Conditional functions in trajectory theory of dynamical systems. Izv. Akad. Nauk SSSR, Ser. Mat. *42*, 927–964 (1978) [Russian]. Zbl. 408.28018. English transl.: Math. USSR, Izv. *13*, 221–251 (1979)

[Z] Zimmer, R.: Ergodic theory, group representations and rigidity. Bull. Am. Math. Soc. New Ser. *6*, 383–416 (1982). Zbl. 532.22009

Additional Bibliography

[Aa] Aaronson, J. An introduction to infinite ergodic theory. AMS Mathematical Surveys and Monographs, vol. 50 (1997)

[BJR] Bellow, A., Jones, R.L., Rosenblatt, J.M.: Harmonic Analysis and ergodic theory. Proceedings of the conference on almost everywhere convergence, Columbus OH, Academic Press, 73–98 (1990)

[Ber] Bergelson, V.: Weakly mixing PET. Ergod. Th. & Dynam. Systems. *7*, 337–349 (1987)

[Bo1] Bourgain, J.: On the maximal ergodic theorems for certain subsets of integers. Israel J. Math. *61*, 29–72 (1988)

[Bo2] Bourgain, J.: On the pointwise ergodic theorem on L^p for arithmetic sets. Israel J. Math. *61*, 73–84 (1988)

[Bo3] Bourgain, J.: Pointwise ergodic theorems for arithmetic sets. Publ. Math. IHES *69*, 5–45 (1989)

[CL] Conze, J.-P., Lesigne, E.: Sur un theoreme ergodique pour des mesures diagonales. C.R. Acad. Sci. *306*, 491–493 (1988)

[dJR] del Junco, A., Rudolph, D.J.: On ergodic actions whose self-joinings are graphs. Ergod. Th. & Dynam. Systems *7*, 531–557 (1987)

[F3] Furstenberg, H.: Disjointness in ergodic theory, minimal sets, and a problem in diophantine approximation. Math. Syst. Theory. *1*, 1–69 (1967)

[F4] Furstenberg, H.: Recurrence in ergodic theory and combinatorial number theory. Princeton Univ. Press, 1981

[Ho] Host, B.: Mixing of all orders and pairwise independent joinings of systems with singular spectrum. Israel J. Math. *76*, 289–298 (1991)

[Kr1] Krengel, U.: Ergodic theorems. De Gruyter Studies in Mathematics, vol. 6 (1985)

[Kr2] Krengel, U.: On Rudolph's representation of aperiodic flows. Ann. Inst. Henri Poincare *B12*, 319–338 (1976)

[L] Lesigne, E.: Theoremes ergodiques pour un translation sur une nilvariete. Ergod. Th. & Dynam. Systems *9*, 115–126 (1989)

[O] Oxtoby, J.C.: Ergodic sets. Bull. Amer. Math. Soc. *58*, 116–136 (1952)

[Ra1] Ratner, M.: Rigidity of horocycle flows. Ann. Math. *115*, 587–614, (1982)

[Ra2] Ratner, M.: Horocycle flows, joinings and rigidity of products. Ann. Math. *118*, 277–313, (1983)

[RW] Rosenblatt, J.M., Wierdl, M.: Pointwise ergodic theorems via harmonic analysis. London Math. Society Lecture Notes Series *205*, 3–151, (1995)

[Ru3] Rudolph, D.J.: Fundamentals of measurable dynamics. Oxford University Press, 1990

[Ru4] Rudolph, D.J.: A joinings proof of Bourgain's return time theorem. Ergod. Th. & Dynam. Systems *14*, 197–203 (1994)

[Ry1] Ryzhikov, V.V.: Connection between the mixing properties of a flow and the isomorphism of the transformations that compose it. Math. Notes *49*, 621–627 (1991)

[Ry2] Ryzhikov, V.V.: Joinings and multiple mixing of the actions of finite rank. Funct. Anal. Appl. *27*, 128–140 (1993)

[Ry3] Ryzhikov, V.V.: Joinings, wreath products, factors and mixing properties of dynamical systems. Russian Acad. Sci. Izv. Math. *42*, 91–114 (1994)

[Th] Thouvenot, J.-P.: La convergence presque sure des moyennes ergodiques suivant certanes sous-suites d'entiers (d'apres Jean Bourgain). Asterisque *189–190*, 133–153 (1990)

[Th1] Thouvenot, J.-P.: Some properties and applications of joinings. London Math. Society Lecture Notes Series *205*, 207–235 (1995)

[W] Wierdl, M.: Pointwise ergodic theorems along prime numbers. Israel J. Math. *64*, 315–336 (1988)

II. Ergodic Theory of Smooth Dynamical Systems

Contents

Chapter 6
Stochasticity of Smooth Dynamical Systems. The Elements of KAM-Theory

Ya.G. Sinai

§ 1. Integrable and Nonintegrable Smooth Dynamical Systems. The Hierarchy of Stochastic Properties of Deterministic Dynamics

A Hamiltonian system with n degrees of freedom and with a Hamiltonian $H(p_1,\ldots,p_n,q_1,\ldots,q_n)$ is called integrable if it has n first integrals $I_1 = H, I_2, \ldots, I_n$ which are in involution. The famous Liouville theorem states (see [Ar1], [KSF]) that if the n-dimensional manifold obtained by fixing the values of the integrals $I_1 = C_1$, $I_2 = C_2$, ..., $I_n = C_n$ is compact, and these integrals are functionally independent in some neighbourhood of the point $(C_1,\ldots,C_n)$ then this manifold is the n-dimensional torus. One could introduce cyclic coordinates φ_1, ..., φ_n such that the equations of motion would obtain a simple form $\dot\varphi_i = F_i(I_1,\ldots,I_n) = \text{const}$, $1 \leqslant i \leqslant n$, and the motion would be quasi-periodic with n frequencies.

From the point of ergodic theory this situation means that a flow $\{S^t\}$ which corresponds to a Hamiltonian system with the Hamiltonian $H(p,q)$ and with the invariant measure $dp\,dq$ is non-ergodic. Its ergodic components are (mod 0) n-dimensional tori and in every such torus an ergodic flow with purely point spectrum is obtained. We shall call a dynamical system integrable one also in a more general case when it is non-ergodic and when on almost every of its ergodic components a dynamical system with purely point spectrum is obtained.

There are many examples of integrable systems: the geodesic flows on surfaces of revolution, the geodesic flow on a triaxial ellipsoid, a billiard in an ellipse, the system of three point eddies in two-dimensional hydrodynamics and so on. A large number of new examples of integrable systems were found in the last decade by the inverse scattering method [T].

Nevertheless one has to consider integrability of a dynamical system not as a rule, but instead as an exclusion. The property of complete integrability of a Hamiltonian system has disappeared already under small sufficiently generic type perturbations of a Hamiltonian. The famous KAM-theory stated that invariant tori with "sufficiently incommensurable" frequencies did not disappear but only shifted slightly in the phase space and the corresponding motion remains quasi-periodic. These invariant tori form a set of positive measure which is not a domain. Its complement consists of stochastic layers where the dynamics is essentially more complicated. The main mechanism whereby such layers occur is connected with the appearance of heteroclinic and homoclinic trajectories (see

Chap. 7, Sect. 2). These layers have a complicated topological structure, in some directions it is the structure of a Cantor set type, and contain an infinite number of periodic trajectories. We shall describe below rigorous mathematical results concerning a structure and properties of trajectories in stochastic layers. Unfortunately these results are dealing yet with trajectories belonging to some subsets of measure zero and do not refer to "generic" trajectories. Even the theorem on positivity of the measure of the set occupied by stochastic layers is not proved yet, but all investigators generally believe that it is true.

The other general mechanism of non-integrability is connected with the occurence of Smale's horseshoe (see Chap. 7, Sect. 2), i.e. of the subspace of a phase space where dynamics has some special properties of instability. By moving away from integrability, a set occupied by invariant tori shrinks and correspondingly a set filled with trajectories with a complex nonintegrable behavior is extended. One can consider it as limiting such dynamical systems which possess the most strong statistical properties in the whole phase space. The most important examples of such systems form geodesic flows on compact manifolds of negative curvature, billiards in domains with convex inwards boundary (see Chaps. 7 and 8) and some one-dimensional mappings (see Chap. 9). The investigation of ergodic properties of such systems is based on the notion of hyperbolicity which will be considered in detail in Sect. 1 of Chap. 7.

We shall now enumerate those properties of a dynamical system, which will be called stochastic properties.

1) The existence of an invariant measure. In many cases, for instance for Hamiltonian systems or for dynamical systems of the algebraic nature, a priori there exists a natural invariant measure. Nevertheless, there are some important situations when such invariant measure is not known a priori but it could be found by analysis of properties of the dynamics. Among these systems there are dynamical systems with hyperbolic (see Chap. 7, Sect. 2 and Sect. 3) and strange (see Chap. 8, Sect. 2) attractors. The natural method of constructing an invariant measure for these systems is to take any initial smooth measure and to study problems of the existence and of the properties of a limit, as $t \to \infty$, of its iterates under the action of the dynamics, and of the dependence of this limit on the choice of the initial measure. In some cases this program can be realized.

2) Ergodicity (see Chap. 1, Sect. 3). When the invariant measure is chosen, it is natural to consider ergodic properties of the dynamical system with respect to this measure. The simplest one is the problem of ergodicity.

3) Mixing (see Chap. 1, Sect. 3). From the point of statistical mechanics, mixing means the irreversibility, i.e. that any initial measure absolutely continuous with respect to the invariant measure converges to it weakly, under the action of dynamics.

4) The K-property (see Chap. 1, Sect. 3). If a dynamical system is a K-system, then it has countably-multiple Lebesgue spectrum, positive entropy and is

mixing of all degrees. The K-property means that a deterministic dynamical system can be coded into a regular stationary random process of probability theory.

5) The Bernoulli property. In the case of discrete time, if a code can be constructed so that the resulting process is a sequence of independent random variables, then the dynamical system is called a Bernoullian one (see Chap. 3, Sect. 4). In the case of continuous time, this definition refers to any automorphism from the flow. It is worthwhile mentioning that usually a code is given by complicated nonsmooth functions and therefore a connection with the smooth structure in the phase space would be lost.

6) The central limit theorem. Let (M, μ, T) be an ergodic automorphism. In view of the Birkhoff-Khinchin ergodic theorem (see Chap. 1, Sect. 2) for any $f \in L^1(M, \mu)$ almost everywhere

$$\lim_{n\to\infty} \left[\frac{1}{n} \sum_{k=1}^{n} f(T^k(x)) - \bar{f} \right] = 0$$

where $\bar{f} = \int f d\mu$. The difference $\left[\frac{1}{n} \sum_{k=1}^{n} f(T^k(x)) - \bar{f} \right]$ is called a time fluctuation with respect to the mean. We shall say that f satisfies the central limit theorem (CLT) of probability theory if there exists a number $\sigma = \sigma(f)$ such that

$$\lim_{n\to\infty} \mu \left\{ x\colon \sqrt{n} \left[\frac{1}{n} \sum_{k=1}^{n} f(T^k(x)) - \bar{f} \right] < a \right\} = \frac{1}{\sqrt{2\pi\sigma}} \int_{-\infty}^{a} e^{-u^2/2\sigma} \, du$$

C.L.T. states that time fluctuations have the order $O(1/\sqrt{n})$. In some cases one has to prove CLT for a possibly wider class of phase functions. It is more difficult as a rule, than to prove the K-property.

7) The rate of correlations decay. A function $f \in L^2(M, \mu)$ with zero mean satisfies the property of exponential decay of correlations if there exist positive numbers $C, \rho < 1$ such that

$$\left| \int f(T^n(x)) f(x) \, d\mu \right| < C\rho^{|n|}$$

This property can be verified easily enough when T is a Markov automorphism and f is a function with a sufficiently simple structure. In the general case, in order to prove the exponential decay of correlations or for an analysis of their decay rate for smooth functions f, one has to construct approximations of the dynamical system under consideration by Markov automorphisms. It is often very complicated to construct such approximations.

In what follows, when referring to stochastic properties of dynamical systems we shall mean the hierarchy 1)–7) enumerated above.

§ 2. The Kolmogorov-Arnold-Moser Theory (KAM-Theory)

We begin to discuss the KAM-theory by considering one important example, and then more general results will be studied. Let us consider the difference equation of stationary sin-Gordon equation type

$$u_{n+1} - 2u_n + u_{n-1} = \lambda V'(u_n) \tag{6.1}$$

where V is a periodic function with period 1 and λ is a parameter. If $V(u) = 1 - \cos 2\pi u$ then we obtain precisely a stationary sin-Gordon difference equation. We shall assume that $V \in C^\infty(R^1)$, while this assumption can be taken sufficiently milder. Let us introduce new variables $z_n = u_n - u_{n-1}$, $\varphi_n = \{u_n\}$ where as usual the fractional part of a number is denoted by $\{\cdot\}$. Then (6.1) takes the form

$$z_{n+1} = z_n + \lambda V'(\varphi_n), \qquad \varphi_{n+1} = \varphi_n + z_{n+1} \pmod 1 \tag{6.2}$$

The space of pairs (z, φ) where $-\infty < z < \infty$, $\varphi \in S^1$, forms a two-dimensional cylinder C. The relations (6.2) show that sequences (z_n, φ_n) are trajectories of the transformation of the cylinder C into itself, where

$$T(z, \varphi) = (z', \varphi'), \quad z' = z + \lambda V'(\varphi), \quad \varphi' = \varphi + z' \pmod 1 \tag{6.3}$$

The transformation T in the case of $V(u) = 1 - \cos 2\pi u$ was introduced by B.V. Chirikov (cf [Chi]) and since then it has been called the Chirikov transformation. A general transformation of type (6.3) will be called a standard transformation.

Standard transformations arise in many problems of the theory of nonlinear oscillations, in plasma physics, in solid state physics (the Frenkel-Kontorova model). Much physical and mathematical literature is devoted to investigations of these transformations. We want to mention [Chi] from the corresponding physical review papers, and from the mathematical works besides the books by V.I. Arnold [Ar1], [Ar2] and by J. Moser [Mos] we should like to mention papers by S. Aubry (see for instance [AP]), by J. Mather [Mat1] and by I. Parseval [P].

In the three last papers a more general situation is also considered. In order to discuss it we shall consider a Lagrangian $L(u, u')$ which satisfies the following conditions

a) $L(u + 1, u' + 1) = L(u, u')$ (periodicity).

b) $-\dfrac{\partial^2 L}{\partial u\, \partial u'} \geqslant \text{const} > 0$ and is continuous (twist condition).

For any set of points $u_0, u_1, \ldots, u_m$ on a line we should define the Lagrangian $L(u_0, u_1, \ldots, u_m) = \sum_{i=1}^m L(u_i, u_{i-1})$. Variational principles of mechanics and of statistical physics force to construct chains $\{u_i\}_{-\infty}^{\infty}$ such that $L(u_s, u_{s+1}, \ldots, u_{s+m})$ takes its minimal value for any $s, m > 0$ and for fixed u_s, u_{s+m}. The corresponding necessary condition takes the following form

$$\frac{\partial L(u_s,\ldots,u_{s+m})}{\partial u_n}=\frac{\partial L(u_{n+1},u_n)}{\partial u_n}+\frac{\partial L(u_n,u_{n-1})}{\partial u_n}=0,\qquad -\infty<n<\infty.$$

Let us introduce a momentum $p=\dfrac{\partial L(u,u')}{\partial u}$ conjugated to u. Then the last chain of equations changes to

$$p_n+\frac{\partial L(u_{n+1},u_n)}{\partial u_n}=0,\qquad p_{n+1}-\frac{\partial L(u_{n+1},u_n)}{\partial u_{n+1}}=0,\qquad -\infty<n<\infty. \tag{6.4}$$

It follows from b) that the first equation can be uniquely solved, and therefore u_{n+1} can be considered as a function of p and u_n. Then p_{n+1} can be found from the second equation. So the transformation T which maps (u_n, p_n) into (u_{n+1}, p_{n+1}) is defined. The condition a) allows us to pass to the phases $\varphi_n\{u_n\}$ and to consider the action of T in the same cylinder C. One could obtain a standard transformation taking $L(u,u')=\frac{1}{2}(u-u')^2+\lambda V(u')$. The useful example of the other type arises in the following way. Let $\Gamma\subset C^\infty$ be a convex curve and it forms a boundary of a convex domain on a plane. We introduce a cyclic variable u on Γ and set

$$L(u,u')=\|\mathbf{r}(u)-\mathbf{r}(u')\|^2$$

where $\mathbf{r}(u)$ is a radius-vector of the point with coordinate u. Then it is easy to demonstrate that sequence $\{u_n\}$ satisfying (6.4), correspond to billiard trajectories inside Γ.

Let us come back to the transformation (6.2). We shall consider it formally in case $\lambda=0$. The corresponding solutions are formed by those sequences $\{z_n = I, \varphi_n=\varphi_0+nI\}$, where I is a constant. In geometrical language it means that every circle $\Gamma_{I,0}=\{(z,\varphi): z=I\}$ is invariant with respect to T and T can be reduced on Γ to the rotation on the angle I. The KAM-theory states that if λ is sufficiently small, the majority of these circles are preserved and suffer only small variation in form on the cylinder C. More precisely, the following theorem is valid (see [Ar1], [Ar2], [Mos]).

Theorem 2.1. *Let a number I be poorly approximated by rationals, i.e. $\min_p |I-p/q|>1/q^{2+\varepsilon}$ for some $\varepsilon>0$ and for all sufficiently large q. Then one can find $\lambda_0(I)=\lambda_0$ such that for all λ, $|\lambda|\leqslant\lambda_0$, a standard transformation T has an invariant curve $\Gamma_{I,\lambda}$ which is C^0-close to $\Gamma_{I,0}$ and T is reduced on $\Gamma_{I,0}$ to the rotation. If λ is sufficiently small, then a set of curves $\Gamma_{I,\lambda}$ has a positive measure.*

Surely this theorem of the KAM-theory is a theorem of perturbation theory. Its essential peculiarity is connected with the fact that constructed invariant curves do not form any domain. According to one of the Birkhoff theorems (cf [Bi]) in any neighbourhood of such a curve, there are periodic trajectories. In a "generic" case, i.e. when a function V belongs to an everywhere dense subset of the second category among these periodic trajectories there are hyperbolic ones and their separatrices intersect transversally. In the case mentioned above in Section 1, stochastic layers arise, which are situated between invariant curves given by the KAM-theory. One can find some information on these layers in Chap. 7, Sect. 2.

The new approach to the KAM-theory considered in the papers by Aubry [AD], Mather [Mat1] and Parseval [P] mentioned above which is developed yet again in the two-dimensional case only. Aubry introduces the following definition.

Definition 2.1. A sequence $\{u_n\}$ is called a *configuration with a minimal energy* (*CME*) if the following property holds: if we consider for any s, t, $s < t$, the sequence $\{u'_n\}$ such that $u'_n = u_n$ if $n \leqslant s$, $n \geqslant t$, then

$$\sum_{n=s+1}^{t} L(u'_n, u'_{n-1}) \geqslant \sum_{n=s+1}^{t} L(u_n, u_{n-1}).$$

In other words, if we consider L as an energy, then any finite perturbation of CME does not decrease the energy of the configuration. Aubry proved that for any CME $\{u_n\}$ there exists $\lim_{n\to\infty} \frac{1}{n} u_n = \omega$ which can be considered as the analogy of the rotation number of a homeomorphism of the circle. Moreover for any ω there exist CME with this ω. If ω is a rational number, $\omega = p/q$, then in a generic case among CME there are hyperbolic periodic trajectories of period q. If ω is an irrational number, then there exists a measurable (mod 0) injective mapping f_ω of the unit circle S^1 into C such that $f_\omega(S^1)$ is invariant with respect to T and $Tf_\omega(x) = f(x + \omega)$, i.e. $T|f_\omega(S^1)$ is the rotation on the angle ω. If the mapping f_ω is continuous, then $f_\omega(S^1)$ is a continuous curve which envelops the cylinder C. If ω is poorly approximable by rational numbers and λ is sufficiently small, then the curve $f_\omega(S^1)$ coincides with a curve given by the KAM theory. If f_ω is discontinuous then $f_\omega(S^1)$ is a Cantor perfect set which is called a cantorus. It is invariant under the transformation T and as before T reduces on it to the rotation. It was proposed in [AD] the geometrical approach to constructing of such invariant sets, which is based on one of the Birkhoff theorems.

For any standard transformation there exists a critical value λ_{cr}, or such that for all $\lambda > \lambda_{cr}$ there are no invariant curves enveloping C. Therefore for such values of λ, all invariant sets $f_\omega(S^1)$ are cantori. Apparently for each value of I mentioned in Theorem 2.1 there exists $\lambda_{cr}(I)$ such that for all $\lambda < \lambda_{cr}(I)$ there is an invariant curve $\Gamma_{I,\lambda}$, and for $\lambda > \lambda_{cr}(I)$ it destroys and transforms into a cantorus, but the nature of this bifurcation is not completely clear yet. If λ increases further, then a projection of any cantorus is concentrated in a small neighbourhood of the minimum of the function V. Aubry proved that for such λ a projection on the φ-axis of a cantorus has zero measure (cf [AD]). Apparently a Hausdorff dimension of this projection is also equal to zero.

The above given description shows how separate ergodic components of a standard transformation can be constructed. Invariant curves $\Gamma_{I,\lambda}$ form a set of positive measure if λ is small, and therefore T has invariant sets of positive measure where it is nonergodic. From the other side in the case of large λ, numerical simulations and considerations on the physical level, using Chirikov's criterion of resonanses overlapping, show that T has invariant sets of large measure where it obeys some properties of stochasticity. Nevertheless there are no corresponding rigorous mathematical results.

We are now going to formulate the main statement of the KAM-theory. Not only concrete results are significant in the KAM-theory but also a method for obtaining these results, because it could be applied to a large variety of problems. Unfortunately we have no opportunity to discuss this question in more detail. The KAM-theory will be considered explicitly in Volume 3 of the given series. We discuss only the main theorem of this theory in its application to the Hamiltonian systems. Let us consider an integrable Hamiltonian system with n degrees of freedom. Then in a phase space there exists an open set $\mathcal{O} = U \times \mathrm{Tor}^n$ where U is a neighbourhood in the n-dimensional space, Tor^n is a n-dimensional torus with coordinates $\varphi_1, \ldots, \varphi_n$, and a Hamiltonian H_0 has the form $H_0 = H_0(I_1, \ldots, I_n)$ where $(I_1, \ldots, I_n)$ are coordinates in U. Let the Hamiltonian $H(I, \varphi) = H_0(I) + \varepsilon H_1(I, \varphi, \varepsilon)$ be a small perturbation of the Hamiltonian H_0. We assume that H is analytic in the domain $\mathcal{O}' = U' \times S$ where U' is a complex neighbourhood of U in $\mathbb{C}^n$ and S is a complex neighbourhood of Tor^n in $\mathbb{C}^n$.

The main theorem of the KAM-theory (see [Ar1], [Ar2], [Mos]). *Let* $\det \left\| \frac{\partial^2 H}{\partial I_i \partial I_j} \right\| \neq 0$ *in* U. *Then for all sufficiently small* $\varepsilon > 0$, *one can find a subset* $K \subset \mathcal{O}$ *such that* $\operatorname{mes} K \to \operatorname{mes} \mathcal{O}$ *as* $\varepsilon \to 0$ *and there exists a measurable partition of K into invariant n-dimensional tori. The Hamiltonian dynamical system with the Hamiltonian H reduces on every such torus to the quasi-periodic motion with pure point spectrum and n basic frequencies.*

This theorem shows that a small perturbation of an integrable system is nonergodic and has an invariant subset of positive measure. Ergodic components belonging to this subset have pure point spectrum. In particular it disproved completely the hypothesis which appeared often in physical works, that a generic multi-dimensional nonlinear Hamiltonian system is ergodic. We would like to mention additionally that tori which are constructed in the KAM-theory depend smoothly on a parameter $u \in K$.

There exist different modifications of this theorem which allow to extend the KAM-theory onto non-Hamiltonian systems, or to consider the milder conditions of smoothness of the system, or to admit some separate degeneracies when $\det \left\| \frac{\partial^2 H}{\partial I_i \partial I_j} \right\| \equiv 0$. We shall not consider these questions here.

Chapter 7
General Theory of Smooth Hyperbolic Dynamical Systems

Ya. B. Pesin

§1. Hyperbolicity of Individual Trajectories

1.1. Introductory Remarks. The idea of studying the global behavior of dynamical systems by analyzing their local properties has emerged as very prolific. The theory of hyperbolic dynamical systems may serve as an impressive illustration of partial realization of this idea.

We will describe the hyperbolicity of an individual trajectory of a dynamical systems S^t by observing the behavior of near-by trajectories. There are three types of behavior in a neighborhood of a given trajectory $S^t(x)$:

a) the trajectory $S^t(x)$ attracts all near-by trajectories as $t \to +\infty$ (complete stability);

b) $S^t(x)$ attracts all near-by trajectories as $t \to -\infty$ (complete instability);

c) $S^t(x)$ attracts some trajectories when $t \to +\infty$ and attracts some others as $t \to -\infty$.

The definition of hyperbolicity is based upon this last type of behavior.

In this chapter we assume that the phase space M is a compact n-dimensional Riemannian C^∞-manifold, and that the transformation S (or the flow S^t) is of class C^∞ unless stated otherwise.

Let S^t, $t \in \mathbb{R}$ be a flow generated by a smooth vector field $X(x)$. This means that the function $x(t) = S^t(x_0)$ is the solution of the differential equation

$$\dot{x} = X(x)$$

on M satisfying the initial condition $x(0) = x_0$. Then $X(x)$ induces a vector field on the tangent space $\mathcal{T}M$, which is called the *vector field of variation of* X and is denoted by $\mathcal{X}(x) = (X(x), X_x(x))$. We denote by $\tilde{S}^t(x, v) = (S^t(x), dS^t_x v)$ the flow generated by $\mathcal{X}(x)$. Each transformation $\tilde{S}^t$ maps the tangent space $\mathcal{T}_x = (\mathcal{T}M)_x$ into $\mathcal{T}_{S^t(x)} = (\mathcal{T}M)_{S^t(x)}$. The corresponding system of differential equations is called the *system of variational equations*

$$\dot{x} = X(x), \qquad \dot{v} = X_x(x)v. \tag{7.1}$$

In the discrete time case we have a diffeomorphism S and the group S^t, $t \in \mathbb{Z}$. Instead of the system of variational equations (7.1), one should deal directly with the group $\tilde{S}^t(x, v) = (S^t(x), dS^t_x v)$, $t \in \mathbb{Z}$, which maps the tangent space $\mathcal{T}_x$ to the tangent space $\mathcal{T}_{S^t(x)}$.

The solutions of (7.1) define a cocycle $\mathfrak{U}(x,t)$ which induces a map $\mathcal{T}_x \to \mathcal{T}_{S^t(x)}$. Suppose that the dynamical system under consideration preserves an ergodic invariant measure ν. Then the multiplicative ergodic theorem is applicable (see Chap. 1, Sect. 2) and it implies the existence of Lyapunov exponents $\chi_1 \leqslant \chi_2 \leqslant \ldots \leqslant \chi_n$. This theorem shows, in particular, that trajectories of variational equations, which are typical with respect to ν, behave in a sufficiently regular way.

In this chapter we deal essentially with the case when there are no zero exponents (when the time is discrete) or there is only one zero exponent corresponding to the flow direction (when the time is continuous). It is believed that such a situation is "typical" in a sense, although rigorous mathematical results in this direction are yet to be established.

1.2. Uniform Hyperbolicity. We begin by considering discrete time dynamical systems. In order to standardize the notation we shall denote by S^1 a diffeomorphism and by $\{S^t\}$, $(t \in \mathbb{Z})$ the group generated by S^1. However, the notation S will also be used when we deal with a single diffeomorphism (but not the group $\{S^t\}$).

Definition 1.1. A trajectory $\{S^t(x)\}$ is called *uniformly completely hyperbolic*[1] if there exist subspaces $E^s(S^t(x))$ and $E^u(S^t(x))$[2] and constants $C > 0$, λ, μ such that

$$0 < \lambda < 1 < \mu, \tag{7.2}$$

and for all t and $\tau \geqslant 0$

$$\mathcal{T}_{S^t(x)} = E^s(S^t(x)) \oplus E^u(S^t(x)), \tag{7.3}$$

$$dS^t E^s(x) = E^s(S^t(x)), \; dS^t E^u(x) = E^u(S^t(x)), \tag{7.4}$$

$$\begin{aligned} &\|dS^\tau v\| \geqslant C\lambda^\tau \|v\|, && v \in E^s(S^t(x)), \\ &\|dS^\tau v\| \leqslant C^{-1}\mu^\tau \|v\|, && v \in E^u(S^t(x)), \\ &\gamma(S^t(x)) \geqslant \text{const.} \end{aligned} \tag{7.5}$$

Here $\gamma(S^t(x))$ is the angle between the subspaces $E^s(S^t(x))$ and $E^u(S^t(x))$.

The subspaces $E^s(x)$ and $E^u(x)$ are respectively called the *stable* and *unstable* subspaces.

If x is a periodic point (in particular, a fixed point) whose trajectory $\{S^t(x)\}$ is uniformly completely hyperbolic then x is called *a hyperbolic periodic (fixed) point*.

One may interpret the splitting of the tangent space $\mathcal{T}_x$ into subspaces $E^s(x)$, $E^u(x)$, by saying that trajectories of the variational equation corresponding to $v \in E^s(x)$ (or $v \in E^u(x)$) approach each other with an exponential rate as

[1] We shall also call it *uniformly hyperbolic* if there is no misunderstanding.

[2] The indices "s" and "u" stand for "stable" and "unstable" respectively.

$t \to \infty$ (respectively, $t \to -\infty$). We shall see below that such a behavior of solutions of the variation equations implies a similar behavior of trajectories of the system $\{S^t\}$.

1.3. Nonuniform Hyperbolicity. There are two ways to weaken the conditions of uniform hyperbolicity. First, the hyperbolicity may be nonuniform (but complete). Second, it may be only partial, i.e. it holds only for a part of the tangent space.

Definition 1.2. A trajectory $\{S^t(x)\}$ is said to be *nonuniformly hyperbolic* if the following three conditions hold. First, there exist subspaces $E^s(S^t(x))$ and $E^u(S^t(x))$ $(t \in \mathbb{Z})$ satisfying (7.3) and (7.4). Second, there exist numbers λ and μ satisfying (7.2). Last, there is a number $0 < \alpha < 1$ such that for any ε, $0 < \epsilon < \min\{\ln \mu, (-\ln \lambda)\}$ one can find a function $C(x, \varepsilon) > 0$ for which

$$C(S^t(x), \varepsilon) \leqslant C(x, \varepsilon) e^{\varepsilon |t|} \quad (t \in \mathbb{Z}) \tag{7.6}$$

and for all t and $\tau \geqslant 0$

$$\begin{aligned} &\|dS^\tau v\| \leqslant C(S^{t+\tau}(x), \varepsilon)\lambda^\tau \|v\|, && v \in E^s(S^t(x)), \\ &\|dS^\tau v\| \geqslant C^{-1}(S^{t+\tau}(x), \varepsilon)\mu^\tau \|v\|, && v \in E^u(S^t(x)), \\ &\gamma(S^t(x)) \geqslant C^{-1}(S^t(x), \varepsilon)\gamma(x). \end{aligned} \tag{7.7}$$

Condition (7.6) may be interpreted in the following way: the estimates (7.7) may get worse along the trajectory (as t increases) but relatively slowly (i.e. with an exponential rate which is small compare with $\max\{\log \mu, \log \lambda^{-1}\}$). Condition (7.6) seems quite technical, but plays a crucial role in the study of nonuniform hyperbolicity. At first glance, one can consider this condition to be rather artificial and restrictive, we shall see below that this is not so and that typical trajectories (with respect to some invariant measure) often satisfy it.

Notice that for a periodic trajectory $\{S^t(x)\}$ the conditions of uniform and nonuniform hyperbolicity coincide.

We obtain the definition of *partial hyperbolicity* if we replace condition (7.2) by

$$0 < \lambda < \min(1, \mu) \tag{7.8}$$

(i.e. the number μ is not necessarily bigger than 1).

Similarly, we differ the concepts of *uniform partial hyperbolicity* and of *nonuniform partial hyperbolicity*.

In examples, partial hyperbolicity usually appears in the following way. The tangent space $\mathcal{T}_{S^t(x)}$ is split into a direct sum of three subspaces invariant under dS^t

$$\mathcal{T}_{S^t(x)} = E^s(S^t(x)) \oplus E^0(S^t(x)) \oplus E^u(S^t(x)), \tag{7.9}$$

where $dS^t|E^s$ is a contraction and $dS^t|E^u$ is an expansion (i.e., (7.5) or (7.7) holds depending on whether hyperbolicity is uniform or not).

The subspace E^0 is called *neutral*. Vectors lying in this subspace may contract and expand but not too fast. More exactly (we restrict ourselves to the case of uniform partial hyperbolicity) there exist C_1, λ_1, and μ_1 such that

$$C_1 > 0, \qquad 0 < \lambda < \lambda_1 \le 1 \le \mu_1 < \mu < \infty, \tag{7.10}$$

and for all t and $\tau \ge 0$

$$C_1^{-1}\lambda_1^{\tau}\|v\| \leqslant \|dS^{\tau}v\| \leqslant C_1\mu_1^{\tau}\|v\|, \qquad v \in E^0(S^t(x)). \tag{7.11}$$

We shall call this version of hyperbolicity *partial hyperbolicity in a narrow sense*.

In a similar fashion, one can define uniform, nonuniform, and partial hyperbolicity for dynamical systems with continuous time; the only difference is that the splitting (7.2) of the tangent space $\mathscr{T}_{S^t(x)}$, $(t \in \mathbb{R})$ should be replaced by

$$\mathscr{T}_{S^t(x)} = E^s(S^t(x)) \oplus E^u(S^t(x)) \oplus X(S^t(x)), \tag{7.12}$$

where $X(S^t(x))$ is the one-dimensional subspace generated by the flow (and obviously invariant under the flow). Comparing (7.9) and (7.12), one sees that any diffeomorphism S^t (which acts as the time t translation along the trajectories of the flow) satisfies the conditions of partial hyperbolicity.

1.4. Local Manifolds. Hyperbolicity conditions allow one to describe the asymptotic behavior of trajectories which start in a small neighborhood of a hyperbolic trajectory. This description is given by the following theorem which is one of the key results in the hyperbolic theory.

Theorem 1.1. *(on local manifold, see* [Pe2]*). Let $\{S^t(x)\}$, $(t \in \mathbb{Z})$ be a nonuniformly (completely or partially) hyperbolic trajectory. Then there exists a local stable manifold (LSM) $V^s(x)$ such that for $y \in V^s(x)$ the distance between $S^t(x)$ and $S^t(y)$ decreases with an exponential rate, i.e. for any t and $\tau \ge 0$*

$$d(S^{t+\tau}(x), S^{t+\tau}(y)) \leqslant KC(S^t(x), \varepsilon)\lambda^{\tau}e^{\varepsilon\tau}d(S^t(x), S^t(y)), \tag{7.13}$$

where d is the distance in M induced by the Riemannian metric and $K > 0$ is a constant.

Let us make some remarks.

Remark 1.1. The local manifold $V^s(x)$ at a point x is constructed via a smooth map

$$\psi(x) : B^s(r) \to E^u(x)$$

which satisfies

$$\psi(0) = 0 \quad \text{and} \quad d\psi(0) = 0. \tag{7.14}$$

Here $B^s(r)$ is the ball of radius r in $E^s(x)$ centered at the origin; the radius $r = r(x)$ is said to be the *size* of LSM. One gets LSM by projecting the graph of ψ into M by the exponential map *exp*, i.e.,

$$V^s(x) = \exp\{\,(x, \psi(x)) : x \in B^s(r)\,\}.$$

It follows from (7.14) that

$$x \in V^s(x), \qquad \mathcal{T}_x V^s(x) = E^s(x). \tag{7.15}$$

Remark 1.2. One can obtain a more refined information on the smoothness of LSM. More precisely, if $\{S^t\} \in C^{r+\alpha}$, $r \geqslant 1$, $0 < \alpha \leqslant 1$, (i.e. the differential of order r, $d^r S^t$ is Hölder continuous with the Hölder exponent α) then $V^s(x) \in C^r$ (cf [Pe2]); in particular, if $\{S^t\} \in C^r$ then $V^s(x) \in C^{r-1}$.

Remark 1.3. One can construct a LSM at any point $y = S^t(x)$. The sizes of LSM at x and at $S^t(x)$ are related by

$$r(S^t(x)) \geqslant K e^{-\varepsilon|t|} r(x), \tag{7.16}$$

where $K > 0$ is a constant. For typical points x the function $r(S^t(x))$ is an oscillating function of t which is of the same order as $r(x)$ for many values of t. Nevertheless, for some t the value $r(S^t(x))$ may become as small as it is allowed by (7.16).

Remark 1.4. In view of 7.13 and 7.16, the size of LSM $V^s(S^t(x))$ may decrease with an exponential rate which is smaller than the rate of convergence of trajectories $\{S^t(x)\}$ and $\{S^t(y)\}$.

Remark 1.5. If a trajectory $\{S^t(x)\}$ is uniformly hyperbolic (completely or partially) then it certainly satisfies the conditions of Theorem 1.1 (since the uniform hyperbolicity is a particular case of the nonuniform one). In the case of complete hyperbolicity, Theorem 1.1. becomes the well-known Hadamard-Perron theorem (cf [An1]). In the case of partial hyperbolicity, it was proved by Brin and Pesin in [BP]. In addition, one can show that in the case of uniform hyperbolicity (complete or partial) Theorem 1.1 holds for $\{S^t\} \in C^1$, and if $\{S^t\} \in C^r$, $r \geq 1$ then $V^s(x) \in C^r$. Moreover one can show that $r(S^t(x)) \geq$ const and $C(S^t(x), \varepsilon) \leq$ const for all t.

Remark 1.6. In the hyperbolic theory there is a symmetry between the objects marked by index "s" and the ones marked by index "u". Namely, when time direction is reversed the statements concerning objects with index "s" go over into the statements about the corresponding objects with index "u". In particular, this allows one to define a *local unstable manifold* (LUM) $V^u(x)$ at a point x as a LSM for S^{-1}. Its properties are similar to those of $V^s(x)$.

Remark 1.7. For a dynamical system with continuous time, LSM and LUM are defined to be those constructed for the diffeomorphism S^1, a time-one map along the trajectories of the flow.

We shall see below that LSM and LUM play a crucial role in the analysis of ergodic and topological properties of hyperbolic dynamical systems. Only in very special situations (for example, for systems with additional symmetry) local manifolds can be defined explicitly (see below). In general, their construction uses a version of the fixed point theorem (cf [Pe2]).

1.5. Global Manifolds. The *global stable manifold* (GSM) at a point $x \in M$ is a smoothly immersed submanifold defined by

$$W^s(x) = \bigcup_{-\infty<t<\infty} S^{-t}(V^s(S^t(x))) \quad (t \in \mathbb{Z}). \tag{7.17}$$

This manifold has the same class of smoothness as the LSM. One can characterize it as follows: a point $y \in W^s(x)$ if and only if

$$d(S^t(x), S^t(y)) \to 0 \quad \text{as } t \to \infty.$$

One can define the *global unstable manifold* (GUM) in a similar fashion.

For dynamical systems with continuous time the GSM at a point x is defined by (7.17) with $t \in \mathbb{R}$. In addition, in this case one can construct a smooth *global weakly stable manifold* by

$$W^{so}(x) = \bigcup_{-\infty<t<\infty} W^s(S^t(x)). \tag{7.18}$$

For any $y \in W^{so}(x)$ the whole trajectory $\{S^t(y)\}$ lies in $W^{so}(x)$. Similarly, one can define the *global unstable manifold* and the *global weakly unstable manifolds*.[3]

§2. Basic Classes of Smooth Hyperbolic Dynamical Systems. Definitions and Examples

2.1. Anosov Systems.

Definition 2.1. A dynamical system is called *Anosov* (respectively, *Anosov diffeomorphism* or *Anosov flow*) if any trajectory of the system is uniformly

[3] A different system of notation and different terminology for systems with continuous time is also in use. GSM is denoted by $W^{ss}(x)$ and is called the strong stable manifold, and the global weakly stable manifold is denoted by $W^s(x)$ and is called the stable manifold. A similar system of notation and terminology is used with respect to unstable manifolds.

completely hyperbolic and the constants C and λ can be chosen independently of the point.[4]

Theorem 2.1. *(cf* [An1], [AS], [KH]*). Let $\{S^t\}$ be a C^1-Anosov system. Then the following statements hold.*

1) The stable and instable subspaces $E^s(x)$, $E^u(x)$ ($x \in M$) generate two continuous invariant distributions (subbundles) of the tangent space $\mathcal{T}M$ (denoted respectively by E^s and E^u). These distributions are transversal (i.e. $E^s(x) \cap E^u(x) = 0$ for any $x \in M$).

2) The distribution E^s is integrable. The corresponding maximal integral submanifolds form a continuous C^1-foliation of M (i.e., the LSM $V^s(x)$ depend continuously on the base point x in the C^1 topology; see [An1], [KH]*). We denote it by W^s. It has the following properties:*

(a) *the leaf of W^s passing through a point x coincides with GSM $W^s(x)$;*

(b) *the foliation W^s is S^t-invariant (i.e. $S^t(W^s(x)) = W^s(S^t(x))$ for any $x \in M$) and contracting, i.e. for any $x \in M$, $y \in W^s(x)$, and $t \geqslant 0$*

$$d^{(s)}(S^t(x), S^t(y)) \leqslant C\lambda^t d^{(s)}(x, y).$$

Here $0 < \lambda < 1$ and $C > 0$ are constants (independent of x and y) and d^s is the distance measured along the leaves of W^s induced by the Riemannian metric on the leaf (which is considered as a smooth submanifold in M).

3) Similar assertions hold for the distribution E^u. We denote the corresponding foliation by W^u. It is invariant under $\{S^t\}$ and is contracting for $t \leqslant 0$. The foliations W^s and W^u are transversal at each point $x \in M$ (i.e. $\mathcal{T}_x = \mathcal{T}_x W^s(x) \oplus \mathcal{T}_x W^u(x)$ and $\mathcal{T}_x W^s(x) \cap \mathcal{T}_x W^u(x) = 0$).

4) In the case of continuous time the distributions $E^s \oplus X$ and $E^u \oplus X$ (see 7.12) are continuous, invariant under dS^t, and integrable. Their maximal integrable submanifolds form continuous S^t-invariant C^1-foliations of M which we denote respectively by W^{so} and W^{uo}. The leaf of W^{so} (respectively of W^{uo}) passing through a point x coincides with $W^{so}(x)$ (respectively with $W^{uo}(x)$).

The distributions E^s and E^u are called respectively the *stable* and *unstable* distributions for the system S^t. Similarly, W^s and W^u are respectively called the stable and unstable foliations.

Let us make some remarks.

Remark 2.1. One can show that if the dynamical system $\{S^t\}$ is of class of smoothness C^r then W^s and W^u are continuous C^r-foliations, i.e., LSM $V^s(x)$ depend continuously on the base point x in the C^r topology for any $1 \leq r \leq \infty$. However, if $\{S^t\}$ is a $C^{1+\alpha}$ system ($\alpha > 0$) then the distribution

[4] There is an interesting question: If any trajectory of a system is uniformly completely hyperbolic (with no additional information on C and λ) is this system Anosov? [Pe2] gives a positive answer to this question for C^2-systems which preserve a measure equivalent to the Riemannian volume.

E^s satisfies the Hölder condition: $\widetilde{d}(E^s(x), E^s(y)) \leqslant Cd(x, y)^\alpha$ where $C > 0$, is a constant independent of x and y and $\widetilde{d}$ is the distance in the Grassman bundle induced by the Riemannian metric. A similar statement holds for E^u.

Remark 2.2. The distributions E^s and E^u (as well as the foliations W^s and W^u) are "typically" not C^1 but can be shown to be Hölder continuous. The most refined estimate of the Hölder exponent was obtained by Hasselblatt in [Has].

Fix $x \in M$ and let $\lambda_1 = \lambda_1(x)$, $\lambda_2 = \lambda_2(x)$ and $\mu_1 = \mu_1(x)$, $\mu_2 = \mu_2(x)$ be numbers for which the following statements hold:

(a) $\lambda_1 < \lambda_2 < 1 < \mu_2 < \mu_1$;

(b) for every $t > 0$,

$$\frac{1}{c}\lambda_1{}^t\|v\| \le \|dS^t(v)\| \le c\lambda_2^t\|v\| \quad \text{for } v \in E^s(x),$$
$$\frac{1}{c}\mu_1{}^{-t}\|v\| \le \|dS^t(v)\| \le c\mu_2^{-t}\|v\| \quad \text{for } v \in E^s(x),$$

where $c > 0$ is a constant independent of x and t.

Let us set

$$\gamma_1 = \inf_{x\in M} (\log\mu_2 - \log\lambda_2)/\log\mu_1, \quad \gamma_2 = \inf_{x\in M} (\log\lambda_2 - \log\mu_2)/\log\lambda_1.$$

It is shown in [Has] that $E^s \in C^{\gamma_1-\varepsilon}$ and $E^u \in C^{\gamma_2-\varepsilon}$ for every $\varepsilon > 0$. Moreover, if γ_1 (respectively, γ_2) is not an integer then $E^s \in C^{\gamma_1}$ (respectively, $E^u \in C^{\gamma_2}$).

We describe some topological properties of Anosov diffeomorphisms (for a more detailed account of topological properties cf [An1], [PS1], [KH]; the case of continuous time is also considered there).

A point $x \in M$ is called *nonwandering* with respect to a homeomorphism S if for any neighborhood U of x there exists $t \neq 0$ such that $S^t(U) \cap U \neq \emptyset$. A homeomorphism S is called *topologically transitive* if it has an everywhere dense trajectory.

Theorem 2.2. *1) If S is a topologically transitive C^2-Anosov diffeomorphism, then $\overline{W^s(x)} = M$ and $\overline{W^u(x)} = M$ for any $x \in M$.*

2) Periodic points of a C^2-Anosov diffeomorphism S are dense in the set of nonwandering points $\Omega(S)$.[5] *The number P_n of periodic points of period $\leqslant n$ is finite and*

$$\lim_{n\to\infty} \frac{\log P_n}{n} = h,$$

where h is the topological entropy of $S|\Omega(S)$ (for the definition of topological entropy see Sect. 3).

[5] The set $\Omega(S)$ is closed and S-invariant.

We now consider some examples of Anosov systems. We denote by Tor^n the n-torus and identify it with the factor space $\mathbb{R}^n/\mathbb{Z}^n$. Let $A : \mathrm{Tor}^n \to \mathrm{Tor}^n$ be an algebraic automorphism given by an integer matrix $A = (a_{ij})$ and let $\lambda_1, \ldots, \lambda_n$ be the complex eigenvalues of A.

Theorem 2.3. *(cf* [An1]*). An algebraic automorphism A of* Tor^n *is Anosov if and only if $|\lambda_i| \neq 1$ for all $i \in [1, n]$ (in this case A is called a hyperbolic toral automorphism).*

Since $\det A = \pm 1$ the map A preserves the Lebesgue measure on Tor^n. The GSM and GUM coincide respectively with the projection of k-dimensional and $(n-k)$-dimensional hyperplanes on Tor^n. Here k is the number of i for which $|\lambda_i| < 1$. The k-dimensional hyperplanes are parallel to the subspace spanned by the eigenvectors corresponding to $|\lambda_i| < 1$ and the $(n-k)$-dimensional hyperplanes are parallel to the subspace spanned by eigenvectors corresponding to $|\lambda_i| > 1$. A small C^1-perturbation of a hyperbolic toral automorphism[6] is an Anosov diffeomorphism of the torus (see Theorem 2.11 below). Moreover, any Anosov diffeomorphism of the torus is topologically conjugate to some hyperbolic automorphism. There exist examples of Anosov automorphisms on nilmanifolds which are not tori. A nilmanifold is a factor of some nilpotent Lie group by a discrete subgroup).

The simplest way to obtain an Anosov flow is to construct a special flow over an Anosov diffeomorphism (see Chap. 1, Sect. 4). Geodesic flows on Riemannian manifolds of negative curvature give another class of interesting examples. Their properties are discussed in Sect. 4.

2.2. Hyperbolic Sets.

Definition 2.2. A set Λ is called *hyperbolic* with respect to a dynamical system $\{S^t\}$ if it is closed and consists of trajectories satisfying the condition of uniform complete hyperbolicity with the same constants C and λ.

If Λ coincides with M then $\{S^t\}$ is an Anosov system. If $\Lambda \neq M$ then Λ contains "holes" i.e. has a structure of a Cantor set. Nevertheless, the local properties of trajectories in Λ are similar to those of Anosov systems.

As in Anosov systems, the subspaces $E^s(x)$ and $E^u(x)$ form two continuous distributions (subbundles) of the tangent bundle $\mathcal{T}\Lambda$ which we denote respectively by E^s and E^u. They are invariant under dS^t, transversal to each other, and satisfy Hölder condition (if $\{S^t\} \in C^{1+\alpha}$). They are also integrable. Moreover, the intersections of Λ with the GSM and GUM (i.e. the sets $\Lambda \cap W^s(x)$ and the sets $\Lambda \cap W^u(x)$) form two C^1*-laminations of* Λ[7] (cf [PS1]) which are tangent to E^s and to E^u respectively. In the case of

[6] i.e. a map $A + \varepsilon S$ where S is a diffeomorphism of the torus, $S(x_1, \ldots, x_n) = (f_1(x_1, \ldots, x_n), \ldots, f_n(x_1, \ldots, x_n))$ and the functions f_i are periodic in $x_1, \ldots, x_n$ with the period 1.

[7] A C^1-lamination of Λ is a partition of Λ induced by the restriction to Λ of a C^1-foliation defined in a neighborhood of Λ.

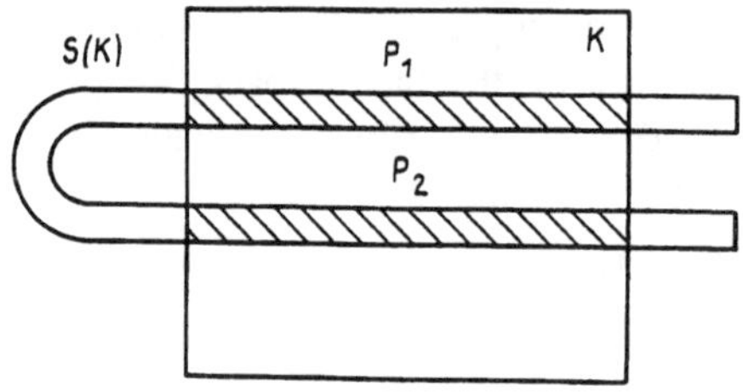

Fig. 1

continuous time there exist two additional C^1-laminations of Λ formed by $\Lambda \cap W^{so}(x)$ and $\Lambda \cap W^{uo}(x)$. The sets

$$W^s(\Lambda) = \bigcup_{x \in \Lambda} W^s(x), \qquad W^u(\Lambda) = \bigcup_{x \in \Lambda} W^u(x)$$

are called respectively the *stable* and *unstable manifolds of the hyperbolic set* Λ.

For a $C^{1+\alpha}$ diffeomorphism ($\alpha \in (0, 1]$) it is known that the Riemannian volume of a hyperbolic set Λ equals 0 or 1 (cf [Bo2]). In the last case, the diffeomorphism is Anosov. This may not hold for C^1-diffeomorphisms; for a counterexample see [Bo2].

A well-known example of a hyperbolic set is a *Smale horseshoe*. Consider a map S of the square $K = [0, 1] \times [0, 1] \subset \mathbb{R}^2$ which acts in the following way. First, the square is strongly stretched in the horizontal direction and compressed in the vertical direction, then it is bent in the form of a horseshoe, and finally this horseshoe is put back on the square such that $K \cap S(K)$ consists of two strips $P_i = [0, 1] \times [a_i, b_i]$ where $i = 1, 2$ and $0 < a_1 < b_1 < a_2 < b_2 < 1$ (see Fig. 1). Each strip P_i contains two substrips $P_{i_1 i_2}$, $i_1, i_2 = 1$ or 2 which are parts of $S^2(K)$ and so on. Continuing in this fashion for each $n > 0$ we obtain 2^n disjoint strips $P_{i_1 \dots i_n}$, $i_j = 1$ or 2 such that $P_{i_1 \dots i_n} \subset P_{i_1 \dots i_{n-1}}$. The width of $P_{i_1 \dots i_n}$ tends to zero as $n \to \infty$. Under the action of S^{-1}, horizontal and vertical directions are interchanged, otherwise the situation is similar. The existence of a C^2-map with the above properties is proved, for example, in [N] (cf also [KH]). It is also proved that the set

$$\Lambda = \bigcap_{-\infty < n < \infty} S^n(K)$$

is hyperbolic.

One can generalize the above construction by forming in the first step any finite number of strips, say l, instead of 2. Then at the second step l^2 strips are formed, and so on (the number of vertical strips corresponding to the map S^{-1} may differ from l). One can extend this construction to the multidimensional case (cf [C]). There are other modifications of the above construction (see [KH], [Pe4]). We point out a smooth realization of Smale horseshoe (see [N]): there exists a C^2-diffeomorphism $\widetilde{S}$ of a two-dimensional sphere which extends the map S constructed above such that $\Omega(\widetilde{S}) = \Lambda \cup p \cup q$ where p is a stable fixed point and q is an unstable one[8].

In general, one call a compact hyperbolic set Λ a *horseshoe* for a C^1 diffeomorphism f (cf [KH]) if there exists s, k, and sets $\Lambda_0, \dots, \Lambda_{k-1}$ such that $\Lambda = \Lambda_0 \cup \dots \cup \Lambda_{k-1}$, $f^k(\Lambda_i) = \Lambda_i$, $f(\Lambda_i) = \Lambda_{i+1} \bmod k$, and $f^k|\Lambda_0$ is conjugate to the full shift on s-symbols.

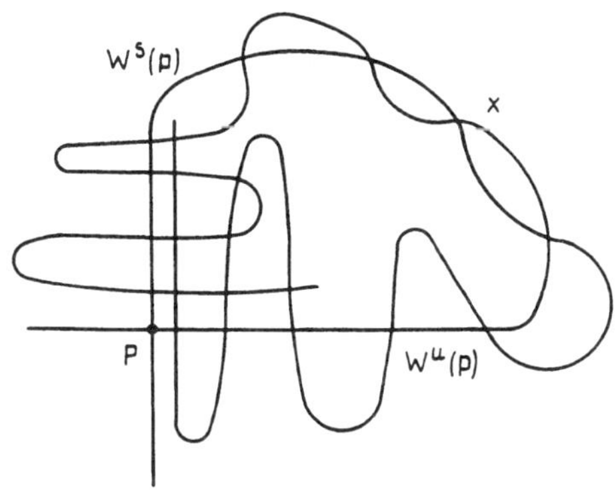

Fig. 2

Other examples of hyperbolic sets are invariant sets in a neighborhood of a homoclinic orbit. Let p be a hyperbolic periodic point of a diffeomorphism S. A point x of intersection between the stable manifold $W^s(p)$ and the unstable manifold $W^u(p)$ which is different from p is called a *homoclinic point*. If the intersection of $W^s(p)$ with $W^u(p)$ is transversal at x (i.e. $\mathcal{T}_x W^s(p) \oplus \mathcal{T}_x W^u(p) = \mathcal{T}_x$ and $\mathcal{T}_x W^s(p) \cap \mathcal{T}_x W^u(p) = 0$) the point x is called a *transversal homoclinic point*. The existence of one transversal homoclinic point implies the existence of an infinite number of homoclinic points with different orbits. Roughly speaking, we have the following situation. Since $W^s(p)$ and $W^u(p)$ are invariant, the trajectory of a homoclinic point consists of homoclinic points. This gives rise to strong oscillations of $W^u(p)$ which stretches along itself when the iterates of x converge to p along $W^s(p)$. A similar picture is valid for $W^s(p)$. Joining the two pictures generates a grid of homoclinic points (see Fig. 2).

Consider now a compact set Γ which is the union of p and the orbit of a transversal homoclinic point x. Choose disjoint open sets $U_0, U_1, \dots, U_n$ in the following way: U_0 is a neighborhood of p which contains all but a finite number of points of the orbit $\{S^n(x)\}$ and each of the remaining points is covered by one of U_i.

[8] First, one can construct a map $S_0 : \mathcal{D}^2 \to \mathcal{D}^2$ of a disk $\mathcal{D}^2$ containing K which extends S and then one modifies S_0 to the desired map $\widetilde{S}$.

Theorem 2.4. *(cf* [A2], [KH]*). For any neighborhood V of the set Γ there exist neighborhoods $U_i \in V$, $i = 0, 1, \ldots, n$ such that*

$$\Lambda = \bigcap_{m=-\infty}^{\infty} S^m \left(\bigcup_{i=1}^{n} U_i \right)$$

is an invariant hyperbolic set[9].

Thus, Λ consists of points y whose orbits $\{S^n y\}$ $(-\infty < n < \infty)$ are contained in U. As U diminishes, Λ tends to Γ.

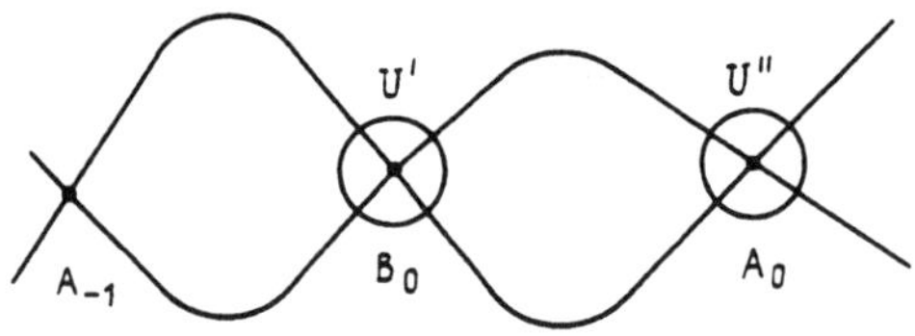

Fig. 3

Let us give a simple argument showing how the appearance of a homoclinic point gives rise to "stochastic trajectories". For simplicity, we consider the case of a C^2-diffeomorphism of a two-dimensional cylinder M which has a fixed hyperbolic point $\mathcal{O}$. Let γ^u and γ^s denote the pieces of $W^u(\mathcal{O})$ and of $W^s(\mathcal{O})$ bounded by $\mathcal{O}$ from one side and by a transversal homoclinic point A_0 from another side. The point $S^{-1}A_0 = A_{-1}$ is also homoclinic. It is clear that there must be at least one other homoclinic point B_0 between A_{-1} and A_0 (see Fig. 3). Consider two sufficiently small neighborhoods $U' \ni B_0$ and $U'' \ni A_0$. Given a point x, we write 1 if an iterate of x falls in U', and 0 if it falls in U''.

Let $\varepsilon = \{\varepsilon_1, \varepsilon_2, \ldots, \varepsilon_n\}$ be a sequence of 0 and 1. Then there exists a point $x \in \Lambda$ such that the iterates of x fall consequently in U' and in U'' according to the sequence ε. Since the majority of sequences ε are "random", we see that random sequences are "realized" by trajectories of S.

Homoclinic points are a particular case of the so-called *heteroclinic points* (a point x is called heteroclinic if there exist two periodic points p_1 and p_2 such that $x \in W^s(p_1) \cap W^u(p_2)$). The preceding constructions and results are still valid for cycles of heteroclinic orbits.

2.3. Locally Maximal Hyperbolic Sets. A hyperbolic set may have a very complicated structure. We shall therefore restrict ourselves to a special though sufficiently large class of hyperbolic sets introduced by Alekseev [A1] and Anosov [An2].

[9] One can show that Λ is a Cantor-like set, i.e. a perfect nowhere dense set.

Definition 2.3. A compact set Λ invariant under a diffeomorphism S is called *locally maximal* (or *isolated*) if there exists a neighborhood $U(\Lambda) \supset \Lambda$ such that any S-invariant set Λ', with $\Lambda \subset \Lambda' \subset U(\Lambda)$ coincides with Λ.

Let Λ be a locally maximal hyperbolic set (LMHS) of a diffeomorphism S. Denote by $\Omega(S|\Lambda)$ the set of nonwandering points and by $\mathrm{Per}(S|\Lambda)$ the set of periodic points for the restriction of S to Λ.

A homeomorphism S of a compact metric space X is called *topologically mixing* if for any sets $U, V \subset X$ there exists N such that $S^n(U) \cap V \neq \emptyset$ for any $n \geqslant N$.

Theorem 2.5. *(Spectral decomposition theorem for LMHS cf* [Bo2],[K1], [KH]*).*

1) The set $\Omega(S|\Lambda)$ may be uniquely decomposed into a disjoint union of closed invariant subsets $\Omega_1, \dots, \Omega_k$, such that the action of S on any Ω_i is topologically transitive.

2) Any Ω_i may be uniquely decomposed into a disjoint union of closed sets $\Omega_i^1, \dots, \Omega_i^{n_i}$ which are cyclically permuted by S and $S^{n_i}|Q_i^1$ is topologically mixing.

3) $\overline{W^s(x)} \supseteq \overline{W^u(x)} \supseteq \Omega_i$ for any $x \in \Omega_i$.

4) $\Omega(S|\Lambda) = \overline{\mathrm{Per}(S|\Lambda)}$.

A similar spectral decomposition theorem of LMHS is true for a dynamical system $\{S^t\}$ with continuous time. In this case, Statements 2 and 3 of Theorem 2.5 should be replaced by the following alternative: either $\overline{W^s(x)} \supset \Omega_i$ and $\overline{W^u(x)} \supset \Omega_i$ for any $x \in \Omega_i$ or $\{S^t\}$ may be represented as a special flow (see Chap. 1, Sect. 4) over a homeomorphism S of some compact set X (cf [Pla]). Although X may have a complicated topological structure (it may contain "holes"), it can be shown that $S|X$ possesses some kind of hyperbolic behavior: more precisely it satisfies the so- called axiom A^* (cf [PS1]; or its version, the axiom $A^\#$, cf [AJ]). A similar alternative is valid for topologically transitive Anosov flows. In this case if $\{S^t\}$ is a special flow over a map $S : X \to X$ one can show that X is a smooth compact Riemannian manifold and S is an Anosov diffeomorphism ([An1]).

2.4. Axiom A-Diffeomorphisms.

Definition 2.4. One says that a diffeomorphism S satisfies *axiom A* (or is an *A-diffeomorphism* (cf [Bo2], [KH])) if its non-wandering set $\Omega(S)$ is hyperbolic and its periodic points are dense in $\Omega(S)$.

One can show that if S satisfies axiom A then $\Omega(S)$ is locally maximal. The components of topological transitivity of $\Omega(S)$ (i.e., the sets Ω_i in Theorem 2.5) are called *basic sets*.

2.5. Hyperbolic Attractors. Repellers.

Definition 2.5. A set Λ is called an *attractor* of a dynamical system $\{S^t\}$ ($t \in \mathbb{Z}$ or $t \in R$) if there exists a neighborhood $U \supset \Lambda$ such that $\overline{S^t(U)} \subset U$ for $t > 0$ and $\Lambda = \bigcap_{t \geqslant 0} S^t(U)$.[10]

One can easily see that Λ is closed, locally maximal and invariant under $\{S^t\}$.

Definition 2.6. A set Λ is called a *hyperbolic attractor* if it is simultaneously an attractor and a hyperbolic set.

Strictly speaking hyperbolic attractors do not appear in real physical systems. However, they can serve as appropriate models for "real" situations.

Theorem 2.6. *(cf* [KH]*). A LMHS Λ is a hyperbolic attractor if and only if $W^u(x) \subset \Lambda$ for any $x \in \Lambda$.*

Thus, a hyperbolic attractor is the union of the unstable manifolds of its points and the complexity of the topological structure of such an attractor is related to the fact that the intersections of unstable manifolds with some transversal submanifold constitute a Cantor-like set.

The first example of a hyperbolic attractor, the *solenoid*, was constructed by Smale and Williams. The corresponding general construction is described, for example, in [N]. We consider only the most simple version of this construction. Let g be a map of the circle S^1 given by $g(x) = 2x \pmod 1$ and ψ_0 the immersion of S^1 into $M = \mathscr{D}^2 \times S^1$ given by $\psi_0(x) = (0, 0, 2x)$. We approximate this immersion in C^1-topology by a smooth immersion $\psi(x) = (\psi_1(x), \psi_2(x), 2x)$. We then extend ψ to a map $S : M \to M$ such that any section $\mathscr{D}^2 \times x_0$ of $\mathscr{D}^2 \times S^1$ is uniformly compressed and then gets mapped inside the section $\mathscr{D}^2 \times \psi(x_0)$. An example of such a map S is

$$S(\alpha, \beta, x) = \left(10^{-10}\alpha + \frac{1}{2}\cos x,\ 10^{-10}\beta + \frac{1}{2}\sin x,\ 2x\right).$$

The set

$$\Lambda = \bigcap_{n \geqslant 0} S^n(M)$$

is a solenoid. Topologically it is a locally trivial fibre bundle with the base space S^1 whose fibers are perfect Cantor-like sets. The set Λ is connected but not locally connected ([N]).

One can obtain another example of a hyperbolic attractor from a hyperbolic automorphism A of the torus Tor^2. More exactly the following result holds.

Theorem 2.7. *(cf* [Pl]*). There exists a C^2-diffeomorphism $S : \mathrm{Tor}^2 \to \mathrm{Tor}^2$ and two open neighborhoods $U_1 \subset U_2$ of the origin $\mathscr{O} = (0, 0)$ such that*

---

[10] If we require beforehand that Λ be closed, then the condition $\overline{S^t(U)} \subset U$ may be dropped.

(1) $S(x) = A(x)$ *for* $x \in \mathrm{Tor}^2 \backslash U_2$*;*
(2) $S(x) = B(x)$ *for* $x \in U_1$

where $B : U_1 \to \mathrm{Tor}^2$ *is a map which has three fixed points: an unstable fixed point* $\mathcal{O}$ *and two hyperbolic points* $\mathcal{O}_1, \mathcal{O}_2 \in W^u(\mathcal{O})$*. In addition, the stable and the unstable manifolds of* $\mathcal{O}_1$ *and* $\mathcal{O}_2$ *are subsets of the corresponding GSM and GUM of these points with respect to* A*.*

S has a hyperbolic attractor which coincides with the closure of the global unstable manifolds of $\mathcal{O}_1$ *and* $\mathcal{O}_2$*.*

Definition 2.7. Let S be a smooth map and J a compact invariant subset (i.e., $S(J) = J$). The dynamical system $\{S^t\}$ is called *expanding* on J and J is a called *repeller* if

(a) there exist $C > 0$ and $\lambda > 1$ such that $\|dS_x^n v\| \geqslant C\lambda^n \|v\|$ for all $x \in J$, $v \in \mathcal{T}_x$, and $n \geqslant 1$ (with respect to the Riemannian metric on M);
(b) there exists an open neighborhood V of J (called a *basin*) such that $J = \{x \in V : S^n(x) \in V \text{ for all } t \geqslant 0\}$.

Obviously, S is a local homeomorphism, i.e., there exists $r_0 > 0$ such that for every $x \in J$ the map $S|B(x, r_0)$ is a homeomorphism onto its image.

Spectral Decomposition Theorem. *(see* [KH]*) The set* $\Omega(S)$ *of non-wandering points can be decomposed into finitely many disjoint closed S-invariant subsets,* $\Omega(S) = J_1 \cup \ldots \cup J_m$*, such that* $S \mid J_i$ *is topologically transitive. Moreover, for each i there exist a number* t_i *and a set* $A_i \subset J_i$ *such that the sets* $S^k(A_i)$ *are disjoint for* $0 \leqslant k < t_i$*, their union is the set* J_i*,* $f^{t_i}(A_i) = A_i$*, and the map* $S^{t_i} \mid A_i$ *is topologically mixing.*

Definition 2.8. A smooth map $S\colon M \to M$ is called *conformal* if for each $x \in M$ we have $dS_x = a(x)\mathrm{Isom}_x$, where Isom_x denotes an isometry of $\mathcal{T}_x$ and $a(x)$ is a scalar. A smooth conformal map f is expanding if $|a(x)| > 1$ for every point $x \in M$. The repeller J for a conformal expanding map is called a *conformal repeller.*

A well-known example of a conformal expanding map is a one-dimensional Markov map (cf [Ra]). Consider disjoint closed intervals $I_1, I_2, \ldots, I_p \subset I$ and a map $S : \bigcup_j I_j \to I$ such that

(a) for every j, there is a subset $P = P(j)$ of indices with $S(I_j) = \bigcup_{k \in P} I_k$ (mod 0);
(b) for every $x \in \bigcup_j \mathrm{int} I_j$, the derivative of S exists and satisfies $\mid S'(x) \mid \geqslant \alpha$ for some fixed $\alpha > 0$;
(c) there exists $\lambda > 1$ and $n_0 > 0$ such that if $S^m(x) \in \bigcup_j \mathrm{int} I_j$, for all $0 \leqslant m \leqslant n_0 - 1$ then $|(S^{n_0})'(x)| \geqslant \lambda$.

Let $J = \{x \in I : S^n(x) \in \bigcup_{j=1}^p I_j \text{ for all } n \in \mathbb{N}\}$. The map S^m is expanding and the set J is the repeller for S.

A multidimensional example of a conformal expanding map is given by a toral endomorphism. This is a map of an n-dimensional torus defined by a diagonal matrix $(k_1, \ldots, k_n)$, where k_i is an integer and $|k_i| > 1$, $i = 1, 2, \ldots, n$.

More complicated examples of conformal expanding maps are provided by rational maps of degree $\geqslant 2$ of the Riemann sphere. For instance, consider the map $z \mapsto z^2 + c$ where $|c| < \frac{1}{4}$. This map is expanding on the Julia set J (which is defined as the closure of the set of repelling periodic points) and J is a repeller.

In a small neighborhood of a hyperbolic set (in particular, of a hyperbolic attractor or repeller) the stochastic properties of the dynamical system are most evident. In many cases, the origin of the stochastic behavior (which is usually caused by some instability of trajectories) is the existence in the phase space of a dynamical system of invariant sets which may be approximately modeled by the Smale horseshoe or by the solenoid of Smale and Williams (or by their modifications).

2.6. Partially Hyperbolic Dynamical Systems.

Definition 2.9. A dynamical system is said to be *uniformly partially hyperbolic* (UPH-system) if any trajectory of this system satisfies the condition of partial hyperbolicity and one can choose the same constants C, λ, and μ for all trajectories.

An invariant set Λ which consists of uniformly partially hyperbolic trajectories with the same constants C, λ, and μ is called a *uniformly partially hyperbolic set* (UPH-set). If an UPH-set is an attractor, then it is called a *uniformly partially hyperbolic attractor* (UPH-attractor). By a narrow sence UPH-system (or a UPH-system in the narrow sence) we mean an UPH-system whose trajectories satisfy the condition of partial hyperbolicity in the narrow sense (with the same constants C, λ, μ and C_1, λ_1, μ_1 independent of trajectory). Similarly, we speak of narrow sence UPH-sets and UPH-attractors (or UPH-sets and UPH-attractors in the narrow sense).

One can construct perhaps the simplest example of an UPH-system in the following way. Let T be an Anosov diffeomorphism of a smooth manifold M, and N a smooth manifold. Then the map $S : M \times N \to M \times N$ defined by $S(x, y) = (T(x), y)$, $x \in M$, $y \in N$, is an UPH-diffeomorphism. The neutral distribution for S is integrable. The corresponding leaves are compact and diffeomorphic to N. Let $\pi : M \times N \to M$ be the projection. Then $\pi(W_S^s(z)) = W_T^s(\pi(z))$, $\pi(W_S^u(z)) = W_T^u(\pi(z))$.

A more general construction is a skew-product over an Anosov diffeomorphism. Any small perturbation of the map S described above is an UPH-diffeomorphism (see Theorem 2.11 below). Moreover, it is conjugate to a skew-product over an Anosov diffeomorphism.

One can construct similar examples of UPH-sets and UPH-attractors: if the base map T has a hyperbolic set (respectively, a hyperbolic attractor) then $\Lambda \times N$ is an UPH-set (respectively, an UPH-attractor) for S and any small perturbation of S has an UPH-set (respectively, an UPH-attractor).

2.7. Mather Theory. A dynamical system $\{S^t\}$ on a smooth compact Riemannian manifold M generates the group of continuous linear operators $\{S^t_*\}$ acting on a Banach space $\Gamma^0(\mathcal{T}M)$ of continuous vector fields $v(x)$ on M by the formula

$$(S^t_* v)(x) = dS^t v(S^{-t}(x)).$$

The spectrum Q of the complexification of S^1_* is called the *Mather spectrum* of the dynamical system $\{S^t\}$. It was introduced by Mather in [Mat1].

Theorem 2.8. *(on the structure of Mather spectrum, Mather* [Mat1], [KSS]*). Let $\{S^t\}$ be a C^1-dynamical system on a smooth compact Riemannian manifold M. Suppose in addition that nonperiodic trajectories of S^t are dense in M. Then the following holds.*

(1) *Any connected component of the spectrum Q coincides with an annulus Q_i around 0 with radii λ_i and μ_i where*

$$0 < \lambda_1 \leq \mu_1 < \ldots < \lambda_p \leq \mu_p;$$

and $p \leq \dim M$.

(2) *The invariant subspace $\mathcal{E}_i \in \Gamma^0(\mathcal{T}M)$ of S^1_* corresponding to the component Q_i of the spectrum is a module over the ring of continuous functions; the collection of the subspaces $E_i(x) = \{v(x) \in \mathcal{T}_x : v \subset \mathcal{E}_i\}$ constitutes a dS^1-invariant continuous distribution on M and $\mathcal{T}_x = \bigoplus_{i=1}^p E_i(x)$, $x \in M$.*

(3) *If the time t is continuous then the unit circle lies in Q.*

(4) *Similar assertions about the spectrum of the complexification of S^t_* are true for any $t \neq 0$.*

We add to this the following statement.

Theorem 2.9. *(cf* [BP]*). Assume that the conditions of Theorem 2.8 hold.*

(1) *If $\mu_k < 1$ then the distribution $F^s_k = \bigoplus_{i=1}^k E_i$ is integrable and the maximal integral manifolds of this distribution generate a continuous C^1-foliation of M which we denote by W^s_k. The leaf $W^s_k(x)$ of W^s_k which passes through a point $x \in M$ is a C^1-immersed submanifold of M*[11]*;*

(2) *If $\lambda_k > 1$ then a similar statement holds about the distribution $F^u_k = \bigoplus_{i=k}^p E_i$ (we shall denote by W^u_k the corresponding foliation and by $W^u_k(x)$ its leaf at a point $x \in M$).*

(3) *The foliation W^s_k is S^t-invariant and contracting, i.e. for any $x \in M$, $y \in W^s_k(x)$, and $t \geqslant 0$*

---

[11] It can be shown that if $\{S^t\} \in C^r$ then $W^s_k(x) \in C^r$.

$$d_k^{(s)}(S^t(x), S^t(y)) \leqslant C(\lambda_k + \varepsilon)^t d_k^{(s)}(x, y),$$

where ε *is any number satisfying* $0 < \varepsilon \leqslant \min\{\lambda_{k+1} - \mu_k, 1 - \mu_k\}$; $C = C(\varepsilon) > 0$ *is a constant independent of* x, y, *and* t; *and where* $d_k^{(s)}$ *is the distance in* $W_k^s(x)$ *induced by the Riemannian metric (*$W_k^s(x)$ *is considered as a smooth submanifold of* M*).*

(4) *The foliation* W_k^u *is* S^t*-invariant and contracting for* $t \leqslant 0$.

Theorems 2.8 and 2.9 imply the existence of two filtrations of distributions

$$F_1^s \subset F_2^s \subset \cdots \subset F_l^s, \qquad F_m^u \supset \cdots \supset F_p^u$$

and the corresponding filtrations of foliations

$$W_1^s \subset W_2^s \subset \cdots \subset W_l^s, \qquad W_m^u \supset \cdots \supset W_p^u,$$

where l and m are such that $\mu_l < 1$ and $\lambda_m > 1$.[12] In general, these filtrations may happen to be trivial (for example, if the spectrum Q consists of the only one annulus which usually contains the unit circle).

Theorems 2.8 and 2.9 remain true if the manifold M is replaced by an S^t-invariant subset Λ (in this case $\Gamma^0(\mathcal{T}M)$ should be replaced by $\Gamma^0(\mathcal{T}\Lambda)$).

One can characterize the classes of dynamical systems considered above using their Mather spectrum.

Theorem 2.10. *(cf* [Mat1]*). 1) A diffeomorphism* S^1 *is Anosov if and only if its Mather spectrum* Q *is contained in a disjoint union of two annuli,* $Q \subset Q_1 \cup Q_2$, *where* Q_1 *lies strictly inside and* Q_2 *strictly outside the unit circle and* $Q \cap Q_1 \neq \emptyset$, $Q \cap Q_2 \neq \emptyset$.

2) A flow $\{S^t\}$ *is Anosov if and only if its Mather spectrum* Q *is contained in a disjoint union of the unit circle and two annuli* Q_1 *and* Q_2 *as above.*

3) A dynamical system is an UPH-system (respectively an UPH-system in the narrow sense) if its Mather spectrum is contained in a disjoint union of two annuli, $Q \subset Q_1 \cup Q_2$, *(respectively, three annuli,* $Q \subset Q_1 \cup Q_2 \cup Q_3$*), where* $Q_i \cap Q_j = \emptyset$ *for* $i \neq j$ *and* $Q \cap Q_i \neq \emptyset$ *for* $i = 1, 2$, *(respectively* $i = 1, 2, 3,$*).*

4) An invariant set Λ *is hyperbolic (respectively an UPH-set) if the Mather spectrum on* Λ *satisfies 1) and 2) (respectively, 3)).*

We denote by $\mathrm{Diff}^r(M)$ (respectively, $\Gamma^r(\mathcal{T}M)$) the space of C^r-diffeomorphisms (respectively, C^r-vector fields) equipped with the C^r-topology. The Mather spectrum is stable under small perturbations of dynamical systems in the following sense.

Theorem 2.11. *(cf* [KSS]*). Let* $\{S^t\}$ *be a dynamical system of class* C^r, $r \geq 1$ *and* $Q = \cup_{i=1}^p Q_i$ *the decomposition of its Mather spectrum into the annuli* Q_i *with radii* λ_i *and* μ_i *where* $0 < \lambda_1 \leq \mu_1 < \ldots < \lambda_p \leq \mu_p$ *and* $p \leq$

---

[12] Note that $m = l + 1$ or $l + 2$.

dim M. *Let also* $\mathcal{T}M = \bigoplus_{i=1}^{p} E_i$ *be the corresponding decomposition of the tangent bundle into* dS^1*-invariant subbundles* E_i, $i = 1, \dots, p$. *Then for any sufficiently small* $\varepsilon > 0$ *there exists a neighborhood* $\mathfrak{U}$ *of* $\{S^t\}$ *(respectively in* $\mathrm{Diff}^r(M)$ *if* $t \in \mathbb{Z}$ *or in* $\Gamma^r(TM)$ *if* $t \in \mathbb{R}$*) such that for any system* $\{\widetilde{S}^t\} \in \mathfrak{U}$ *its Mather spectrum* Q *is a union of disjoint components* $\widetilde{Q}_i$ *each is contained in an annulus with radii* $\widetilde{\lambda}_i, \widetilde{\mu}_i$ *satisfying* $|\lambda_i - \widetilde{\lambda}_i| \leqslant \varepsilon$ *and* $|\mu_i - \widetilde{\mu}_i| \leqslant \varepsilon$. *Moreover, the distribution* $\widetilde{E}_i$ *corresponding to the component* Q_i *satisfies* $\widetilde{d}(E_i, \widetilde{E}_i) \leq \varepsilon$.

Remark 2.1. While each component Q_i of the spectrum of $\{S^t\}$ is an annulus, the corresponding component Q_i of the spectrum of $\{\widetilde{S}^t\}$ may be a union of several annuli.

It follows from Theorem 2.11 that Anosov systems and UPH-systems form open subsets respectively in $\mathrm{Diff}^r(M)$ (in the case of diffeomorphisms) and in $\Gamma^r(\mathcal{T}M)$ (in the case of flows), $r \geqslant 1$. One can prove a stronger result about Anosov diffeomorphisms. Namely, they are *structurally stable* (cf [An1]). More generally, if Λ is a LMHS of a C^r-diffeomorphisms S^1, then any C^r-diffeomorphism $\widetilde{S}^1$ which is sufficiently closed to S^1 in C^r-topology has a LMHS $\widetilde{\Lambda}$, and there exists a homeomorphism $h : \Lambda \to \widetilde{\Lambda}$ which is close to identity in C^0-topology (in particular, $\widetilde{\Lambda}$ lies in a small neighborhood of Λ) such that $h \circ S^1|\Lambda = \widetilde{S}^1 \circ h|\Lambda$ (cf [KSS]).

2.8. Nonuniformly Hyperbolic Dynamical Systems. Lyapunov Exponents. We fix a Borel measure ν which is invariant under a dynamical system $\{S^t\}$.

Definition 2.10. A system $\{S^t\}$ is said to be *nonuniformly completely hyperbolic* (NCH), respectively *nonuniformly partially hyperbolic* (NPH), if there exists an invariant set Λ of positive ν-measure, consisting of trajectories satisfying the conditions of nonuniform complete hyperbolicity (respectively nonuniform partial hyperbolicity). In addition we assume that the functions $C(S^t(x), \varepsilon)$, $\gamma(S^t(x))$ and constants λ, μ, ε, which are defined for each trajectory $\{S^t(x)\}$, $x \in \Lambda$ (see (7.6), (7.7)), are the restrictions to this trajectory of some measurable functions $C(x)$, $\gamma(x)$, $\lambda(x)$, $\mu(x)$, and $\varepsilon(x)$ on Λ.

We shall now give another definition of NCH and NPH-systems. As we saw in the beginning of this chapter, to any smooth dynamical system there corresponds a cocycle $\mathfrak{U}(X, t)$ which describes the behavior of solutions of the variational equations. This cocycle satisfies the conditions stated in Sect. 2 of Chap. 1. Therefore, it defines a Lyapunov characteristic exponent $\chi(x, v)$ on the tangent bundle TM (where $x \in M$, $v \in T_x$).

Suppose that $\{S^t(x)\}$ is a regular trajectory (see Chap. 1, Sect. 2).

Theorem 2.12. *(cf* [Pe2]*). A trajectory* $\{S^t(x)\}$ *is nonuniformly (completely) hyperbolic if*

$$\chi(x,v) \neq 0 \begin{cases} \text{for any } v \neq \alpha X(x) & \text{if } t \in R \\ \text{for any } v \in \mathcal{T}_x & \text{if } t \in Z \end{cases} \tag{7.19}$$

(here $X(x)$ is the vector field corresponding to the flow $\{S^t\}$ and $\alpha \in \mathbb{R}$). A trajectory $\{S^t(x)\}$ is nonuniformly partially hyperbolic, if

$$\chi(x,v) < 0 \quad \text{forsome } v \in \mathcal{T}_x. \tag{7.20}$$

It follows from this theorem and the multiplicative ergodic theorem (see Chap. 1, Sect. 2) that NCH (respectively, NPH)-systems can be defined in the following way: A set Λ which consists of trajectories satisfying 7.19 (respectively 7.20) has full measure with respect to some invariant Borel measure. That is why NCH-systems are also called the *systems with non-zero Lyapunov exponents*.

For NCH and NPH-systems the distribution E^s is, in general, only measurable. In order to obtain more exact information let us consider the sets

$$\Lambda_l = \left\{x \in M : C(x) \le l,\ \gamma^{-1}(x) \le l,\ \lambda(x) \le \frac{l-1}{l} < \frac{l+1}{l} \le \mu(x)\right\}. \tag{7.21}$$

The sets Λ_l are measurable and form a nested system of subsets:

$$\Lambda_l \subset \Lambda_{l+1}, \quad \bigcup_l \Lambda_l = \Lambda \pmod 0.$$

The estimates (7.6) and (7.7) are uniform over $x \in \Lambda_l$ but may get worse as l grows. As shown in [Pe2], the sets Λ_l are closed and the distribution E^s is continuous on Λ_l. Moreover, the sizes of LSM are uniformly bounded from below on Λ_l and LSM $V^s(x)$ continuously depend on $x \in \Lambda_l$. For this reason, the sets Λ_l are called *sets with uniform estimates*[13].

We shall use mes to denote the Riemannian volume on M. When mes $\Lambda > 0$, the LSM on Λ_l have an important property known as *absolute continuity* (we do **not** assume here that mes is an invariant measure). Fix $x \in \Lambda$ and $l > 0$ and choose a small neighborhood $U(x)$ of x (whose size depends only on l). For $y \in \Lambda_l \cap U(x)$ consider the LSM $V^s(y)$. Take two smooth submanifolds W_1, $W_2 \subset U(x)$ which are transversal to $V^s(y)$ and define $A_i = \{z : z = W_i \cap V^s(y)$ for some $y \in \Lambda_l \cap U(x)\}$, $i = 1, 2$. Let $H^s : A_1 \to A_2$ be the map which moves a point $z = W_1 \cap V^s(y) \in A_1$ to the point $p(z) = W_2 \cap V^s(y)$. H^s is called the *holonomy map* (induced by local stable manifolds).

Let μ_{W_i} denote the measure induced on W_i, $i = 1, 2$ by the restriction of the Riemannian volume. If l is sufficiently large we have that $\mu_{W_1}(A_1) > 0$.

Theorem 2.13. *(on the absolute continuity of LSM, see* [Pe2]*). The measure $H^s_*\mu_{W_1}$ is absolutely continuous with respect to the measure μ_{W_2} (for any*

[13] or *regular sets* or *Pesin sets*.

choice of the point x, neighborhood $U(x)$, number l, and transversals W_1 and W_2).

Let ξ_l be the partition of $U(x) \cap \Lambda_l$ generated by $V^s(y)$ where $y \in \Lambda_l \cap U(x)$, $x \in \Lambda_l$. It follows from Theorem 2.13 that for almost every element $C_{\xi_l}(y) = V^s(y) \cap U(x) \cap \Lambda_l$ of ξ_l the conditional measure mes $(\cdot|C_{\xi_l})$ is absolutely continuous with respect to $\mu_{V^s(y)}$. This fact plays a crucial role in the study of ergodic properties of hyperbolic dynamical systems. It allows one to use information on local properties of dynamics to study the global behavior of its trajectories. The property of absolute continuity for Anosov systems was introduced by Anosov in [An1] (see also [AS]) where the proof of Theorem 2.13 in this case was given. In [BP], Brin and Pesin extended this property to UPH-systems. The case of NPH-systems was studied by Pesin in [Pe2]. In the case of Anosov systems and UPH-systems one can simplify the formulation of this property since the sizes of LSM are "equal" and the holonomy map is well-defined for all $z \in W_1$.

In [Pe1], Pesin constructed an NCH-flow of class C^2 without fixed points by modifying a C^2-Anosov flow. Katok [K2] constructed an NCH-diffeomorphism of class C^2 on any two-dimensional manifold. Brin, Feldman, and Katok [BFK] extended this result to multidimensional case and constructed a C^2-diffeomorphism with all but one non-zero Lyapunov exponents. In these examples the Lebesgue measure is invariant and the corresponding diffeomorphisms are isomorphic to Bernoulli shift with respect to this measure. In particular, this provides examples of Bernoulli diffeomorphisms on any compact manifolds.

§3. Ergodic Properties of Smooth Hyperbolic Dynamical Systems

3.1. u-Gibbs Measures. In this section we describe a special class of invariant measures for hyperbolic dynamical systems which determine statistical properties of the "majority" of trajectories as well as stochastic properties of the whole system. This class includes in particular, smooth invariant measures.

In order to simplify the exposition we restrict ourselves to the case of diffeomorphisms; the necessary modifications in the case of continuous time will be made at the end of this section. We shall always assume that the dynamical system under consideration is of class C^2. Let $x \in M$ be a point whose trajectory $\{S^n(x)\}$ is nonuniformly completely hyperbolic (see Sect. 1). We shall describe a measure on GUM $W^u(x)$ which is absolutely continuous with respect to the Riemannian volume on $W^u(x)$ and is uniquely defined (up to a constant factor). For $k \geq 1$, consider GUM $W^u(S^{-k}(x))$. For any $w \in W^u(S^{-k}(x))$ denote by $J^{(u)}(w)$ the Jacobian of the map $W^u(S^{-k}(x)) \xrightarrow{S} W^u(S^{-k+1}(x))$ at w.

Fix $y \in W^u(x)$. For $n \geq 1$, consider the fraction

$$\psi_n(x, y) = \prod_{k=1}^{n} \frac{J^{(u)}(S^{-k}(x))}{J^{(u)}(S^{-k}(y))} . \tag{7.22}$$

Since $\mathrm{dist}(S^{-k}(x), S^{-k}(y)) \to 0$ exponentially fast as $k \to \infty$, the factors corresponding to large k tend to 1 with exponential rate. Thus, there exists a positive limit $\lim_{n\to\infty} \psi_n(x, y) = \psi(x, y)$. It follows that for any $z, y \in W^u(x)$

$$\psi(x, z) \cdot \psi(z, y) = \psi(x, y).$$

Let ν_x^u denote the Lebesgue measure on $W^u(x)$ generated by the Riemannian metric (we view $W^u(x)$ as a smooth submanifold of M). Let also μ_x^u be a measure on $W^u(x)$ whose Radon-Nikodim derivative is $d\mu_x^u(y)/d\nu_x^u(y) = \psi(x, y)$.

Suppose that a dynamical system has a hyperbolic attractor Λ (see Sect. 2). As we saw for every $x \in \Lambda$ the GUM $W^u(x)$ is contained in Λ. Fix a small neighborhood $U(x)$ of x in M and consider LUM $V^u(y)$ for $y \in U(x) \cap \Lambda$. These manifolds generate a partition $\xi^u = \xi^u_{U(x)\cap\Lambda}$ of the set $U(x) \cap \Lambda$. Let λ be a finite Borel measure on Λ. Since the partition ξ^u is measurable λ induces a conditional measure $\lambda(\cdot|C_{\xi^u})$ on λ-almost every C_{ξ^u}.

Definition 3.1. A measure λ is said to be an *u-Gibbs measure* if $\lambda(\cdot|C_{\xi^u})$ may be written as $d\lambda(z|C_{\xi^u}) = \Xi \cdot \psi(z, y) d\nu_y^u(z)$, $z \in C_{\xi^u}$ where Ξ is the normalizing factor

$$\Xi = \int_{C_{\xi^u}} \psi(z, y) d\nu_y^u(z) \quad (C_{\xi^u} = V^u(y)).$$

It is clear that $\lambda(\cdot|C_{\xi^u})$ is uniquely defined, i.e. it does not depend on the choice of the neighborhood $U(x)$. We use the notation Ξ to follow the standard notation for the statistical sum in statistical mechanics.

Theorem 3.1. *(Bowen, Ruelle, Sinai,* [KH]*). An u-Gibbs measure exists and is invariant under S; if $S|\Lambda$ is topologically transitive, then u-Gibbs measure is unique.*

We shall see below that this measure has many "nice" properties. There is a natural way to construct such a measure. Let $A \subset M$ be an open set whose boundary is piecewise smooth. Consider LUM $V^u(x)$ and its images under S^n. The leaf $S^n(V^u(x))$ is approximatively equidistributed along the attractor. Set

$$\lambda_n(A) = \frac{\int_{A\cap S^n(V^u(x))} \psi(z, S^n(x)) d\nu_x^u(z)}{\int_{S^n(V^u(x))} \psi(z, S^n(x)) d\nu_x^u(z)} . \tag{7.23}$$

In "good situations" $\lambda_n(A)$ tend to a limit, as $n \to \infty$, which does not depend on the initial leaf $V^u(x)$. This limit is the value of $\lambda(A)$. The proof of this assertion and the analysis of the properties of λ is based on the special symbolic representation of $S|\Lambda$ which also reveals the profound connection between the theory of hyperbolic dynamical systems and the equilibrium statistical mechanics of lattice systems.

3.2. Symbolic Dynamics. We say that a smooth dynamical system admits a symbolic representation if almost every trajectory (in the topological sense or with respect to some invariant measure) can be encoded by means of a finite or a countable alphabet, such that the dynamical system gets associated with a subshift acting on some subset of the space of two-side infinite sequences. If the coding is "good" then the structure of the corresponding subset in the space of sequences is relatively simple. Good coding requires special partitions.

Let Λ be LMHS. We say that a subset $R \subset \Lambda$ is a *rectangle* if the diameter of R is sufficiently small, $R = \overline{\mathrm{Int} R}$, and $V^s(x) \cap V^u(y) \in R$ for any $x, y \in R$.

The definition does not exclude that rectangles are full of holes and look like a direct product of Cantor sets. Any rectangle may be obtained using the following procedure. Take some $z \in \Lambda$ and choose open subsets $U^s \subset V^s(z) \cap \Lambda$ and $U^u \subset V^u(z) \cap \Lambda$. If we construct $V^s(y)$ for any $y \in U^u$ and $V^u(x)$ for any $x \in U^s$, (where it can be done) then $R = \bigcup_{x,y}(V^u(x) \cap V^s(y))$ is a rectangle.

Definition 3.2. A *Markov partition* is a finite cover of Λ by rectangles $R_1, R_2, \ldots, R_k$ such that

(1) $\mathrm{Int} R_i \cap \mathrm{Int} R_j = \emptyset$ for $i \neq j$;
(2) if $x \in \mathrm{Int} R_i$, $S(x) \in \mathrm{Int} R_j$, then $S(V^u(x) \cap R_i) \supset V^u(S(x)) \cap R_j$ and $S^{-1}(V^s(x) \cap R_i) \supset V^s(S^{-1}(x)) \cap R_j$.

The Markov condition (2) in the above definition imposes strong restrictions on the arrangement between the "boundaries" of elements of the Markov partition. For example, if we consider a hyperbolic automorphism of the two-torus, then the intersection $R_i \cap S(R_j)$ must be regular (see Fig. 4).

Theorem 3.2. *For any $\varepsilon > 0$ there exists a Markov partition of the LMHS Λ whose elements have diameter less than ε.*

Sinai elaborated the first general approach to the construction of Markov partitions for Anosov systems in [Si2], [Si3]. Bowen [Bo2] made important improvements and extended it to hyperbolic sets. Adler and Weiss [AW] described a simple construction of Markov partitions for hyperbolic toral automorphisms. This partition consists of five elements which are projections on the torus of "real" parallelograms in the covering plane.

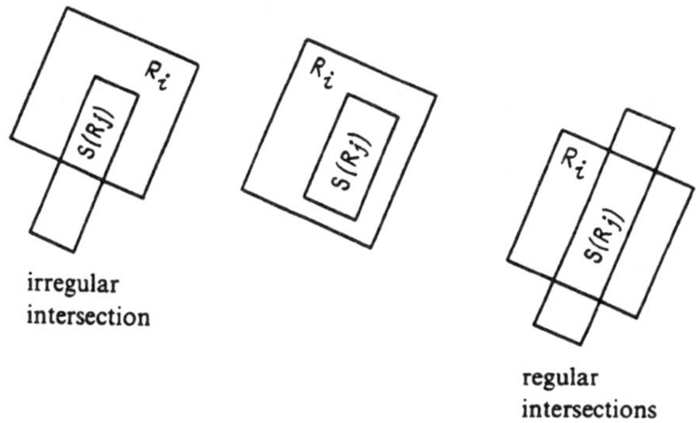

Fig. 4

Starting from a finite partition $\xi = \{R_1, \dots, R_k\}$ of a set Λ one can construct the coding map ψ from Λ into the space Σ_k of two-sided infinite sequences of k symbols. Namely, set $\psi(x) = \{\omega_n\}$, where ω_n is such that $S^n(x) \in R_{\omega_n}$. The map ψ is a conjugacy between S and the shift σ on some subset of Σ_k. The triple $(\psi, \psi(\Lambda), \sigma)$ is called a *symbolic representation* of the map $S|\Lambda$. If ξ is a Markov partition then the map ψ constructed above is one-to-one everywhere except for point whose trajectories hit the boundaries of rectangles, i.e., the set

$$\bigcup_{n=-\infty}^{\infty} f^n\left(\bigcup_{i=1}^{k} \partial^i R_i\right).$$

The main advantage of using Markov partitions is that the closure of the image $\psi(\Lambda)$ is a subshift of finite type (Σ_A, σ) where $A = (a_{ij})$ is the *transfer* (or *transition*) matrix:

$$a_{ij} = \begin{cases} 1, & \text{if } \operatorname{Int} R_i \cap S^{-1}(\operatorname{Int} R_j) \neq \emptyset \\ 0, & \text{otherwise} \end{cases}$$

(see Chap. 3, Sect. 6). More exactly the following holds.

Theorem 3.3. *(cf* [Bo2]*). For every $\omega = \{\omega_n\} \in \Sigma_A$ the intersection*

$$\bigcap_{n \in \mathbb{Z}} S^n(R_{\omega_n})$$

is nonempty and consists of exactly one point. Thus, the map ψ^{-1} which is defined on Σ_A by the formula

$$\psi^{-1}(\omega) = \bigcap_{n \in \mathbb{Z}} S^n(R_{\omega_n}), \quad \omega = (\omega_n)$$

is continuous and onto. It is one-to-one except for a set of the first category in the sense of Bair.

3.3. Measures of Maximal Entropy. Theorem 3.3 allows us to reduce several questions about topological and stochastic properties of hyperbolic attractors to some problems of statistical mechanics of u-Gibbs measures.

Let Λ be a LMHS of a diffeomorphism S, and suppose that $S|\Lambda$ is topological transitive. Let also(Σ_A, σ) be a symbolic representation of Λ constructed using some Markov partition ξ. Consider a stationary Markov chain with the transition probabilities $p_{ij} = a_{ij}z_i/\lambda(A)z_i$, where $\lambda(A)$ is the maximal positive eigenvalue of the transfer matrix A and $\{z_i\}$ the corresponding eigenvector (see [AJ]). We denote by μ_0^* the Markov measure corresponding to this Markov chain and by μ_0 the push forward of μ_0^* under ψ^{-1}. As shown in [AJ] μ_0 is the *measure of maximal entropy* for $S|\Lambda$ (for the definition see Sect. 3.5 below). We state some important properties of μ_0.

Denote by P_k the number of periodic points in Λ of period k and by $P_k(B)$ the number of such points lying in a subset $B \subset \Lambda$.

Theorem 3.4. *(cf* [AJ], [KH]*). If $B \subset A$ satisfies $\mu_0(B\backslash \text{Int}B) = 0$, then*

$$\mu_0(B) = \lim_{n\to\infty} \frac{P_n(B)}{P_n} .$$

This theorem shows that periodic points form a kind of "uniform lattice" inside any LMHS. A modification of Theorem 3.4 for flows and its application to geodesic flows are stated in Sect. 4 (see Theorem 4.2).

Fix $x \in \Lambda$ and denote by μ_0^s and μ_0^u the conditional measures induced by μ_0 on $W^s(x)$ and on $W^u(x)$ respectively.

Theorem 3.5. *(cf* [Mar]*). The following statements hold*

$$\mu_0^s(S(A)) = K^{-1}\mu_0^s(A) \qquad \text{for any } A \in W^s(x),$$
$$\mu_0^u(S(A)) = K\mu_0^u(A) \qquad \text{for any } A \in W^u(x),$$

where $K = \exp h > 1$ and $h = h(S)$ is the topological entropy of $S|\Lambda$ (for the definition see Sect. 3.5 below).

Thus, the measure of maximal entropy has the important property that it induces conditional measures on GSM (and on GUM) which are uniformly contracted (respectively, expanded) under the action of S. It can be shown that μ_0 is the unique measure satisfying this property.

3.4. Construction of u-Gibbs Measures. We present a method for constructing an u-Gibbs measure which is inspired by some constructions in statistical physics. It was developed by Sinai in [Si6]. By Theorem 2.5 we may assume without loss of generality that $S|\Lambda$ is topologically transitive (otherwise we consider a component of topological transitivity). Let $\xi = \{R_1, \dots, R_k\}$ be a Markov partition of Λ and ψ: $\Lambda \to \Sigma_A$ the corresponding symbolic representation. For $x \in M$, consider

$$\varphi^u(x) = -\log|\mathrm{Jac}(dS|E^u(x))|. \tag{7.24}$$

Then $\varphi_*^u(\omega) = \varphi^u(\psi^{-1}(\omega))$ is a continuous function on Σ_A. Consider a sequence of measures $\mu_n^* = \rho_n\mu_0^*$ on Σ_A, where μ_0^* is the measure of maximal entropy constructed above and the densities ρ_n are given by

$$\rho_n = \frac{\exp\sum_{k=0}^{n}\varphi_*^u(\sigma^k(\omega))}{\int\limits_{\Sigma_A}\exp\sum_{k=0}^{n}\varphi_*^u(\sigma^k(\omega))d\mu_0^*}.$$

The function $\varphi^u(x)$ is Hölder continuous (since so is the distribution $E^u(x)$). This implies that

$$|\varphi_*^u(\omega^1) - \varphi_*^u(\omega^2)| \leqslant d_\beta(\omega^1, \omega^2)^\alpha \tag{7.25}$$

for any $\omega^1, \omega^2 \in \Sigma_A$[14]. It follows from Sect. 6 of Chap. 3 that there exists a unique weak limit of the measures ρ_n which we denote by μ^*.

Theorem 3.6. *(cf* [Si6]*). The preimage of μ^* under ψ coincides with the weak limit of the measures λ_n* (cf. (7.23)) *and is an u-Gibbs measure on the attractor Λ.*

3.5. Topological Pressure and Topological Entropy. In this section we describe another approach to the construction of u-Gibbs measures due to Bowen [Bo2]. This approach is based on the thermodynamic formalism and its adaptation to the theory of dynamical systems. The mathematical foundation of the thermodynamic formalism, i.e., the formalism of equilibrium statistical physics, has been led by Ruelle [Ru1]. Bowen, Ruelle, and Sinai have used the thermodynamic approach to study ergodic properties of smooth hyperbolic dynamical systems (cf references and discussion in [KH]). The main constituent components of the thermodynamic formalism are the following.

(a) The topological pressure of a continuous function φ (this function determines the "potential of the system").
(b) The variational principle for the topological pressure, which establishes the variational property of the "free energy" of the system (the free energy is defined as the sum of the measure-theoretic entropy and the integral of φ with respect to a probability distribution in the phase space of the system).
(c) The existence, uniqueness, and ergodic properties of equilibrium measures (which are extremes of the variational principle).

Ruelle's version of the thermodynamic formalism is based on the classical notion of topological pressure for compact invariant sets which we now describe.

[14] Recall that the distance in Σ_A is defined by $d_\beta(\omega^1, \omega^2) = \beta^N$, where $0 < \beta < 1$ and N is the maximal nonnegative integer for which $\omega_n^1 = \omega_n^2$ for $|n| < N$

Consider a continuous map $S : X \to X$ of a compact metric space X, a continuous function φ on X, and a finite open cover $\mathfrak{U}$ of X. Denote by $\mathfrak{W}_m(\mathfrak{U})$ the set of all strings $\mathbb{U} = U_{i_1} U_{i_2} \dots U_{i_n}$ of length $m = m(\mathbb{U})$ consisting of elements of $\mathfrak{U}$. Set

$$X(\mathbb{U}) = \{x \in X : S^k(x) \in U_{i_k},\ k = 0, 1, \dots, m-1\},$$

$$S_m\varphi(\mathbb{U}) = \sup\left\{\sum_{k=0}^{m-1} \varphi(S^k(x)) : x \in X(\mathbb{U})\right\}.$$

If $X(\mathbb{U}) = \emptyset$ we set $S_m\varphi(\mathbb{U}) = -\infty$. We say that $\Gamma \subset \mathfrak{W}_m(\mathfrak{U})$ covers X if $X = \bigcup_{\mathbb{U}\subset\Gamma} X(\mathbb{U})$. Consider

$$Z_m(\varphi, \mathfrak{U}) = \inf_{\Gamma} \sum_{\mathbb{U}\in\Gamma} \exp S_m\varphi(\mathbb{U}),$$

where the infimum is taken over all elements Γ of $\mathfrak{W}_m(\mathfrak{U})$ which cover X. It can be shown (cf [Bo2]) that the limit

$$P(\varphi, \mathfrak{U}) = \lim_{m\to\infty} \frac{1}{m} \log Z_m(\varphi, \mathfrak{U})$$

exists and is finite. One can also show that the following limit

$$P(\varphi) = \lim_{\operatorname{diam}\mathfrak{U}\to 0} P(\varphi, \mathfrak{U})$$

exists (cf [Bo2]).

Definition 3.3. The quantity $P(\varphi)$ is called the *topological pressure of the function* φ (with respect to the map S).

We will also use the more explicit notation $P_S(\varphi)$.

The following statement is known as the variational principle for topological pressure.

Theorem 3.7. *(Ruelle* [Ru1], *Walters* [W1], *Bowen* [Bo2]*).*

$$P_S(\varphi) = \sup_{\mu} \left(h_\mu(S) + \int \varphi d\mu \right),$$

where the supremum is taken over all S-invariant Borel measures.

Definition 3.4. An S-invariant Borel measure μ_φ is called an *equilibrium state* corresponding to a function $\varphi \in C(X)$ if

$$h_{\mu_\varphi}(S) + \int \varphi\, d\mu_\varphi = P_S(\varphi).$$

Equilibrium states may not exist (cf [PS1] for more details). Thus the following general criterion for the existence of equilibrium states is important.

A homeomorphism S is said to be *expansive* if, for arbitrary $\varepsilon > 0$ and any $x, y \in X$, there exists an integer n such that $\rho(S^n x, S^n y) \geq \varepsilon$ (here ρ is the distance in X). For example, the full shift σ on Σ_k or a subshift of finite type σ on Σ_A can be shown to be expansive.

Theorem 3.8. *(cf* [Bo2]*). If S is expansive, then for any $\varphi \in C(X)$ there exists an equilibrium state μ_φ.*

A given function φ may have many equilibrium states. From the point of view of Statistical Physics this case corresponds to the presence of phase transitions (cf [PS1]).

Definition 3.5. The topological pressure $P_S(0)$ of the function $\varphi(x) \equiv 0$ is called the *topological entropy* of the map S and is denoted by $h(S)$.

There is an equivalent definition of the topological entropy. Fix $h > 0$ and consider a new distance $\rho_n(x, y)$ in X defined as

$$\rho_n(x, y) = \max_{0 \leqslant i \leqslant n} \rho(S^i(x), S^i(y)).$$

Given $\varepsilon > 0$, let $N(n, S, \varepsilon)$ be the minimal number of ε-balls (in the ρ_n-distance) which are needed to cover X. It can be shown (cf [K3]) that

$$h(S) = \lim_{\varepsilon \to 0} \overline{\lim_{n \to \infty}} \frac{\log N(n, S, \varepsilon)}{n} = \lim_{\varepsilon \to 0} \underline{\lim_{n \to \infty}} \frac{\log N(n, S, \varepsilon)}{n}.$$

Theorem 3.7 implies the following *variational principle for topological entropy* (cf [Bo2]):

$$h(S) = \sup_{\mu} h_\mu(S),$$

where the supremum is taken over all S-invariant Borel probability measures. *Measures of maximal entropy* are those for which the supremum is attained. They are equilibrium states for $\varphi = 0$.

Remark 3.1. One can extend the notions of topological pressure and topological entropies to continuous maps acting on noncompact subsets of compact metric spaces. See Sect. 6.3 for details. One can use this extension to develop the thermodynamic formalism for discontinuous maps which can alway be considered to be continuous on the noncompact subset consisting of all the points whose trajectories do not fall into the discontinuity set.

We now describe Bowen's approach to the construction of u-Gibbs measures. Consider the function $\varphi^u(x)$ on a hyperbolic attractor Λ defined by (7.24). Let $\varphi_*^u(\omega) = \varphi^u(\psi^{-1}(\omega))$ be the pull back of this function to Σ_A by the coding map ψ. Let also μ^* be the corresponding equilibrium state. By (7.25) the function $\varphi_*^u(\omega)$ is Hölder continuous (in the metric d_β). Since the shift map σ on Σ_A is expansive the measure μ^* is uniquely defined.

Theorem 3.9. *(cf* [Bo2]*). The image $\psi^{-1}\mu^*$ is an u-Gibbs measure on the hyperbolic attractor Λ.*

It follows from this theorem that the u-Gibbs measure is the single equilibrium state of the function $\varphi^u(x)$.

3.6. Properties of u-Gibbs Measures. They play a crucial role in the study of statistical properties of "typical" trajectories in the neighborhood of a hyperbolic attractor. More exactly the following holds.

Theorem 3.10. *(cf* [Bo2], [PS1]*). If Λ is a hyperbolic attractor of a diffeomorphism S and $S|\Lambda$ is topologically transitive, then there exists a neighborhood U of Λ, such that for any continuous function φ on U and for almost every point x (with respect to the Riemannian volume)*

$$\lim_{n\to\infty} \frac{1}{n}\sum_{k=0}^{n-1} \varphi(S^k(x)) = \int \varphi\, d\mu,$$

where μ is the unique u-Gibbs measure on Λ.

Such measures μ are called *Sinai-Ruelle-Bowen measures* (or SRB-measures).

Consider a measure ν which is absolutely continuous with respect to the Riemannian volume with the support inside the neighborhood U of Λ. Following the general approach of Bogolyubov and Krilov let us consider a sequence of measures (See Chap. 1, Sect. 1).

$$\mu_n = \frac{1}{n}\sum_{k=0}^{n-1} S_*^k \nu. \tag{7.26}$$

It follows from Theorem 3.10 that

$$\mu_n \to \mu \text{ as } n \to \infty \tag{7.27}$$

in the weak* topology. One can interpret this fact by saying that in a neighborhood of a hyperbolic attractor the system "forgets an initial distribution" during its evolution. This can be considered as one of the most crucial indications of the stochastic behavior.

For hyperbolic attractors, a stronger statement than (7.27) holds. Namely, the sequence of measures $S_*^n \nu$ converges to μ in the weak* topology. However, for more general attractors one should expect only the convergence in the mean.

Another approach to the construction of u-Gibbs measures was developed by Pesin and Sinai in [PS2]. Its main advantage is that it does not use Markov partitions and allows one to prove directly that the measures μ_n converge weakly to an u-Gibbs measure on Λ. This approach turns out to be effective in studying UPH-attractors (where other methods fail to work).

Theorem 3.11. *(cf* [PS2]*). Let Λ be an UPH-attractor of a C^2-diffeomorphism S. Then any limit measure for the sequence μ_n is an u-Gibbs measure on Λ.*

Since the topological structure of UPH-attractors is significantly more complicated than that of hyperbolic attractors the sequence of measures μ_n may not converge even in the case when S is topologically transitive.

In conclusion we point out that Bowen's approach to the construction of equilibrium states may be generalized to arbitrary LMHS. Although in this case we cannot define an u-Gibbs measure, the equilibrium state corresponding to φ^u is of some interest (cf [Bo2]). An analog of u-Gibbs measures for Smale horseshoes was constructed in [C].

u-Gibbs measures have rich ergodic properties which determine their important role in hyperbolic theory. Let μ be an u-Gibbs measure on a hyperbolic attractor Λ of a diffeomorphism S. Then the results of Section 3.4 and Theorem 3.12 stated below imply that on any component of topological transitivity on Λ the dynamical system (S, μ) is ergodic, some power of S is Bernoulli and has exponential decay of correlations; for a wide class of smooth or piecewise smooth functions the Central Limit Theorem of probability theory holds. The entropy of $S|\Lambda$ with respect to μ can be calculated by the formula (7.28) given below.

3.7. Small Stochastic Perturbations. We consider a family of probability distributions $q(\cdot|x, \varepsilon)$ on M where ε is a parameter. We suppose that $q(\cdot|x, \varepsilon)$ continuously depends on x and for any fixed ρ

$$\min q(U_\rho(x)|x, \varepsilon) \to 1 \quad \text{as} \quad \varepsilon \to 0,$$

where $U_\rho(x)$ is the ball of radius ρ centered at x.

Let S be a homeomorphism of M. We construct a family of Markov chains Π_ε where a random point x moves according to the following rule: first x gets mapped to $S(x)$ and then it moves to a random point y chosen according to the distribution $q(\cdot|S(x), \varepsilon)$. A family of Markov chains Π_ε defined as above is called a (small) *stochastic perturbation* of the homeomorphism S. It is not difficult to prove that if $I_\varepsilon = \{\pi_\varepsilon\}$ is a family of Π_ε-invariant measures, then for $\varepsilon \to 0$ any limit measure (in the sense of weak* convergence) for the family I_ε coincides with some S-invariant measure.

In the case when S is a C^2-diffeomorphism of a smooth manifold M and Λ is a hyperbolic attractor, Kifer (cf [Kif]) found some conditions which imply that the sequence π_ε converges to the unique u-Gibbs measure on Λ. In the particular case when S is an Anosov diffeomorphism of a smooth compact manifold, this result was obtained earlier by Sinai in [Si6].

3.8. Equilibrium States and Their Ergodic Properties. Apart from the u-Gibbs measures described above, some other equilibrium measures are of great interest in the hyperbolc theory.

Let Λ be a LMHS of a C^2 diffeomorphism S, and let φ be a continuous function on Λ. It is easy to show that S is expansive on Λ and hence there exists an equilibrium state μ_φ (see Sect. 3.5). In general, μ_φ is not unique. To ensure the uniqueness one should assume that φ satisfies some sufficiently weak conditions of smoothness, for example, that φ is Hölder continuous. In this case its pull back to Σ_A by the coding map ψ (i.e., the function $\phi(\omega) = \varphi(\psi^{-1}(\omega)))$ is continuous and satisfies (7.25)[15]. Thus (cf [AJ]) there exists a unique equilibrium state μ_ϕ on Σ_A for ϕ and its push forward by ψ is the unique equilibrium state for φ.

Since ergodic properties of equilibrium states for subshifts of finite type corresponding to the functions satisfying (7.25) are well understood (see Chap. 3 Sect. 6), this approach allows one to obtain detailed information about the ergodic properties of the equilibrium measure μ_φ.

Theorem 3.12. *(cf* [Bo2]*). Let φ be a Hölder continuous function on Λ. Suppose that S is topologically transitive. Then there exists a unique equilibrium state μ_φ which has the following properties.*

(1) *The measure μ_φ is positive on open subsets of Λ.*

(2) *S is ergodic with respect to μ_φ.*

(3) *There exist $m \geqslant 1$ and a set $\Lambda_0 \subset \Lambda$ such that $\Lambda = \bigcup_{k=0}^{m-1} S^k(\Lambda_0)$, $S^m(\Lambda_0) = \Lambda_0$, and $S^i(\Lambda_0) \cap S^j(\Lambda_0) = \emptyset$ for $i \neq j$, $0 \leqslant i, j \leqslant m$.*

(4) *The system $(S^m|\Lambda_0, \mu_\varphi)$ is isomorphic to a Bernoulli automorphism (in particular, it is mixing and has K-property).*

(5) *The exponential decay of correlations and the Central Limit Theorem (see Chap. 6, Sect. 1) hold for the system $(S^m|\Lambda_0, \mu_\varphi)$ with respect to the class of Hölder continuous functions.*

If S is not topologically transitive then the above statements hold on any component of topological transitivity (see Theorem 2.5). Several results about the uniqueness and the properties of equilibrium states for functions satisfying some weaker assumptions than the Hölder continuity were obtained in [Si6].

We now consider the case when S is a $C^{1+\alpha}$-expanding map. Let J be a repeller for S (see Section 2.5). A finite cover of J by rectangles $R_1, R_2, \ldots, R_k$ is called a Markov partition if

(1) $\mathrm{Int}R_i \cap \mathrm{Int}R_j = \emptyset$ for $i \neq j$;

(2) if $x \in \mathrm{Int}R_i$ and $S(x) \in \mathrm{Int}R_j$, then $R_j \subset S(R_i)$.

One can build Markov partitions for the map S of arbitrarily small diameter. Such a partition generates a symbolic model of the repeller by a subshift of finite type (Σ_A^+, σ), where Σ_A^+ is the space of one-sided infinite sequences

[15] Recall that (ψ, Σ_A, σ) is a symbolic representation of S on Λ.

admissible with respect to the transfer matrix A (see Chap. 3). For every Hölder continuous function φ on J there exists the unique equilibrium state for φ that has properties described in Theorem 3.12. The equilibrium state corresponding to the function $\varphi^u(x) = \text{Jac}(dS)$ is the unique u-Gibbs measure for S.

3.9. Ergodic Properties of Dynamical Systems with Nonzero Lyapunov Exponents. The definitive reason for the existence and strong ergodic properties of u-Gibbs measures is the high degree of instability of the trajectories on hyperbolic attractors. Here we shall consider a situation when the instability of the trajectories of a dynamical system is sufficiently weak.

Let S be a C^2-diffeomorphism of a smooth Riemannian manifold M. We denote by Λ the set of biregular points[16] with nonzero values $\chi_i(x)$ of the characteristic Lyapunov exponent $\chi^+(x, v)$. The diffeomorphism $S|\Lambda$ is NCH-diffeomorphism.

Definition 3.6. A measure μ is said to be a *measure with nonzero exponents* or a *hyperbolic measure*, if μ is S-invariant and $\mu(\Lambda) = 1$.

We introduce a class of measures with nonzero exponents which are analog of u-Gibbs measures.

Definition 3.7. A measure μ is called a *Sinai-Ruelle-Bowen measure* (or SRB- measure) on Λ if μ has nonzero Lyapunov exponents and the family of LUM (or the family of LSM) has the property of absolute continuity with respect to μ (see Sect. 2.8).

Generally speaking the set Λ is not an attractor, but if it admits a SRB-measure, then for μ-almost every x the LUM $V^u(x)$ (or the LSM $V^s(x)$) is almost contained in Λ.

No general result about the existence of SRB-measures is known except for the case when μ is the Liouville measure (see Chap. 1, Sect. 1): in [Pe2], Pesin proved that if μ is hyperbolic then it is an SRB-measure. A characterization of SRB-measures in terms of their entropies is given in [LY] (see below).

SRB-measures have rich ergodic properties described below. Pesin established these properties for NCH systems preserving the Liouville measure in [Pe2]; Ledrappier extended these results to an arbitrary SRB-measure in [L3]. We begin with the well-known entropy formula which expresses the entropy of a diffeomorphism via its Lyapunov exponents.

Theorem 3.13. ([Pe2], [K3], [L3]). *For a $C^{1+\varepsilon}$ diffeomorphism S ($\varepsilon > 0$) the entropy with respect to a SRB-measure (in particular, with respect to the Liouville measure with nonzero Lyapunov exponents) is given by the formula*

---

[16] For the definition of biregular points and of characteristic Lyapunov exponents, see Sect. 2 of Chap. 1

$$h_\mu(S) = \int_M \sum_{i=1}^{k(x)} q_i(x)\chi_i(x)d\mu(x). \tag{7.28}$$

The formula (7.28) was first proved by Pesin (cf [Pe2]) for diffeomorphisms preserving the Liouville measure and extended by Ledrappier to arbitrary SRB-measures (cf [L3]). For an upper bound see Theorem 3.17 below.

The following statement describes the partition of SRB-measures into ergodic components.

Theorem 3.14. *(cf* [Pe2], [L3]*). Let μ be an SRB-measure for a $C^{1+\varepsilon}$ diffeomorphism ($\varepsilon > 0$). Then the invariant sets Λ_i, $i = 0, 1, 2, \ldots$ exist, such that*

(1) $\bigcup_{i\geq 0} \Lambda_i = \Lambda$, $\Lambda_i \cap \Lambda_j = \emptyset$ *for* $i \neq j$;
(2) $\mu(\Lambda_0) = 0$, $\mu(\Lambda_i) > 0$ *for* $i > 0$;
(3) $S|\Lambda_i$ *is ergodic for* $i > 0$.

Some sufficient conditions for the ergodicity of $S|\Lambda$ are given in [Pe2]: if μ is a smooth measure and the foliation W^u (or the foliation W^s) is locally continuous (cf [Pe2] for definition), then any ergodic component Λ_i is open (mod 0). The topological transitivity of $S|\Lambda$ therefore implies the ergodicity of $S|\Lambda$. For smooth systems the local continuity of W^u is similar to a property stated in the main theorem for dispersed billiards (see Theorem 1.5 of Chap. 8, Sect. 1).

The following statement describes the partition into K-components.

Theorem 3.15. *(cf* [Pe2], [L3]*). Any set Λ_i, for $i > 0$, is a disjoint union of sets $\Lambda_i^{(j)}$, $j = 1, \ldots, n_i$ such that*

(1) $S(\Lambda_i^{(j)}) = \Lambda_i^{(j+1)}$ *for* $j = 1, \ldots, n_i - 1$, *and* $S(\Lambda_i^{(n_i)}) = \Lambda_i^{(1)}$;
(2) $S^{n_i}|\Lambda_i$ *is isomorphic to a Bernoulli automorphism.*

Let $\pi(S)$ be the π partition for S (i.e. the maximal partition with zero entropy; cf [Ro]). We shall use ξ^- (respectively ξ^+) to denote the partition of M by $W^s(x)$ (respectively by $W^u(x)$). For $x \in M\setminus\Lambda$, we set $W^s(x) = W^u(x) = x$.

Theorem 3.16. *(cf* [Pe2]*). There exists a measurable partition η of M with the following properties.*

(1) *For almost every $x \in \Lambda$ the element $C_\eta(x)$ is an open* (mod 0) *subset of $W^s(x)$.*
(2) $S\eta \geqslant \eta$.
(3) $\bigvee_0^\infty S^k\eta = \varepsilon$ *and* $\bigwedge_{-\infty}^0 S^k\eta = \nu(\xi^-) = \nu(\xi^+) = \nu(\xi^-) \wedge \nu(\xi^+) = \pi(S)$[17].
(4) $h_\mu(S) = h_\mu(S, \eta)$.

[17] Recall that the measurable hull of ξ is denoted by $\nu(\xi)$.

We describe some ergodic properties of general hyperbolic measures. We begin with a general result which gives an upper estimate for the entropy of any C^1-diffeomorphism.

Theorem 3.17. *(cf [K3]). The entropy of a C^1-diffeomorphism S with respect to a Borel invariant measure μ satisfies the following inequality*

$$h_\mu(S) \leqslant \int_M \sum_{i=1}^{k(x)} q_i(x)\chi_i(x)d\mu(x).$$

Here $\chi_1(x) \geqslant \cdots \geqslant \chi_{k(x)}(x) \geqslant 0 \geqslant \chi_{k+1}(x) \geqslant \cdots \geqslant \chi_n(x)$ are the ordered values of the characteristic Lyapunov exponent at the point x, $n = \dim M$, and $q_i(x)$ are the multiplicities of the corresponding values.

Margulis proved this inequality for diffeomorphisms preserving the Liouville measure (unpublished) and Ruelle [Ru3] established it in the full generality.

It follows from Theorem 3.17 that if for almost every point (with respect to the measure μ) the characteristic Lyapunov exponents are zero, then $h_\mu(S) = 0$. If we consider arbitrary Borel measures μ, then the preceding inequality may be strict (cf [K3]).

In [LY], Ledrappier and Young showed that if the equality occurs (i.e., (7.28) holds) the measure μ is an SRB-measure and thus the formula (7.28) characterizes SRB-measures. Moreover, they extended the formula (7.28) to arbitrary Borel hyperbolic probability measure μ invariant under a $C^{1+\varepsilon}$-diffeomorphism: one should replace the numbers $q_i(x)$ by the Hausdorff dimension of the conditional measure $\mu^u(x)$ induced by μ on the LUM $V^u(x)$ (see the definition of the Hausdorff dimension of a measure in Sect. 6.2). Moreover, the numbers $q_i(x)$ are integers if and only if μ is an SRB-measure.

Katok [K2, K3], showed that a diffeomorphism with non-zero Lyapunov exponents can be approximated by a horseshoe (cf Sect. 2.2). More precisely, let f be a $C^{1+\alpha}$ diffeomorphism of a compact Riemannian manifold preserving an ergodic hyperbolic measure μ of positive entropy. Then for any $\varepsilon > 0$ there exists a horseshoe Λ which is contained in an ε-neighborhood of $\operatorname{supp}\mu$ for which

$$h_\mu(f) - \varepsilon < h_{\text{top}}(f|\Lambda).$$

As a consequence of this result one obtains at least an exponential growth of the number of periodic points of f. More precisely, the following estimate holds (cf [K2, K3]):

$$0 < h_\mu(f) \leq \overline{\lim_{n\to\infty}} \frac{\log P_n}{n}.$$

3.10. Ergodic Properties of Anosov Systems and of UPH-Systems. The theorems stated above are fully applicable to Anosov systems if one sets $\Lambda = M$. The following assertion is a consequence of Theorem 3.10.

Theorem 3.18. *Let S be a C^2-Anosov diffeomorphism. Suppose that S is topologically transitive. Then*

(1) *the sequence of measures μ_n defined by (7.26) (with ν to be a smooth measure) weakly* converges as $n \to +\infty$ to an u-Gibbs measure μ^u which is the unique equilibrium state for the function φ^u (see (7.24));*

(2) *the sequence of measures μ_n weakly converges as $n \to -\infty$ to an s-Gibbs measure*[18] *μ^s which is the unique equilibrium state for the function*

$$\varphi^s(x) = \log |\mathrm{Jac}(dS|E^s(x))|.$$

Sinai and Lifschitz proved (cf [Si6]) that $\mu^u = \mu^s$ if and only if for any periodic point x of period p

$$|\mathrm{Jac}(dS^p(x))| = 1.$$

This is a very uncommon property: Anosov diffeomorphisms with $\mu^u = \mu^s$ lie in the complement in $\mathrm{Diff}^2(M)$ to a set of the second category in the sense of Bair (cf [Si6]). Nevertheless this class of diffeomorphisms include some interesting examples of automorphisms of algebraic origin. Thus we suppose that $\mu = \mu^u = \mu^s$. This implies that μ is absolutely continuous with respect to the Liouville measure. The following result is a corollary of Theorem 3.12.

Theorem 3.19. *(cf* [An1], [AS]*). Suppose that a C^2-Anosov diffeomorphism S is topologically mixing and preserves a measure which is absolutely continuous with respect to the Liouville measure. Then S is isomorphic to a Bernoulli automorphism (in particular, it is ergodic, mixing, has the K-property and positive entropy). Moreover, it has exponential decay of correlations and satisfies the Central Limit Theorem for Hölder continuous functions.*

The major part of Theorem 3.19 (up to the K-property) was proved by Anosov (cf [An1]).

The existence of u-Gibbs measures and s-Gibbs measures for UPH-diffeomorphisms follows from Theorem 3.11 (cf [PS2]). These measures may be obtained as limit measures for the sequence μ_n (see (7.26)). They have positive entropy (cf [PS2]).

We describe ergodic properties of UPH-diffeomorphisms (in the narrow sense) preserving the Liouville measure. The following notion of transitivity of the pair of foliations $\{W^u, W^s\}$ plays a crucial role.

Definition 3.8. Consider two points $x, y \in M$. A sequence of points $x_1, \dots, x_n$ is called an (R, N)-*Hopf chain* if the following holds: $x_1 = x$, $x_N = y$, the point x_i, $i = 2, \dots, N$ lies on the stable or on the unstable manifold of the point x_{i-1}, and the distance (measured along this manifold) between x_i and x_{i-1} does not exceed R. For $x \in M$ we define the (R, N)-set

[18] The definition of an s-Gibbs measure is similar to the definition of an u- Gibbs measure.

of x as the set of all $y \in M$ which are connected with x by an (R, N)-Hopf chain.

For Anosov diffeomorphisms one can find R and N such that for any $x \in M$ the (R, N)-set of x coincides with the whole manifold. The proof of ergodicity for Anosov diffeomorphisms is based on this fact. In the case of UPH-diffeomorphisms the sum of dimensions of stable and unstable manifolds is less than the dimension of M. Therefore, it is quite possible that for any $x \in M$ the (R, N)-set of x has measure zero. For example, if the pair of foliations $\{W^u, W^s\}$ is integrable[19] and generates a foliation W, then the (R, N)-set of x coincides with $W(x)$. Certainly, in such a case S is not ergodic.

Definition 3.8. A pair of foliations $\{W^u, W^s\}$ is called *locally transitive* if for any $\varepsilon > 0$, there exists an N such that (ε, N)-set of any point $x \in M$ contains a neighborhood of x. UPH-*diffeomorphism* is called *locally transitive* if the pair of foliations $\{W^u, W^s\}$ is locally transitive.

Brin and Pesin showed that a small perturbation of a sufficiently smooth locally transitive UPH-diffeomorphism (in the narrow sense) is again locally transitive and UPH (cf [BP]). It is also proved in [BP] that if S is a C^2-locally transitive UPH-diffeomorphism (in the narrow sense) preserving the Liouville measure, then under some additional conditions on W^u and W^s the map S has K-property.

A modification of the notion of (local) transitivity was introduced by Pugh and Shub [PSh].

Definition 3.9. A UPH-diffeomorphism f is said to be *globally transitive* or to have the *accessibility property* if for any $x, y \in M$ there exists an (R, N)-Hopf chain which connect them.

Pugh and Shub showed that a C^2 UPH-diffeomorphism f (in the narrow sense) preserving the Liouville measure and with the accessibility property is ergodic provided that: a) stable and unstable distributions have the Hölder exponent sufficiently close to one, b) f is *dynamically coherent* (i.e., the neutral distribution E^0 is integrable and for each leaf L of the corresponding continuous C^1 foliation $\mathscr{L}$ the sets $W^s(L)$ and $W^u(L)$ are unions of leaves of $\mathscr{L}$).

Let f be a C^2 diffeomorphism of a compact Riemannian manifold M which preserves the Liouville measure.

Definition 3.10. f is said to be *stably ergodic* if there exists a neighborhood U of f in the space of C^2 diffeomorphisms of M preserving the Liouville measure such that every $g \in U$ is ergodic.

---

[19] i.e., the distribution $T_x W^u(x) \oplus T_x W^s(x)$ is integrable; an example is the direct product of an Anosov diffeomorphism and the identity map (see Sect. 2).

In [PSh] sufficient conditions are given for an UPH-diffeomorphism f (in the narrow sense) preserving the Liouville measure which guarantee that f is stably ergodic[20].

3.11. Continuous Time Dynamical Systems. The definitions of u-Gibbs measures, measures with nonzero Lyapunov exponents, and SRB-measures may be transferred to the continuous time dynamical systems (the zero exponent corresponding to the flow direction must be excluded). The definition and the construction of Markov partitions and the corresponding symbolic model for flows on hyperbolic sets require some substantial modifications (cf [Bo2]). Theorems 3.1, 3.2, 3.10–3.12 (with exception of Statements 3 and 4), 3.13–3.15, 3.17 and the corresponding consequences of these theorems are literally transferred to this case (one should only assume that $n \in \mathbb{R}$). Theorems 3.4 and 3.5 may be transferred with obvious modifications (cf Theorem 4.2 and [AJ]). The analog of Statements 3 and 4 of Theorem 3.12 is: if the first possibility of the alternative (see 2.3) is realized, then the flow $\{S^t\}$ is Bernoulli with respect to μ_φ (cf [Bo1]).

§4. Hyperbolic Geodesic Flows

For a long time geodesic flows have played an important stimulating role in the development of hyperbolic theory. When Hadamard and Morse in the beginning of the 20th century were studying the statistics of geodesics on surfaces of negative curvature, they pointed out that the local instability of trajectories gave rise to some global properties of dynamical systems such as ergodicity and topological transitivity. The further study of geodesic flows has later inspired the introduction of different classes of hyperbolic dynamical systems (Anosov systems, UPH-systems, and NCH-systems preserving the Liouville measure). On the other hand, geodesic flows always were a touchstone for applying new advanced methods of the general theory of dynamical systems. This in particular, has led to some new interesting results in differential and Riemannian geometry. In this chapter we will describe some of these results. The reader can find the connection of geodesic flows with classical mechanics in Chap. 1, Sect. 1[21].

4.1 Manifolds of Negative Curvature. Let Q be a compact p-dimensional Riemannian *manifold of negative curvature*. This means that for any $x \in Q$ and any two linearly independent vectors $v_1, v_2 \in \mathcal{T}_x$ the sectional curvature $K_x(v_1, v_2)$ satisfies

[20] These conditions include the accessibility property, smoothness of invariant distributions, as well as that f is dynamically coherent and df is sufficiently bunched.

[21] For detailed references see [Pe3].

$$K_x(v_1, v_2) \leqslant -k, \tag{7.29}$$

where $k > 0$ is a constant independent of x, v_1, v_2. The geodesic flow $\{g^t\}$ on the unit tangent bundle $M = SQ$ is defined in Chap. 1, Sect. 1. The key statement which allows one to investigate its topological and ergodic properties is the following theorem of Anosov (cf [An1]).

Theorem 4.1. *The geodesic flow on a compact Riemannian manifold of negative curvature is Anosov.*

The proof of this theorem is based on the study of solutions of the variational equations (7.1) corresponding to the flow $\{g^t\}$. One can show that these equations along a given trajectory of the flow are the Jacobi equation along the corresponding geodesic $\gamma(t)$ that is well-known in the Riemannian geometry:

$$Y''(t) + R_{XY}Y = 0. \tag{7.30}$$

Here $Y(t)$ is a vector field along $\gamma(t)$, $X(t) = \dot\gamma(t)$, and R_{XY} is the curvature tensor (cf [Pe3]). More precisely, the relation between the variational equations (7.1) and the Jacobi equation (7.30) can be described as follows. Fix a vector $v \in M$ and consider the projection $\pi : M \to Q$ and the connection map $K : M \to \mathcal{T}_{\pi(v)}Q$ which is induced by the Riemannian metric (i.e. K is the Levi-Civita connection). Take $\xi \in \mathcal{T}_vM$ and let $Y_\xi(t)$ be the solution of the Jacobi equation with the initial conditions $Y_\xi(0) = d\pi\xi$ and $Y'_\xi(0) = K\xi$. The map $\xi \overset{\chi}{\to} Y_\xi(t)$ is an isomorphism and $d\pi dS^t\xi = Y_\xi(t)$, $KdS^t\xi = Y'_\xi(t)$. The above identification χ allows us to say that (7.30) are the variational equations corresponding to the flow S^t.

Fix $v \in M$ and denote by $\gamma_v(t)$ the geodesic given by the vector v. Using the Fermi coordinates $\{e_i(t)\}$, $i = 1, \dots, p$ along $\gamma_v(t)$[22] we can rewrite (7.30) in the matrix form

$$\frac{d^2}{dt^2}A(t) + K(t)A(t) = 0, \tag{7.31}$$

where $A(t) = (a_{ij}(t))$ and $K(t) = (k_{ij}(t))$ are matrix functions and $K_{ij}(t) = K_{\gamma_v(t)}(e_i(t), e_j(t))$. Using (7.29) it can be shown (cf [An1]) that the boundary value problem for (7.31) has a unique solution. Thus given s, there exists a unique solution $A_s(t)$ satisfying the boundary conditions $A_s(0) = I$ (where I is the identity matrix) and $A_s(s) = 0$. Furthermore, there exists a limit

$$\lim_{t\to\infty} \dot A_s(t)\big|_{t=0} = A^+.$$

A *positive limit solution* $D^+(t)$ of the equation (7.31) is defined by its initial conditions $D^+(t)\big|_{t=0} = I$ and $\dot D^+(t)\big|_{t=0}A^+$. It is nondegenerate (i.e. $\det(D^+(t)) \neq 0$ for every $t \in \mathbb{R}$) and $D^+(t) = \lim_{s\to+\infty} A_s(t)$. Moreover,

[22] $\{e_i(t)\}$ is obtained by the time t parallel translation along $\gamma_v(t)$ of an orthonormal basis $\{e_i(0)\}$ in $\mathcal{T}_{v(0)}$ where $e_1(t) = \dot\gamma(t)$.

$$D^+(t) = C(t)\int_t^\infty C^{-1}(s)(C^{-1}(s))^* ds, \tag{7.32}$$

where $C(t)$ is the solution of (7.31) corresponding to the initial conditions $C(0) = 0$ and $\frac{d}{dt}C(t)\big|_{t=0} = I$. Similarly, considering $s \to -\infty$ we can define a *negative limit solution* $D^-(t)$ of (7.31). The vectors $\chi^{-1}(D^\pm(0)e_i(0))$, $i = 1, \dots, p$ are linearly independent and span a subspace $E^\pm(v) \subset \mathcal{T}_v M$. One can show that the subspaces $E^+(v)$ and $E^-(v)$ generate two g^t-invariant distributions satisfying the conditions of uniform hyperbolicity. This proves Theorem 4.1.

One can show that for geodesic flows on manifolds of negative curvature, only the second possibility of the alternative for Anosov flows takes place (see Sect. 2.3). Namely, the flow $\{g^t\}$ is mixing. This fact may be deduced from a theorem due to Arnold (cf [An1]), or may be proved directly (cf [Si1]). The description of topological and ergodic properties of geodesic flows is now a consequence of general results on Anosov flows.

Given a Borel set $A \subset M$, $t > 0$, and $\varepsilon > 0$, set

$$N_{t,\varepsilon}(A) = \sum_{\gamma\in Q(\varepsilon,t)} \omega_\gamma(A) \Big/ \sum_{\gamma\in Q(\varepsilon,t)} \tau(\gamma),$$

where $Q(\varepsilon, t)$ is the set of periodic orbits γ of the period (not necessarily the minimal one) lying between $t-\varepsilon$ and $t+\varepsilon$, $\tau(\gamma)$ is the minimal period of γ, $\omega_\gamma(\cdot)$ is the measure on M which is the image of the Lebesgue measure on $[0, \tau(\gamma))$ under the map $t \to g^t(x)$, $x \in \gamma$.

Theorem 4.2. *Let $\{g^t\}$ be a geodesic flow on a compact Riemannian manifold on negative curvature. Then the following hold.*

(1) *The flow $\{g^t\}$ is isomorphic to a Bernoulli flow; in particular $\{g^t\}$ is ergodic, mixing, has continuous spectrum, positive entropy, and K-property. It is also isomorphic to a Bernoulli flow.*

(2) *The flow $\{g^t\}$ is topologically mixing; in particular, it is topologically transitive.*

(3) *Periodic orbits of $\{g^t\}$ are dense in M; the number $P(T)$ of periodic orbits of period $\leqslant T$ is finite and*

$$\lim_{T\to\infty} \frac{hTP(T)}{e^{hT}} = 1,$$

where $h = h(g^1)$ is the topological entropy of g^1.

(4) *The flow $\{g^t\}$ admits a uniquely defined measure of maximal entropy μ_0; given a Borel set $A \subset M$ satisfying $\mu_0(A \setminus \operatorname{Int} A) = 0$ and any $\varepsilon > 0$, we have*

$$\lim_{t\to\infty} N_{t,\varepsilon}(A) = \mu_0(A).$$

Consider the stable foliation W^s and the unstable foliation W^u for the geodesic flow $\{g^t\}$. The leaf $W^s(v)$ (respectively $W^u(v)$) passing through a linear element v is called the *stable* (respectively *unstable*) *horosphere*[23]. It can be shown that for any $v \in M$, the corresponding stable and unstable horospheres are everywhere dense in M.

Theorem 4.2 is an illustration of how the methods of dynamical systems theory work to produce important information on geometric structure of compact manifolds of negative curvature. This theorem establishes such pure geometric properties as the density of closed geodesics, their asymptotic number, and their distribution in the phase space, the existence of everywhere dense geodesics, and the density of stable and unstable horospheres.

We describe now another more geometrical approach to the construction of stable and unstable horospheres. This approach turns out to be useful in obtaining some additional information on geometry of horospheres. It will also help us to transfer some of the above results to a broader class of metrics (see 4.2).

We denote by H the universal Riemannian cover of Q. Recall that H is a simply connected complete Riemannian manifold such that $Q = H/\Gamma$ where Γ is a discrete subgroup of the group G of isometries of H. According to the Hadamard-Cartan theorem, any two points in H are joint by the single geodesic and for any $x \in H$ the exponential map $\exp_x : \mathbb{R}^p \to H$ is a diffeomorphism. Hence, the map

$$\varphi_x(y) = \exp_x[(1 - \|y\|^{-1}]$$

is a homeomorphism of the open unit ball B onto H.

Two geodesics $y_1(t)$, $y_2(t)$ in H are called *asymptotic* for $t > 0$ if

$$\rho(\gamma_1(t), \gamma_2(t)) \leqslant \text{const}$$

for all $t > 0$ (ρ is the distance in H induced by the Riemannian metric). It can be shown that geodesics which are asymptotic for $t > 0$ do not intersect and that for any point x there is a single geodesic passing through x and asymptotic for $t > 0$ to a given geodesic. This is an equivalence relation, and the equivalence classes are called *points at infinity*. The set of these classes is denoted by $H(\infty)$ and is called the *ideal boundary* of H (sometimes the *absolute*). One can define a topology τ on $H \cup H(\infty)$ such that $H \cup H(\infty)$ becomes a compact metric space and the restriction of τ on H coincides with the topology induced in H by the Riemannian metric.

The map φ_x may be extended to a homeomorphism (still denoted by φ_x) of the closed ball $\overline{B} = B \cup S^{p-1}$ (where S^{p-1} is the $(p-1)$-dimensional sphere in $\mathbb{R}^p$) onto $H \cup H(\infty)$ by the equality

$$\varphi_x(y) = \gamma_y(+\infty), \qquad y \in S^{p-1}.$$

[23] The modern terminology is somewhat different from the traditional one which uses horospheres (horocycles in the two dimensional case) to denote the projections of $W^s(v)$ and $W^u(v)$ in Q. Nowadays the latter are called limit spheres (respectively limit circles) and horospheres are the framing of limit spheres (see below).

In particular, φ maps S^{p-1} homeomorphically onto $H(\infty)$. The image of all the asymptotic geodesics under φ generates the partition of B into disjoint curves "tending" to the same point $y \in S^{p-1}$. A *limit sphere* is a submanifold in H which is orthogonal to the set of all the asymptotic geodesics (corresponding to y). More exactly, it means that through any point of this submanifold passes a single geodesic which is directed orthogonally to the submanifold. It can be shown that limit spheres exist and have the following properties.

(1) Any limit sphere is uniquely defined by a point $q \in H(\infty)$ corresponding to the bundle under consideration and a point $x \in M$. (For this reason, it will be denoted by $L(q, x)$).

(2) $\bigcup_{-\infty<t<\infty} L(q, \gamma(t)) = H$.

(3) $L(q, \gamma(t_1)) \cap L(q, \gamma(t_2)) = \emptyset,\ t_1 \neq t_2$.

(4) $\rho(L(q, \gamma(t_1)), L(q, \gamma(t_2))) = |t_2 - t_1|$.

Here $\gamma(t)$ is a geodesic satisfying $\gamma(+\infty) = q$. The last property expresses the equidistance of limit spheres.

The set $\varphi^{-1}(L(q, x)) \cup \varphi^{-1}(q) \subset \overline{B}$ is homeomorphic to the $(p-1)$-dimensional sphere in $\overline{B}$ which is tangent to S^{p-1} at the single point $\varphi^{-1}(q)$.

There is another method to construct a limit sphere as a limit of some spheres. Namely, consider the geodesic $\gamma(t)$ joining q and x and for any t the sphere in H passing through x and centered at $\gamma(t)$. The limit of these spheres as $t \to +\infty$ coincides with $L(q, x)$[24].

The *positive* (respectively, *negative*) *framing* of a limit sphere $L(q, x)$ is the set of all orthonormal vectors pointing inside (respectively outside) the sphere. It can be shown that the positive (respectively, negative) framing coincides with the stable horosphere $W^s(v)$ (respectively with the unstable horosphere $W^u(v)$) where v is defined by $\pi(v) = x$, $\gamma_v(+\infty) = q$.

Consider the geodesic flow $\{g^t\}$ on a compact connected n - dimensional Riemannian manifold M. Let H be the universal cover of M. One can show that for any $x \in H$ there exists the limit

$$h_V = \lim_{R\to\infty} \frac{1}{R} \log \operatorname{vol}(B(x, R))$$

which does not depend on x (cf [Ma3]). The number h_V is called the *volume growth* specified by the Riemannian metric. In general, h_V does not exceed the topological entropy $h_T = h(S^1)$, i.e.,$h_T \geq h_V$. In the case of metrics of non-positive curvature, one can show that they coincide, i.e., $h_T = h_V$.

4.2. Riemannian Metrics Without Conjugate (or Focal) Points. Two points $x, y \in Q$ are *conjugate* along the geodesic $\gamma(t)$ (which joins them) if for some $t_1, t_2 \in \mathbb{R}$, we have $x = \gamma(t_1)$, $y = \gamma(t_2)$, and there exists a

[24] This justifies the name "limit sphere".

nonzero solution $Y(t)$ of the Jacobi equation (7.30) along $\gamma(t)$ satisfying $Y(t_1) = Y(t_2) = 0$[25].

Two points $x_1 = \gamma(t_1)$ and $x_2 = \gamma(t_2)$ are called *focal* along the geodesic $\gamma(t)$ if there exists a nonzero solution $Y(t)$ of the Jacobi equation (7.30) along $\gamma(t)$ satisfying $Y(t_1) = 0$, $Y'(t_1) \neq 0$, and $\frac{d}{dt}(\|Y(T)\|^2)\,|_{t=t_2} = 0$.

A Riemannian metric is said to have *no conjugate* (respectively, *focal*) *points* if there are no conjugate (respectively, focal) points along any geodesic.

A Riemannian metric is said to be of *nonpositive curvature* if for any two linearly independent vectors $v_1, v_2 \in \mathcal{T}_x$ the sectional curvature $K_x(v_1, v_2)$ satisfies

$$K_x(v_1, v_2) \leqslant 0.$$

One can show that if the Riemannian metric is of nonpositive curvature then it has no focal points. The latter implies there are no conjugate points. On the other hand a Riemannian metric with no focal points may have sectional curvature which is strictly positive on some subset.

Eberlein [Eb] proved that if the Riemannian metric has no conjugate points one can construct positive and negative limit solutions of (7.31) along $\gamma(t)$ as well as two distributions of subspaces $E^+(v)$, $E^-(v)$, $v \in SH$.

Under some additional assumptions one can show that the distributions E^+ and E^- are integrable. To explain this let us consider the universal Riemannian cover H of Q (so that $Q = H/\Gamma$, where Γ is a discrete subgroup of the group of isometries of H) and lift the distributions E^+ and E^- to SH. We say that the Riemannian metric satisfies the *Axiom of Asymptoticity* if the following condition holds. For any $x \in H$ and a vector v, let $x_n \in H$, $x_n \to x$, vectors $v_n \to v$, and numbers $t_n \to \infty$. Let also γ_n be the geodesic that joins the points x_n and $\gamma_{v_n}(t_n)$. Since the sequence of vectors $\gamma_n'(0)$ is compact the sequence of geodesics γ_n has a limit[26]. The Axiom of Asympoticity requires that any limit geodesic for the sequence of geodesics γ_n is asymptotic to γ[27]. Note that if the Riemannian metric has no focal points it satisfies the Axiom of Asymptoticity.

Pesin proved that if the Riemannian metric has no conjugate points and satisfies the Axiom of Asympoticity then the limit spheres (defined as the limits of spheres) exist. Their (positive or negative) framing (i.e., horospheres) are the integral manifolds for the distributions E^+ and E^- (cf [Pe3] for more details).

If for any $v \in M$

$$E^-(v) \oplus E^+(v) \oplus X(v) = \mathcal{T}_v M \tag{7.33}$$

($X(v)$ is the one-dimensional subspace corresponding to the flow direction), then $\{g^t\}$ can be shown to be an Anosov flow (cf [Eb]).

---

[25] For a more geometric definition of conjugate points see [Pe3].

[26] A sequence of geodesics γ_n converges to a geodesic γ if $\gamma_n(0) \to \gamma(0)$ and $\gamma_n'(0) \to \gamma'(0)$.

[27] It is not known that the sequence of geodesics γ_n in fact converges to γ.

The condition (7.33) holds for metrics of negative curvature, but it may also hold for metrics which have some domains of zero or even positive curvature. This gives rise to the notion of a *manifold of Anosov type* which admits a Riemannian metric such that the corresponding geodesic flow is Anosov (an important subclass consists of *manifolds of hyperbolic type* which admit a Riemannian metric of negative curvature; in particular, this class contains any surface of genus > 0). For these manifolds one gets information about topological and ergodic properties of geodesic flows with respect to any metric without conjugate points.

We state a sufficient condition which guarantees that (7.33) holds along a geodesic. Let $v, w \in M$ and w be orthogonal to v. Let also $Y_w(t)$ be a positive limit solution of (7.31) determined by w (i.e. $Y_w(0) = D^+(0)w$ and $Y_w'(0) = \dot{D}^+(0)w$). We use $K_w(t)$ to denote the sectional curvature at $\gamma_v(t)$ defined by $Y_w(t)$ and $\gamma_v(t)$.

Theorem 4.3. *(cf* [Pe3]*). Assume that for any w,*

$$\overline{\lim_{t\to+\infty}} \frac{1}{t} \int_0^t K_w(s)\, ds < 0. \tag{7.34}$$

Then

(1) *condition (7.33) holds along the geodesic $\gamma_v(t)$;*

(2) $\chi^+(\xi, v) > 0$ *for any* $\xi \in E^+(v)$ *and* $\chi^+(\xi, v) < 0$ *for* $\xi \in E^-(v)$ *(where $\chi^+(\xi, v)$ is the characteristic Lyapunov exponent of ξ and v).*

We denote by Λ the set of vectors $v \in M$ for which (7.34) holds (for any $w \in M$ which is orthogonal to v). If this set has positive Liouville measure, then $\{g^t\}|\Lambda$ is an NCH-flow. This allows one to describe ergodic properties of the flow $\{g^t\}|\Lambda$. Indeed, in this case one can obtain a more refined information on the flow provided some additional assumptions hold.

Following Eberlein [Eb] we say that a Riemannian metric satisfies the *Visibility Axiom* if for any $\varepsilon > 0$ there exists $R = R(\varepsilon)$ such that for any $x \in H$ and any geodesic interval $\gamma(t)$, $t_0 \le t \le t_1$ in H, for which $d(x, \gamma) \ge R$, we have

$$\angle_x(\gamma(t_0), \gamma(t_1)) \le \varepsilon.$$

A Riemannian metric of negative curvature satisfies the Visibility Axiom while a Riemannian metric of nonpositive curvature satisfies this axiom provided the manifold does not admit a global geodesically complete embedding of the Euclidean plane $\mathbb{R}^2$. The Visibility Axiom characterizes the whole class of Riemannian metrics without conjugate points: if a Riemannian metric without conjugate points satisfies the Visibility Axiom then so does any other Riemannian metric without conjugate points. In particular, a two-dimensional compact Riemannian manifold M whose Euler characteristic $\chi(M) < 0$ with Riemannian metric without conjugate points satisfies the Visibility Axiom.

An important role of the Visibility Axiom in the study of geodesic flows is illustrated by the remarkable result of Eberlein.

Theorem 4.4. *(cf* [Eb]*). The geodesic flow on a compact Riemannian manifold satisfying the Visibility Axiom is topologically transitive.*

We now state the result describing ergodic properties of the geodesic flow on the set Λ.

Theorem 4.5. *Let $\{g^t\}$ be the geodesic flow on a compact Riemannian manifold without focal points satisfying the Visibility Axiom. Assume that $\mu(\Lambda) > 0$ where μ is the Liouville measure. Then*

(1) *the set Λ is open and everywhere dense;*
(2) *the flow $\{g^t | \Lambda\}$ is isomorphic to a Bernoulli flow.*

We describe some additional conditions on the Riemannian metric of a two-dimensional manifold which guarantee that Λ is of positive Liouville measure.

Theorem 4.6. *Let $\{g^t\}$ be the geodesic flow on a compact two-dimensional Riemannian manifold of negative Euler characteristic. Then the set Λ has positive Liouville measure if any one of the following conditions holds*

(1) *the Riemannian metric has no focal points;*
(2) *the entropy of the geodesic flow is positive.*

4.3. Entropy of Geodesic Flows. Let $v \in M$. We define $v^\perp$ as the set of vectors $w \in M$ which are orthogonal to v. Consider the linear map $S_v : v^\perp \to v^\perp$ defined by the equality: $S_v w = K_\xi(w)$, where $\xi(w)$ is the vector in $E^-(v)$ such that $d\pi\xi(w) = w$.

Theorem 4.7. *(cf* [Pe3]*). For a Riemannian metric of class C^4 without focal points S_v is a linear self-adjoint operator of the second quadratic form for the limit sphere $L(\pi(v), \gamma_v(+\infty))$ at the point $\pi(v)$ (which is a submanifold in H of class C^2).*

Using formula (7.32) we can represent S_v in a form similar to the continued fraction decomposition for the operator $B(x)$ from Sect. 1 of Chap. 8.

Denote by $\{e_i(v)\}$, $i = 1, \dots, p-1$, the orthonormal basis in $v^\perp$ consisting of eigenvectors of S_v. Let $K_i(v)$ be the corresponding eigenvalues. The numbers $K_i(v)$ are called the *principal curvatures* and the directions determined by the vectors $e_i(v)$ the *directions of principal curvatures* for the limit sphere at $\pi(v)$.

Theorem 4.8. *(cf* [Pe3]*). Under the assumptions of Theorem 4.3 we have the following formula for the entropy of the geodesic flow*

$$h_\mu(g^1) = -\int_M \sum_{i=1}^{p-1} K_i(v)d\mu(v) = -\int_M \operatorname{tr} S_v d\mu(v),$$

where $\operatorname{tr} S_v$ *denotes the trace of* S_v.

This result is analogous to the result of [SC] for dispersed billiards. Some estimates from below for the entropy of geodesic flow were obtained in [OS].

4.4. Riemannian Metrics of Nonpositive Curvature. As we saw in the previous section geodesic flows on manifolds without conjugate (or focal) points may exhibit some hyperbolic properties which are similar to (but somewhat weaker than) those of geodesic flows on manifolds of negative curvature. Notice however, that while the property of the Riemannian metric to be of negative curvature is local and is stable with respect to small perturbations,[28] the absence of conjugate points is a global property which is not stable and usually is difficult to verify. Nevertheless, there is an important class of Riemannian metrics to which the above results are applied. These are metrics of nonpositive curvature which lie on the boundary of the set of metrics of negative curvature.

A simple example of the geodesic flow on an n-torus with the standard metric, which is not ergodic and has zero entropy, shows that this situation is rather different. This example is not typical however. Indeed, if the universal cover H of a compact manifold M of nonpositive curvature satisfies the Visibility Axiom then the set Λ (which consists of geodesic satisfying (7.34)) has positive Liouville measure, the geodesic flow $g^t|\Lambda$ is an NCH-flow, and Theorem 4.4 holds. In particular, $g^t|\Lambda$ is isomorphic to a Bernoulli flow.

Let us note that for metrics of nonpositive curvature the Visibility Axiom is equivalent to the property that the universal cover H does not admit any geodesically complete embedding of the plane.

It is an open problem whether the set Λ has full measure[29]. For geodesic flows on compact surfaces one can show that the curvature along any geodesic which does not belong to Λ is identically zero. Roughly speaking the zero curvature reveals the possibility of a geodesically complete embedding into H of an infinite strip of zero curvature which consists of geodesics joining two given points on the ideal boundary[30]. It is conjectured that the total measure of all such strips is equal to zero.

In the mid 80th, significant progress was made in the study of the geometric structure and classification of manifolds of non-positive curvature (cf the survey [BK]). We briefly outline main results.

---

[28] Any sufficiently close metric also has negative curvature.

[29] The argument in the proof of Theorem 9.1 in [Pe5] contains a gap.

[30] One can show that any two geodesics on the universal cover of a compact manifold without focal points which are asymptotic both for $t > 0$ and for $t < 0$ (i.e., they both join two distinct points on the ideal boundary) bound a flat strip (this statement is known as the *flat strip theorem* (cf [Pe3], [BKn]).

Given a unit vector v the *rank of* v, rank(v), is the dimension of the space of parallel Jacobi fields along the geodesic γ_v[31]. The *rank of the manifold* M, rank(M), is the minimal rank of a unit vector. This definition agrees with the usual definition of rank for locally symmetric spaces. A vector v (and the corresponding geodesic γ_v) are called *regular* if rank(v) = rank(M). A k-*flat* is a totally geodesically and isometrically immersion of a k-dimensional Euclidean space in M. A k-flat F is called *regular* if k = rank(M) and F contains a regular geodesic.

Theorem 4.9.

(1) *The set of regular vectors is open and dense in the space SM of unit tangent vectors.*

(2) *Every regular vector is tangent to a unique k-flat.*

(3) *Periodic regular vectors are dense in SM and if M is compact the k-flat to which such a vector is tangent is compact.*

The following theorem provides a classification of complete Riemannian manifolds of negative curvature and of higher rank. It was proved by Ballman [Ba] and independently (for compact manifolds) by Burns and Spatzier [BuSp].

Theorem 4.10. *If $k \geq 2$, the de Rham factors of the universal cover H of M are (a) Euclidean, or (b) of rank 1, or (c) symmetric spaces. If in addition H has no Euclidean factor, M has a finite cover which is a product of locally symmetric spaces and rank 1 manifolds.*

§5. Geodesic Flows on Manifolds of Constant Negative Curvature

One can use the powerful methods of the theory of unitary representations of Lie groups to develop a profound study of ergodic properties of geodesic flows on manifolds of constant negative curvature. The idea of describing these flows in an algebraic way is due to Gelfand and Fomin (cf [GGV]). Sometimes the dynamical systems to which the Gelfand-Fomin method is applicable are called systems of algebraic origin. Several results concerning these systems are described in the survey [KSS]. Here we shall consider only the geodesic flows on manifolds of constant negative curvature. We shall use the Poincaré model on the upper half plane $H = \{z = (x + iy) : y > 0\}$. The line $y = 0$ is the ideal boundary (and is denoted by $H(\infty)$).

The geodesics in H (with respect to its metric) coincide with the semicircles[32] centered on the ideal boundary or with the rays orthogonal to the

[31] Recall that γ_v is the geodesic that passes through the point $\pi(v)$ in the direction v.

[32] Here and below semicircles, circles, straight lines are understood in the Euclidean sense.

ideal boundary. The Riemannian metric with the curvature $-k$ is given as the inner product $\langle\, ,\,\rangle_L$ at a point $z = x + iy \in H$ by $\langle\, ,\,\rangle_L = \frac{k}{y^2}\langle\, ,\,\rangle$, where $\langle\, ,\,\rangle$ is the Euclidean scalar product.

An *asymptotic bundle* is the set of all oriented lines which reach the same point of the ideal boundary. Limit circles orthogonal to some asymptotic bundle are the circles tangent to the ideal boundary and the lines parallel to it.

The *isometries* in *Lobachevsky geometry* are the fractional linear transformations

$$z \to \frac{a + bz}{c + dz}$$

which preserve the upper half plane. Here a, b, c, d are real which are normalized such that $\det \begin{vmatrix} a & b \\ c & d \end{vmatrix} = 1$.

The matrices $\begin{vmatrix} a & b \\ c & d \end{vmatrix}$ and $\begin{vmatrix} a' & b' \\ \gamma' & d' \end{vmatrix}$ define the same transformation if $\begin{vmatrix} a & b \\ c & d \end{vmatrix} = \begin{vmatrix} a' & b' \\ \gamma' & d' \end{vmatrix}$. The matrices $\begin{vmatrix} 1 & 0 \\ 0 & 1 \end{vmatrix}$ and $\begin{vmatrix} -1 & 0 \\ 0 & -1 \end{vmatrix}$ form the center Z of $SL(2, \mathbb{R})/Z$. An isometry $g \in G$ is called

a)) *elliptic* if g is conjugate with the transformation $z \to \lambda z$, $|\lambda| = 1$;
b) *hyperbolic* if g is conjugate with the transformation $z \to \lambda z$, $|\lambda| > 1$;
c) *parabolic* if g is conjugate with the transformation $z \to z + 1$.

We are especially interested in hyperbolic transformations. Consider some $g_0 : g_0 z = \lambda z$, $\lambda > 1$. The line $\gamma = \{\operatorname{Re} z = 0\}$ is invariant under g_0, i.e. $g_0\gamma = \gamma$ and any point $z \in \gamma$ moves under the action of g_0 by the non-Euclidean distance $\ln \lambda$. If g is conjugate with g_0 then there exists a g-invariant line with the same property. Since $g = g_1 g_0 g_1^{-1}$ for some $g_1 \in G$ we find that $\lambda > 1$ is the eigenvalue of g. We shall use M_0 to denote the unit tangent bundle SH. Then $G = M_0 = SL(2, \mathbb{R})/Z$, where N consists of two elements $\begin{vmatrix} 1 & 0 \\ 0 & 1 \end{vmatrix}$ and $\begin{vmatrix} -1 & 0 \\ 0 & -1 \end{vmatrix}$, because any isometry is uniquely defined by its action on the unit tangent vector based at $z = i$ and directed vertically upwards. A surface Q of constant negative curvature is the factor space $Q = H/\Gamma$ where Γ is a discrete subgroup of G. The phase space of the geodesic flow, i.e., the unit tangent bundle over Q coincides with the factor space $M = \Gamma\backslash G$, the measure μ on M is induced by the Haar measure on G. The two following observations play an important role in the Gelfand-Fomin construction:

1) The right action of G induces in the Hilbert space $L^2(M, \mu)$ the unitary representation by the formula

$$U_g f(x) = f(xg), \qquad x \in \Gamma\backslash G, \qquad g \in G.$$

2) The geodesic flow $\{S^t\}$ is generated by the one-parameter subgroup $g_t = \begin{pmatrix} e^t & 0 \\ 0 & e^{-t} \end{pmatrix}$, i.e. $S^t(x) = xg_t$.

One can find the spectrum of the geodesic flow by considering irreducible representations of G and defining the spectrum of $\{U_t\}$ for any irreducible representation (this was first done in [GGV]).

For the surfaces of constant negative curvature there is a close relation between the lengths of closed geodesics and the eigenvalues of hyperbolic isometries $g \in \Gamma$. Consider a geodesic γ in H which is invariant under g. Take some $q \in \gamma$ and a linear element x tangent to γ at q. Then gx and x are identified. From the geometric point of view this means that the interval along γ between x and gx is a closed geodesic. It can be shown that all closed geodesics may be obtained in such a way. Thus we see that the lengths of simple closed geodesics are equal to the logarithms of eigenvalues larger than 1 of the transformations $g \in \Gamma$.

There is a remarkable formula due to Selberg (cf [GGV]). It relates the eigenvalues of $g \in \Gamma$ to the eigenvalues of the Laplacian operator on Q. Thus there is a relation between the eigenvalues of the Laplacian operator on the surface of constant negative curvature and the lengths of closed geodesics. There is no generalization of this result to surfaces of nonconstant negative curvature. This problem seems to be one of the most interesting in the area.

We describe an example of a noncompact surface of constant negative curvature and of finite volume. Consider the *modular subgroup* Γ of $SL(2, \mathbb{R})$ which consists of integer matrices $\begin{vmatrix} a & b \\ c & d \end{vmatrix}$ with determinant 1. We shall use the theorem of Artin to describe the ergodic properties of the geodesic flow $\{S^t\}$ on $Q = H/\Gamma$.

Consider a geodesic γ in Q and one of its liftings $\widetilde{\gamma}$ to H. Denote $\widetilde{x} = \widetilde{\gamma}(-\infty)$ and $\widetilde{y} = \widetilde{\gamma}(+\infty)$. Suppose, by way of example, that $\widetilde{x} > 0$ and $\widetilde{y} < 0$ and let $\widetilde{x} = [\widetilde{n}_1, \widetilde{n}_2, \ldots]$ and $-\widetilde{y} = [\widetilde{m}_1, \widetilde{m}_2, \ldots]$ be the continued fraction expansions of $\widetilde{x}$ and $\widetilde{y}$ with $\widetilde{n}_i > 0$ and $\widetilde{m}_i > 0$. Let $\widehat{\gamma}$ be another lifting of γ in H and $\widehat{x} = \widehat{\gamma}(-\infty)$, $\widehat{y} = \widehat{\gamma}(+\infty)$. Then $\widehat{\gamma} = g\widetilde{\gamma}$ for some $g \in \Gamma$. Let $\widehat{x} = [\widehat{n}_1, \widehat{n}_2, \ldots]$ and $-\widehat{y} = [\widehat{m}_1, \widehat{m}_2, \ldots]$ be the corresponding continued fraction expansions (again we suppose that $\widehat{x} > 0$ and $\widehat{y} < 0$). The Artin Theorem says that $(\widetilde{x}, \widetilde{y})$ and $(\widehat{x}, \widehat{y})$ define the same geodesic in M if and only if $\sigma^k(\ldots, \widetilde{m}_2, \widetilde{m}_1, \widetilde{n}_1, \widetilde{n}_2, \ldots) = (\ldots, \widehat{m}_2, \widehat{m}_1, \widehat{n}_1, \widehat{n}_2, \ldots)$ for some integer k, where σ is the shift in the space Σ of two-sided infinite sequences of positive integers. Thus, we obtain a *coding* map $\psi : SQ \to \Sigma$.

One can show that the ergodicity of $\{S^t\}$ with respect to the Riemannian volume μ on SQ is equivalent to the ergodicity of σ with respect to the measure $\psi_*\mu$. This, in turn, is equivalent to the ergodicity of σ in the space Σ_+ of one-sided infinite sequences of positive integers (note that Σ is the natural extension of Σ_+) with respect to the measure ν which is the projection of $\psi_*\mu$.

In order to study the ergodicity of σ let us consider the *Gauss map* T of the unit interval $[0, 1]$ into itself given by

$$T(x) = \frac{1}{x} - \left[\frac{1}{x}\right],$$

where $[y]$ is the integer part of y. It is easy to see that if $x = [n_1, n_2, \ldots]$, then $T(x) = [n_2, n_3, \ldots]$ so that T is conjugate to the shift σ in Σ_+. Let χ be the corresponding conjugacy map. The measure $\chi_*^{-1}\nu$ coincides with the Gauss measure on $[0, 1]$ with the density $\frac{dx}{\log 2(1+x)}$ which is T-invariant. The ergodicity of Gauss measure with respect to T is well-known (cf [KSF]). A generalization of the above construction for some other discrete groups is given in [S].

§6. Dimension-like Characteristics of Invariant Sets for Dynamical Systems

6.1. Introductory Remarks. As we mentioned in Sect. 2, the topological structure of hyperbolic sets (including strange attractors) may be rather complicated and resembles in a sense, the classical Cantor set. This allows to describe the geometric structure of a hyperbolic set using certain characteristics of dimension type. In many cases understanding of dynamics need not require detailed description of (usually subtle) topological behavior of its individual trajectories but instead may enjoy the study of only "typical" trajectories with respect to an ergodic measure, i.e., the study of the ergodic properties of the system.

The dimension-like characteristics occupy, so to say, an intermediate place between topological and ergodic characteristics, and are closely related to them. In recent physics literature many dimension-like characteristics of invariant sets[33] for dynamical systems have been introduced. They turn out to be effective in the numerical study of dynamical systems. The rigorous mathematical study of these characteristics has recently shaped a new area in the theory of dynamical systems which we briefly outline in this section.

6.2. Hausdorff Dimension. Let X be a complete metric space and $\mathcal{F}$ the set of all open sets in X. Fix $\alpha \geq 0$. The α-Hausdorff measure of a subset $Y \subset X$ is defined as

$$m_H(Y, \alpha) = \lim_{\varepsilon \to 0} \inf_{\mathcal{G} \subset \mathcal{F}} \left\{ \sum_{S \in \mathcal{G}} (\operatorname{diam} S)^\alpha : \bigcup_{S \in \mathcal{G}} S \supset Y, \quad \operatorname{diam} S \leq \varepsilon \right\}, \quad (7.35)$$

[33] these invariant sets are not hyperbolic but have much in common with them; cf [FOY].

where $\mathcal{G}$ is a finite or countable subset of $\mathcal{F}$. It is easy to see that the above limit exists. For a fixed α, $m_H(\cdot, \alpha)$ is a regular, σ-additive Borel outer measure (cf [Bil]). For a fixed Y, the function $m_H(Y, \cdot)$ has the following property: there exists $\alpha_0 \geq 0$ such that $m_H(Y, \alpha) = \infty$ for $\alpha < \alpha_0$ and $m_H(Y, \alpha) = 0$ for $\alpha > \alpha_0$. The *Hausdorff dimension* of Y is defined by

$$\dim_H Y = \alpha_0 = \inf\{\alpha : m_H(Y, \alpha) = 0\} = \sup\{\alpha : m_H(Y, \alpha) = \infty\}.$$

We state some basic properties of the Hausdorff dimension:

(1) $\dim_H \left(\bigcup_i Y_i\right) = \sup_i \dim_H Y_i$ for any countable union of subsets $Y_i \subset X$;
(2) $\dim_H(Y \times Z) \geqslant \dim_H Y + \dim_H Z$ for any Borel (in particular, for any compact) subsets $Y, Z \subset X$; the strict inequality can occur (there exists a subset $Y \subset \mathbb{R}^2$ for which $\dim_H(Y \times Y) > 2 \dim_H Y$);
(3) if Y is a Borel subset of X and B an open set in $\mathbb{R}^n$, then $\dim_H(Y \times B) = \dim_H Y + n$.

We describe some results concerning the calculation and the estimates of Hausdorff dimension for invariant sets of dynamical systems. Let Λ be an invariant subset of a C^1-diffeomorphism S of a compact Riemannian manifold M. Denote by $M_S(\Lambda)$ the set of all S-invariant Borel probability ergodic measures on Λ. For $\mu \in M_S(\Lambda)$, consider the set G_μ which consists of points $x \in \Lambda$ satisfying

$$\frac{1}{n}\sum_{k=0}^{n-1} \varphi(S^k(x)) \to \int_\Lambda \varphi \, d\mu$$

for any continuous function φ on Λ. An equivalent definition is the following: G_μ consists of those points $x \in \Lambda$ for which the sequence of measures

$$\frac{1}{n}\sum_{k=0}^{n-1} \delta(S^k(x))$$

($\delta(x)$ is the point measure at x) weakly converges to μ. Consider the ordered characteristic Lyapunov exponents $\chi^1(x) \geq \cdots \geq \chi^n(x)$ ($n = \dim M$) at x. The functions $\chi^i(x)$ are S-invariant; thus for $\mu \in M_S(\Lambda)$ and μ-almost all $x \in M$ we have $\chi^i_\mu \overset{\text{def}}{=} \chi^i(x) = \text{const}$, where $\chi^1_\mu \geq \cdots \geq \chi^n_\mu$. Set

$$\alpha(\mu) = \begin{cases} 0, & \text{if } \chi^1_\mu < 0 \\ \sup\left\{\alpha : 0 \leq \alpha \leq n, \ \sum_{i=1}^{[\alpha]} \chi^i_\mu + (\alpha - [\alpha])\chi^{[\alpha]+1}_\mu \geq 0\right\}, & \text{if } \chi^1_\mu \geq 0. \end{cases}$$

Defined in this way, $\alpha(\mu)$ is called the *Lyapunov dimension* of μ. A modified version of a hypothesis by Mori [Mor] may be stated as follows: $\dim_H \Lambda = \alpha(\mu)$ for some measure $\mu \in M_S(\Lambda)$. This hypothesis was supported in several physics papers (for example, see [FOY]). A modification of this hypothesis is proved in the two dimensional case (cf [Y2]). In the multidimensional case similar results hold if, instead of the Hausdorff dimension, we consider another dimension-like characteristic (cf Sect. 6.3 below). Apparently, there is no formula which relates the Hausdorff dimension only with the characteristic Lyapunov exponents. Therefore, the following estimate of Hausdorff dimension obtained by Ledrappier in [L2] is very interesting.

Theorem 6.1.

$$\dim_H \Lambda \leqslant \sup_{\mu \in M_S(\Lambda)} \alpha(\mu).$$

If Λ is a LMHS of a $C^{1+\alpha}$-diffeomorphism S, then it is possible to obtain more exact results. Fix $x \in \Lambda$ and a small neighborhood $U(x)$ of x in M. Since Λ is locally maximal we may assume that $\Lambda \cap U(x) = (\Lambda \cap U_1) \times (\Lambda \cap U_2)$, where $U_1 \subset V^s(x)$ and $U_2 \subset V^u(x)$ are open sets. It turns out that in the case $\dim M = 2$, the Hausdorff dimension of Λ equals the Hausdorff dimension of $\Lambda \cap U(x)$ (which does not depend on the choice of $x \in \Lambda$ and of $U(x)$). The latter equals the sum of the Hausdorff dimensions of $\Lambda \cap U_1$ and $\Lambda \cap U_2$ (which also do not depend on x). A similar situation arises if instead of Λ we consider the set G_μ for some $\mu \in M_S(\Lambda)$ (recall that $\mu(G_\mu) = 1$). We now make these ideas precise.

Consider the function $\psi^u(t) = P(t\varphi^u)$, where $\varphi^u(x)$ is defined by (7.24) and $P(t\varphi^u)$ is the topological pressure of the function $t\varphi^u(x)$ on Λ. It is easy to see that $\psi^u(0) = P(0) = h(S) \geq 0$. It can also be shown that $\psi^u(t)$ is strictly decreasing (cf [W1]). The results of Bowen [Bo2] imply $\psi^u(1) \leqslant 0$. Therefore, there exists a unique root t^u of the equation

$$P(t\varphi^u) = 0.$$

This equation was first introduced by Bowen (cf [Bo3]) and is known now as *Bowen's equation*. Similarly considering the function $\psi^s(t) = P(t\varphi^s)$, one can prove the existence of a unique root t^s of Bowen's equation $P(t\varphi^s) = 0$.

Theorem 6.2. *(cf* [Ma1]*). Suppose that* $\dim M = 2$, Λ *is a LMHS for a* $C^{1+\alpha}$*-diffeomorphism* S *and* $S|\Lambda$ *is topologically transitive. Then for any* $\mu \in M_S(\Lambda)$ *and* $x \in \Lambda$

$$\begin{gathered}\dim_H G_\mu = \dim_H(G_\mu \cap V^u(x)) + \dim_H(G_\mu \cap V^s(x)),\\ \dim_H(G_\mu \cap V^s(x)) = h_\mu/\chi^1_\mu,\\ \dim_H(G_\mu \cap V^u(x)) = h_\mu/|\chi^2_\mu|,^{34}\end{gathered} \tag{7.36}$$

where h_μ *is the metric entropy of* S.

We sketch the proof of (7.36). Let ξ be a Markov partition of Λ and $\xi_n = \bigvee_{k=0}^{n-1} S^k\xi$. It follows from the Shannon-McMillan-Breiman theorem (see Chap. 3, Sect. 3) and from the properties of characteristic Lyapunov

[34] As $\dim M = 2$ and Λ is a hyperbolic set, we have according to our notation $\chi^1_\mu > 0 > \chi^2_\mu$.

exponents that for any $\varepsilon > 0$ there exists a set $G_\mu^\varepsilon \subset G_\mu$ satisfying the following conditions: 1) $\mu(G_\mu^\varepsilon) \geq 1 - \varepsilon$; 2) the number of elements of ξ_n which cover the set $V^u(x) \cap G_\mu^\varepsilon$ is proportional to $\exp(h_\mu \pm \varepsilon)n$; 3) the diameter of any element of ξ_n intersecting G_μ^ε is proportional to $\exp(\chi_\mu^2 \pm \varepsilon)n$. One can then show that ξ_n is the "best" cover, i.e. it almost realizes the infinum in (7.35). For this reason $\dim_H(V^u(x) \cap G_\mu^\varepsilon) \approx (h_\mu \pm \varepsilon)/(|\chi_\mu^2| \pm \varepsilon)$. Now take the limit as $\varepsilon \to 0$.

Theorem 6.3. *(cf* [Ma1]*). Under the conditions of Theorem 6.2 for any* $x \in \Lambda$ *the following hold:*

(1) $\dim_H \Lambda = \dim_H(\Lambda \cap V^u(x)) + \dim_H(\Lambda \cap V^s(x))$;

(2) $\dim_H(\Lambda \cap V^u(x)) = t^u = \sup_{\mu \in M_S(\Lambda)}(G_\mu \cap V^u(x)) = \dim_H(G_{\mu_1} \cap V^u(x))$, *where* μ_1 *is the Gibbs measure corresponding to the function* $t^u \varphi^u$;

(3) $\dim_H(\Lambda \cap V^s(x)) = t^s = \sup_{\mu \in M_S(\Lambda)}(G_\mu \cap V^s(x)) = \dim_H(G_{\mu_2} \cap V^s(x))$, *where* μ_2 *is the Gibbs measure corresponding to the function* $t^s \varphi^s$.

Let Λ be a hyperbolic attractor. Then $V^u(x) \subset \Lambda$ for any $x \in \Lambda$; thus $\dim_H(V^u(x) \cap \Lambda) = 1$. Consequently, $\dim_H \Lambda = 1 + t^s$. In addition, if μ is the Sinai-Ruelle-Bowen measure on Λ, then $\dim_H(G_\mu \cap V^s(x)) = \chi_\mu^1/|\chi_\mu^2|$ (see Section 3) which implies $\dim_H G_\mu = 1 + \chi_\mu^1/|\chi_\mu^2|$.

6.3. Other Dimension Characteristics. Let X be a complete metric space, $Y \subset X$. We denote by $N(\varepsilon)$ the minimal number of balls of radius ε needed to cover Y. The *upper* (respectively, *lower*) *box dimension* of Y is defined by

$$\overline{C}(Y) = \overline{\lim_{\varepsilon \to 0}} \frac{\log N(\varepsilon)}{\log(1/\varepsilon)} \left(\text{respectively, } \underline{C}(Y) = \underline{\lim_{\varepsilon \to 0}} \frac{\log N(\varepsilon)}{\log(1/\varepsilon)}\right).$$

It is not difficult to show that $\dim_H Y \leqslant \underline{C}(Y) \leqslant \overline{C}(Y)$. Let μ be a Borel measure on X. Then

$$\begin{aligned} \dim_H \mu &= \inf\{\dim_H Y : Y \subset X,\ \mu(Y) = 1\}, \\ \overline{C}(\mu) &= \underline{\lim_{\delta \to 0}} \inf\{\overline{C}(Y) : Y \subset X,\ \mu(Y) \geq 1 - \delta\}, \\ \underline{C}(\mu) &= \underline{\lim_{\delta \to 0}} \inf\{\underline{C}(Y) : Y \subset X,\ \mu(Y) \geq 1 - \delta\} \end{aligned}$$

are called respectively the *dimension of the measure*, the *upper* and the *lower box dimension* of the measure.

Given ε and $\delta > 0$ we use $N(\varepsilon, \delta)$ to denote the minimal number of balls of radius ε needed to cover a set of μ-measure $\geq 1 - \delta$. Then

$$\underline{C}_L(\mu) = \overline{\lim_{\delta \to 0}} \underline{\lim_{\varepsilon \to 0}} \frac{\log N(\varepsilon, \delta)}{\log(1/\varepsilon)}, \quad \overline{C}_L(\mu) = \overline{\lim_{\delta \to 0}}\, \overline{\lim_{\varepsilon \to 0}} \frac{\log N(\varepsilon, \delta)}{\log(1/\varepsilon)},$$

are called respectively the *lower* and the *upper Ledrappier box dimension* of μ. It is easy to see that

$$\dim_H \mu \le \underline{C}(\mu) \le \overline{C}(\mu), \quad \overline{C}_L(\mu) \le \overline{C}(\mu), \quad \underline{C}_L(\mu) \le \overline{C}_L(\mu).$$

One can also show that $\dim_H \mu \le \underline{C}_L(\mu)$.

Let ξ be a measurable partition of X. Fix $\varepsilon > 0$ and set

$$H_\mu(\varepsilon) = \inf\{H_\mu(\xi) : \operatorname{diam} \xi \le \varepsilon\}$$

($H_\mu(\xi)$ is the entropy of ξ; see Chap. 3, Sect. 1). Then

$$\overline{R}(\mu) = \overline{\lim_{\varepsilon\to 0}} \frac{H_\mu(\varepsilon)}{\log(1/\varepsilon)}, \qquad \underline{R}(\mu) = \underline{\lim}_{\varepsilon\to 0} \frac{H_\mu(\varepsilon)}{\log(1/\varepsilon)}$$

are called respectively the *upper* and the *lower information dimension* of μ (or the *upper* and the *lower Renyi dimension* of μ).

Theorem 6.4. *(cf [Y2]). Let μ be a Borel probability measure on a complete metric space X. Suppose that for μ-almost every x*

$$d_\mu \stackrel{\text{def}}{=} \lim_{\rho\to 0} \frac{\log \mu(B(x,\rho))}{\log \rho} = \alpha. \tag{7.37}$$

($B(x,\rho)$ is the ball centered at x of radius ρ). Then

$$\dim_H \mu = \overline{C}(\mu) = \underline{C}(\mu) = \overline{C}_L(\mu) = \underline{C}_L(\mu) = \overline{R}(\mu) = \underline{R}(\mu) = \alpha.$$

Measures for which the limit (7.37) exists almost everywhere and is constant are called *exact dimensional*. They constitute a class of measures for which virtually all the known characteristics of dimension type coincide. This determines their crucial role in applications of dimension theory to dynamical systems.

In the mid 80th Eckmann and Ruelle [ER] conjectured that hyperbolic measures are exact dimensional. This statement has since become known as the Eckmann-Ruelle conjecture. It was partly based on the following result in two dimensions by Young.

Theorem 6.5. *(cf [Y2]). Let S be a $C^{1+\alpha}$-diffeomorphism of a smooth compact surface and μ a Borel ergodic hyperbolic measure with characteristic Lyapunov exponents $\chi_\mu^1 > 0 > \chi_\mu^2$ (see Sect. 3). Then*

$$d_\mu = h_\mu(S)\left(\frac{1}{\chi_\mu^1} - \frac{1}{\chi_\mu^2}\right).$$

Ledrappier proved the Eckmann-Ruelle conjecture for hyperbolic SRB-measures and Pesin and Yue extended his approach to arbitrary Gibbs measures (cf [Pe4]). The argument is based upon local product structure of Gibbs measures. Barreira, Pesin, and Schmeling showed that any hyperbolic measure has "nearly" local product structure. Using this result they obtained a complete proof of the Eckmann-Ruelle conjecture.

Theorem 6.6. *(cf* [BPS]*). Let S be a $C^{1+\alpha}$-diffeomorphism of a smooth compact Riemannian manifold and μ a Borel ergodic hyperbolic measure. Then μ is exact dimensional.*

Neither of the assumptions in the Eckmann–Ruelle conjecture can be omitted. Ledrappier and Misiurewicz constructed an example of a smooth map of an interval preserving an ergodic measure with zero Lyapunov exponent which is not exact dimensional and Pesin and Weiss presented an example of a Hölder homeomorphism whose measure of maximal entropy is not exact dimensional (cf [Pe4]).

One can use the dimension-like characteristics introduced above to describe the geometric structure of invariant sets for dynamical systems. In some cases one can compute these characteristics precisely or obtain some sharp estimates.

Theorem 6.7. *(cf* [Y1]*). Let Λ be an invariant set of $C^{1+\alpha}$ diffeomorphism S of a smooth compact manifold M, μ a Sinai-Ruelle-Bowen measure (see 3.9), and $\chi_\mu^1 \geq \ldots \geq \chi_\mu^k > 0 > \chi_\mu^{k+1} \geq \ldots \geq \chi_\mu^k$ the corresponding characteristic Lyapunov exponents. Then*

$$\overline{C}(\operatorname{supp}\mu) \geq k + (\chi_\mu^1 + \cdots + \chi_\mu^k)/|\chi_\mu^{k+1}| = A .$$

Corollary. *(cf* [Y1]*). If Λ is a hyperbolic attractor of a $C^{1+\alpha}$ diffeomorphism S, $S|\Lambda$ is topologically transitive, and μ is the Sinai-Ruelle-Bowen measure on Λ, then $\underline{\overline{C}}(\Lambda) \geq A$.*

Consider a hyperbolic periodic point p of a C^2-diffeomorphism S of a smooth surface M, and let x be a transversal homoclinic point corresponding to p (see Sect. 2). Fix some small $\varepsilon > 0$ and consider a LMHS Λ_ε which is contained in the ε-neighborhood of the trajectory $\{S^n(x)\}$ (see Theorem 2.4). Let λ and γ be the eigenvalues of $DS(p)$, $0 < \lambda < 1 < \gamma$.

Theorem 6.8. *(cf* [AP]*). There exist constants $C_1 > 0$ and $C_2 > 0$ such that*

$$\beta \leq \dim_H(\Lambda_\varepsilon \cap V^u(x)) \leq \alpha,$$

where $\alpha = \alpha(\varepsilon)$ and $\beta = \beta(\varepsilon)$ coincide with the unique roots of the equations

$$\alpha \ln \frac{C_1\gamma}{\varepsilon} = -\ln(\gamma^\alpha - 1), \qquad \beta \ln \frac{C_2\gamma}{\varepsilon} = -\ln(\gamma^\beta - 1).$$

Corollary. *As $\varepsilon \to 0$ the function $\varphi(\varepsilon) = \dim_H(\Lambda_\varepsilon \cap V^u(x))$ behaves asymptotically as*

$$\varphi(\varepsilon) = \ln\ln\frac{1}{\varepsilon}\Big/\ln\frac{1}{\varepsilon} - \ln\ln\ln\frac{1}{\varepsilon}\cdot\ln\gamma\Big/\ln\frac{1}{\varepsilon}.$$

A similar result (with γ replaced by λ^{-1}) holds for $\dim_H(\Lambda_\varepsilon \cap V^s(x))$.

6.4. Carathéodory Dimension Structure. Carathéodory Dimension Characteristics. We describe a construction which is a generalization of the classical Carathéodory construction in the general measure theory. It was elaborated by Pesin in [P6] (cf also [Pe4]) to produce various characteristics of dimension type.

The starting point for the construction is a space X which is endowed with a special structure. We introduce this structure axiomatically by describing its basic elements and relations between them. It is called the Carathéodory dimension structure (or, briefly, C-structure). This structure enables one to yield the Carathéodory dimension characteristics of subsets of X and measures on X. C-structures can be generated by other structures on X associated with metrics, measures, functions, or dynamical systems acting on X.

Let us point out that Carathéodory dimension characteristics are invariants of an isomorphism which preserves the C-structure. In the case when the C-structure is generated by a dynamical system on X, the Carathéodory dimension characteristics become invariants of the dynamics and can be used to characterize invariant sets and invariant measures. There are deep relations between them and other important invariants of dynamics such as entropy (which characterizes the complexity of the dynamics) and Lyapunov exponents (which characterize the instability of trajectories).

We now proceed with the formal description. Let X be a set and $\mathscr{F}$ a collection of subsets of X. Assume that there exist two set functions $\eta, \psi : \mathscr{F} \to \mathbb{R}^+$ satisfying the following conditions:

A1. $\emptyset \in \mathscr{F}$; $\eta(\emptyset) = 0$ and $\psi(\emptyset) = 0$; $\eta(U) > 0$ and $\psi(U) > 0$ for any $U \in \mathscr{F}$, $U \neq \emptyset$;

A2. for any $\delta > 0$ one can find $\varepsilon > 0$ such that $\eta(U) \leqslant \delta$ for any $U \in \mathscr{F}$ with $\psi(U) \leqslant \varepsilon$.

A3. for any $\varepsilon > 0$ there exists a finite or countable subcollection $\mathscr{G} \subset \mathscr{F}$ which covers X (i.e., $\bigcup_{U \in \mathscr{G}} U \supset X$) and $\psi(\mathscr{G}) \stackrel{\text{def}}{=} \sup\{\psi(U) : U \in \mathscr{G}\} \leqslant \varepsilon$.

Let $\xi : \mathscr{F} \to \mathbb{R}^+$ be a set function. We say that the collection of subsets $\mathscr{F}$ and the set functions ξ, η, ψ, satisfying Conditions $A1$, $A2$, and $A3$, introduce the *Carathéodory dimension structure* or *C-structure* τ on X and we write $\tau = (\mathscr{F}, \xi, \eta, \psi)$.

Given a set $Z \subset X$ and numbers $\alpha \in \mathbb{R}$, $\varepsilon > 0$, we define

$$M_C(Z, \alpha, \varepsilon) = \inf_{\mathscr{G}}\{\sum_{U \in \mathscr{G}} \xi(U)\eta(U)^\alpha\},$$

where the infimum is taken over all finite or countable subcollections $\mathscr{G} \subset \mathscr{F}$ covering Z with $\psi(\mathscr{G}) \leqslant \varepsilon$. By Condition $A3$ the function $M_C(Z, \alpha, \varepsilon)$ is correctly defined. It is non-decreasing as ε decreases. Therefore, there exists the limit

$$m_C(Z, \alpha) = \lim_{\varepsilon \to 0} M_C(Z, \alpha, \varepsilon).$$

The function $m_C(Z,\alpha)$ has the following properties: for any $\alpha \in \mathbb{R}$

(1) $m_C(\emptyset,\alpha)=0$ for $\alpha>0$;
(2) $m_C(Z_1,\alpha)\leqslant m_C(Z_2,\alpha)$ if $Z_1\subset Z_2\subset X$;
(3) $m_C(\bigcup_{i\geqslant 0} Z_i,\alpha)\leqslant \sum_{i\geqslant 0} m_C(Z_i,\alpha)$, where $Z_i\subset X, i=0,1,2,\dots$.

If $m_C(\emptyset,\alpha)=0$ (this holds true for $\alpha>0$ but can also happen for some negative α) the set function $m_C(\cdot,\alpha)$ becomes an outer measure on X. In this case we call it the *Carathéodory outer measure* (specified by the collection of subsets $\mathscr{F}$ and the set functions ξ,η,ψ). According to the general measure theory the outer measure induces a σ-additive measure on the σ-field of measurable sets in X which is called the *Carathéodory measure*. Note that this measure is not necessarily σ-finite.

Consider the function $m_C(Z,\cdot)$ for a fixed set Z. There exists a critical value α_C, $-\infty\leqslant\alpha_C\leqslant+\infty$ such that $m_C(Z,\alpha)=\infty$ for $\alpha<\alpha_C$ and $m_C(Z,\alpha)=0$ for $\alpha>\alpha_C$. We define the *Carathéodory dimension of a set* $Z\subset X$ by

$$\dim_C Z=\alpha_C=\inf\{\alpha: m_C(Z,\alpha)=0\}=\sup\{\alpha: m_C(Z,\alpha)=\infty\}.$$

The Carathéodory dimension clearly depends on the choice of C-structure $\tau=(\mathscr{F},\xi,\eta,\psi)$ on X. Its basic properties are as follows:

(1) $\dim_C\emptyset\leqslant 0$;
(2) $\dim_C Z_1\leqslant\dim_C Z_2$ if $Z_1\subset Z_2\subset X$;
(3) $\dim_C(\bigcup_{i\geqslant 0} Z_i)=\sup_{i\geqslant 0}\dim_C Z_i$, where $Z_i\subset X$, $i=0,1,2,\dots$.

One can show that the Carathéodory dimension is invariant under an isomorphism which preserves the C-structure.

We now modify the above construction to produce another type of Carathéodory dimension characteristics. We shall assume that the following Condition $A3'$ holds which is stronger than Condition $A3$:

A3′. There exists $\epsilon>0$ such that for any $\epsilon\geqslant\varepsilon>0$, one can find a finite subcollection $\mathscr{G}\subset\mathscr{F}$ covering X such that $\psi(U)=\varepsilon$ for any $U\in\mathscr{G}$.

Given $\alpha\in\mathbb{R}$, $\varepsilon>0$, and a set $Z\subset X$, define

$$R_C(Z,\alpha,\varepsilon)=\inf_{\mathscr{G}}\left\{\sum_{U\in\mathscr{G}}\xi(U)\eta(U)^\alpha\right\},$$

where the infimum is taken over all finite or countable subcollections $\mathscr{G}\subset\mathscr{F}$ covering Z such that $\psi(U)=\varepsilon$ for any $U\in\mathscr{G}$. According to $A3'$, $R_C(Z,\alpha,\varepsilon)$ is correctly defined. We set

$$\underline{r}_C(Z,\alpha)=\varliminf_{\varepsilon\to 0} R_C(Z,\alpha,\varepsilon),\quad \overline{r}_C(Z,\alpha)=\varlimsup_{\varepsilon\to 0} R_C(Z,\alpha,\varepsilon).$$

There exist two critical values $\underline{\alpha}_C$ and $\overline{\alpha}_C\in\mathbb{R}$ such that

(1) $\underline{r}_C(Z,\alpha) = \infty$ for $\alpha < \underline{\alpha}_C$ and $\underline{r}_C(Z,\alpha) = 0$ for $\alpha > \underline{\alpha}_C$ (while $\underline{r}_C(Z,\alpha_C)$ can be 0, ∞, or a finite positive number);
(2) $\overline{r}_C(Z,\alpha) = \infty$ for $\alpha < \overline{\alpha}_C$ and $\overline{r}_C(Z,\alpha) = 0$ for $\alpha > \overline{\alpha}_C$ (while $\overline{r}_C(Z,\alpha_C)$ can be 0, ∞, or a finite positive number).

Given a subset $Z \subset X$, we define the *lower* and *upper Carathéodory capacities of the set* Z by

$$\underline{\mathrm{Cap}}_C Z = \underline{\alpha}_C = \inf\{\alpha : \underline{r}_C(Z,\alpha) = 0\} = \sup\{\alpha : \underline{r}_C(Z,\alpha) = \infty\},$$
$$\overline{\mathrm{Cap}}_C Z = \overline{\alpha}_C = \inf\{\alpha : \overline{r}_C(Z,\alpha) = 0\} = \sup\{\alpha : \underline{r}_C(Z,\alpha) = \infty\}.$$

Their basic properties are as follows:

(1) $\dim_C Z \leqslant \underline{\mathrm{Cap}}_C Z \leqslant \overline{\mathrm{Cap}}_C Z$ for any $Z \subset X$;
(2) $\underline{\mathrm{Cap}}_C Z_1 \leqslant \underline{\mathrm{Cap}}_C Z_2$ and $\overline{\mathrm{Cap}}_C Z_1 \leqslant \overline{\mathrm{Cap}}_C Z_2$ for any $Z_1 \subset Z_2 \subset X$;
(3) for any sets $Z_i \subset X$, $i = 1, 2, \dots$

$$\underline{\mathrm{Cap}}_C\Big(\bigcup_{i\geqslant 0} Z_i\Big) \geqslant \sup_{i\geqslant 0} \underline{\mathrm{Cap}}_C Z_i, \quad \overline{\mathrm{Cap}}_C\Big(\bigcup_{i\geqslant 0} Z_i\Big) \geqslant \sup_{i\geqslant 0} \overline{\mathrm{Cap}}_C Z_i.$$

For any $\varepsilon > 0$ and any set $Z \subset X$ let us set

$$\Lambda(Z,\varepsilon) = \inf_{\mathcal{G}} \left\{ \sum_{U\in\mathcal{G}} \xi(U) \right\},$$

where the infimum is taken over all finite or countable subcollections $\mathcal{G} \subset \mathcal{F}$ covering Z for which $\psi(U) = \varepsilon$ for all $U \in \mathcal{G}$.

Let us assume that the set function η satisfies the following condition:

A4. $\eta(U_1) = \eta(U_2)$ for any $U_1, U_2 \in \mathcal{F}$ for which $\psi(U_1) = \psi(U_2)$.

Provided this condition holds, the function $\eta(\varepsilon) = \eta(U)$ if $\psi(U) = \varepsilon$ is correctly defined. Then the lower and upper Carathéodory capacities admit the following description:

$$\underline{\mathrm{Cap}}_C Z = \varliminf_{\varepsilon\to 0} \frac{\log \Lambda(Z,\varepsilon)}{\log(1/\eta(\varepsilon))}, \quad \overline{\mathrm{Cap}}_C Z = \varlimsup_{\varepsilon\to 0} \frac{\log \Lambda(Z,\varepsilon)}{\log(1/\eta(\varepsilon))}.$$

6.5. Examples of C-structures and Carathéodory Dimension Characteristics. We now provide some examples of C-structures and Carathéodory dimension characteristics generated by them.

1. Hausdorff dimension and box dimension. We begin with a C-structure on the Euclidean space $\mathbb{R}^m$ generated by a metric ρ which is equivalent to the standard metric. Let $\mathcal{F}$ be the collection of all **open** subsets of $\mathbb{R}^m$. For $U \in \mathcal{F}$ define

$$\xi(U) = 1, \quad \eta(U) = \psi(U) = \operatorname{diam} U.$$

It is easy to see that the collection of subset $\mathcal{F}$ and set functions η and ψ satisfy Conditions $A1, A2, A3, A3'$, and $A4$ and hence determine a C-structure $\tau = (\mathcal{F}, \xi, \eta, \psi)$ on $\mathbb{R}^m$. The corresponding Carathéodory set function $m_C(\cdot, \alpha)$ is the Hausdorff measure of Z.

Further, for a subset Z, the above C-structure generates the Carathéodory dimension of Z which coincide with the Hausdorff dimension of Z. Moreover, the lower and upper Carathéodory capacities of Z are lower and upper box dimensions of Z.

In the above definition of the C-structure τ one can choose $\mathcal{F}$ to be the collection of all **closed** subsets of $\mathbb{R}^m$ or even **all** subsets of $\mathbb{R}^m$ and obtain the same value of the Hausdorff measure. If, instead, one chooses $\mathcal{F}$ to be the collection of all open or closed *balls* in $\mathbb{R}^m$ then the value of the corresponding Hausdorff measure can change but the Hausdorff dimension and lower and upper box dimensions of Z remain the same.

2. Dimension spectra. Another example are C-structures on the Euclidean space $\mathbb{R}^m$ induced by measures and metrics on $\mathbb{R}^m$.

Let μ be a Borel finite measure on $\mathbb{R}^m$. Given numbers $q \geqslant 0$ and $\gamma > 0$, we introduce a C-structure on $\mathbb{R}^m$ generated by the metric ρ and the measure μ. Namely, let $\mathcal{F}$ be the collection of open balls in $\mathbb{R}^m$. For any ball $B(x, \varepsilon) \in \mathcal{F}$, we define

$$\xi(B(x,\varepsilon)) = \begin{cases} \mu(B(x,\gamma\varepsilon))^q & \text{if } q > 0 \\ 1 & \text{if } q = 0 \end{cases}, \quad \eta(B(x,\varepsilon)) = \psi(B(x,\varepsilon)) = \varepsilon.$$

It is easy to verify that the collection $\mathcal{F}$ and the set functions ξ, η, and ψ satisfy conditions $A1, A2, A3$, and $A3'$. Hence they define a C-structure in $\mathbb{R}^m$, $\tau_{q.\gamma} = (\mathcal{F}, \xi, \eta, \psi)$. The corresponding Carathéodory set function $m_C(Z, \alpha)$, (where $Z \subset \mathbb{R}^m$ and $\alpha \in \mathbb{R}$) is called the (q, γ)-set function. We denote it by $m_{q.\gamma}(Z, \alpha)$. If $m_{q.\gamma}(\emptyset, \alpha) = 0$ (this holds true for $\alpha > 0$ but can also happen for some negative α) the set function $m_{q.\gamma}(\cdot, \alpha)$ becomes an outer measure on $\mathbb{R}^m$. It induces a σ-additive measure on $\mathbb{R}^m$ which we call the *(q, γ)-measure*.

Further, the above C-structure produces the Carathéodory dimension of Z which is called the *(q, γ)-dimension of the set* Z and is denoted by $\dim_{q.\gamma} Z$. It generates the lower and upper Carathéodory capacities of Z which are called the *lower* and *upper (q, γ)-box dimensions of the set* Z and are denoted by $\underline{\dim}_{q.\gamma} Z$ and $\overline{\dim}_{q.\gamma} Z$ respectively.

We remark that any Lipschitz continuous homeomorphism of $\mathbb{R}^m$ with a Lipschitz continuous inverse that moves the measure μ into an equivalent measure is an isomorphism of the C-structure $\tau_{q.\gamma}$ and thus, preserves the (q, γ)-dimension and lower and upper (q, γ)-box dimensions.

Further, we define the *q-dimension of a set* Z and *lower* and *upper q-box dimensions of a set* Z by

$$\dim_q Z = \inf_{\gamma>1} \dim_{q,\gamma} Z = \lim_{\gamma\downarrow 1} \dim_{q,\gamma} Z,$$
$$\underline{\dim}_q Z = \inf_{\gamma>1} \underline{\dim}_{q,\gamma} Z = \lim_{\gamma\downarrow 1} \underline{\dim}_{q,\gamma} Z,$$
$$\overline{\dim}_q Z = \inf_{\gamma>1} \overline{\dim}_{q,\gamma} Z = \lim_{\gamma\downarrow 1} \overline{\dim}_{q,\gamma} Z.$$

There is an essential difference between the lower and upper $(q,1)$-box dimensions and the lower and upper q-box dimensions: for any $q \geqslant 0$ and any set $Z \subset \mathbb{R}^m$, we have that $\underline{\dim}_{q,1} Z \leqslant \underline{\dim}_q Z$ and $\overline{\dim}_{q,1} Z \leqslant \overline{\dim}_q Z$.

Following [Fe] we call a Borel finite measure μ *diametrically regular* if it satisfies the following condition:

there exist constants $\gamma_0 > 1$ and $C_0 > 0$ such that for any point x and any $r > 0$

$$\mu(B(x, \gamma_0 r)) \leqslant C_0 \mu(B(x, r)). \tag{7.38}$$

Such a measure is sometimes called a *Federer measure*; in harmonic analysis it is also known as a *doubling measure*. Given $\gamma > 1$, choose the least positive integer $n = n(\gamma)$ such that $\gamma\gamma_0^{-n} < 1$ then

$$\mu(B(x, \gamma r)) \leqslant C_0^n \mu(B(x, r))$$

for any point x and $r > 0$. This immediately implies that if μ is a diametrically regular measure on $\mathbb{R}^m$, then

$$\underline{\dim}_{q,\gamma} Z = \underline{\dim}_q Z = \underline{\dim}_{q,1} Z \quad \text{and} \quad \overline{\dim}_{q,\gamma} Z = \overline{\dim}_q Z = \overline{\dim}_{q,1} Z$$

for any $Z \subset \mathbb{R}^m$, $q \geqslant 0$, and $\gamma > 1$.

Although the assumption (7.38) is sufficiently strong one can show that it holds in many interesting cases: for example, for equilibrium measures (corresponding to Hölder continuous functions) for conformal expanding maps, two-dimensional axiom A diffeomorphisms, and symbolic dynamical systems. If Assumption (8.15) is violated the $(q,1)$-box dimension of a set Z may be strictly less than the q-box dimension of Z.

Using the formula (7.37) for the Carathéodory capacities one can obtain formulae for lower and upper q-box dimensions of a set Z. Consider the function

$$\varphi_q(r) = \int_X \mu(D(x, r))^{q-1}\, d\mu(x), \tag{7.39}$$

where $D(x, r)$ is the *closed* ball of radius r centered at x and X is the support of the measure μ. One can show (cf [Pe4]) that for any set Z of full measure

$$\underline{\dim}_q Z = \varliminf_{\varepsilon\to 0} \frac{\log \varphi_q(r)}{\log(1/r)}, \quad \overline{\dim}_q Z = \varlimsup_{\varepsilon\to 0} \frac{\log \varphi_q(r)}{\log(1/r)}. \tag{7.40}$$

One can easily see that $\underline{\dim}_q Z \leqslant \overline{\dim}_q Z$ for every compact set Z and every $q \geqslant 1$. On the other hand one can build an example of a measure μ for which the strict inequality occurs on an arbitrary interval in q.

We introduce the *HP-spectrum for dimensions* (after Hentschel and Procaccia; cf [HP], and also [Pe4]), specified by a measure μ on $\mathbb{R}^m$ with the support X. It is the one-parameter family of pairs of quantities $\left(\underline{HP}_q(\mu), \overline{HP}_q(\mu)\right)$, $q > 1$, where

$$\underline{HP}_q(\mu) = \frac{1}{q-1} \varliminf_{r\to 0} \frac{\log \varphi_q(r)}{\log(1/r)}, \quad \overline{HP}_q(\mu) = \frac{1}{q-1} \varlimsup_{r\to 0} \frac{\log \varphi_q(r)}{\log(1/r)}.$$

It follows directly from (7.40) that

$$\underline{HP}_q(\mu) = \frac{1}{q-1}\underline{\dim}_q X, \quad \overline{HP}_q(\mu) = \frac{1}{q-1}\overline{\dim}_q X.$$

Let S be a dynamical system acting in a domain $U \subset \mathbb{R}^m$ and μ an invariant ergodic Borel measure. For integer $q = 2, 3, \ldots$ the values $\underline{HP}_q(\mu)$ and $\overline{HP}_q(\mu)$ yield special invariants of S which play an important role in the numerical study of the orbit distribution generated by S. Grassberger, Hentschel, and Procaccia suggested an approach to obtain information related to invariant ergodic measures of a dynamical system by observing individual trajectories. The approach is based on computing correlations between q-tuples of points in the orbit distribution for $q = 2, 3, \ldots$. This led to characteristics known as correlation dimensions of order q which capture information on the global behavior of "typical" trajectories by observing a single one. These characteristics proved to be experimentally the most accessible and offered a substantial advantage over the other characteristics of dimension type that were used in the numerical study of dynamical systems with chaotic behavior.

The formal definition of the correlation dimension is as follows. Let $S: \mathbb{R}^m \to \mathbb{R}^m$ be a continuous map. Given $x \in \mathbb{R}^m$, $n > 0$, $r > 0$, and an integer $q \geqslant 2$, define the *correlation sum of order* q as

$$C_q(x, n, r) = \frac{1}{n^q}\operatorname{card}\Big\{(i_1 \ldots i_q) \in \{0, 1, \ldots, n\}^q : \rho(S^{i_j}(x),\ S^{i_k}(x)) \leqslant r \quad \text{for any } 0 \leqslant j, k \leqslant q\Big\},$$

where ρ is the metric in $\mathbb{R}^m$ and $\operatorname{card}\{A\}$ denotes the cardinality of the set A. We call the quantities

$$\underline{\alpha}_q(x) = \frac{1}{q-1} \varliminf_{r\to 0} \lim_{n\to\infty} \frac{\log C_q(x, n, r)}{\log(1/r)},$$
$$\overline{\alpha}_q(x) = \frac{1}{q-1} \varlimsup_{r\to 0} \lim_{n\to\infty} \frac{\log C_q(x, n, r)}{\log(1/r)}$$

the *lower* and *upper correlation dimensions of order* q *at the point* x respectively. We shall first discuss the existence of the limit as $n \to \infty$. Consider the function $\varphi_q(r)$ (see (7.39)). It is clearly non-decreasing in r and hence may have only a finite or countable set of discontinuity points. Thus, it is right-continuous; it is continuous at r if and only if $\mu(S(x,r)) = 0$ for μ-almost every $x \in X$ ($S(x,r)$ is the sphere of radius r centered at x). In [PT], Pesin and Tempelman proved the following result.

Theorem 6.10. *For any integer $q \geqslant 2$ there exists a set $Z \subset X$ of full measure such that for any $\varepsilon > 0$, $R > 0$, and any $x \in Z$ one can find a positive integer $N = N(x, \varepsilon, R)$ for which the inequality*

$$|C(x,n,r) - \varphi_q(r)| \leqslant \varepsilon$$

holds for every $n \geqslant N$ and $0 < r \leqslant R$. In other words, $C(x,n,r)$ tends to $\varphi_q(r)$ as $n \to \infty$ for μ-almost every $x \in X$ uniformly over r for $0 < r \leqslant R$.

We stress that the lower and upper q-correlation dimensions do not depend either on the dynamical system f or on the point x for μ-almost every x (provided μ is ergodic). Instead, they are completely specified by the measure μ.

There is another approach to the description of orbit distributions which was suggested by Tél. Consider a hypercube K in $\mathbb{R}^m$ and a cover of K by a uniform grid of mesh size r. Let p_i be an average number of visits of a "typical" orbit to a given box B_i of a grid. We define *Rényi spectrum for dimensions* as the one-parameter family of pairs of quantities $(\underline{R}_q(\mu), \overline{R}_q(\mu))$, $q > 1$

$$\underline{R}_q(\mu) = \frac{1}{q-1} \varliminf_{r\to\infty} \frac{\log \sum_{i=1}^N p_i^q}{\log(1/r)}, \quad \overline{R}_q(\mu) = \frac{1}{q-1} \varlimsup_{r\to\infty} \frac{\log \sum_{i=1}^N p_i^q}{\log(1/r)},$$

where $N = N(r)$ is the total number of boxes B_i of the grid with $p_i > 0$. One can show (see [Pe4]) that for any $q > 1$,

$$\underline{R}_q(\mu) = \underline{HP}_q(\mu), \quad \overline{R}_q(\mu) = \overline{HP}_q(\mu).$$

The collection of numbers $\{p_i\}$ determines the orbit distribution corresponding to the scale level r (recall that r is the mesh size of the grid). Define *scaling exponents* α_i by $p_i = r^{\alpha_i}$. In the seminal paper [HJKPS], the authors suggested using the limit distribution of numbers α_i as $r \to 0$ as a quantitative characteristic of orbit distributions. More precisely, given a point $x \in X$, define the *pointwise dimension of the measure μ at x* by

$$d_\mu(x) = \lim_{x\to 0} \frac{\log(\mu(D(x,r))}{\log r}$$

(assuming that the limit exists). It generates the *multifractal structure* of X as follows

$$X = \hat{X} \cup \left(\bigcup_\alpha X_\alpha \right).$$

Here X_α is the set of points where the pointwise dimension takes on the value α (the *level set*) and $\hat{X}$ is the set of points with *no* pointwise dimension (the *irregular part*). The above decomposition can be characterized by the *dimension spectrum for pointwise dimensions* of the measure μ or $f_\mu(\alpha)$*-spectrum (for dimensions)*, where $f_\mu(\alpha) = \dim_H X_\alpha$. The $f_\mu(\alpha)$-spectrum provides a description of the fine-scale geometry of the set X (more precisely, the part of it where the measure μ is concentrated) whose constituent components are the sets X_α. The study of the $f_\mu(\alpha)$-spectrum is a main part of the *multifractal analysis* of dynamical systems. Another crucial component is to establish the *Legendre transform relation* between $f_\mu(\alpha)$ and the Hentschel-Procaccia and Rényi spectra for dimensions. The first result in this direction was obtained by P. Collet, J.L. Lebowitz, and A. Porzio [CLP]. Pesin and Weiss [PW1, PW2] effected the complete multifractal analysis for conformal dynamical systems (see below).

3. Topological pressure and topological entropy. We describe a C-structure on a metric space which is generated by a dynamical system. The corresponding Carathéodory dimension characteristics provide invariants of dynamics which are general versions of the classical topological pressure and topological entropy (see Sect. 3.5). This reveals the "dimension" nature of these important invariants.

Let (X, ρ) be a compact metric space with metric ρ, $S\colon X \to X$ a continuous map, and $\varphi\colon X \to \mathbb{R}$ a continuous function. We follow the notations in Section 3.5. Given a finite open cover $\mathscr{U}$ of X, define the collection of subsets

$$\mathscr{F} = \mathscr{F}(\mathscr{U}) = \{X(\boldsymbol{U}) : \boldsymbol{U} \in \mathscr{W}_m(\mathscr{U}),\ m \geq 1\}$$

and three functions $\xi, \eta, \psi\colon \mathscr{S}(\mathscr{U}) \to \mathbb{R}$ as follows

$$\xi(\boldsymbol{U}) = \exp\left(\sup_{x \in X(\boldsymbol{U})} \sum_{k=0}^{m(\boldsymbol{U})-1} \varphi(S^k(x))\right),$$
$$\eta(\boldsymbol{U}) = \exp(-m(\boldsymbol{U})), \qquad \psi(\boldsymbol{U}) = m(\boldsymbol{U})^{-1}.$$

The collection of subsets $\mathscr{F}$, and the functions η, ξ, and ψ determine a C-structure $\tau = \tau(\mathscr{U}) = (\mathscr{S}, \mathscr{F}, \xi, \eta, \psi)$ on X[35]. The corresponding Carathéodory function $m_C(Z, \alpha)$ (where $Z \subset X$ and $\alpha \in \mathbb{R}$) depends on the cover $\mathscr{U}$ (and the function φ).

Given a set $Z \subset X$, the C-structure τ generates the Carathéodory dimension of Z and lower and upper Carathéodory capacities of Z specified by the cover $\mathscr{U}$. We denote them by $P_Z(\varphi, \mathscr{U})$, $\underline{CP}_Z(\varphi, \mathscr{U})$, and $\overline{CP}_Z(\varphi, \mathscr{U})$ respectively. One can show that for any set $Z \subset X$ the following limits exist:

[35] Some modifications are needed in the case when $X(\boldsymbol{U}_1) = X(\boldsymbol{U}_2)$ while $m(\boldsymbol{U}_1) \neq m(\boldsymbol{U}_1)$; cf [Pe4] for details.

$$P_Z(\varphi) \stackrel{\text{def}}{=} \lim_{|\mathscr{U}|\to 0} P_Z(\varphi, \mathscr{U}),$$

$$\underline{CP}_Z(\varphi) \stackrel{\text{def}}{=} \lim_{|\mathscr{U}|\to 0} \underline{CP}_Z(\varphi, \mathscr{U}),$$

$$\overline{CP}_Z(\varphi) \stackrel{\text{def}}{=} \lim_{|\mathscr{U}|\to 0} \overline{CP}_Z(\varphi, \mathscr{U}),$$

where $|\mathscr{U}| = \max\{\operatorname{diam} U_i : U_i \subset \mathscr{U}\}$ is the diameter of the cover $\mathscr{U}$.

We call the quantities $P_Z(\varphi)$, $\underline{CP}_Z(\varphi)$, and $\overline{CP}_Z(\varphi)$, respectively the *topological pressure* and *lower* and *upper capacity topological pressures* of the function φ on the set Z (with respect to S). We emphasize that the set Z can be arbitrary and need not be compact or invariant under the map S (compare to Sect. 3.5).

Obviously, the functions η and ψ satisfy Condition $A4$ and hence, for any $Z \subset X$

$$\underline{CP}_Z(\varphi) = \lim_{|\mathscr{U}|\to 0} \varliminf_{N\to\infty} \frac{1}{N} \log \Lambda(Z, \varphi, \mathscr{U}, N),$$

$$\overline{CP}_Z(\varphi) = \lim_{|\mathscr{U}|\to 0} \varlimsup_{N\to\infty} \frac{1}{N} \log \Lambda(Z, \varphi, \mathscr{U}, N),$$

where

$$\Lambda(Z, \varphi, \mathscr{U}, N) = \inf_{\mathscr{G}} \left\{ \sum_{U \in \mathscr{G}} \exp\left(\sup_{x \in X(U)} \sum_{k=0}^{N-1} \varphi(S^k(x)) \right) \right\}$$

and the infimum is taken over all finite or countable collections of strings $\mathscr{G} \subset \mathscr{S}(\mathscr{U})$ such that $m(U) = N$ for all $U \in \mathscr{G}$ and $\mathscr{G}$ covers Z.

This formula can be used to show that for any S-invariant compact subset $Z \subset X$ the topological pressure and lower and upper capacity topological pressures coincide and the common value yields the classical topological pressure (see Section 3.5). It is worth pointing out that this common value is a topological invariant (i.e., it is preserved under a homeomorphism which commutes with S). This means that the pressure does not depend on the metric on X.

If a set Z is neither invariant nor compact one has three, in general distinct, quantities: the topological pressure, $P_Z(\varphi)$, and lower and upper capacity topological pressures, $\underline{CP}_Z(\varphi)$ and $\overline{CP}_Z(\varphi)$. The latter coincide if the set Z is invariant and may not otherwise (cf [Pe4]). In view of the variational principle (see below), the topological pressure $P_Z(\varphi)$ seems more adapted to the case of non-compact sets and plays a crucial role in the thermodynamic formalism.

We state now a more general version of the variational principle (see Theorem 3.7) by considering topological pressure on non-compact sets.

Denote by $\mathfrak{M}(X)$ the set of all S-invariant Borel ergodic measures on X. Given an S-invariant (not necessarily compact) set $Z \subset X$, denote also by

$M_S(Z) \subset M_S(X)$ the set of measures μ for which $\mu(Z) = 1$. For each $x \in X$ and $n \geqslant 0$ we define a probability measure $\mu_{x.n}$ on X by

$$\mu_{x.n} = \frac{1}{n} \sum_{k=0}^{n-1} \delta_{S^k(x)},$$

where δ_y is the δ-measure supported at the point y. Denote by $V(x)$ the set of limit measures (in the weak* topology) of the sequence of measures $(\mu_{x.n})_{n \in \mathbb{N}}$. It is easy to see that $\emptyset \neq V(x) \subset M_S(X)$ for each $x \in X$. Put

$$\mathscr{L}(Z) = \{x \in Z : V(x) \cap M_S(Z) \neq \emptyset\}.$$

It is easy to check that $\mathscr{L}(Z)$ is a Borel S-invariant set.

The following statement establishes the variational principle for the topological pressure on non-compact sets. It was proved by Pesin and Pitskel' in [PP].

Theorem 6.11. *Let $Z \subset X$ be an S-invariant set. Then for any continuous function φ on X,*

$$P_{\mathscr{L}(Z)}(\varphi) = \sup_{\mu \in M_S(Z)} \left(h_\mu(S) + \int_Z \varphi \, d\mu \right).$$

We consider the special case $\varphi = 0$. Given a set $Z \subset X$, we call the quantities

$$h_Z(S) \stackrel{\text{def}}{=} P_Z(0), \quad \underline{Ch}_Z(S) \stackrel{\text{def}}{=} \underline{CP}_Z(0), \quad \overline{Ch}_Z(S) \stackrel{\text{def}}{=} \overline{CP}_Z(0)$$

respectively, the *topological entropy* and *lower* and *upper capacity topological entropies* of the map S on Z (compare to Sect. 3.5). We stress again that the set Z can be arbitrary and need not be compact or invariant under S.

For an invariant set Z one can show that

$$\begin{aligned} Ch_Z(S) &= \lim_{|\mathscr{U}| \to 0} \underline{\lim}_{N \to \infty} \frac{1}{N} \log \Lambda(Z, 0, \mathscr{U}, N) \\ &= \lim_{|\mathscr{U}| \to 0} \overline{\lim}_{N \to \infty} \frac{1}{N} \log \Lambda(Z, 0, \mathscr{U}, N), \end{aligned}$$

where $\Lambda(Z, 0, \mathscr{U}, N)$ is the smallest number of strings $\boldsymbol{U}$ of length N, for which the sets $X(\boldsymbol{U})$ cover Z. This formula reveals the meaning of the quantity $Ch_Z(S)$: it is the exponential rate of growth in N of the smallest number of strings $\boldsymbol{U}$ of length N, for which the sets $X(\boldsymbol{U})$ cover Z.

For a compact invariant set Z we have that $h_Z(S) = \underline{Ch}_Z(S) = \overline{Ch}_Z(S)$.

6.6. Multifractal Formalism. A recent innovation in the theory of dynamical systems is the multifractal analysis of invariant sets and measures.

The main constituent component of this analysis is the notion of multifractal spectra which capture integrated information about various multifractal structures generated by dynamical systems. The f_α-spectrum described above is an example of a multifractal spectrum; it characterizes the multifractal decomposition associated with the pointwise dimension. The general concept of multifractal spectra was introduced by Barreira, Pesin, and Schmeling (cf in [BPS] cf also [Pe4]). The formal description follows.

Let X be a set, $Y \subset X$ a subset, and $g: Y \to [-\infty, +\infty]$ a function. The *level sets*

$$X_\alpha^g = \{x \in X : g(x) = \alpha\}, \quad -\infty \leq \alpha \leq +\infty$$

are disjoint and produce a *multifractal decomposition* of X,

$$X = \bigcup_{-\infty \leq \alpha \leq +\infty} K_\alpha^g \cup \hat{X},$$

where the set $\hat{X} = X \setminus Y$ is called the *irregular part*.

Now, let G be a set function, i.e., a real function that is defined on subsets of X. Assume that $G(Z_1) \leqslant G(Z_2)$ if $Z_1 \subset Z_2$. We introduce the *multifractal spectrum* specified by the pair of functions (g, G) (or simply the (g, G)-*multifractal spectrum*) as the function $\mathscr{F}: [-\infty, +\infty] \to \mathbb{R}$ defined by

$$\mathscr{F}(\alpha) = G(K_\alpha^g).$$

The function g generates a special structure on X, called the *multifractal structure*, and the function $\mathscr{F}$ captures important information about this structure.

We consider the case when X is a complete separable metric space. Let $S: X \to X$ be a continuous map. There are two *natural* set functions on X. The first one is generated by the metric structure on X; namely

$$G_D(Z) = \dim_H Z.$$

The second one is generated by the dynamics on X; namely

$$G_E(Z) = h_Z(S).$$

The multifractal spectrum generated by the function G_D is called the *dimension spectrum* while multifractal spectrum generated by the function G_E is called the *entropy spectrum*.

There are also three **natural** ways to choose the function g.

(1) Let μ be a Borel finite measure on X. Consider the subset $Y \subset X$ consisting of all points $x \in X$ for which the limit

$$d_\mu(x) = \lim_{r \to 0} \frac{\log \mu(B(x, r))}{\log r}$$

exists (i.e., the lower and upper pointwise dimensions coincide). We set

$$g(x) = d_\mu(x) \stackrel{\text{def}}{=} g_D(x), \quad x \in Y.$$

If μ is a hyperbolic measure for $C^{1+\alpha}$ diffeomorphism then by Theorem 6.7 $\mu(Y) = 1$. We obtain two multifractal spectra $\mathscr{D}_D = \mathscr{D}_D^{(\mu)}$ and $\mathscr{D}_E = \mathscr{D}_E^{(\mu)}$ specified respectively by the pairs of functions (g_D, G_D) and (g_D, G_E). We call them the *multifractal spectra for (pointwise) dimensions* (note that $\mathscr{D}_D$ is just the $f_\mu(\alpha)$-spectrum studied above). We stress that these spectra do not depend on the map S.

(2) Assume that the measure μ is invariant with respect to S. Consider a finite measurable partition ξ of X. For every $n > 0$, we write $\xi_n = \xi \vee S^{-1}\xi \vee \cdots \vee S^{-n}\xi$, and denote by $C_{\xi_n}(x)$ the element of the partition ξ_n that contains the point x. Consider the set $Y = Y_\xi \subset X$ consisting of all points $x \in X$ for which the limit

$$h_\mu(S, \xi, x) = \lim_{n\to\infty} -\frac{1}{n} \log \mu(C_{\xi_n}(x))$$

exists. We call $h_\mu(S, \xi, x)$ the *local entropy* of S at the point x (with respect to ξ). Clearly, Y is S-invariant and $h_\mu(S, \xi, S(x)) = h_\mu(S, \xi, x)$ for every $x \in Y$. By the Shannon–McMillan–Breiman theorem, $\mu(Y) = 1$. In addition, if ξ is a generating partition and μ is ergodic then

$$h_\mu(S) = h_\mu(S, \xi, x)$$

for μ-almost all $x \in X$. Set

$$g(x) = h_\mu(S, \xi, x) \stackrel{\text{def}}{=} g_E(x), \quad x \in Y.$$

We emphasize that g_E may depend on ξ. We obtain two multifractal spectra $\mathscr{E}_D = \mathscr{E}_D^{(\mu)}$ and $\mathscr{E}_E = \mathscr{E}_E^{(\mu)}$ specified respectively by the pairs of functions (g_E, G_D) and (g_E, G_E). These spectra are called *multifractal spectra for (local) entropies*. These spectra provide integrated information on the deviation of local entropy in the Shannon–McMillan–Breiman theorem from its mean value that is the entropy of the map.

(3) Let X be a differentiable manifold and $S: X \to X$ a C^1-map. Consider the subset $Y \subset X$ consisting of all points $x \in X$ for which the limit

$$\lambda(x) = \lim_{n\to+\infty} \frac{1}{n} \log \|d_x S^n\|$$

exists. The function $\lambda(x)$ is measurable and invariant under S; it relates to characteristic Lyapunov exponents at x. By Kingman's sub-additive ergodic theorem, if μ is an S-invariant Borel probability measure, then $\mu(Y) = 1$. We set

$$g(x) = \lambda(x) \stackrel{\text{def}}{=} g_L(x), \quad x \in Y.$$

This produces two multifractal spectra $\mathscr{L}_D$ and $\mathscr{L}_E$ specified respectively by the pairs of functions (g_L, G_D) and (g_L, G_E). These spectra are called *multifractal spectra for Lyapunov exponents*. It is worth emphasizing that these spectra do not depend on the measure μ.

If μ is ergodic the function $\lambda(x)$ is constant almost everywhere. Let λ_μ denote its value. Multifractal spectra for Lyapunov exponents provide integrated information on the deviation of Lyapunov exponent from its mean value λ_μ.

We consider some examples. First, we analyze four multifractal spectra $\mathscr{D}_D$, $\mathscr{D}_E$, $\mathscr{E}_D$, and $\mathscr{E}_E$ for a subshift of finite type (Σ_A^+, σ). The space Σ_A^+ is endowed with the metric

$$d_\beta(\omega, \omega') = \sum_{k=1}^{\infty} \frac{|\omega_k - \omega'_k|}{\beta^k},$$

where $\beta \geqslant 2$.

Denote by $\mathscr{P}$ the class of finite partitions of Σ_A^+ into disjoint cylinder sets (not necessarily all at the same level). Clearly, each $\xi \in \mathscr{P}$ is a generating partition. We use it to define the spectra for entropies $\mathscr{E}_D$ and $\mathscr{E}_E$.

The following theorem establishes the relations between the multifractal spectra for dimensions and entropies.

Theorem 6.12. *(cf* [BPS], [Pe4]*) For every $\alpha \in \mathbb{R}$, we have*

$$\mathscr{E}_E(\alpha) = \mathscr{E}_D(\alpha) \log \beta = \mathscr{D}_E(\alpha / \log \beta) = \mathscr{D}_D(\alpha / \log \beta) \log \beta.$$

In particular, the common value is independent of the partition $\xi \in \mathscr{P}$.

We now provide a complete description of the spectrum $\mathscr{D}_D$ (and hence, all other spectra) for a subshift of finite type. Let φ be a Hölder continuous function on Σ_A^+ and $\nu = \nu_\varphi$ the corresponding equilibrium state. Define the function ψ such that $\log \psi = \varphi - P_{\Sigma_A^+}(\varphi)$. Clearly ψ is a Hölder continuous function on Σ_A^+ such that $P_{\Sigma_A^+}(\log \psi) = 0$ and ν is the unique equilibrium state for $\log \psi$.

Define the one-parameter family of functions φ_q, $q \in (-\infty, \infty)$ on Σ_A^+ by

$$\varphi_q(\omega) = -T(q) \log \beta + q \log \psi(\omega), \quad \omega \in \Sigma_A^+,$$

where $T(q)$ is chosen such that $P_{\Sigma_A^+}(\varphi_q) = 0$. One can show that for every $q \in \mathbb{R}$ there exists only one number $T(q)$ with the above property. It is obvious that the functions φ_q are Hölder continuous on Σ_A^+. We denote by ν_q the unique equilibrium state corresponding to φ_q.

The following statement effects the complete multifractal analysis of equilibrium states corresponding to Hölder continuous functions for subshifts of finite type.

Theorem 6.13. *(cf* [Pe4]*).*

(1) *The pointwise dimension $d_\nu(\omega)$ exists for ν-almost every $\omega \in \Sigma_A^+$ and*

$$d_\nu(\omega) = -\frac{\int_{\Sigma_A^+} \log \psi(\omega)\, d\nu(\omega)}{\log \beta} = \frac{h_\nu}{\log \beta}.$$

(2) *The function $T(q)$ is real analytic for all $q \in \mathbb{R}$; $T(0) = \dim_H \Sigma_A^+$ and $T(1) = 0$; $T'(q) \leqslant 0$ and $T''(q) \geqslant 0$.*
(3) *The function $\alpha(q) = -T'(q)$ takes on values in an interval $[\alpha_1, \alpha_2]$, where $0 \leqslant \alpha_1 = \alpha(\infty) \leqslant \alpha_2 = \alpha(-\infty) < \infty$; moreover, $f_\nu(\alpha(q)) = T(q) + q\alpha(q)$.*
(4) *For any $q \in \mathbb{R}$ we have $\nu_q(X_{\alpha(q)}) = 1$ where $X_{\alpha(q)} = \{\omega \in \Sigma_A^+ : d_\nu(\omega) = \alpha(q)\}$.*
(5) *If ν is not the measure of maximal entropy then the functions $f_\nu(\alpha)$ and $T(q)$ are strictly convex and form a Legendre transform pair.*
(6) *For $q > 1$,*

$$\frac{T(q)}{1-q} = \underline{HP}_q(\nu) = \overline{HP}_q(\nu) = \underline{R}_q(\nu) = \overline{R}_q(\nu).$$

We describe four multifractal spectra $\mathscr{D}_D$, $\mathscr{E}_E$, $\mathscr{L}_D$, and $\mathscr{L}_E$ for equilibrium states supported on conformal repellers of smooth expanding maps.

Let J be a repeller of a conformal $C^{1+\alpha}$-expanding map S. Consider a Markov partition ξ of J. It is clear that ξ is a generating partition. The same holds true for any partition of J by rectangles which are obtained from ξ (not necessarily all at the same level) and correspond to disjoint cylinder sets in Σ_A^+. We denote the class of such partitions by $\mathscr{P}_S$.

Let φ be a Hölder continuous function on J and ν the corresponding equilibrium state. Write $\log \psi = \varphi - P_J(\varphi)$.

For each q, $p \in \mathbb{R}$ consider the functions on J

$$\varphi_{D.q} = -T_D(q) \log |a| + q \log \psi, \quad \varphi_{E.p} = -T_E(p) + p \log \psi,$$

where the numbers $T_D(q)$ and $T_E(p)$ are chosen such that

$$P_J(\varphi_{D.q}) = P_J(\varphi_{E.p}) = 0.$$

One can show that the functions $T_D(q)$ and $T_E(p)$ are real analytic decreasing and convex. Set

$$\alpha_D(q) = -T_D'(q), \quad \alpha_E(p) = -T_E'(p).$$

Note that $\mathscr{D}_D(\alpha) = f_\nu(\alpha)$ is the dimension spectrum for pointwise dimensions. Its complete description provides a complete multifractal analysis of conformal repeller. It was established by Pesin and Weiss in [PW1] (cf also [PW2] and [Pe4]).

Theorem 6.14.

(1) *The pointwise dimension $d_\nu(x)$ exists for ν-almost every $x \in J$ and*

$$d_\nu(x) = \frac{h_\nu(S)}{-\int_J \log|a(x)|\,d\nu(x)}.$$

(2) *The function* $f_\nu(\alpha)$ *is defined on the interval* $[\alpha_1, \alpha_2]$ *which is the range of the function* $\alpha_D(q)$*; this function is real analytic and* $f_\nu(\alpha_D(q)) = T(q) + q\alpha_D(q)$.

(3) *If* ν *is not the measure of full dimension then the functions* $f_\nu(\alpha)$ *and* $T_D(q)$ *are strictly convex and form a Legendre transform pair.*

(4) *The function* $T_D(q)$ *satisfies Statement 6 of Theorem 6.13.*

Using a Markov partition and the corresponding symbolic representation of the conformal repeller one can try to deduce Theorem 6.14 from Theorem 6.13. Note that under the coding map the Riemannian metric pulls back into a metric on the symbolic space which is **not** the standard metric d_β. In order to overcome this obstacle one should modify the Markov partition and build another partition known as a *Moran cover* of the repeller. One obtains its elements by moving elements of the Markov partition each on its own appropriate time. The coding map associated with the Moran cover pulls back the Riemannian metric into a metric on the symbolic space which is equivalent to the standard metric d_β.

The spectrum $\mathscr{L}_D$ is the dimension spectrum for Lyapunov exponents. Its complete description was obtain by Weiss [We] (cf also [Pe4]).

Theorem 6.15. *If the measure of maximal entropy* $\nu_{\max}$ *is not the measure of full dimension then*

$$\mathscr{L}_D(\beta) = f_{\nu_{\max}}\left(\frac{h_J(S)}{\beta}\right)$$

is a real analytic strictly convex function on an open interval $\lfloor\beta_1, \beta_2\rfloor$ *containing the point* $\beta = h_J(S)/s$. *Otherwise the spectrum is a delta function.*

Since the coding map preserves the entropy the complete description of the spectrum $\mathscr{E}_E$ (the entropy spectrum for the local entropies) follows immediately from Theorem 6.13. More precisely, the following statement holds.

Theorem 6.16.

(1) *There exists a set* $S \subset J$ *with* $\nu(S) = 1$ *such that for every partition* $\xi \in \mathscr{P}_\xi$ *and every* $x \in S$*, the local entropy of* ν *at* x *exists, does not depend on* x *and* ξ*, and*

$$g_E(x) = h_\nu(S, \xi, x) = -\int_J \log\psi\,d\nu.$$

(2) *The function* $T_E(p)$ *is real analytic, and satisfies* $T_E'(p) \leqslant 0$ *and* $T_E''(p) \geqslant 0$ *for every* $p \in \mathbb{R}$. *In addition, we have* $T_E(0) = h_J(S)$ *and* $T_E(1) = 0$.

(3) *The domain of the function* $\alpha \mapsto \mathscr{E}_E(\alpha)$ *is a closed interval in* $[0, +\infty)$ *and coincides with the range of the function* $\alpha_E(p)$. *For every* $p \in \mathbb{R}$, *we*

have

$$\mathscr{E}_E(\alpha_E(p)) = T_E(p) + p\alpha_E(p).$$

(4) *If ν is not the measure of maximal entropy then $\mathscr{E}_E$ and T_E are analytic strictly convex functions, and hence $(\mathscr{E}_E, T_E)$ form a Legendre transfer pair.*

(5) *If ν is the measure of maximal entropy then $\mathscr{E}_E$ is a delta function.*

We now give a complete description of the spectrum $\mathscr{L}_E$ – the entropy spectrum for Lyapunov exponents.

Theorem 6.17. *(cf* [Pe4]*) For every $\alpha \in \mathbb{R}$ we have*

$$\mathscr{L}_E(\alpha) = \mathscr{E}_E^{(m)}(\alpha \dim_H J),$$

where m is the measure of full dimension.

One can establish similar results for two-dimensional LMHS (cf [Pe4]).

§7. Coupled Map Lattices

Coupled map lattices (CML) form a special class of infinite-dimensional dynamical systems. They are described by the equation of the form

$$u_{\bar{j}}(n+1) = f(u_{\bar{j}}(n)) + \varepsilon g_{\bar{j}}(\{u_{\bar{i}}(n)\}_{|\bar{i}-\bar{j}|\leqslant s}).$$

Here $n \in \mathbb{Z}$ is the discrete time coordinate, $\bar{j} = (j_k)$, $k = 1, \dots, d$ is the discrete space coordinate, and $u(\bar{j}, n) = u_{\bar{j}}(n)$ is a characteristic of the medium (for example, its density, or distribution of the temperature, etc.). Furthermore, $f : \mathbb{R}^d \to \mathbb{R}^d$ and $g : (\mathbb{R}^d)^{2s+1} \to \mathbb{R}^d$ are smooth functions; f is called the *local map* and g the *interaction* of size s (where s can be infinite). Finally, ε is a parameter which is assumed to be sufficiently small.

A natural source of CML are discrete versions of partial differential equations of evolution type. They arise while modeling partial differential equations by a computer. In general, no information on the global behavior of solutions of a partial differential equation can be derived from the study of its discrete versions even when steps of discretization are small.

In [Ka], Kaneko, developed a new point of view on CML as phenomenological models which can be used to describe the behavior of an unbounded medium with high level of energy pumping (corresponding to large Reynolds numbers; cf also [Bu7]). In this case one can observe particle-like localized structures, i.e., distinct spatial structures obeying individual dynamics and interacting with nearest neighbors. Moreover, if the medium is space-homogeneous the individual dynamics are identical. Thus, the behavior of the medium obeys the above equation (with the local map f representing the individual dynamics). The discovery by Kaneko, has drown attention of many physicists and mathematicians to CML and exposed a great interest to this area.

Bunimovich and Sinai initiated the rigorous mathematical study of coupled map lattices in [BuSi]. They constructed an analog of SRB-measures assuming that the local map is an expanding circle map and that the interaction is of finite range and preserves the unique fixed point of the local map. In [BKu1, BKu2], Bricmont and Kupiainen extended these results to general expanding circle maps. In [KK], Keller and Künzle studied the case when the local map is a piecewise smooth map of an interval.

The first attempt to analyze coupled map lattices with multidimensional local maps of hyperbolic type was made by Pesin and Sinai in [PS]. They constructed conditional distributions for the SRB-measure on unstable local manifolds assuming that the local map possesses a hyperbolic attractor. Finally, in [JP], Jiang and Pesin established the existence and uniqueness of Gibbs distributions for arbitrary CML whose local map possesses a hyperbolic set.

One can view CML as infinite-dimensional dynamical systems. In order to proceed with this description let us notice that every solution of a CML is uniquely determined if one fixes an *initial condition* $u_{\bar{j}}(0)$ and also prescribes a *boundary condition*. Since the medium is unbounded the latter corresponds to a fixed rate of grows (or decay) of solutions along the space coordinate. We consider solutions which are bounded. In this case, the CML admits the following description as an infinite-dimensional dynamical systems.

Let M be a smooth compact Riemannian manifold and f a C^r-map of M, $r \geq 1$. Let also $\mathbb{Z}^d$, $d \geqslant 1$ be the d-dimensional integer lattice. Set $\mathcal{M} = \otimes_{i \in \mathbb{Z}^d} M_i$, where M_i are copies of M. The space $\mathcal{M}$ admits the structure of an infinite-dimensional Banach manifold with the Finsler metric induced by the Riemannian metric on M, i.e.,

$$\|\bar{v}\| = \sup_{i \in \mathbb{Z}^d} \|v_i\|.$$

We define the *direct product* map on $\mathcal{M}$ by $F = \otimes_{i \in \mathbb{Z}^d} f_i$, where f_i are copies of f.

Consider a map G on $\mathcal{M}$ which is C^r-close to the identity map id. Set $\Phi = F \circ G$. The map G is said to be an *interaction* between points (*space sites*) of the lattice $\mathbb{Z}^d$. Iterates of the map Φ generate a $\mathbb{Z}$-action on $\mathcal{M}$ called *time translations*.

We also consider the group action of the lattice $\mathbb{Z}^d$ on $\mathcal{M}$ by *spatial translations* S^k. Namely, for any $k \in \mathbb{Z}^d$ and any $\bar{x} = (x_i) \in \mathcal{M}$, we set $\left(S^k(\bar{x})\right)_i = x_{i+k}$.

If G commutes with the spatial translations S^k, i.e., $S^k \circ G = G \circ S^k$, we call G *spatial translation invariant*. In this case the pair $\tau = (\Phi, S)$ generates a $\mathbb{Z}^{d+1}$-action which **represents a CML on** $\mathcal{M}$.

Following Pesin and Sinai [PS3] we say that a CML (Φ, S) displays:

1) *temporal chaos* if there exists a measure μ invariant under the $\mathbb{Z}^1$-action $\{\Phi^l\}$ which is mixing;

2) *spatial chaos* if there exists a measure μ invariant under the $\mathbb{Z}^d$-action $\{S^k\}$ which is mixing;

3) *spatio-temporal chaos* if there exists a measure μ invariant under the $\mathbb{Z}^{d+1}$-action $\{\Phi^l, S^k\}$ which is mixing.

We consider a special class of *CML of hyperbolic type*, i.e. a CML whose local map possesses a hyperbolic set Λ. One can see that the map F possesses an infinite-dimensional hyperbolic set

$$\Delta_F = \otimes_{i\in\mathbb{Z}^d}\Lambda_i,$$

where Λ_i is a copy of Λ. Moreover, for each point $\bar{x} = (x_i) \in \Delta_F$ the tangent space $T_{\bar{x}}\mathcal{M}$ admits the splitting $T_{\bar{x}}\mathcal{M} = E^s(\bar{x}) \oplus E^u(\bar{x})$, where the *stable* and *unstable subspaces* are

$$E^s(\bar{x}) = \otimes_{i\in\mathbb{Z}^d}E^s(x_i), \quad E^u(\bar{x}) = \otimes_{i\in\mathbb{Z}^d}E^u(x_i).$$

Furthermore, for each point $\bar{x} = (x_i) \in \Delta_F$ the *local stable* and *unstable manifolds* passing through $\bar{x}$ are

$$V_F^s(\bar{x}) = \otimes_{i\in\mathbb{Z}^d}V_i^s(x_i), \quad V_F^u(\bar{x}) = \otimes_{i\in\mathbb{Z}^d}V_i^u(x_i),$$

where $V_i^s(x_i)$ and $V_i^u(x_i)$ are the local stable and unstable manifolds at x_i respectively. If the hyperbolic set Λ is locally maximal, so is Δ_F.

The construction of invariant measures for CML is based upon an extension of the thermodynamic formalism to the infinite-dimensional case. This extension faces some obstacles. The most crucial one is non-compactness of the hyperbolic set Δ_F. To overcome this obstacle we introduce a new metric on $\mathcal{M}$ known as a *metric with weights* with respect to which the space becomes compact. This metric is defined as follows: given $0 < q < 1$ and $\bar{x}, \bar{y} \in \mathcal{M}$, we set

$$\rho_q(\bar{x}, \bar{y}) = \sup_{i\in\mathbb{Z}^d} q^{|i|}d(x_i, y_i),$$

where $|i| = |i_1| + |i_2| + \cdots + |i_d|$, $i = (i_1, i_2, \cdots, i_d) \in \mathbb{Z}^d$.

For different $0 < q < 1$ the metrics ρ_q induce the same compact (Tychonov) topology in $\mathcal{M}$.

Although working with ρ_q-metrics gives one some advantages in studying invariant measures for the maps F and Φ, it also introduces some new problems. For example, the set $\mathcal{M}$ is no longer a differential manifold and the maps F and Φ, while being continuous, need not be differentiable. Therefore, we restrict the class of perturbations to so-called short range maps (the concept introduced by Bunimovich and Sinai in [BuSi]).

We say that the interaction G is *short ranged* if G is of the form $G = (G_i)_{i\in\mathbb{Z}^d}$, where $G_i : \mathcal{M} \to M_i$ satisfy the following condition: for any fixed

$k \in \mathbb{Z}^d$ and any points $\bar{x} = (x_j)$, $\bar{y} = (y_j) \in \mathcal{M}$ with $x_j = y_j$ for all $j \in \mathbb{Z}^d$, $j \neq k$ we have

$$d(G_i(\bar{x}), G_i(\bar{y})) \leqslant C\theta^{|i-k|} d(x_k, y_k),$$

where C and θ are constants and $C > 0$, $0 < \theta < 1$.

If G is spatial translation invariant then G can be shown to be short ranged with a decay constant θ, if and only if

$$d(G_0(\bar{x}), G_0(\bar{y})) \leqslant C\theta^{|k|} d(x_k, y_k)$$

for any $\bar{x} = (x_j)$, $\bar{y} = (y_i) \in \mathcal{M}$ with $x_j = y_j$ for all $j \in \mathbb{Z}$, $j \neq k$.

The following theorem shows that CML of hyperbolic type are structurally stable.

Theorem 7.1. *(cf* [JP]*). For any $\varepsilon > 0$ there exists $\delta > 0$ such that, if* $\text{dist}_{C^1}(\Phi, F) \leqslant \delta$, *then there is a unique homeomorphism $h : \Delta_F \to \mathcal{M}$ satisfying $\Phi \circ h = h \circ F|_{\Delta_F}$ with* $\text{dist}_{C^0}(h, id) \leqslant \varepsilon$. *In particular, the set $\Delta_\Phi = h(\Delta_F)$ is hyperbolic and locally maximal.*

Provided the interaction G is short ranged one gets a crucial additional information on the conjugacy map h.

Theorem 7.2. *(cf* [JP]*). For any $0 < \theta < 1$ there exists $\delta > 0$ such that if G is a C^2-spatial translation invariant short range map with a decay constant θ and* $\text{dist}_{C^1}(G, id) \leqslant \delta$, *then the conjugacy map h is Hölder continuous with respect to the metric ρ_q, $0 < q < 1$.*

We describe equilibrium measures for CML. For any continuous function φ on $\mathcal{M}$, the variational principle of statistical mechanics claims that

$$P_\tau(\varphi) = \sup_\nu \Big(h_\tau(\nu) + \int \varphi d\nu\Big),$$

where $\tau = (\Phi, S)$ is the $\mathbb{Z}^{d+1}$-action generated by the CML, $P_\tau(\varphi)$ is the topological pressure of the action corresponding to φ, and $h_\tau(\nu)$ is the entropy of the action with respect to the invariant measure ν (cf [Ru1] for definitions).

A measure μ is called an *equilibrium measure* for φ with respect to a $\mathbb{Z}^{d+1}$-action τ if

$$P_\tau(\varphi) = h_\tau(\mu) + \int \varphi d\mu.$$

The action τ is said to be *expansive* if there exists $\varepsilon > 0$ such that for any $\bar{x}, \bar{y} \in \mathcal{M}$,

$$d(\tau^k \bar{x}, \tau^k \bar{y}) \leqslant \varepsilon \text{ forall } k \in \mathbb{Z}^{d+1} \text{ implies } \bar{x} = \bar{y}.$$

In [Ru1], Ruelle showed that expansiveness of a $\mathbb{Z}^{d+1}$-action implies the upper semi-continuity of the metric entropy $h_\tau(\mu)$ with respect to μ. Therefore, it also implies the existence of equilibrium measures for continuous functions. One can easily check that the action (F, S) is expansive on Δ_F in the ρ_q-metric. The expansiveness of the action (Φ, S) on Δ_Φ is a direct consequence of the structural stability theorem. Thus, we have the following result.

Theorem 7.3. *Let $\tau = (\Phi, S)$ be a $\mathbb{Z}^{d+1}$-action on Δ_Φ. Assume that G is short ranged spatial translation invariant and sufficiently C^1-close to identity. Then for any $0 < q < 1$ and any continuous function φ on (Δ_Φ, ρ_q) there exists an equilibrium measure μ_φ for φ with respect to τ. The measure μ_φ does not depend on q.*

While this theorem guarantees the existence of equilibrium measures for continuous functions (with respect to ρ_q-metrics), it does not tell us anything about uniqueness and ergodic properties of these measures. Uniqueness of equilibrium measures implies their ergodicity and usually some stronger ergodic properties (mixing, etc.).

Ruelle [Ru1] obtained the following general result about uniqueness which is a direct consequence of the convexity of the topological pressure on the Banach space $C^0(\Delta_\Phi)$ of all continuous functions in the ρ_q-metric.

Theorem 7.4. *Assume that the map f is topologically mixing. Then for a residual set of continuous functions in $C^0(\Delta_\Phi)$, the corresponding equilibrium measures are unique.*

Ruelle's theorem does not specify the class of functions for which the uniqueness takes place. One of the major recent advances in the theory of CML is to show that uniqueness takes place for Hölder continuous functions with sufficiently small Hölder constant. The main tool is the thermodynamic formalism applied to symbolic models corresponding to CML.

One of the main manifestations of the structural stability theorem is that the conjugacy map h is Hölder continuous in ρ_q-metric. Therefore, the study of existence, uniqueness, and ergodic properties of an equilibrium measure μ_φ corresponding to a (Hölder) continuous function φ on Δ_Φ for the perturbed map Φ is equivalent to the study of these properties for the equilibrium measure $\mu_{\varphi\circ h}$ for the unperturbed map F.

We shall assume that f is topologically mixing. For any $\varepsilon > 0$ there exists a Markov partition of Λ of "size" ε. Let A be the corresponding transfer matrix.

Consider $\Sigma_A^{\mathbb{Z}^d}$ as a subset of the direct product $\Omega^{\mathbb{Z}^{d+1}}$, where $\Omega = \{1, 2, \ldots, m\}$. Let σ_t and σ_s be the time and space translations on $\Sigma_A^{\mathbb{Z}^d}$ defined as follows: for $\bar{\xi} = (\xi_i) \in \Sigma_A^{\mathbb{Z}^d}$, $\xi_i = \xi_i(\cdot) \in \Sigma_A$,

$$(\sigma_t^k \bar{\xi})_i(j) = \xi_i(j + k),\ k \in \mathbb{Z}; \quad (\sigma_s^k \bar{\xi})_i = \xi_{i+k},\ k \in \mathbb{Z}^d.$$

We define the coding map $\bar{\pi} = \otimes_{i\in\mathbb{Z}^d}\pi : \Sigma_A^{\mathbb{Z}^d} \to \Delta_F$. It is a semi-conjugacy between the uncoupled map lattice and the symbolic dynamical system, i.e., the following diagram is commutative:

$$\begin{array}{ccc} \Delta_F & \overset{(F,S)}{\rightarrow} & \Delta_F \\ \uparrow \bar{\pi} & & \uparrow \bar{\pi} \\ \Sigma_A^{\mathbb{Z}^d} & \overset{(\sigma_t,\sigma_s)}{\rightarrow} & \Sigma_A^{\mathbb{Z}^d} \end{array}$$

The action (σ_t, σ_s) on $\Sigma_A^{\mathbb{Z}^d}$ is called the *lattice spin system* and provides the symbolic model for the CML. Let us point out that the lattice spin systems corresponding to CML are of a special type and have not been studied in the framework of the "classical" statistical mechanics until recently. The study of Gibbs distributions for these special lattice spin systems required new and advanced technique which was developed in [JM] and [BKu2]. We give a concise description of the corresponding results.

We remind the reader the concept of Gibbs states for lattice spin systems of statistical physics.

An element $\bar{\xi} \in \Sigma_A^{\mathbb{Z}^d} \subset \Omega^{\mathbb{Z}^{d+1}}$ is called a *configuration*. Let φ be a Hölder continuous function on $\Sigma_A^{\mathbb{Z}^d}$ with respect to the ρ_q-metric. For each finite subset $X \subset \mathbb{Z}^{d+1}$ define the function $p_X(\bar{\xi})$ on $\Sigma_A^{\mathbb{Z}^d}$ by

$$p_X(\bar{\xi}) = \frac{1}{\sum_{\bar{\eta}, \bar{\eta}(\hat{X})=\bar{\xi}(\hat{X})} \exp\left(\sum_{x\in\mathbb{Z}^{d+1}} \varphi(\tau^x\bar{\eta}) - \varphi(\tau^x\bar{\xi})\right)},$$

where τ^x is the action $(\sigma_t)^i \circ (\sigma_s)^j$, $\hat{X} = \mathbb{Z}^{d+1} \setminus X$, and $x = (i, j)$, $i \in \mathbb{Z}^d$, $j \in \mathbb{Z}$.

A probability measure μ on $\Sigma_A^{\mathbb{Z}^d}$ is called a *Gibbs state corresponding to the function* φ if for any finite subset $X \subset \mathbb{Z}^{d+1}$,

$$\mu_X(\bar{\xi}_X) = \int_{\Omega_{\hat{X}}} p_X(\bar{\xi}) d\mu_{\hat{X}},$$

where μ_X and $\mu_{\hat{X}}$ are the probability measures on Ω_X and $\Omega_{\hat{X}}$ respectively that are induced by natural projections. This equation is known as the *Dobrushin-Ruelle-Lanford equation.*

The relation between translation invariant Gibbs states and equilibrium measures can be stated as follows (cf [Ru1]).

Theorem 7.5. *If the transfer matrix A is aperiodic then μ is an equilibrium measure for φ if and only if it is a translation invariant Gibbs state corresponding to the function φ.*

In statistical mechanics Gibbs states are usually defined for potentials rather than for functions. We briefly describe this approach.

A *potential U* is a collection of functions defined on the family of all finite configurations, i.e.,

$$U = \{U_X : X \subset \mathbb{Z}^{d+1},\ U_X : \Omega_X \to \mathbb{R}\}.$$

Gibbs states for a potential U are defined as the convex hull of the thermodynamic limits of the *conditional Gibbs distributions*:

$$\mu_{\bar{\eta}^*.X}(\bar{\xi}(X)) = \frac{\exp(\sum_{V\cap X\neq\emptyset} U_V(\bar{\xi}(X) + \bar{\eta}^*(\hat{X}))}{\sum_{\bar{\eta}.\bar{\eta}(\hat{X})=\bar{\eta}^*(\hat{X})} \exp(\sum_{V\cap X\neq\emptyset} U_V(\bar{\eta}))},$$

where $\bar{\eta}^*$ is a fixed configuration. We describe potentials corresponding to Hölder continuous functions. Let φ be such a function. We write φ in the form of a series $\varphi = \sum_{n=0}^{\infty} \varphi_n$ where the value of φ_n depends only on configurations inside the $(d+1)$-dimensional cube Q_n centered at the origin of side $2n \times \cdots \times 2n$. We also set $Q_0 = (0, 0)$ and define the functions φ_n as follows. Fix a configuration η^* and set

$$\varphi_0(\bar{\xi}) = \varphi\big(\bar{\xi}(Q_0) + \bar{\eta}^*(\hat{Q}_0)\big).$$

Continuing inductively define

$$\varphi_{n+1}(\bar{\xi}) = \varphi\big(\bar{\xi}(Q_{n+1}) + \bar{\eta}^*(\hat{Q}_{n+1})\big) - \varphi\big(\bar{\xi}(Q_n) + \bar{\eta}^*(\hat{Q}_n)\big), \quad n = 1, 2, \ldots.$$

It is easy to see that $\|\varphi_n\| \to 0$ exponentially fast as $n \to \infty$. We define the potential U_φ associated with the function φ on Q_n by setting

$$U_\varphi(\bar{\xi}(Q_n)) = \varphi_n(\bar{\xi}(Q_n)).$$

For other $(d+1)$-dimensional cubes that are translations of Q_n we assign the same value of U_φ. For other finite subsets of $\mathbb{Z}^{d+1}$ we define the potential to be zero. Thus, we obtain a translation invariant potential whose values on finite volumes decrease exponentially when the diameter of the volume grows.

Let us set

$$\mathscr{P}(q, \epsilon) = \{U : \sup_{n\geq 1} q^{-n} \|U_{Q_n}\| \leqslant \epsilon\}.$$

The definition of Gibbs states corresponding to potentials is consistent with the one corresponding to functions. More precisely, Gibbs distributions corresponding to Hölder continuous functions φ are exactly the Gibbs distributions.

A potential U_0 on $\Sigma_A^{\mathbb{Z}}$ is called *longitudinal* if it is zero everywhere except for configurations on vertical finite intervals of the lattice. A potential U_0 is said to be *exponentially decreasing* if

$$|U_0(\bar{\xi}(I))| \leq Ce^{-\lambda|I|},$$

where $C > 0$ and $\lambda > 0$ are constants, I is a vertical interval (i.e., in the time direction), $|I|$ is its length, and $\bar{\xi}(I)$ is a configuration over I. Exponentially deceasing longitudinal potentials correspond to those potential functions whose values depend only on the configuration $\bar{\xi}(0, j)$, $j \in \mathbb{Z}$.

Theorem 7.6. *(Uniqueness and Mixing Property of Gibbs States; cf* [JP]*) For any exponentially deceasing longitudinal potential* U_0 *and every* $0 < q < 1$*, there exists* $\epsilon > 0$ *such that the Gibbs state for any potential* $U = U_0 + U_1$ *with* $U_1 \in \mathscr{P}(q, \epsilon)$ *is unique and exponentially mixing.*

The well-known Ising model provides an example where the Gibbs states are not unique even for potentials of finite range (cf [JP]).

Theorem 7.6 enables one to obtain the following result about uniqueness and mixing property of equilibrium measures for CML.

Theorem 7.7. *Let* (Φ, S) *be a coupled map lattice and* $\varphi = \varphi_0 + \varphi_1$ *a function on* Δ_Φ*, where* φ_0 *is a Hölder continuous function depending only on the coordinate* x_0 *and* φ_1 *is a Hölder continuous function with a small Hölder constant in the metric* ρ_q*. Then there exists a unique equilibrium measure* μ_φ *on* Δ_Φ *corresponding to* φ*. This measure is exponentially mixing and is positive on open sets.*

The chaotic behavior described above is essentially infinite-dimensional. There are other cases where the chaotic behavior is essentially finite-dimensional. This means that there exists a finite-dimensional (often smooth) submanifold in the infinite-dimensional phase space which is invariant with respect to time translations or space translations or both and which supports an invariant mixing measure. Such submanifolds are usually associated with special classes of solutions.

It may also happen that such a submanifold is *stable* in the infinite-dimensional phase space, i.e., solutions which start in a small neighborhood of this submanifold approach it with time.

In this case chaotic behavior is persistent and thus is *physically observable*. Otherwise, chaotic behavior occurs on a "tiny" finite-dimensional submanifold and is "invisible". In some cases the invariant submanifold is stable in a weaker sense: it possesses an infinite-dimensional separatrix which is everywhere dense in the infinite dimensional phase space. In this case the chaotic behavior should also be considered as physically observable. However, it is essentially unstable (with respect to small perturbations of initial data) and hence, is significantly more difficult to study.

In order to illustrate finite-dimensional chaotic behavior consider a CML of *finite size s* and its solutions which can grow with a fixed exponential rate along the space coordinate. Such a CML admits a description as an infinite-dimensional dynamical system $(\mathcal{M}, \Phi)$ where $\mathcal{M}$ is a Banach space endowed with a ρ_q-metric (for an appropriate q).

If the local map is hyperbolic (i.e., possesses a LMHS) one can observe spatial-temporal chaos (for the evolution operator Φ) in the space of *traveling wave solutions*, i.e., solution of the form:

$$u(\bar{j}, n) = \psi((\bar{l}, \bar{j}) + mn),$$

where m is an integer, $\bar{l} = (l_i) \in \mathbb{Z}^p$ an integer p-tuple, and $\psi : \mathbb{Z}^p \to \mathbb{R}^d$ is a function (cf [AfPe], [AfPeTe])[36]. This chaos is physically observable and persistent in a weak sense.

If the local map is of Morse-Smale type so is the dynamics of the evolution operator in the space of traveling waves (cf [OP]). This is a very general result which allows one to describe the behavior of solutions of the discrete versions of some well-known PDE's (for example, Kolmogorov-Petrovsky-Piskunov equation, Ginzburg-Landau equation, Huxley equation, etc.).

One can also observe other types of chaotic behavior (temporal and spatial) which are also finite-dimensional, and physically observable (see [OP]).

Additional References

[AKM] R. L. Adler, A. G. Konheim, M. H. McAndrew, Topological entropy, Trans. Amer. Math. Soc., 114 (1965) 309–319

[AfPe] V. S. Afraimovich, Ya. Pesin, Travelling waves in lattice models of multi-dimensional and multi-component media: I, Nonlinearity 6 (1993) 429-455.

[AfPeTe] V. S. Afraimovich, Ya. Pesin, A. A. Tempelman, Travelling waves in lattice models of multi-dimensional and multi-component media: II. Ergodic properties and dimension, Chaos 3:2 (1993) 233–241

[Ba] W. Ballman, Non-positively curved manifolds of higher rank, Ann. of Math. 122 (1985) 597–609

[BPS] L. Barreira, Ya. Pesin, J. Schmeling, On a general concept of multifractality: multifractal spectra for dimensions, entropies, and Lyapunov exponents. Multifractal rigidity, Chaos, 7:1 (1997) 27–38

[Bo3] R. Bowen, Hausdorff dimension of quasi-circles, Publ. Math. IHES, 50 (1979) 259–273

[BK1] J. Bricmont and A. Kupiainen, Coupled analytic maps, Nonlinearity, 8 (1995) 379–396

[BK2] J. Bricmont and A. Kupiainen, High temperature expansions and dynamical systems, Comm. Math. Phys., 178 (1996) 703–732

[Bu7] L. Bunimovich, Coupled map lattices: one step forward and two steps back, Physica D, 86 (1995) 248–255

[BuSi] L. Bunimovich and Ya.G. Sinai, Space-time chaos in coupled map lattices, Nonlinearity, 1 (1988) 491–516

[BuSp] K. Burns, R. Spatzier, Manifolds of nonpositive curvature and their buildings, Publ. Math. IHES 65 (1987) 35–59

[CLP] P. Collet, J. L. Lebowitz, A. Porzio, The Dimension spectrum of some dynamical systems, J. Stat. Phys., 47 (1987) 609–644

[Di] E. I. Dinaburg, On the relations among various entropy characteristics of dynamical systems, Math. of the USSR, Izvestia, 5 (1971) 337–378

[Eb] P. Eberlein, Geodesic flows on certain manifolds without conjugate points, Trans. Amer. Math. Soc. 167 (1972) 151–170

[ER] J. P. Eckmann, D. Ruelle, Ergodic theory of chaos and strange attractors, Rev. Mod. Phys., 57 (1985) 617–656

[Fe] H. Federer, Geometric measure theory, Springer-Verlag, Berlin–New York, 1969

[36] The numbers m and l_i determine the *velocity* of the wave. It is assumed that the numbers l_i are relatively prime and $m > s\sum_{i=1}^{p} l_i$, i.e., the velocity of the wave is large.

[HJKPS] T. C. Halsey, M. Jensen, L. Kadanoff, I. Procaccia, B. Shraiman, Fractal measures and their singularities: the characterization of strange sets, Phys. Rev. A, 33:2 (1986) 1141–1151

[Ha] B. Hasselblatt, Regularity of the Anosov splitting and of horospheric foliations, Ergod. Theory and Dyn. Syst., 14 (1994) 645–666

[HP] H. G. E. Hentschel, I. Procaccia, The Infinite number of generalized dimensions of fractals and strange attractors, Physica D, 8 (1983) 435–444

[JM] M. Jiang, A. Mazel, Uniqueness of Gibbs states and exponential decay of correlation for some lattice models, Journal of Statistical Physics, 82:3-4 (1995) 797–821

[JP] M. Jiang, Ya. Pesin, Equilibrium measures for coupled maplLattices: existence, uniqueness, and finite-dimensional approximations, Comm. Math.-Phys., 193:3 (1998) 675–711

[Ka] K. Kaneko, editor, Theory and applications of coupled map lattices, Wiley, New York, 1993

[KH] A. Katok, B. Hasselblatt, Introduction to the Modern Theory of Dynamical Systems, Encyclopedia of Mathematics and its Applications, Cambridge University Press, 1994

[KK] G. Keller, M. Künzle, Transfer operators for coupled map lattices, Ergod. Theor. & Dyn. Systems, 12 (1992) 297–318

[Ma3] A. Manning, Topological entropy for geodesic flows, Ann. Math. 110 (1979) 567– 573

[OP] D. R. Orendovici, Ya. B. Pesin, Chaos in traveling waves of lattice systems of unbounded media, Proceedings of the "IMA Volumes in Math. and Appl.", v. 119, Springer-Verlag, 1999, 327–359

[PS3] Ya. Pesin, Ya. Sinai, Space-time chaos in chains of weakly interacting hyperbolic mappings, Advances in Soviet Mathematics, 3 (1991) 165–198

[Pe4] Ya. Pesin, Dimension Theory in Dynamical Systems: Contemporary View and Applications, Chicago Lectures in Mathematics, Univ. of Chicago Press, 1997

[Pe5] Ya. Pesin, Geodesic flows on closed Riemannian manifolds without focal points, Math. USSR Izvestia, 11:6 (1977) 1195–1228

[Pe6] Ya. Pesin, Dimension type characteristics for invariant sets of dynamical systems, Russian Math. Surveys, 43:4 (1988) 111–151

[PT] Ya. Pesin, A. Tempelman, Correlation dimension of measures invariant under group actions, Random and Computational Dynamics, 3:3 (1995) 137–156

[PW1] Ya. Pesin, H. Weiss, A Multifractal analysis of equilibrium measures for conformal expanding maps and Markov Moran geometric constructions, J. Stat. Phys., 86:1-2 (1997) 233–275

[PW2] Ya. Pesin, H. Weiss, The Multifractal analysis of Gibbs measures: motivation, mathematical foundation, and examples, Chaos, 7:1 (1997) 89–106

[PSh] C. Pugh, M. Shub, Stably ergodic dynamical systems and partial hyperbolicity, J. Complexity, 13 (1997) 125–179

[Ra] D. Rand, The Singularity spectrum for Cookie-cutters, Ergod. Theory and Dyn. Syst., 9 (1989) 527–541

[Ru3] D. Ruelle, An inequality for the entropy of differentiable maps, Boletim da Sociedade Brasileira Matemática, 9 (1978) 83–87

[We] H. Weiss, The Lyapunov Spectrum of Equilibrium Measures for Conformal Expanding Maps, J. Stat. Phys. (1999)

Chapter 8
Billiards and Other Hyperbolic Systems

L.A. Bunimovich

There are many important problems in science, (most notably in physics), where hyperbolic dynamical systems with singularities have arisen. Moreover, a Poincaré map for a smooth (and even analytic) flow often possesses singularities. One should also mention that the representation of a flow as a special flow (see Chap. 1, Sect. 4) and the passage to the corresponding induced (Poincaré) map is one of the most effective tools to study ergodic properties of dynamical systems with continuous time.

§1. Billiards

Dynamical systems with elastic reflections (or billiards) form the most important class of dynamical systems with singularities both in the general theory and in applications. By billiard we mean a dynamical system corresponding to the motion due to inertia of a point mass within a region which has a piecewise smooth boundary. The reflections from the boundary are taken to be elastic. (We refer the readers to the recent books on billiards [KT], [Ta]). Here, we concentrate almost entirely on hyperbolic billiards which are essentially beyond the content of these books.

1.1. The General Definition of a Billiard. Let Q be a compact Riemannian manifold with piecewise smooth boundary, $\dim Q = d$. The boundary ∂Q consists of a finite number of smooth compact submanifolds $\Gamma_1, \ldots, \Gamma_r$ of codimension 1. A point $q \in \bigcup_{i=1}^{r} \left(\Gamma_i \backslash \bigcup_{i \neq j} \Gamma_j\right)$ and a set $\tilde{\Gamma}_i = \Gamma_i \backslash \bigcup_{i \neq j} \Gamma_j$ will be respectively called a *regular point* and a *regular component of the boundary*. A point which is not regular will be called *singular*.

A smoothness of the boundary plays a crucial role. The traditional assumption is that the regular components of the boundary are of class C^3, i.e., the curvature is continuously differentiable. This assumption ensures that there are no trajectories which have an infinite number of reflections in a finite time [H]. However, often C^2-smoothness will suffice, especially for the analysis of topological properties of billiards.

Let $\mathcal{T}_q$ be the tangent space of the submanifold Γ_i at the point q, and $n(q)$ the unit normal vector at q directed toward the interior of Q. If q is a regular point then $n(q)$ is uniquely defined. At a singular point there can be several vectors $n(q)$.

We denote by M a unit tangent bundle over Q, $M = \{x : x = (q, v), q \in Q, v \in S^{d-1}\}$ where $S^{d-1} \subset \mathbb{R}^{d-1}$ is the unit sphere. A point $x = (q, v) \in M$ is called a line element with a footpoint q. Let $\pi : M \to Q$ be the natural projection of M onto Q, i.e., $\pi(q, v) = q$. If Q is a compact smooth manifold with piecewise smooth boundary, then M is also a manifold with the boundary $\partial M = \pi^{-1}(\partial Q)$, consisting of a finite number of regular components $\partial M_i = \pi^{-1}(\Gamma_i)$. Obviously, $\dim M = 2d - 1$.

We introduce a measure μ on M by setting $d\mu = d\rho(q) d\omega_q$, where $d\rho(q)$ is the Riemannian volume element on Q at the point q and ω_q is the Lebesgue measure on the sphere $S^{d-1} = \pi^{-1}(q)$.

We now consider the geodesic flow (see Chap. 1, Sect. 1) on the space M. It is determined by the vector field $X = \{X(x), x \in M\}$ where $X(x)$ is a tangent vector to M at the point x. The vector field X characterizes the motion of points along geodesic lines with unit speed. Let N_{ij} be a set of inner points $x = (q, v) \in M$ such that the segment of the geodesic drawn along the direction[1] X intersects ∂Q in a point of the set $\Gamma_i \cap \Gamma_j$. It is clear that N_{ij} is a closed submanifold of codimension 1 and therefore $\mu\left(\bigcup_{i \neq j} N_{ij}\right) = 0$.

We shall assume that for almost every point $x \in \operatorname{Int} M$ (with respect to μ) the geodesic drawn along the direction X intersects with the boundary. This property is fulfilled for all known interesting examples of billiards. Let s be the smallest positive number such that the geodesical segment of length s drawn along the direction x ends in a regular point of a boundary ∂Q. (It can be shown that the set of points x, such that the corresponding geodesics hits singular points of the boundary, has measure 0.) Let y be the tangent vector to Q which arose from x by parallel displacement along the geodesic of length s. We shall now construct the new tangent vector $y' = y - 2(n(q), y)n(q)$ at the point $q = \pi(y)$. This means that the vector y is reflected from the boundary according to the law of elastic reflections: "the angle of incidence equals the angle of reflection." Then we take a segment of the geodesic drawn along the direction y' up to the next intersection with the boundary and so on. It can be shown that the set of trajectories which have an infinite number of reflections from the boundary within a finite time has measure 0 if the regular components of the boundary ∂Q are C^3. We shall also assume that for almost all points x, all the geodesical segments that are generated by the process just described have finite lengths.

Therefore one can define a one-parameter group of transformations $\{T^t\}$ on a subset $M' \subset M$ of full measure by associating to any $x \in M'$ and t, $-\infty < t < \infty$, the tangent vector $T^t x$, obtained by the parallel displacement of x along the direction defined by its geodesic for a distance t. If t is a moment of reflection at the boundary, we set $T^t x = \lim_{t' \to t+0} T^{t'} x$. It is convenient to

[1] By definition the geodesic passing through q along the direction of the vector v will be called the geodesic drawn along the direction $x = (q, v)$.

identify a point y of the boundary ∂M with the point $y' = y - 2(n(q), y)n(q)$. The result set will also be denoted M'.

Definition. The group of transformations $\{T^t\}$ is called a *billiard* in Q.

It can be proved that the group $\{T^t\}$ preserves the measure μ (cf. [KSF]). Therefore, a billiard is a measure preserving flow and can be examined with the tools of ergodic theory. Usually $M(Q)$ is called the phase space (the configuration space) of a billiard.

It can be easily seen from this definition that billiards are geodesic flows on nonclosed manifolds with reflection from the boundary according to the law "the angle of incidence equals the angle of reflection." A billiard in a region Q can be also defined as the Hamiltonian system with a potential $V(q) = 0$ if $q \in Q \backslash \partial Q$ and $V(q) = \infty$ if $q \in \partial Q$.

A billiard $\{T^t\}$ has the following natural special representation (see Chap. 1, Sect. 4). Consider the following transformation T_1 of the set $M_1 = \{x \in \partial M : (n(q), x) \geq 0, q = \pi(x)\}$ into itself. Draw the geodesic along the direction x up to its first intersection with the boundary. A vector y, equal to the reflection of the tangent vector at that intersection point, will be denoted $T_1 x$. We thus obtain the special representation of the flow $\{T^t\}$ generated by the transformation T_1 and the function $\tau(x)$, where $\tau(x)$ is the length of the geodesic segment under consideration. T_1 preserves the measure ν which is the projection of μ onto ∂M.

1.2. Billiards in Polygons and Polyhedrons. Let $Q \subset \mathbb{R}^d$ be a convex polyhedron, i.e., a closed bounded set $Q = \{q \in \mathbb{R}^d : f_i(q) \geq 0, i = 1, \dots, r\}$ where the functions $f_1, \dots, f_r$ are linear. In this case, regular components of ∂Q are faces Γ_i, $i = 1, \dots, r$, of the polyhedron Q. We denote by n_i the unit vector orthogonal to Γ_i which is directed inside Q.

It follows from the definition that the trajectories of billiards in domains contained in Euclidean space are broken lines. Let us consider the isometric mapping $\sigma_j : S^{d-1} \to S^{d-1}$ acting on every point $x = (q, v)$, $q \in \Gamma_i$, according to the formula $\sigma_i(v) = v - 2(n_i, v)n_i$. We assume that some trajectory of the billiard in Q has vertices in faces with numbers $i_1, i_2, \dots$. Then by means of successive reflections of Q, we can obtain a straight line instead of the broken one. This straight line intersects with the polyhedrons $Q, Q_{i_1}, Q_{i_1,i_2}, \dots$. Here $Q_{i_1,\dots,i_k}$ is the result of successive reflections of Q, relative to faces $\Gamma_{i_1}, \dots, \Gamma_{i_k}$, where Γ_{i_l} is a face of $Q_{i_1,\dots,i_{l-1}}$.

Consider a point $x_0 = (q_0, v_0) \in M$. The vector $v_0 \in S^{d-1}$ defines the initial velocity of the billiard trajectory originating from the point $q_0 \in Q$. The velocity vector becomes $v_k = (\sigma_{i_k}\sigma_{i_{k-1}} \cdots \sigma_1)v_0$ between the kth and $(k+1)$th reflections.

Let G_Q be a subgroup of the group of all isometries of S^{d-1} generated by $\sigma_1, \dots, \sigma_r$.

Theorem 1.1. *If G_Q is a finite group, then the billiard in the polyhedron Q is nonergodic. Moreover, to every orbit $\Omega = \Omega(v_0) = \{gv_0 \in S^{d-1} : g \in G_Q, v_0 \in S^{d-1}\}$ of G_Q acting on S^{d-1}, there corresponds a set A_Q invariant with respect to the action of the flow $\{T^t\}$ consisting of all points $x = (q, v) \in M$, such that $v \in \Omega$.*

The geometric meaning of this theorem is as follows: if the group G_Q is finite, then only finite sets of directions can arise from the initial direction during the motion along trajectories of a billiard. For $d = 2$, the finiteness of the group is equivalent to the commensurability of all angles of a polygon Q.

The simplest example in the class under consideration is a billiard in a rectangle. In this case, G_Q contains four elements: **Id**, $\sigma_1, \sigma_2, \sigma_2\sigma_3$. The phase space of this billiard can be decomposed into $\{T^t\}$-invariant sets A_Ω, where $\Omega = \Omega(v)$. It is easy to see that if $\sigma_1 v \neq v$ and $\sigma_2 v \neq v$, then A_Ω is a two-dimensional torus. The flow $\{T^t_\Omega\}$ induced on A_Ω by $\{T^t\}$ is a one-parameter group of shifts on a torus. Hence (see Chap. 1, Sect. 3), the flow $\{T^t_\Omega\}$ is ergodic if the number v_2a_1/v_1a_2 is irrational where v_1 and v_2 are the projections of the velocity along the directions of the sides of the rectangle, and where a_1, a_2 are the lengths of these sides. So the decomposition of the phase space into invariant tori A_Ω coincides with its decomposition into ergodic components. Analogously, the ergodic components of the billiard in a rectangular parallelepiped are, except for a set of measure zero, one-parameter groups of shifts acting on d-dimensional invariant tori.

The general problem of the ergodic properties of billiards in an arbitrary polyhedra is still open. However, there are two main results on this subject (cf. [BKM], [KSF]).

Theorem 1.2. *The entropy of a billiard in an arbitrary (not necessarily convex) polyhedron is equal to zero.*

Theorem 1.3. *If angles of a polygon Q are commensurable, then almost all trajectories of the corresponding billiard are dense in Q.*

The 2D case is much better understood. Namely, a typical billiard in a polygon is ergodic.

Theorem 1.4. [KMS] *There is a dense G_σ subset in the space of polygons consisting of polygons for which the billiard flow is ergodic.*

However, the first explicit examples of polygons which generate ergodic billiards were only recently found [V].

It was conjectured that every trajectory of a billiard in a convex polygon must be periodic or everywhere dense. G. A. Galperin has proved however that this conjecture is wrong (cf. [Ga]). In this example there is a trajectory which is everywhere dense in some subdomain of Q. Apparently this situation often occurs.

Some problems in classical mechanics can be reduced to the investigation of billiards in polygons and polyhedrons. Consider $n \geq 2$ point particles moving along a straight segment, colliding elastically between them and also with the ends of the segment. The order of particles in this segment is preserved. Therefore the configuration space of this dynamical system is a simplex. It is easy to show that the elastic collisions of the particles correspond to the reflections of trajectories from the boundary of this simplex according the law "the angle of incidence equals the angle of reflection" (cf. [KSF]).

1.3. Billiards in Domains with Smooth Convex Boundary. Let $Q \subset \mathbb{R}^2$ be a domain bounded by a smooth convex curve $\Gamma = \partial Q$. In the simplest case, Γ is a circle. It is easy to see that every link of a broken line which corresponds to an arbitrary trajectory in the configuration space of a billiard is tangent to some circle concentrix to Γ but with a smaller radius. A billiard in a circle is therefore a completely integrable Hamiltonian system. Let Γ now be an ellipse $\Gamma = \Gamma_c = \{q \in \mathbb{R}^2 : \mathrm{dist}(q, A_1) + \mathrm{dist}(q, A_2) = c\}$ with foci A_1 and A_2. One can verify that all links of any trajectory of a billiard in Q are tangent to one and the same ellipse Γ_{c_1}, $c_1 < c$, confocal with Γ_c or to a hyperbola $H_{c_1} = \{q \in \mathbb{R}^2 : \mathrm{dist}(q, A_1) - \mathrm{dist}(q, A_2) = c_1\}$ confocal with Γ_c. (In the latter case, the points of tangency may belong not only to links of a broken line but also to their continuations.) Therefore a billiard in an ellipse is nonergodic.

The investigation of the ergodic properties of billiards is also important as well for some problems in the theory of partial differential equations.

Consider the following problem in the domain Q

$$\begin{aligned} (\Delta + \lambda)u(x, y) &= 0 \\ u|_{\partial Q} &= 0 \end{aligned} \tag{8.1}$$

where $(x, y) \in Q$, Δ is Laplace operator and λ is a spectral parameter. This problem arises in the study of small oscillations of a membrane with fixed boundary. It also arises in quantum mechanics, when investigating the eigenvalues and eigenfunctions of the Schrödinger equation, with a potential equal to zero inside Q and equal to infinity on the boundary. It is thus natural to expect that solutions of the quantum problem are connected with solutions of the corresponding problem in classical mechanics, i.e., the billiard in Q.

Definition. A *caustic* for a billiard in Q is a smooth closed curve $\gamma \subset Q \subset \mathbb{R}^2$, such that if one link of the billiard trajectory is tangent to γ, then every other link of this trajectory is also tangent to γ.

There is a unique family of caustics for a circle, namely the concentric circles. But for ellipses there are two such families (confocal ellipses and hyperbolas).

One of the main challenges in the theory of two-dimensional billiards is to prove G. Birkhoff's conjecture which, in the modern terminology, is formulated as follows:

If a neighborhood of a strictly convex smooth boundary ∂Q of a billiard is foliated by caustics, then the curve ∂Q is an ellipse.

Recently M. Bialy proved the elegant theorem leading in this direction.

Theorem 1.5. [Bi] *If a billiard table is foliated by smooth closed convex caustics so that almost every trajectory is tangent to a caustic, then its boundary is a circle.*

Another important result in this area was obtained by Bolotin who described a broad class of integrable planar billiards [Bo]. In his work he considered a family of confocal ellipses and hyperbolas together with two limit lines. The boundary of an integrable planar billiard of the first type consists of pieces of these curves. A boundary of an integrable billiard of the second kind is formed by pieces of the confocal parabolas and by pieces of the corresponding limit line. The classes of integrable billiard found in [Bo] include also the billiards on the surfaces of constant curvature. (The planar billiards correspond to zero curvature.)

V. F. Lazutkin has proved that there exists an uncountable set of caustics, of positive measure in Q, if the boundary $\Gamma = \partial Q$ is convex and sufficiently smooth (cf. [La1], [La2]). Moreover Γ is a limit point of this set and the measure of the set of all line elements tangent to caustics is positive in M. Therefore a billiard inside a sufficiently smooth convex curve in the plane is non-ergodic.

J. Mather showed [Ma] that the billiard map has no caustics if the curvature of a C^2 smooth convex billiard curve ∂Q vanishes at some point.

This result has been strengthened in [GK] in the following way. Denote by L, D, and A respectively the length of a convex smooth curve ∂Q, the diameter of ∂Q and area of Q. Let $K_{\min}$ and $K_{\max}$ be the minimum and the maximum values of a curvature of ∂Q.

Theorem 1.6. [GK] *If $\sqrt{2}D^2 K_{\min} K_{\max} \leq 1$, then the billiard table contains a convex region, free of convex caustics, whose area is not less than $A - \sqrt{2}K_{\min} L D^2$.*

The assumption that ∂Q is C^2 is essential. A billiard doesn't have caustics in a neighborhood of the boundary ∂Q if a curvature of ∂Q is discontinuous at some point [Hu].

It was shown in [La2] that with the help of the invariant sets of a billiard, one can construct "quasi"–eigenfunctions (quasimodes) and the corresponding "quasi"–eigenvalues for the Dirichlet problem in Q. The support of any such eigenfunction is localized in a neighborhood of one of the invariant sets of a billiard defined by caustics.

In a sense, the opposite holds for the ergodic billiards and for the ergodic geodesic flows [Sh], [CdV], [Ze]. Let $f \in C^\infty(M)$ be non-negative. Then there exists a subsequence $\{\psi_k\}$ of density one of the eigenfunctions of the Laplace operator on M such that

$$\lim_{k\to\infty} \int_M f|\psi_k|^2 d\mu = \int_M f d\mu$$

1.4. Dispersing or Sinai Billiards. Consider the billiard in the domain Q illustrated in Fig. 5. We shall assume that all regular components of the boundary ∂Q are smooth curves (of class C^3) which are convex inwards and have a strictly positive curvature with respect to a framing of ∂Q in each point $q \in \partial Q$ by inner unit normal vectors $n(q)$. The properties of such billiards are quite different from those described above.

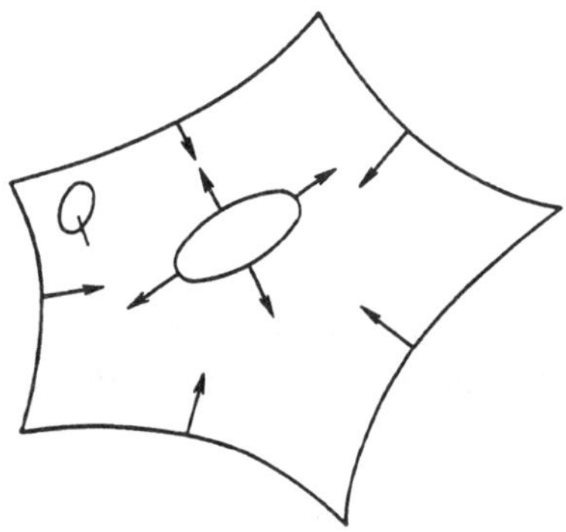

Fig. 5

It is easy to see that the difference between this example and the billiards considered in Sect. 1.2 and 1.3 stems from the structure of the boundary ∂Q. In Sect. 1.2, ∂Q consisted of pieces of hyperplanes (straight lines), in Sect. 1.3, it was convex outwards Q, and in this example it is convex inwards Q.

The ergodic properties of billiards in domains in Euclidean spaces ($Q \subset \mathbb{R}^d$) or on a torus with the Euclidean metric are completely defined by the structure of the boundary ∂Q. In particular, a billiard in a domain Q with an inwardly convex boundary is a hyperbolic dynamical system.

The rigorous mathematical studies of billiards satisfying a condition of hyperbolicity were started by Ya. G. Sinai (cf. [Si4], [Si8]). In [Si2] he introduced the important class of dispersing billiards, which are now called Sinai billiards.

Let us recall that a boundary ∂Q is framed by the unit normal vectors $n(q)$ directed towards the interior of Q. Therefore for any regular point $q \in \partial Q$, one can define a linear, self-adjoint operator, associated to the second quadratic form $K(q)$, which acts in the tangent space $\mathcal{T}_q$ to the boundary ∂Q at q.

Definition. A billiard is called *dispersing* if $K(q) > 0$ for any regular point $q \in \partial Q$.

In the case $d = 2$, a billiard is dispersing if the curvature of ∂Q is positive. We shall now explain why dispersing billiards are the natural analog of smooth dynamical systems of hyperbolic type. Consider again the billiard in the domain represented in Fig. 5. We will assume for simplicity that the regular components of the boundary ∂Q intersect transversally at their endpoints.

Take a smooth curve $\tilde{\gamma} \subset Q$ and a continuous set γ of unit vectors normal to $\tilde{\gamma}$. Then γ is a smooth curve in M. It is evident that two curves γ correspond to every curve $\tilde{\gamma}$ according to the choice of a field of normal vectors. After fixing the curve γ, we can speak about the curvature of the curve $\tilde{\gamma}$. Therefore it is more correct, in general, to speak about the curvature of the curve γ. We shall say that the curve γ is *convex* when its curvature is positive everywhere. Fig. 6 shows a convex curve γ.

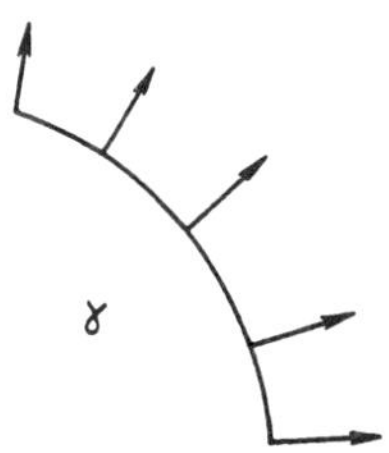

Fig. 6

We denote by $\kappa(x_0)$ the curvature of γ at the point x_0. Let t be so small that no point of the curve γ could reach the boundary ∂Q during the time interval $[0, t]$. It is easy to calculate that the curvature of the curve $\gamma_t = T^t\gamma$ at the point $x_t = T^t x_0$ is given by the equality $\kappa(x_t) = \kappa(x_0)(1 + t\kappa(x_0))^{-1}$. Therefore as the motion proceeds inwards Q, the length of γ_t increases locally linearly with respect to t and its curvature decreases according to a $1/t$ law. From the equality given above it follows that if $\kappa(x_0) > 0$, then $\kappa(x_t) > 0$, that is, if γ_0 is a convex curve, then γ_t is a convex curve too.

We shall now consider what happens upon reflections from the boundary ∂Q. Let $\tau(x_0) > 0$ be a moment of the first reflection from ∂Q of a point x_0. It can be shown, using elementary geometrical arguments, that

$$\kappa(x_{\tau+0}) = \kappa(x_{\tau-0}) + \frac{2k(q_\tau)}{\cos\phi_\tau}. \tag{8.2}$$

Here $k(q_\tau)$ is the curvature of the boundary at the point of reflection and $\phi(x_\tau)$ is the angle between the reflected ray and the normal vector with

$\cos\phi(x_\tau) \geq 0$. Thus, reflection from a boundary which is inwardly convex turns every convex beam that approaches ∂Q with a decreased curvature into a beam with a curvature not less than $2\min_{q\in\partial Q} k(q)$. From this fact one can already derive the local exponential instability (sensitive dependence on initial conditions) of dispersing billiards. In fact, from the equalities given above it follows that the length of a convex curve γ under the action of the differential of a billiard flow between its ith and $(i+1)$th reflections from the boundary increases with a coefficient $1 + \tau_{i+1}(x_0)k_{i+1}(x_0)$. Here, $x_0 \in \gamma$, $\tau_{i+1}(x_0)$ is the time between the ith and $(i+1)$th reflections from ∂Q and $k_{i+1}(x_0)$ is the curvature of ∂Q at the point of the $(i+1)$th reflection. Finally it is easy to see that for a "generic" trajectory, the number of reflections from the boundary increases linearly with time.

Thus dispersing billiards are very similar to geodesic flows in Riemannian manifolds of negative curvature (see Chap. 7, Sect. 4). The role of a negative curvature is played here by a boundary convex inwards Q. Nevertheless there is an essential difference between these classes of hyperbolic dynamical systems. It will be shown below that dispersing billiards are not uniformly hyperbolic, but rather nonuniformly hyperbolic dynamical systems (see Chap. 7, Sect. 1 for a definition).

In view of the local exponential instability one could try to construct local stable and unstable manifolds (LSM and LUM) for dispersing billiards (see Chap. 7, Sect. 1). The corresponding theorem for such billiards was proved in [Si4] (see [Si8] also).

Consider the formulae which describe the evolution of a smooth curve in M under the action of billiard dynamics. These relations allow one to write differential equations for the vector fields tangent to LSM and LUM. Let $0 < t_1 < t_2 < \cdots < t_n < \cdots$ be the moments of the successive reflections of the forward trajectory of the point $x \in M$ from the boundary, $t_n \to \infty$ as $n \to \infty$. Let $\tau_i = t_i - t_{i-1}$, $t_0 = 0$. Denote by $q_i \in \partial Q$, the point of the boundary where the i-th reflection occurs. Let v_i^- and v_i^+ be the velocities directly before and after the i-th reflection, and define ϕ_i by $\cos\phi_i = -(v_i^+, n(q_i))$. Let K_i be the operator of the second fundamental form of the boundary ∂Q at the point q_i. Let U_i be the isometric operator which maps in the direction parallel to the vector $n(q_i)$ which is normal to the hyperplane $A_i^- \subset \mathbb{R}^d$ which contains the point q_i and is orthogonal to the vector v_i^- onto the hyperplane A_i^+, which contains q_i also and is orthogonal to v_i^+. Let V_i denote the operator which maps A_i^- in the direction parallel to the vector v_i^- onto the hyperplane $A_i \subset \mathbb{R}^d$, which is tangent to the boundary ∂Q at the point q_i. Denote by V_i^* the operator adjoint to V_i. Consider the following infinite operator-valued continued fraction

$$B(x) = \cfrac{I}{\tau_i I + \cfrac{I}{2\cos\phi_1 V_1^* K_1 V_1 + U_1^{-1}\cfrac{I}{\tau_2 I + \cfrac{I}{2\cos\phi_2 V_2^* K_2 V_2 + \cdots}} U_1}} \tag{8.3}$$

where I is the identity operator. If $d = 2$ (see Fig 5), then the operator K_i in (8.3) must be replaced by the curvature of ∂Q at the point q_i and the operators U_i, U_i^{-1}, V_i, V_i^* must be replaced by the identity operators. It can be shown that the operator $B(x)$ defines the plane tangent to the local stable manifold of the point x in the phase space of the billiard (cf. [Si8]).

Definition. A smooth $(d-1)$-dimensional submanifold $\gamma \subset \mathbb{R}^d$ is called *convex* or *concave* if the operator of the second fundamental form is positively or negatively defined correspondingly at every point $q \in \gamma$.

Theorem 1.7. *For almost every point $x \in M$ there exist convex and concave $(d-1)$-dimensional submanifolds $\gamma^{(u)}(x) \ni x$ and $\gamma^{(s)}(x) \ni x$ such that*

$$\lim_{t\to\infty}\left[-\frac{\ln \operatorname{diam} T^{-t}(\gamma^{(u)}(x))}{t}\right] > 0$$

and

$$\lim_{t\to\infty}\left[-\frac{\ln \operatorname{diam} T^{t}(\gamma^{(s)}(x))}{t}\right] > 0$$

It was mentioned above that there exist essential differences between the dispersing billiards and the geodesic flows on manifolds of negative curvature. The main one is that for dispersing billiards the flow $\{T^t\}$ is defined on M only almost everywhere and is nonsmooth. The corresponding singularities arise from trajectories falling into singular points of the boundary ∂Q and from trajectories tangent to ∂Q (see Fig. 7). This implies that the LUM and LSM do not exist for points in this set of measure zero. Furthermore the LUM and LSM do not exist at points such that their trajectories come sufficiently often too close to singular points of ∂Q or to trajectories tangent to ∂Q.

Thus the dispersing billiards are nonuniformly hyperbolic dynamical systems (see Chap. 7, Sect. 1). It can be shown [KS] that these local manifolds satisfy the property of absolute continuity (see Chap. 7, Sect. 3). Thus the general theory of NCH-systems implies that ergodic components of a dispersing billiard have positive measure, its entropy is positive and the flow $\{T^t\}$ is the K-flow on almost every ergodic component.

Ya. G. Sinai [Si2, 6] proved that for some classes of dispersing billiards an analog of the entropy formula for geodesic flows on manifolds of negative curvature (see Chap. 7, Sect. 4). Then in the papers [SC], [Ch2, 6], [CM] this result has been extended to much more general classes of hyperbolic billiards. We formulate here the result from [Ch6].

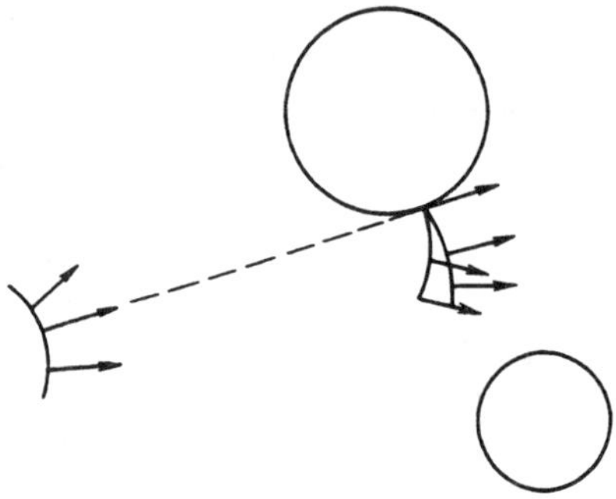

Fig. 7

The nonfocusing assumption says that for almost every point $x \in M_1$ the trajectory segment $\{T^t x, 0 \leq s \leq \tau(x)\}$ has no focusing points. Equivalently, for a.e. $x \in M_1$ and every $0 \leq t \leq \tau(x)$, we have $\det(I + tB(x)) \neq 0$, where $B(x)$ is defined by (8.3).

Theorem 1.8. [Ch6] *Assume the nonfocusing condition stated above. Then the KS entropy of the billiard flow* $\{T^t\}$ *with respect to measure* μ *is given by*

$$h(T^t) = \int_M \operatorname{tr} B(x) d\mu(x).$$

The analogous formulas for entropy of some ergodic focusing billiards were derived in [CM], [Ch6].

Due to the existence of the singular set, the global stable and unstable manifolds of dispersing billiards are submanifolds consisting of a countable number of smooth components. Fig. 8 shows a typical picture of a part of such a global unstable manifold in case $d = 2$. The cusps on this curve correspond to trajectories tangent to the boundary, and break-points to trajectories falling onto singular points of the boundary ∂Q.

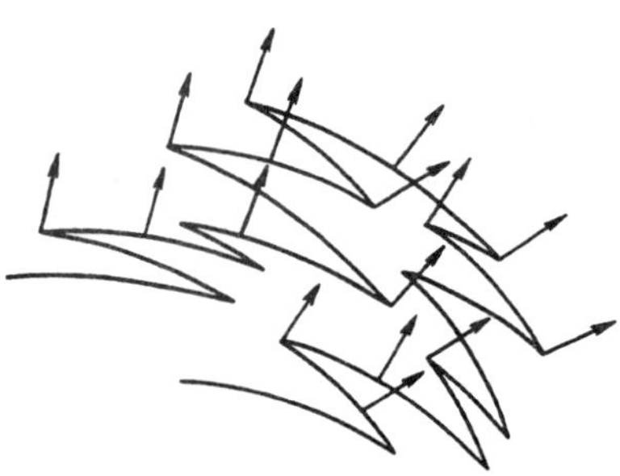

Fig. 8

Therefore, the investigation of ergodic properties of dispersing billiards is much more complicated in comparison with the smooth uniformly completely hyperbolic dynamical systems (see Chap. 7, Sect. 3). For UCH-systems, one can prove ergodicity immediately. The proof is based on the approach which was first used by E. Hopf to prove the ergodicity of the geodesic flows on surfaces of constant negative curvature. According to this approach, for almost any pair of points x_1 and x_2 of the phase space of the dynamical system under consideration, one has to construct the so-called Hopf chain. That is, a finite set $W_1^s, W_1^u, W_2^s, \dots, W_n^s$ which consists of LSM and LUM such that $W_i^s \cap W_j^u \neq \emptyset$, where $j = i \pm 1$. Then one can derive easily from the Birkhoff-Khinchin ergodic theorem (see Chap. 1, Sect. 2) that both points x_1 and x_2 belong to the same ergodic component. However for the dispersing billiards, such a chain cannot be constructed in an analogous manner because a global manifold changes its direction in a singular point "almost" to the opposite direction. Therefore in order to construct the Hopf chain, one has to prove that regular components of LSM and LUM which pass through the majority of points of M are sufficiently large. The needed statement was formulated and proved in [Si4]. After that it was generalized several times and the proof was modified (cf. [BS1], [Si8], [Si7], [KSS1]). A proof of an analogous assertion is necessary before the proof of ergodicity for all billiards which are completely hyperbolic systems (see for instance [Bu1], [Si8], [Bu2], [KSS1]). Thus it is natural to call it the fundamental theorem of ergodic theory of billiards. We shall formulate this theorem in the simplest variant in order to demonstrate clearly its meaning and the idea of the proof.

First we introduce some notation. We shall consider throughout this paragraph a dynamical system generated by the transformation T_1 (see Sect. 1.1). We shall assume for simplicity that Q belongs to a two-dimensional torus and its boundary ∂Q consists of one connected component only. We define on the phase space M_1 a system of coordinates (l, ϕ), where the parameter l denotes the arc length on ∂Q and ϕ is an angle between a line element $x = (q, v)$ and the normal vector $n(q)$, $-\frac{\pi}{2} \leq \phi \leq \frac{\pi}{2}$. It is easy to see that, relative to these coordinates, M_1 is a cylinder. Let S_0 be the set of all line elements tangent to the boundary. The transformation T_1 is a discontinuous one. It has discontinuities on the set $S_{-1} = T_1^{-1} S_0$ (see Fig. 7). This set consists of a denumerable number of smooth curves on the cylinder M_1. In the coordinates (l, ϕ), each curve can be defined by a monotonically increasing function $\phi(l)$ (see Fig. 9) which satisfies the differential equation $\frac{d\phi}{dl} = k(l) + \frac{\cos\phi}{\tau(l, \phi)}$, where $k(l)$ is a curvature of the boundary at the point l and $\tau(l, \phi)$ is the time up to the next reflection from the boundary of the trajectory of the line element (l, ϕ). The limit points of the set S_{-1} are the points which have neighborhoods where the function $\tau(l, \phi)$ is unbounded. To these points correspond periodic trajectories of the billiard which are always tangent to the boundary ∂Q where, at the points of tangency, the boundary lies on one side of

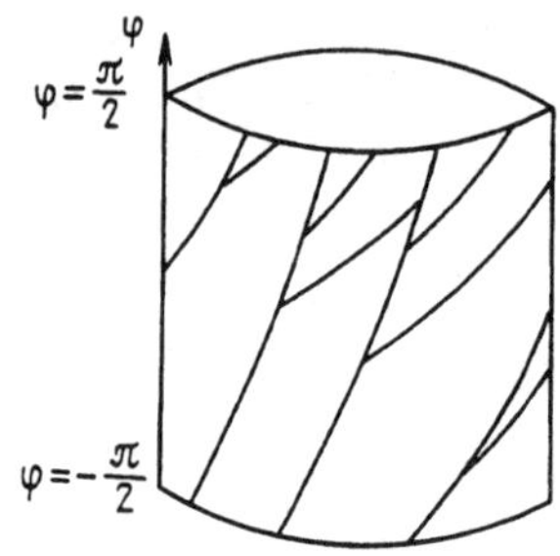

Fig. 9

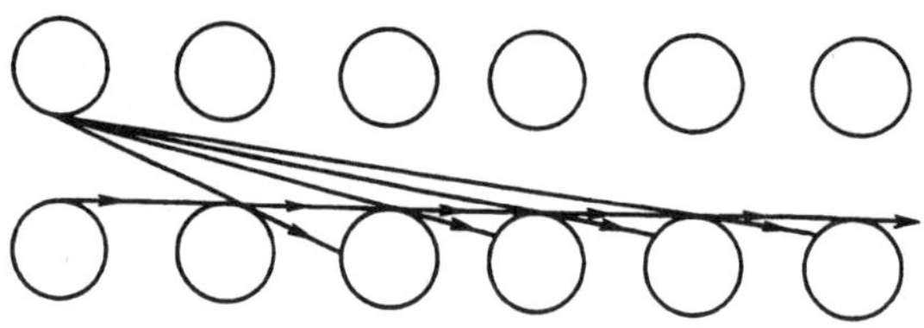

Fig. 10

the trajectory (see Fig. 10). There are only a finite number of such trajectories. The set of discontinuities of the transformation T_1^{-1} has an analogous structure; the only difference is that in the coordinates (l, ϕ) their regular components are decreasing curves. We shall call a piecewise smooth curve $\phi = \phi(l)$ *increasing* (respectively *decreasing*) if $a_i \le \frac{d\phi}{dl} \le a_2$ (respectively $-b_1 \le \frac{d\phi}{dl} \le -b_2$), where $0 < a_1 < a_2$, $0 < b_2 < b_1$ are some constants which depend on geometrical properties of the domain Q. It can be demonstrated that LSM (respectively LUM) are increasing (respectively decreasing) curves on the cylinder M_1. By a *quadrilateral*, we mean a domain $G \subset M_1$ whose boundary consists of four piecewise continuously differentiable curves and one pair of curves not intersecting each other consists of increasing curves while the curves of the other pair are decreasing (Fig. 11). We shall denote by $l(\gamma)$ the length of the piecewise smooth curve $\gamma \subset M_1$. We can formulate, for the example under consideration, the fundamental theorem of ergodic theory of billiards.

Theorem 1.9. *Let $x_0 \in M_1$ be a point such that its positive semitrajectory $T^i x_0$, $i = 0, 1, \dots$ never hit the singular points of the boundary. Then for every α $(0 < \alpha \le 1)$ and an arbitrary number C $(0 < C < \infty)$ there exists $\epsilon = \epsilon(x_0, \alpha, C)$ such that the ϵ-neighborhood U_ϵ of the point x_0 satisfies the following condition: for any increasing curve $\gamma_0 \subset U_\epsilon$, $l(\gamma_0) = \delta_0$, there exists a quadrilateral G_1 (G_2) for which γ_0 is the upper (lower) side such that, if $G_i' = \{x : x \in G_i$, through x passes a regular segment of* LSM

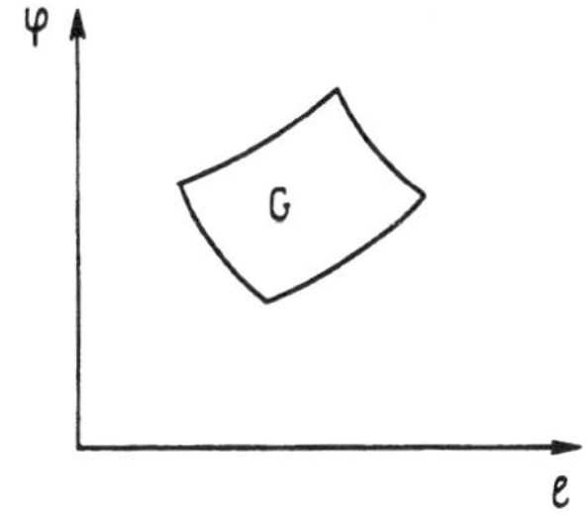

Fig. 11

joining the upper and lower sides of G_i, $l(\gamma^{(s)}(x)) > C\delta_0\}$, $i = 1, 2$, *then* $\nu(G_i') \geq (1-\alpha)\nu(G_i)$, $i = 1, 2$.

The idea of the proof of this theorem is the following. Consider a quadrilateral G for which the increasing curves joining the upper and the lower sides of G have lengths greater than $C\delta_0$. If ϵ is sufficiently small, one can take a large number k_0 such that $T^{k_0}|U_\epsilon$ is a smooth transformation. Then the image $T^{k_0}G$ becomes a very narrow and elongated quadrilateral, which is split up into connected components by curves which belong to S_{-1}. It is easy to see that the preimages of points lying in a small neighborhood of these curves do not belong to G_1' (G_2'). To every connected component obtained by intersection $T_1^{k_0}$ with S_{-1} we apply T_1 and consider the intersection of its image with S_{-1}. Then G_1' (G_2') does not contain the preimages of points which belong to smaller neighborhoods of the curves of discontinuity. We apply again the transformation T_1 to the connected components arising from the intersection $T_1^{k_0+1}G \cap S_{-1}$ and repeat the same arguments, etc. Since in the direction of decreasing curves contraction occurs at each successive step, one can throw out more and more narrow neighborhoods of curves which belong to S_{-1}. Thus the total area of the sets thrown out is relatively small. The analogous theorem holds if one changes LSM onto LUM and the upper and the lower sides of G onto its left and right sides.

In order to prove the desired statement, we can now use a Hopf chain whose links are formed by sets of type G_1' and G_2' constructed in the theorem 1.5, instead of the individual LSM and LUM. However, this construction still provides only a local ergodicity. Indeed, the singularity manifolds have codimension one and therefore they can separate the phase space into different pieces. The general strategy to prove the global ergodicity, i.e., to prove that there is the unique ergodic component, has been developed by Ya. G.-Sinai and N. I. Chernov [SC2]. The central statement of their theory is the, so-called, Sinai-Chernov Ansatz. It says that there exists a LSM (LUM) passing through almost every point x of a positive T^kS, $k > 0$ (negative T^{-k}, $k > 0$) shift of the singularity manifold S. Observe, that here we refer to the natural induced measure in T^kS, k is an integer. (The phase volume of T^kS

equals zero.) One can then use the local stable or unstable manifolds to cross singularities while constructing a Hopf chain.

Theorem 1.10. *A dispersing billiard is ergodic and a K-system.*

It can also be shown (cf. [GO]) that a dispersing billiard is isomorphic to a Bernoulli shift.

1.5. The Lorentz Gas and Hard Spheres Gas. We shall consider in this section two examples of billiards which are among the most popular models of statistical mechanics.

In connection with the problem of description of electronic motion in metals, H. Lorentz introduced in 1905 the dynamical system which is now called the Lorentz gas. We shall consider here its simplest version. The general definition of the Lorentz gas will be given in Sect. 1.10.

Let D be a compact domain with a piecewise smooth boundary in the Euclidean space $\mathbb{R}^d$, $d \geq 1$, $B_1, \ldots, B_r$ be a collection of nonintersecting d-dimensional balls in D which are called scatterers. By definition, the Lorentz gas is a billiard in the domain $Q = D \setminus \bigcup_{i=1}^r B_i$.

We shall now consider the gas of hard, (i.e., absolutely elastic), spheres. Assume that r hard spheres with radius and mass 1 are moving inside a compact domain $D \subset \mathbb{R}^d$, $d \geq 1$, and collide elastically with each other and with the boundary ∂D.

The position of the ith sphere is determined uniquely by coordinates $q_j^{(i)}$, $1 \leq i \leq r$, $1 \leq j \leq d$, of its center. Let $D^- \subset D$ consist of all points which are situated not closer to the boundary ∂D than at the distance ρ. Consider a direct product $D^{(r)} = \underbrace{D^- \times D^- \times \ldots \times D^-}_{r \text{ times}} \subset \mathbb{R}^{dr}$ and exclude from $D^{(r)}$ all inner points of the sets

$$C_{i_1, i_2} = \left\{ q \in D^{(r)} : \sum_{j=1}^{d} (q_j^{(i_1)} - q_j^{(i_2)})^2 \leq (2\rho)^2 \right\}, \quad 1 \leq i_1, i_2 \leq r, \quad i_1 \neq i_2.$$

It is easy to see that the set C_{i_1, i_2} is a product of the $(d-1)$-dimensional sphere and the Euclidean space $\mathbb{R}^{d(r-2)}$, i.e., it is a cylinder. The resulting set $Q \subset \mathbb{R}^{dr}$ is the domain with piecewise smooth boundary. To each configuration of r spheres of radius ρ in D corresponds some point $q \in Q$. Thus the motion of spheres described above induces a group of transformations of the set Q. It can be easily checked that the laws of elastic collisions of spheres between themselves and with the boundary ∂Q correspond to elastic reflections of a moving point q from the boundary ∂Q, i.e., a billiard arises in Q.

1.6. Semi-dispersing Billiards and Boltzmann Hypotheses. It was shown in Sect. 1.5 that in the phase space of a dispersing billiard, the condition of complete hyperbolicity holds for a set of full measure. It is known from the general theory of hyperbolic dynamical systems (see Chap. 7, Sect. 3) that systems which satisfy some weaker conditions of hyperbolicity could possess strong statistical properties as well. It happens that the situation for billiards is in some sense analogous. The corresponding classes of billiards will be considered in this and in the following sections.

Definition. A billiard is called *semi-dispersing* if, for any regular point $q \in \partial Q$, the second fundamental form $K(q) \geq 0$.

In other words, a billiard is semi-dispersing if a boundary ∂Q is not strictly convex inside Q but its curvature equals zero in some directions. Semi-dispersing billiards can serve as analogs of partially hyperbolic dynamical systems. The most important example of semi-dispersing billiards is the gas of hard spheres. One could expect that a semi-dispersing billiard possesses strong ergodic properties only in the case when plane directions at different points of the boundary are nonparallel. For instance, the system of two disks elastically interacting in the torus $\mathbb{T}^2$ has the total momentum as an additional integral of motion, and thus is non-ergodic, while the same system in a square is a K-system (cf. [Si4]). It is easy to see that in the first case the directions with zero curvature are parallel, while in the latter one they are not.

As usual, the first step in the studying of hyperbolic billiard systems is connected with the investigation of the corresponding operator $B(x)$ (see (8.3)). It was proved in [Si8] that a continued fraction converges almost everywhere in a phase space M and it defines a symmetrical non-negatively defined operator which acts in a hyperplane $J(x)$, $x = (q, v)$ which is orthogonal to a vector v. Therefore $J(x)$ can be decomposed into the direct sum of two $B(x)$-invariant, zero and positive subspaces, correspondingly $J_0(x)$ and $J_+(x)$.

In the case of dispersing billiards, the dimension of both these foliations equals $d-1$, where d is the dimension of the phase space. In the case of semi-dispersing billiards, the dimension of the foliations depends on the geometrical properties of the boundary ∂Q. It was shown in [Ch1] that local transversal fibers exist for all points of a set $\widehat{M} \subset M$ containing an open everywhere dense subset of M. Besides, a function which equals the dimension of an expanding (contracting) fiber is equal to a constant on the negative (positive) semitrajectory of the point x. From this result, it follows in particular that the entropy of a semi-dispersing billiard is positive.

We would like to mention that it was N. S. Krylov who first pointed out the exponential instability of the hard spheres gas (cf. [Kr]).

The celebrated Boltzmann hypotheses says that a gas of N hard spheres confined in a torus is ergodic on any submanifold defined by fixing of all classical integrals of motion.

In [S4], Sinai proved the Boltzmann hypotheses for $N = 2$. The very important extension to $N = 3$ has been made in [KSS2] followed by the proof of ergodicity for $N = 4$ [KSS3].

One of the major difficulties on the way towards the proof of Boltzmann hypotheses appears because not always all particles (spheres) may interact with each other. Indeed, sometimes the particles can be partitioned into groups so that the interactions are only within these groups. Certainly, the set of such *insufficient* trajectories has measure zero. Furthermore, any insufficient trajectory is nonhyperbolic because along some directions in the phase space there is neither expansion nor contraction.

The recent elegant paper [SS] deals with the gas of N *nonidentical* particles, i.e., the radii and the masses of the particles can vary. The insufficiency condition for a trajectory can be represented as some algebraic equations. Therefore, these equations do not hold for generic values of the radii and the masses. This idea allowed to prove the nonvanishing of Lyapunov exponents for the generic system of N hard spheres [SS].

Another interesting system of elastically interacting particles was studied in [Wo3]. At first the motion of a billiard ball in a wedge with gravity was investigated numerically in [LM]. Then Wojtkowski showed that this system is equivalent to the motion of two balls on a vertical line subjected to a constant vertical acceleration (the balls fall down) and considered an arbitrary (finite) number of balls. The balls collide elastically with each other and in addition the lowest one collides elastically with the floor as well.

If the masses are equal then this Hamiltonian system is integrable. However the Lyapunov exponents are nonzero almost everywhere if the mass is a nondecreasing function of the particle number (starting with the floor) and if the potential $V(q)$ satisfies the conditions $V'(q) > 0$ and $V''(q) < 0$ where q is the vertical coordinate [Wo4]. For instance, one can take the standard potential $V(q) = -1/q$ for the gravitational field. The conjecture that the system is actually ergodic under these conditions has been proven only for two particles [Ch3].

However, there is already an example of a system of N elastically interacting spherical particles where ergodicity has been proven for any N [BLPS]. In this example the domain where particles move (a billiard table) has the special shape. It is interesting to mention that unfolding of this model gives a new model of statistical mechanics. This model is intermediate between the Lorentz gas and the Boltzmann gas of hard spheres. Imagine a periodic configuration of scatterers. Let there be exactly one particle at each cell of the complement to the union of scatterers. These identical particles have a nonzero radius (on contrary to the Lorentz gas). Suppose that this radius is so big that a particle cannot leave its cell but also the radius is so small that a particle can collide with the particles at the neighboring cells. The model studied in [BLPS] is one (finite, but with arbitrarily long length) strip with such a configuration.

1.7. Billiards in Domains with Boundary Possessing Focusing Components. In this section, we shall consider billiards in domains Q, on the plane or on the two-dimensional torus, such that the boundary ∂Q contains only convex-outwards components. We shall call these focusing. Analogously convex-inwards components of ∂Q will be called dispersing.

In accordance with the traditional ideology, the stochastic properties of billiards are generated by the scattering of trajectories which results from collisions with the convex-inwards boundary of a domain Q. Therefore the first intuitive idea was that if a billiard is dispering and the angles of ∂Q were "smoothed," then a generic trajectory would spend the most time on the dispersing part of the boundary and hit the focusing part very rarely. Hence, the resulting billiard still could have good statistical properties.

However, the situation turns out to be essentially different. Namely, it has been discovered another mechanism to produce hyperbolicity which has been called the mechanism of defocusing. It is different from the mechanism of dispersing.

In order to understand what really happens, let us consider the evolution of a curvature of a smooth curve $\gamma \subset M$ which has a series of consecutive reflections from some focusing component of ∂Q.

The main difficulty of the corresponding analysis is to study the properties of the continued fraction (8.3). In the case under consideration ($d = 2$), elements of the continued fraction are not operators but numbers, and in contrast to the dispersing billiards these elements have different signs. Thus, it is harder to study the problem of its convergence. The following proposition is fundamental for studying reflections from the focusing part of the boundary (cf. [Bu2], [Bu1]).

Theorem 1.11. *Suppose that a bundle of parallel trajectories corresponding to a curve $\gamma \subset M_1$ of zero curvature* (Fig. 12) *has n successive reflections from a focusing component $\Gamma \subset \partial Q$ with constant curvature. Then for any line element $x \in \gamma$ and for any number m, $1 \le m \le n$, the following inequalities hold: $\kappa_-(T_1^m x_0) < 0$, $\kappa_+(T_1^{m-1} x_0) > 0$, $\kappa_+(T_1^{m-1} x_0) > |\kappa_-(T_1^m x_0)|$ while*

$$\lim_{\substack{m\to\infty \\ n\to\infty}} \frac{\kappa_+(T_1^{m-1} x_0)}{|\kappa_-(T_1^m x_0)|} = 1$$

where $x_0 = T^{\tau(x)+0}x$, $\tau(x)$ is the nearest positive moment of reflection from the boundary by the trajectory of the point x, $\kappa_-(T_1^m x_0)$ is the curvature of the image of γ in the point $T_1^m x_0$ before – and $\kappa_+(T_1^m x_0)$ after the mth reflection from the boundary.

Qualitatively, this theorem has the following meaning. A bundle of parallel trajectories with a plane front (i.e., with zero curvature) after reflection from a focusing component $\Gamma \subset \partial Q$ becomes contracting (i.e., has negative curvature) but on its way between any two successive reflections from Γ it

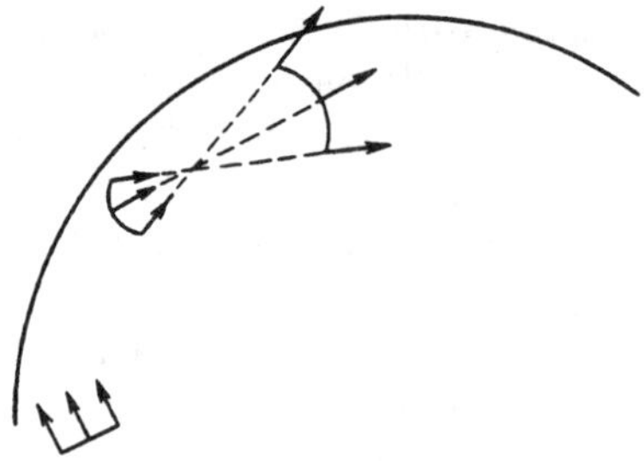

Fig. 12

passes through a conjugate point and comes to the boundary before the following reflection as an expanding bundle. Moreover, the time during which the bundle under consideration has positive curvature (expands) is more than half of the whole time interval between any two successive reflections from Γ in the considered series (see Fig. 12). During this series of reflections the curvature of the bundle under consideration stays bounded from above in its absolute value. So, by successive reflections from any focusing component of a constant curvature, the length of the curve γ in the phase space increases linearly. Of course, to ensure hyperbolicity it is enough that expansion would dominate contraction only on average over many reflections.

The first classes of ergodic billiards in domains with focusing components of the boundary were studied in [Bu1] and [Bu2].

In [Bu1], it was shown that a billiard in a domain whose boundary consists of focusing as well as dispersing components is a K-system if the following conditions hold:

1) the curvature of each focusing component $\Gamma \subset \partial Q$ is constant;

2) there is no pair of focusing components which are arcs of one and the same circle;

3) a complement of each focusing component up to the whole circle is contained entirely inside the domain Q.

We would like to mention that these conditions do not force the focusing part of the boundary ∂Q to be small. In fact, its length could be much greater than the length of the dispersing part of the boundary (Fig. 13).

In [Bu2], the K-property was proved for some classes of billiards in the case when the boundary ∂Q has no dispersing components. Besides, ∂Q has to have at least one focusing component and each focusing component has to have a constant curvature. Could the conditions on the boundary ∂Q be relaxed? Observe, that it is rather delicate question. In fact, a 2D billiard with all boundary components convex inwards is always ergodic. Actually, the ergodicity of a hyperbolic billiard implies by a standard technique [S3], [GO], [CH] that it is also a K system and Bernoulli system. However, billiards

Fig. 13

with boundary components convex outwards may be Bernoulli as well as integrable.

After the discovery of the new mechanism of hyperbolicity (the mechanism of defocusing), the natural conjecture was that it still works for small perturbations of arcs of a circle. This conjecture is natural because in the course of consecutive reflections from a circle nearby trajectories still diverge (linearly). But what we really need to ensure the mechanism of defocusing work? Obviously, just noncontraction in a series of consecutive reflections from a focusing component also suffices. Moreover, we can allow even contraction in the passage through a focusing component, provided that the coefficient of contraction is bounded from below by some positive number. After such series exponential divergence may appear if a trajectory jumps from one regular component of a boundary onto another one or if it returns to the same focusing component after hitting neutral or dispersing components. Thus, we have some "extra" divergence at our disposal that may allow us to make small perturbations of a circular arc and still keep nonconvergence. The arguments why this works were given in [Bu6] and then the proof was given in [M2]. The paper [Do3] contains a simpler proof of this fact, but the class of perturbations allowed is narrower than in [M2]. The simplest proof and allowing the most general (C^3) perturbations was given in [Bu8].

The general question about focusing components of the boundary in chaotic billiards was formulated in [Bu6]. It asks about the conditions that allow a focusing component to be a part of a boundary of some chaotic billiard.

[Bu6] conjectures an answer to this question. It speculates that any focusing component of a boundary of a chaotic billiard must also be absolutely focusing.

The notion of absolutely focusing seems to be new to geometrical optics. A focusing arc (mirror) Γ is called **absolutely focusing** if any infinitesimal parallel beam of rays becomes focusing after the last reflection in the series of consecutive reflections from Γ.

To understand better this notion, observe that a focusing arc focuses any parallel beam of rays after the first reflection from it. This property can be taken as the definition of a focusing arc. In the definition of an absolutely focusing arc we look instead at the beams of rays fallen onto this arc just before the first and after the last reflection from such arc. So, from the physical point of view our definition is even more natural than the definition of a focusing arc (mirror). Indeed, we consider what happens before (a "signal") and after (a "response") a ray passes through a mirror.

Observe that it doesn't follow from the definition that any absolutely focusing arc is also a focusing one. But it is obviously the case. On the other hand there are focusing arcs that are not absolutely focusing. For instance, cut an ellipse along its shortest axis. Then corresponding semiellipses are absolutely focusing if the ratio of the axis is less than $\sqrt{2}$; otherwise a semiellipse is not absolutely focusing [Do3], [Bu8]. The important fact is that any sufficiently short focusing arc is absolutely focusing [M2], [Do3], [Bu8].

The notion of an absolutely focusing arc is global in the sense that it involves only the entrance of the ray to the arc and the exit from it. However, such information is shown to be enough to characterize what is happening in any reflection in a series of consecutive reflections from an absolutely focusing arc [Bu8].

Theorem 1.12. *Suppose that a parallel* (*before the first reflection*) *beam of rays has a sequence of consecutive reflections from an absolutely focusing arc. Then such beam goes through a conjugate point between any two consecutive reflections in this series.* (*In other words, the picture between any two consecutive reflections is as the one depicted in* Fig. 12.)

This condition, which is equivalent to the absolutely focusing, is more convenient for the analysis because of its local nature. For instance, it has been used in [Bu7] where the sufficient conditions were considered for ergodicity of the two-dimensional billiards which may contain all/any of three types of regular boundary components. The local condition of defocusing was independently introduced in [Do3]. V. J. Donnay proved in this paper that the local condition of defocusing holds for any narrow parallel beam of rays which is sufficiently close to a tangent to a focusing arc. The conditions from [Bu7] are too long to formulate here. Basically, the general result of this paper says that a billiard in a 2D region Q is a K-system (and Bernoulli [CH]) if all focusing components of the boundary ∂Q are absolutely focusing, if they are situated sufficiently far from each other, and if any two adjacent regular components of ∂Q intersect properly.

It has been conjectured that absolutely focusing arcs are also the only focusing arcs which can be contained in a boundary of a chaotic billiard [Bu9]. The "sufficient" part of this conjecture has already been proved [Bu7], while the "necessary" part is proven only partially. Let Γ be a focusing arc. Then the rays (trajectories) need a sufficient time to defocus after leaving Γ to make the defocusing mechanism work. However, an arbitrary large time does not suffice if Γ is not absolutely focusing [Bu9]. Denote by $\tau(Q, \Gamma)$ a minimal free path (a time between the two consecutive reflections from the boundary) that a trajectory may have after leaving Γ, i.e., after the last reflection in a series of consecutive reflections from Γ. Then for any $L > 0$ there exists a domain Q, such that $\partial Q \supset \Gamma$, $\tau(Q, \Gamma) > L$ and a billiard in Q has a linearly stable periodic trajectory.

The first class of focusing arcs with nonconstant curvature which may belong to boundaries of chaotic billiards has been considered in [W1]. These arcs satisfy the condition $d^2R/ds^2 \leq 0$, where R is a radius of curvature and s is the length parameter. It was shown later in [M1] that the same is true for the focusing arcs which obey the inequality $d^2(R^{1/3})/ds^2 > 0$. Both [W1] and [M1] use the general, but different approaches. The paper [W1] explores the invariant cones technique, while [M1] works with invariant quadratic forms.

However, the dynamics of billiard systems can be explicitly described in purely geometric terms by the infinite continued fraction (8.3). In case of 2D billiards, each odd element in (8.3) corresponds to the time between two consecutive reflections while each even element is proportional to the ratio of the curvature of the boundary at the point of reflection and the cosine of the angle of incidence.

Therefore, while all elements of a continued fraction have the same sign in dispersing billiards the signs of these elements alternate in billiards with only focusing components. It is easy to see why the problem of hyperbolicity of semi-dispersing 2D billiards can be solved completely while for billiards with at least one focusing component this problem is rather complicated. Indeed, there exists a well known Seidel-Sterne criterium of convergence of continued fractions with elements of constant sign. However, there doesn't exist a criterium of convergence for other classes of continued fractions. (It is interesting that the geometric description of continued fractions in terms of billiard trajectories allowed to obtain some new results on their convergence [Bu10].)

The approach based on the analysis of continued fractions allows to see that the two classes of focusing arcs considered in [W1] and [M1] are dual to each other. Indeed, the sufficient condition of convergence of continued fractions with alternating signs of elements a_i can be written [Bu10], as

$$|a_{n+2j-r}| \geq |a_{n+2j-(r+1)}|^{-1}\left(2 - \frac{\delta_j}{1+\delta_j}\right) + |a_{n+2j-(r-1)}|^{-1}(2 + \delta_{j+1}) \quad (8.4)$$

where $\delta_j \geq -1$ and $r = 1$ (case of [W1]) or $r = 0$ (case of [M1]).

After the appearance of the examples of ergodic billiards in domains with focusing components V. M. Alexeev formulated in 1973 at the Seminar on Dynamical Systems in Moscow University the following problem. Can one construct ergodic geodesic flows on S^2 and $\mathbb{T}or^2$? According to the Gauss-

Bonnet theorem the mean curvature over the two-dimensional surface of genus g equals $2\pi(2-2g)$. Therefore only for $g \geq 2$ a curvature can be negative everywhere in a surface (with respect to the natural metrics). The example has been constructed by R. Osserman who attached to S^2 three semispheres (caps) along to some circles (equators). The geodesic flow outside the caps is the same as in the case when we removed the caps and identified the opposite points of the boundary circle of a half-sphere. The difficult task of smoothing this example to C^∞ was performed by Donnay [Do1,2]. He also constructed ergodic geodesic flows on $\mathbb{T}or^2$. A real analytic Bernoulli geodesic flow on S^2 has been constructed in [BG]. In all these examples the same mechanism of defocusing has been explored. Namely, the geodesics entering a cap pass through the conjugate points and leave the cap being defocused.

Can the mechanism of defocusing work in higher dimensions? This question was answered affirmatively only recently [BR1–3].

First, it was claimed in [Bu6] that the defocusing can generate hyperbolic and ergodic focusing billiards in higher dimensions provided that the focusing components of the boundary are small enough. The 3D billiard was constructed in [W2] which consists of semispheres and planes. In this example the minimal recurrence time to a focusing boundary may be arbitrarily large but the corresponding billiard still has linearly stable periodic orbits. While this result doesn't ensure that this billiard is not ergodic (a linearly stable orbit in higher dimensions generally is not stable) it shows, however, that the construction of hyperbolic focusing billiards in more than 2 dimensions is rather delicate issue.

This fact has been known for centuries in the geometrical optics because of the phenomenon which is called astigmatism. The astigmatism means that the focusing of rays in different planes may be (and for focusing mirrors it is) essentially different. To be more precise let us consider an infinitesimal beam of rays that is reflected from a spherical mirror C. Take the entire series of consecutive reflections from C by a given ray (the "central" ray in the beam of rays under consideration). Then obviously, the corresponding piece of the trajectory of this ray belongs to the two-dimensional plane which contains this ray as well as the center of the sphere. The evolution of curvatures of beams of trajectories in this plane is given via the formula (8.3).

However, in the orthogonal plane the corresponding relations is [Co]

$$x^{\perp}(x_{\tau+0}) = x^{\perp}(x_{\tau-0}) + 2k(q)\cos\phi \tag{8.5}$$

where τ is a moment of reflection $k = 1/R$ is a curvature of a sphere (R is its radius) and ϕ is the angle of incidence.

The relations (8.3) and (8.5) show that focusing in different planes is different (astigmatism). Moreover, in some planes the focusing is so weak (when ϕ is close to $\pi/2$) that the mechanism of defocusing seemingly has no chances to generate the hyperbolicity.

However, it is still the case [BR1–3] if the focusing components of ∂Q are not very big pieces of spheres. For the sake of brevity, we formulate the corresponding result for a rather narrow class of multidimensional focusing domains which can be viewed as multidimensional stadia [BR3].

Theorem 1.13. *Let a region $Q \in R^n$ consist of a rectangular box with spherical caps attached to any of its faces and all spheres defined by the spherical caps lie inside Q and do not intersect each other. Suppose also that internal angles of all spherical caps are less than $\pi/3$. Then the billiard in Q is a Bernoulli system.*

It is not known whether or not focusing hyperbolic billiards exist with spherical caps with internal angles bigger than $\pi/2$. Another challenging question is whether or not spherical caps can be perturbed and what is the analog of absolutely focusing arcs in higher dimensions. The conjecture is that the same definition must carry over.

Thus billiard regions in Theorem 1.13 are formed by two spherical caps attached to opposite faces of (sufficiently long, but not necessarily too long) rectangular box. Some more general classes of focusing domains were proven to confine hyperbolic and ergodic billiards. However, there is a gap between the conditions on smallness of spherical caps in these two cases. Namely, an internal angle of a cap must be less than $\pi/2$ to ensure hyperbolicity and less than $\pi/3$ (so even less than it was required in [Bu6]) for ergodicity. It is not known whether this is just a technical issue.

Some examples of domains which generated billiards satisfying the K-property are shown in Fig. 14. The most popular example is the stadium (Fig. 14a), i.e., a convex domain whose boundary consists of two identical half-circles and two parallel segments. From the other side a billiard in a domain bounded by a sufficiently smooth convex curve is non-ergodic (see Sect. 1.2). The boundary of a stadium is of class C^1. Therefore one of the most interesting problems in this subject is to study how one could smooth out a boundary of a stadium in order to preserve the K-property. The conjecture is that the smoothness of class C^2 is critical.

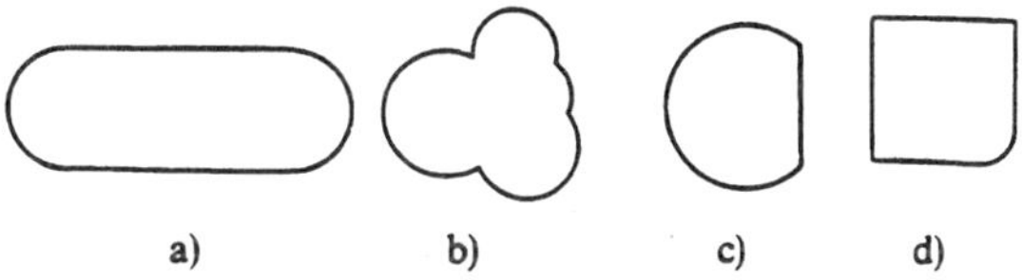

Fig. 14

1.8. Hyperbolic Dynamical Systems with Singularities (a General Approach). The billiards considered in the previous sections are hyperbolic dynamical systems, which act on a manifold with boundary and have singularities (for instance discontinuities) on some set of Lebesgue measure zero.

The proofs of the theorems obtained for the corresponding billiard systems are based on a detailed analysis of the structure of the singularity set. It is natural to ask what type of results could be obtained, having information only on the general structure of this set. This problem was studied by A. Katok and J.-M. Strelcyn [KS].

Let M be a smooth compact Riemannian manifold with boundary ∂M, ρ be the metric in M induced by the Riemannian metric, $N \subset M$ a union of a finite number of compact submanifolds, $f : M\backslash N \to f(M\backslash N)$ a diffeomorphism of class C^2, and μ an invariant Borel measure. The pair (f, N) is said to be a discontinuous dynamical system if the following conditions are satisfied:

1) $\int_M \ln^+ \|df_x\| d\mu < +\infty$, $\int_M \ln^+ \|df_x^{-1}\| d\mu < +\infty$ where df_x is the differential of f at a point x, $\|\cdot\|$ is a norm in the tangent space, $\ln^+ x = \max(\ln x, 0)$;

2) for every $\epsilon > 0$ there exists a $C_1 > 0$ and a_1, $0 < a_1 \leq 1$, such that $\mu(U_\epsilon(N)) \leq C_1 \epsilon^{a_1}$, where $U_\epsilon(N)$ denotes a ϵ-neighborhood of the set N;

3) there exist $C_2 > 0$, $C_3 > 0$, a_2, a_3, $0 < a_2, a_3 \leq 1$, such that for every $x \in M\backslash N$

$$\|df_x\| < C_2 \rho(x, N)^{-a_2},$$
$$\|d^2 f_x\| < C_3 \rho(x, N)^{-a_3}.$$

It follows from condition 1) and Theorem 2.7 (see Chap. 1, Sect. 2) that μ-almost every point $x \in M\backslash N$ will be regular in the sense of Lyapunov (see Chap. 7, Sect. 2). From condition 2) it follows that $\mu(N) = 0$. The conditions 1)–3) enable one to construct LSM and LUM at every regular point $x \in M\backslash N$ at which the characteristic Lyapunov exponents are different from zero, to prove the property of absolutely continuity and to obtain an exact upper estimation of the entropy of the dynamical system under consideration.

The proofs repeat in great detail the proofs of the corresponding statement considered in Chap. 7. We shall formulate the main results obtained in this way.

Theorem 1.14. *Let (f, N) be a discontinuous dynamical system, preserving the measure μ, equivalent to the Riemannian volume, the set $\Lambda \subset M\backslash N$ consists of points with nonzero characteristic Lyapunov exponents and $\mu(\Lambda) > 0$. Then*

1) *the ergodic components of the automorphism $f|\Lambda$ have positive measure;*
2) *if the automorphism $f|\Lambda$ has continuous spectrum, then it is isomorphic to a Bernoulli shift;*
3) *the corresponding formula of* Chap. 7 *is valid for the entropy of the mapping f.*

Recently, general approaches for studying the ergodic properties of nonuniformly hyperbolic systems were developed in [LW] and [M2]. These approaches have already been successfully applied to the study of some billiard type systems.

An example of conservative hyperbolic dynamical systems with singularities, besides billiard systems, is toral linked twist mappings. Let $\mathbb{T}^2 = \mathbb{R}^2/Z^2$ be the standard torus and let P, Q be closed annuli in $\mathbb{T}^2$ defined by

$$P = \{(x, y) \in \mathbb{T}^2 : y_0 \leq y \leq y_1, |y_1 - y_0| \leq 1\}$$

and

$$Q = \{(x, y) \in \mathbb{T}^2 : x_0 \leq x \leq x_1, |x_1 - x_0| \leq 1\}.$$

Let $f : [y_0, y_1] \to \mathbb{R}$, $g : [x_0, x_1] \to \mathbb{R}$ be C^2-functions such that $f(y_0) = g(x_0) = 0$, $f(y_1) = k$, $g(x_1) = l$ for some integers k and l.

Define mappings F_f and G_g of the set $P \cup Q$ onto itself by $F_f(x, y) = (x + f(y), y)$, $G_g(x, y) = (x, y + g(x))$ on P and Q respectively and $F_f|Q\backslash P = \mathbf{Id}$, $G_g|P\backslash Q = \mathbf{Id}$. We shall be interested in the composition of these mappings $H = H_{f,g} = G_g \circ F_f$. It is easy to see that H preserves the Lebesgue measure on $P \cup Q$ and has singularities on the set $\partial P \cup \partial Q$.

Suppose that $\frac{df}{dy} \neq 0$ and $\frac{dg}{dx} \neq 0$ for all $y \in [y_0, y_1]$, $x \in [x_0, x_1]$ and denote $\alpha = \inf\left\{\frac{df}{dy} : y \in [y_0, y_1]\right\}$, $\beta = \inf\left\{\frac{dg}{dx} : x \in [x_0, x_1]\right\}$. We shall then call $F|P$ a (k, α)-twist, $G|Q$ a (l, β)-twist and H a toral linked twist mapping.

F. Przytycki [Pr1] has proved many results concerning the properties of this class of mappings.

Theorem 1.15. *Let H be a toral linked twist mapping composed from (k, α)- and (l, β)-twists.*
If $\alpha\beta > 0$, then H is isomorphic to a Bernoulli shift
If $\alpha\beta < -4$, then H is almost hyperbolic. First of all this means that all Lyapunov exponents are almost everywhere (with respect to Lebesgue measure) different from zero. Second, the set $P \cup Q$ can be decomposed into a countable family of invariant pairwise disjoint sets Λ_i with a positive measure such that for every i $(i = 1, 2, \ldots)$ $H|\Lambda_i$ is ergodic and $\Lambda_i = \bigcup_{j=1}^{j(i)} \Lambda_i^j$, where $\Lambda_i^{j'} \cap \Lambda_i^{j''} = \emptyset$ for $j' \neq j''$, $H|\Lambda_i$ permutes Λ_i^j and for each j a mapping $H^{j(i)}|\Lambda_i^j$ is isomorphic to a Bernoulli shift.

1.9. Markov Approximations and Symbolic Dynamics for Hyperbolic Billiards. For dispersing billiards, a Markov partition can be constructed (see the definition in Chap. 7, Sect. 3). Nevertheless if for smooth hyperbolic systems (for instance A-systems) there exists a Markov partition with a finite number of elements, one could not expect to construct such a partition for billiards. In fact, as the transformation T_1 is discontinuous, then the regular components of the global stable and unstable manifolds can be arbitrarily small. Hence, a Markov partition has to contain elements of arbitrarily small sizes.

A Markov partition was constructed in [BS2] for two-dimensional dispersing billiard, such that the transformations T_1 and T_1^{-1} have a finite number of curves of discontinuity and satisfy some additional technical conditions. This construction was later simplified and generalized [BSC1]. The construction applied equally well to hyperbolic, nondispersing billiards.

The existence of a Markov partition with a countable number of elements does not allow us to obtain immediately the same consequences as hold for a smooth hyperbolic dynamical system which has a finite Markov partition. In the case of an infinite partition, an induced symbolic dynamics has to satisfy some additional conditions, i.e., a union of its elements with small sizes must have a small measure and images of small elements have to transform under the action of $\{T_1^n\}$ mainly into elements with larger sizes.

Images of every LUM under the action of a dynamical system will have large (exponential in time) sizes and therefore, after a sufficiently large time, images of all LUMs would fill the whole phase space almost uniformly. Hyperbolic billiards are discontinuous systems. Hence, simultaneously with the expansion of LUM it is also decomposed when its images intersect the manifolds of discontinuity of the transformation T_1. Using some additional properties of the Markov partition constructed in [BS2] it was proved that in ergodic hyperbolic billiards, expansion is stronger than the decomposition due to discontinuities and that there exists a subset of a large measure (but not of full measure) as in the case of smooth hyperbolic systems, which consists of LUMs, such that its images under the action of large iterates of T_1 are sufficiently dense and uniformly fill the whole phase space of a billiard under consideration.

The properties of this type are needed in order to study questions more delicate than ergodicity, mixing, Bernoulli. Namely one would like to study the rate of correlations decay (rate of mixing), or the Central Limit Theorems.

The corresponding results will be discussed later. Here we only make the important point [BSC2] that one actually doesn't need a Markov Partition to study these properties. By this we mean an approach that is based on some consistent dynamical partitioning of the phase space. (There are other powerful approaches, such as the one based on the Perron-Frobenius-Ruelle operator [Ru1], to study the statistical properties of dynamical systems.) In fact, one considers only finite (but long) segments of trajectories in order to estimate a rate of correlations decay or to establish some other statistical properties. Therefore, the first thing that you do after constructing a Markov Partition is to throw away all of its small elements. (For instance, if we look at the trajectories of length T than certainly all elements with diameters less than $C_1 \exp(-T)$, $C_1 > 0$ must be thrown away because they remain to be small after a time T.)

Thus one needs a sequence of partitions ξ_T, $T \to \infty$, of some *subsets* of a phase space with a big measure (say more than $1 - C_2 \exp(-T)$, $C_2 > 0$) rather than one (Markov) partition of entire phase space.

A sequence of such partitions ξ_T N. I. Chernov called a Markov sieve [BSC2]. (We don't discuss here the definition of a Markov sieve because it is too long and too technical.) Chernov has developed a general theory which deals with the applications of Markov sieves to the study of statistical properties of dynamical systems [Ch5]. In this paper he also constructed Markov sieves for high dimensional dispersing billiards.

1.10. Statistical Properties of Dispersing Billiards and of the Lorentz Gas. The properties of symbolic dynamics stated in [BS2] allow us to prove for dispersing billiards in two-dimensional domains some assertions which are analogous to limit theorems in probability theory.

Let $h(\omega)$ be a function defined on the space of sequences Ω such that $|h(\omega)| < C_1$, where C_1 is a constant and let there also exist a number λ, $0 < \lambda < 1$, satisfying the property that for all n sufficiently large one can find functions $h_n(\omega) = h_n(\omega_{-n} \dots \omega_n)$, $\int_\Omega h_n d\mu_0 = 0$, depending only on coordinates ω_i with $|i| \leq n$, such that $\sup_\omega |h(\omega) - h_n(\omega)| < \lambda^n$. We define by T_0 the shift in the space Ω which corresponds to the transformation T_1 by the symbolic representation $\phi : M \to \Omega$ of the dynamical system under consideration. For those dispersing billiards (in particular, with a finite number of curves of discontinuity) for which a Markov partition was constructed in [BS2] ([BSC1]) the following theorem was proved in [BS4] ([BSC2]).

Theorem 1.16. *Let μ_0 be a measure which satisfies conditions* 1)–3), *stated in* [BS2] (*see* Lemmas 6.2, 6.5, 6.6). *If $\int_\Omega h d\mu_0 = 0$, then there exists a number α, $0 < \alpha < 1$, such that $|\int_\Omega h(T_0^n\omega)h(\omega)d\mu_0| < \exp(-n^\alpha)$ for all n sufficiently large.*

The question whether or not one can actually put $\alpha = 1$ in Theorem 1.16 was around since [BS2]. The answer was not obvious at all. Indeed the cutting of regular local unstable manifolds by singularities may, in principle, slow down a rate of correlations decay. L. S. Young [Y3,Y4] has developed a technique which allows one to address this delicate question for general hyperbolic dynamical systems with singularities. To apply this technique, one needs to show that a number of images of singularity manifolds that are passing through an arbitrary point of the phase space grows slower than exponentially. Such fact has been proven in [BSC2] for dispersing billiards with smooth boundary (i.e., for Sinai billiards). Therefore [Y3] finally settled the question on exponential decay of correlations in a periodic Lorentz gas with a bounded free path.

One can prove (cf. [BS4], [BSC2]) that for functions satisfying conditions of Theorem 1.16 the central limit theorem of the probability theory holds.

The most important example of a dispersing billiard is the *Lorentz gas*. We shall give now the general definition of this dynamical system. Let an infinite number of balls (scatterers) be randomly distributed in the d-dimensional Euclidean space. In the complement of the set of scatterers, an infinite number of point particles are distributed according to some probabilistic or regular law. Every particle moves with a constant velocity and reflects from scatterers according to the law "the angle of incidence equals the angle of reflection." The dynamical system which corresponds to the motion of this infinite ensemble of particles is called the Lorentz gas.

In view of the absence of interactions between moving particles, one can consider dynamical system generated by the motion of a single particle. The dynamical system generated by the motion of an infinite ensemble of particles will be considered in Chap. 10.

The Lorentz gas is one of the most popular models in the non-equilibrium statistical mechanics. In this context, it is of great interest to study the problem of the existence of transport coefficients and the related problem of the slow decay of correlation functions. Among transport coefficients there are coefficients of diffusion, viscosity, heat conductivity, electroconductivity and so on. Since momentum is not conserved in the Lorentz gas, the diffusion coefficient D is a single transport coefficient for this model. According to the Einstein formula,

$$D = \frac{2}{d} \int_0^\infty \int_M (v(x(0)) \cdot v(x(t))) d\mu(x(0)) dt$$

where $v(x(0))$ is the velocity of the particle at the moment $t = 0$.

We shall consider a periodic configuration of scatterers. One can then study the motion of the point particle in a torus $\mathbb{T}^d$ with the Euclidean metric. We shall say that the Lorentz gas has a *finite horizon* if the free path length (i.e., a path between two successive reflections from scatterers) of the point particle is uniformly bounded from above by some constant. In this case, the number of curves of discontinuity of the transformation T_1 is finite.

It was proved in [BS4] (see also [BSC2]) that for the two-dimensional periodic Lorentz gas with a finite horizon, the diffusion coefficient exists and is positive. The generalization to higher dimensions was proved in [Ch5].

Let the rectangle $\Pi = \{q = (q_1, q_2) : 0 \le q_1 < B_1, 0 \le q_2 < B_2\}$ be a fundamental domain which corresponds to the configuration of scatterers in the plane. We shall consider the set $M \cap (\Pi \times S^1) = \widehat{M}$ where M is the phase space of the Lorentz gas. Let μ be a probability measure concentrated on $\widehat{M}$ which is absolutely continuous with respect to the Lebesgue measure on M and its density $f(x) \in C^1$. Then, a point $x \in M$ can be considered as a random variable, distributed according to the measure μ. If $T^t x = x(t) = (q(t), v(t))$, then $q(t)$, $v(t)$ are also random variables. For every t we put $q_t(s) = \frac{1}{\sqrt{t}} q(st)$, $0 \le s \le 1$. The measure μ induces a probability distribution μ_t on the set of all possible trajectories $q_t(s)$, $0 \le s \le 1$, which are points of the space $C_{[0,1]}(\mathbb{R}^2)$ of continuous functions defined on the unit segment with values in $\mathbb{R}^2$.

The following statement which is an analog of Donsker's invariance principle in the theory of random processes was proved in [BS4] for the dynamical system under consideration.

Theorem 1.17. *The measures μ_t converge weakly to a Wiener measure.*

It was the first rigorous result on the convergence of a purely deterministic dynamical system without any random mechanism to the random process of Brownian motion. Usually the motion of a heavy particle under collisions with a gas of light particles that do not interact between themselves is considered as a "physical" image of the random walk. Nevertheless the representation of the random walk as of the motion of a light particle in the field of heavy immobile scatterers with elastic reflections from their boundaries seems to be even more natural. Recently N. I. Chernov proved the same results for multidimensional periodic Lorentz gas with a finite horizon [Ch5].

If the length of a free path of the particle is not bounded from above, then the rate of correlation decay in the case of continuous time is powerlike, but in the case of discrete time it is of the same type as for bounded free path (see the Theorem 1.10) (cf. [Bu5]). The reason for that is the existence of singular periodic trajectories which are tangent to the boundary ∂Q at every point of reflection (Fig. 10).

This is the reason why the diffusion coefficient should not exist for a periodic Lorentz gas with an infinite horizon. Many numerical simulations confirm this claim but there is no proof (see the detailed discussion in [Bl]).

Many generalizations of the periodic Lorentz gas have been considered. Several papers were devoted to the study of periodic Lorentz gas with the nonoverlapping "soft" scatterers, i.e., with some localized potential fields. The strongest results were obtained in [DL] where ergodicity was proven for some examples with repulsing as well as with attracting potentials. (The first (second) case corresponds to the dispersing (focusing) billiards.)

The ergodicity and integrability of some models of periodic Lorentz gas in the magnetic field were shown in [BK].

The paper [CELS] deals with a periodic Lorentz gas in a weak electric field. In this model the dynamics is non-Hamiltonian but time-reversible. Besides, the kinetic energy is conserved. The technique developed in [BSC2] allowed the proof of the existence of SRB-measure, the estimation of the dimension of this measure, and the proof of Ohm's law.

The important problem concerns the behavior of a mean free path of the particle when the radii of scatterers tend to zero. The calculation of a mean free path is also important for any hyperbolic billiard because it allows the establishment of relations between Lyapunov exponents and KS entropy of a billiards flow and those of a billiard map.

Following [BGW], we denote $\mathbb{Z}_\epsilon = \{x \in \mathbb{R}^d : \text{dist}(x, \epsilon\mathbb{Z}^d) > \epsilon^\gamma\}$ for all $0 < \epsilon < 1/2$ and $\gamma \geq 1$. For any $x \in \bar{\mathbb{Z}}_\epsilon$ and $v \in S^{d-1}$ the free path length is defined as $\tau_\epsilon(x, v) = \inf\{t > 0 : x + tv \in \partial\mathbb{Z}_\epsilon\}$. Assume that the phase space of the periodic Lorentz gas $(\mathbb{Z}_\epsilon/\epsilon\mathbb{Z}^d) \times S^{d-1}$ is equipped with the natural Borel probability measure $d\mu_\epsilon(x, v)$. Consider the distribution function F_ϵ of τ_ϵ with respect to μ_ϵ which is the shift of the measure μ_e on time τ_ϵ.

The geometric mean free path for ergodic billiards is given via the following formula (see e.g., [Ch2, 6])

$$l_\epsilon = \int_{M_1} \tau_\epsilon(q, v) d\nu_\epsilon(q, v). \tag{8.6}$$

It has been shown in [Ch2] that

$$l_\epsilon = \frac{1}{|B^{d-1}|} \epsilon^{d-\gamma(d-1)} + O(\epsilon^\gamma) \tag{8.7}$$

where $|B^{d-1}|$ is the volume of a unit ball in $\mathbb{R}^{d-1}$. It follows from (8.7) that $\gamma_c = d/(d-1)$ is the critical value of γ.

Recall that the weak topology on the space of Borelian probability measures on $\mathbb{R}^d$ is defined by the family of seminorms $\mu \to |(\mu, f)|$ for all bounded continuous functions f, while the vague topology on this space is the one defined by the subfamily of these seminorms corresponding to continuous functions with compact support.

Theorem 1.18.
1) *If* $\gamma > \gamma_0$ *then* $F_\epsilon \to 0$ *vaguely as* $\epsilon \to 0$.
2) *If* $1 \le \gamma < \gamma_0$, *then* $F_\epsilon \to \delta_0$ *weakly as* $\epsilon \to 0$.
3) *If* $\gamma = \gamma_0$ *then any vague limit point* F *of the family* $\{F_\epsilon\}$ *is a probability measure and satisfies* $\limsup_{t\to\infty} tF([t, +\infty]) < \infty$.
4) *If* $\gamma = \gamma_c$ *and* $d = 2$ *then any vague limit point* F *of the family* $\{F_\epsilon\}$ *satisfies* $\liminf_{t\to\infty} tF(t, +\infty)) > 0$.

The first and second statements in Theorem 1.18 (for $d = 2$) were proved in [DDG]. The three, four and two (for $d > 2$) were proved in [BGW].

We would like to mention in conclusion that one of the most interesting and important problems in this subject is the study of ergodic properties of the Lorentz gas with a random configuration of scatterers. This problem is closely related to the yet unsolved classical problem of the random walk in a random environment. A natural hypothesis is that in this system, the velocity autocorrelation function decays as const $t^{-(d/2+1)}$.

1.11. Transport Coefficients for the Simplest Mechanical Models. One of the main challenges of statistical mechanics is to prove the existence of finite and nondegenerate transport coefficients for a system of particles governed by Newton's equations of motion. There is no surprise that the Lorentz gas was the first mechanical model for which the existence and nondegeneracy of a transport coefficient was proven from the first principles.

Indeed, the diffusion coefficient is the simplest one in the hierarchy of transport coefficients and the Lorentz gas is the simplest possible model where a diffusion coefficient may exist. Obviously, at least one particle should move in order to ensure the transport of mass in a system. It is exactly what is happening in a Lorentz gas. And diffusion is the only process that is going on in this system. To prove the existence and nondegeneracy of other transport coefficients one should look for the simplest models of interacting particles where these coefficients may exist.

Such a hierarchy of models was considered in [BSp]. It was proven there that both shear and bulk viscosity coefficients do exist and are nondegenerate for the two-disk periodic fluid. The Green-Kubo formulas provide the formal expressions for the transport coefficients as the integrals of the time correlations of certain functions on the phase space. The Einstein's formula is the first in this hierarchy.

In three-dimensional physical space, a fluid in thermal equilibrium has five locally conserved fields: the particle density $n^{(0)}(x,t)$, the three components of the momentum density $n^{(\alpha)}(x,t)$, $\alpha = 1,2,3$, and the energy density $n^{(4)}(x,t)$, which depends on location $x \in \mathbb{R}^3$ and time $t \in \mathbb{R}$. (These are the distributions on phase space indexed by x, t.)

By the local conservation law we have, in a distributional sense,

$$\frac{\partial}{\partial t} n^{(i)}(x,t) + \operatorname{div} j^{(i)}(x,t) = 0, \tag{8.8}$$

where $i = 0, 1, \ldots, 4$, with the local currents $j^{(i)}$. (Since the interaction between particles has some range, the local currents are not uniquely defined. However, the total (space integrated) currents always are (see e.g., [Sp]).)

The Green-Kubo formulas for the transport coefficients have the form

$$D^{ij}_{\alpha\beta} = \int dt \int dx \left(\langle j^{(i)}_{\alpha}(x,t) j^{(i)}_{\beta}(0,0) \rangle - \langle j^{(i)}_{\alpha}(0,0) \rangle \langle j_{\beta}(0,0) \rangle \right) \tag{8.9}$$

where $\alpha, \beta = 1,2,3$, $i, j = 0, 1, \ldots, 4$, and $\langle \cdot \rangle$ denotes the equilibrium ensemble average. Because of rotation invariance and by time reversal symmetry, only three out of 15×15 coefficients do not vanish. They can be expressed as a linear combination of shear and bulk viscosity and by the thermal conductivity (see the details e.g., in [Sp]).

Observe that the Einstein formula has the same structure. Indeed, the diffusion coefficient is represented there as the integral of the time correlation functions of the current of mass $mv(x,t)$. It occurs that one needs at least two particles to ensure a transport of momentum. The unfolding of the system of two elastic disks on torus was studied in [BSp]. This system can be reduced to a billiard by the natural canonical transformation [Si2]. This billiard is a periodic Lorentz gas with infinite horizon if the radius of disks is small. The corresponding billiard is a dispersing one in a domain whose boundary is not smooth but contains singular points if this radius is not too small.

The next model in the hierarchy is the unfolding of a system of three elastic disks on a torus. This is the simplest model where a coefficient of heat conductivity may exist. However, in spite of the proof of ergodicity for this system [KSS2], there are still many new difficulties that appear on the way to the proof of the existence and nondegeneracy of the coefficient of heat conductivity in this model.

It is important to observe that at any higher level in the hierarchy of systems of N elastically interacting particles, there appear the new types of singularities which make the analysis much harder.

§2. Strange Attractors

2.1. Definition of a Strange Attractor. The notion of a "strange attractor" was introduced in the work by D. Ruelle and F. Takens [RT]. They proposed to use this term for invariant attracting sets of dynamical systems which were not manifolds and have in some cross-sections the structure of the perfect Cantor set. According to their general idea, sets with such a structure arise in the dynamical system corresponding to the Navier-Stokes equations. Although this picture has not yet been verified this term has become very popular, especially among physicists since such strange attractors had been found, usually on the basis of computer simulations, in various physical systems and one has begun to connect the existence of a strange attractor with the chaotic behavior of trajectories of a dynamical system. Besides, it is a general belief that trajectories on a strange attractor are (in some sense) unstable which is the cause of stochasticity of the corresponding dynamical system. However, there are only a few rigorous mathematical results which support this point of view. Nevertheless, it became in some sense a fashionable to find a strange attractor. The role of a mathematician is to formulate and to prove rigorous results on the behavior of the studied dynamical system on the basis of available results of computer simulations.

At present many works devoted to the calculation of different dimension-like characteristics of invariant sets of dynamical systems have appeared. The corresponding basic idea is that among these sets only strange attractors can have a fractional dimension. The general theorems concerning this subject can be found in Chap. 7, Sect. 6.

In accordance with the general idea of this book, we shall consider only the rigorous mathematical results related to the ergodic theory of strange attractors. First of all we would like to change slightly the corresponding terminology. Let the flow $\{S^t\}$ be induced by a smooth vector field on a smooth compact manifold M.

Definition. An invariant closed set A is called an *attractor* if there exists a neighborhood $U_0 \supset A$ such that $U_t = S^t U_0 \subset U_0$ as $t > 0$ and $\bigcap_t U_t = A$.

The simplest examples of attractors are a stable stationary point and a stable periodic trajectory. In Chap. 7, Sect. 2, hyperbolic attractors of smooth dynamical systems and, in particular, the Smale-Williams attractor, which has a more complex structure, were considered. We shall introduce the notion of a stochastic attractor taking as a basic point the existence of an invariant measure which is analogous to the u-Gibbs measure $\mu_{\phi^{(u)}}$ for the smooth hyperbolic systems.

Definition. An attractor A will be called *stochastic* if the following conditions hold:

1) for any absolutely continuous measure μ_0 with the support in U_0, its shifts μ_t converge weakly to an invariant measure λ which does not depend on μ_0;

2) the dynamical system $(A, \lambda, \{S^t\})$ is mixing.

The choice of the absolutely continuous measures as a class of initial measures μ_0 has a number of advantages beyond the analytical advantages. This class of measures is stable with respect to the passage to a discrete version of the dynamical system, to numerical simulations on computers, and to small random perturbations.

The hyperbolic attractors are stochastic ones (see Chap. 7, Sect. 2.5). We shall now consider the structure and properties of the stochastic attractor arising in the famous Lorenz model.

2.2. The Lorenz Attractor. Let us consider the following system of three ordinary differential equations

$$\begin{cases} \dfrac{dx}{dt} = -\sigma x + \sigma y \\ \dfrac{dy}{dt} = rx - y - xz \\ \dfrac{dz}{dt} = -bz + xy \end{cases} \tag{8.10}$$

where σ, b, r are some parameters with positive values. The system (8.10) was obtained by Galerkin approximation from the Rayleigh-Benard problem describing the convection in a layer of fluid of uniform depth between two parallel smooth planes with a constant temperature difference between the upper and the lower surfaces (cf. [Lor]). The same system as (8.10) arises in the theory of lasers and in geophysics as the simplest model of a geomagnetic "dynamo."

It was E. Lorenz who first studied the system (8.10) with the help of computer simulation and discovered the rather irregular behavior of its trajectories when $r = 28$, $\sigma = 10$ and $b = 8/3$. The topological structure of the attractor arised in this model was investigated by R. Williams [Wi].

We shall now describe the simplest properties of the system (8.10):

i) The divergence of the right-hand side of (8.10) is equal to $-(\sigma+1+b) < 0$ and therefore any volume in the phase space of the Lorenz system shrinks exponentially with time.

ii) Infinity is an unstable point. Any trajectory of the system will reach a compact subset of the phase space and will remain there at all subsequent times.

iii) The system (8.10) is invariant with respect to the transformation $x \to -x$, $y \to -y$, $z \to -z$.

iv) If $\sigma = 10$, $b = 8/3$, $r > \sigma(\sigma + b + 3)(\sigma - b - 1)^{-1} = 24, 73, \ldots$, then there are only three stationary points in the phase space of the system $O = (0, 0, 0)$, $O_1 = (\sqrt{b(r-1)}, \sqrt{b(r-1)}, r-1)$ and $O_{-1} = (-\sqrt{b(r-1)}, -\sqrt{b(r-1)}, r-1)$ which are all hyperbolic; the stable separatrix $W^{(s)}$ of the point O is two-dimensional and the stable separatrices of the points O_1 and O_2 are both one-dimensional. The branches of the unstable separatrix at point O will be denoted by Γ_1 and Γ_{-1} (see Fig. 15).

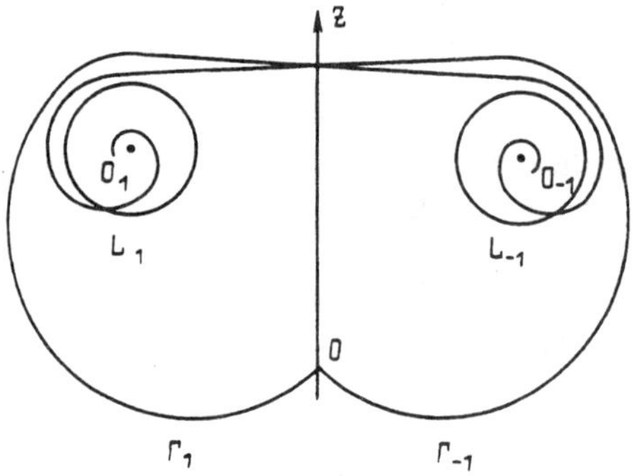

Fig. 15

The sequence of bifurcations leading to the formation of a stochastic attractor in the Lorenz model when $r = 28$, $b = 8/3$ and σ varies between 0 and 10 was considered by V. S. Afraimovich, V. V. Bykov and L. P. Shilnikov (cf. [ABS1]) (see also paper [ABS2] which contains the detailed proofs for a more general class of dynamical systems). The results of [ABS1] were obtained by the combination of the rigorous mathematical analysis of qualitative behavior of the system (8.10) with the quantitative estimations obtained by computer simulations.

It was shown in [ABS1] that Γ_1 and Γ_{-1} become doubly asymptotic when $r = 28$, $b = 8/3$ and $\sigma = \sigma_1 \approx 3.42$. This means that Γ_1 and Γ_{-1} belong locally to the stable separatrix of the point 0. From the results of [ABS1], it follows that two periodic hyperbolic trajectories (L_1 and L_{-1}) appear from the loops Γ_1 and Γ_{-1} when σ crosses the value σ_1. Moreover the branch Γ_1 (resp., Γ_{-1}) is attracted to the point O_{-1} (O_1) when $\sigma_1 < \sigma < 5.87$. Another bifurcation takes place at $\sigma = \sigma_2 \approx 5.87$. As a result the branch Γ_1 is attracted to a periodic motion L_{-1} and symmetrically Γ_{-1} to L_1 (Fig. 15). The limiting set which is called the Lorenz attractor, appears in the phase space of the system (8.10) when σ crosses the value σ_2.

Let us consider the Poincaré map of the plane $P = \{z = r - 1\}$ which contains the stable points O_1 and O_{-1}. We denote by z_i^*, $i = 1, -1$, the points of the first intersection of the branches Γ_i, $i = 1, -1$, with the plane P. It is assumed that Γ_i, $i = 1, -1$, intersects P at the point z_i^* from above to below; we denote by z_i^0, $i = 1, -1$, the points of intersection of the hyperbolic cycles L_i, $i = 1, -1$, with P and by S the set $W^{(s)} \cap P$. (The set S is, generally speaking, not locally closed since the separatrix $W^{(s)}$ is strongly bending and does not divide the space locally. Thus the unstable branch Γ_1 (Γ_{-1}) can come close to the point O_{-1} (O_1) without intersection with $W^{(s)}$. The Poincaré mapping T associates to any point $x \in P$ the point $y = Tx \in P$ where the integral curve originated at x at first intersects the plane P from above to below. It is obvious that T is discontinuous on the set S. Actually points which belong to P and lie on opposite sides of $W^{(s)}$ move along different branches (Γ_1 and Γ_{-1}) of the unstable separatrix. In [ABS1] the properties of the mapping T were formulated axiomatically based on the results of the computer simulations and of the qualitative analysis of the system (8.10) in the following manner.

Let $\Pi = \{|x| \le 1, |y| \le 1\}$ be the rectangle in the plane. We denote $\Pi_1 = \{|x| \le 1, 0 < y \le 1\}$, $\Pi_{-1} = \{|x| \le 1, -1 \le y < 0\}$. A mapping T is defined by the following conditions:

1) $T_i(x, y) = (f_i(x, y), g_i(x, y))$, $i = 1, -1$, where f_i, g_i are functions of class C^2.

2) A mapping T_i has a stationary point $z_i^0 \in K_i = \{|x| < 1, y = i\}$ of saddle type.

3) The segment K_i belongs to the stable separatrix of the point z_i^0, $i = 1, -1$.

4) The functions f_i and g_i can be defined on the segment S so that $\lim_{y\to 0} f_i(x, y) = x_i^*$, $\lim_{y\to 0} g_i(x, y) = y_i^*$ and $\operatorname{sgn} x_i^* = \operatorname{sgn} i$, $\operatorname{sgn} y_i^* = -\operatorname{sgn} i$, $i = \pm 1$. We denote $z_i^* = (x_i^*, y_i^*)$.

5) The mappings T_1 and T_{-1} are uniformly completely hyperbolic (see Chap. 7, Sect. 1) on the set $\Pi \setminus \left(\bigcup_{k=0}^{\infty} T_1^{-k} S \cup \bigcup_{k=0}^{\infty} T_{-1}^{-k} S\right)$. They are expanding in the vertical direction and contracting in the horizontal one.

6) $f_1(x, y) = -f_{-1}(-x, -y)$, $g_1(x, y) = -g_{-1}(-x, -y)$.[2]

Let us define the mapping T by letting $T = T_i|\Pi_i \cup S$, $i = 1, -1$, (see Fig. 16).

It is easy to see that the image $T\Pi$ consists of two curvilinear triangles (Fig. 16). In [SV] it was shown with the help of computer simulations that when $\sigma = 6$, $b = 8/3$, $r = 28$, there exist two narrow triangles with vertices

[2] Condition 6) is not necessary, in the sense that all results which will be formulated below for the mapping T also hold without it.

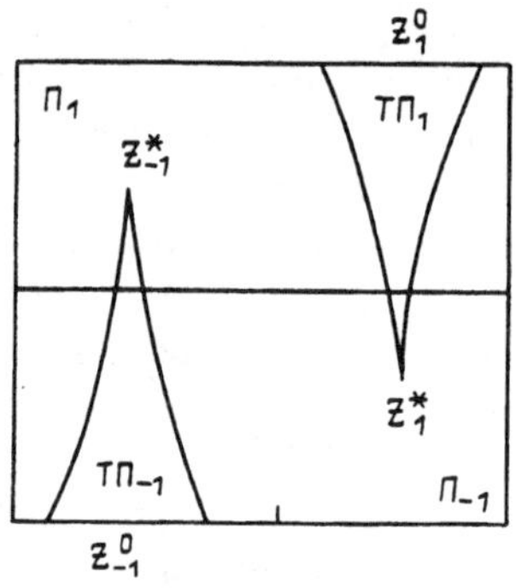

Fig. 16

z_1^* and z_{-1}^* situated on P in the opposite sides of S such that the Poincaré mapping of the plane P in the phase space of the system (8.10) satisfies on these triangles the property of hyperbolicity.

The following assertion was proved in [ABS1][3]

Theorem 2.1. *If the mapping T is topologically transitive* (*see* Chap. 7, Sect. 2.2), *then there is a single stable limiting set of T in Π which has the form $\Lambda = \bigcap_{n\geq 0} T^n \Pi$ and is a one-dimensional hyperbolic set with the following properties:*
i) *Λ consists of two connected components,*
ii) *Λ is the closure of the set of periodic points.*

The attractor Λ contains a continuum of smooth curves which can be projected uniquely on the y-axis. Any such curve is a regular segment of an expanding fiber. We shall denote by ξ the partition of Λ into such maximal expanding curves and by $C_\xi(x)$ the element of ξ which contains the point x.

We therefore have the discontinuous hyperbolic mapping of the square Π into itself. Thus the situation is in some respects similar to the case of Sinai billiards considered in the preceding section. However, the Lorenz system has some essential differences with respect to Sinai billiards. First of all, it has no natural invariant measure.

Theorem 2.2. *Let v be a measure on Π which is absolutely continuous with respect to the Lebesgue measure and the corresponding density is continuously differentiable. Then the sequence of measures $T_*^n v$ converges weakly to an invariant with respect to T measure μ. Moreover the conditional measure $\mu(\cdot|C_\xi)$ induced by μ on regular expanding fibers is absolutely continuous with respect to the Lebesgue measure on C_ξ.*

3 In [ABS1] some other results on the properties of the mapping T were obtained too, but we shall not use them in what follows.

One could prove a stronger assertion which states, in analogy to the corresponding one in Chap. 7, Sect. 3, that convergence is strong almost everywhere.

Nevertheless it occurs that under the above conditions on the mapping T, the attractor Λ does not have to be stochastic; namely, the restriction of T onto Λ does not have to be mixing. In this case, any element of ξ contains lacunae[4]. Therefore there is a cyclic component in a spectrum of the dynamical system under consideration. In order for T to be mixing it is necessary that in condition five an expansion stronger than the usual condition of hyperbolicity holds. Under this condition, the following theorem is true (cf. [BS3]).

Theorem 2.3. *Let $f(x)$ be a continuous function defined in some neighborhood U, $\Lambda \subset U \subset \Pi$, and let ν be an absolutely continuous measure on Π whose support is contained in U. Then, for almost every point $x \in \Pi$ with respect to ν.*

$$\lim_{n\to\infty} \int_{\Pi} f(T^k x) d\nu = \int_{\Lambda} f(x) d\mu.$$

In [Bu4] the central limit theorem (see Chap. 6, Sect. 1) was proved for a wide class of functions on Π, also for sufficiently large n and for some positive number $\gamma < 1$, the following estimation of the rate of mixing was obtained

$$\left| \int_{\Lambda} f(x) g(T^n x) d\mu - \int_{\Lambda} f(x) d\mu \int_{\Lambda} g(x) d\mu \right| < \text{const}\exp(-n^{\gamma}) \quad (8.11)$$

The recent results by L.-S. Young [Y3] allow one to take $\gamma = 1$.

We would like to mention that the mapping T has a global stable foliation which is constructed in an analogous fashion to the partition of the square Π into segments $y = \text{const}$ (cf. [ABS1]). It allows us to represent T as a skew product (see Chap. 1, Sect. 4) over a monotonic transformation of the unit segment which has a single point of discontinuity, and to use the theory of one-dimensional mappings (see Chap. 9, Sect. 1). However, in order to prove mixing, one has to use a smoothness of the stable foliation which apparently does not hold for the dynamical system under consideration.

Let us discuss some available generalizations of the considered example. First of all it is clear that all results hold in case when the mapping T has discontinuities of the first class not on one but on a finite number of curves (compare with dispersing billiards). The phase space may not be two-dimensional but can have any finite dimension. In the latter case, manifolds of discontinuity have to have codimension one. In the phase space of the corresponding dynamical system on the n-dimensional cube, a one-dimensional stochastic

[4] This phenomenon is analogous to the arising of lacunae in piecewise monotonic one-dimensional mappings (see Chap. 9).

attractor exists as well. Hence, one could use a factorization along contracting fibers in order to pass to a one-dimensional mapping.

2.3. Some Other Examples of Hyperbolic Strange Attractors. V. N. Belykh considered the two-dimensional mapping which has a hyperbolic attractor and cannot be reduced to a one-dimensional one (cf. [Be]). This example also appeared in the study of concrete dynamical systems in physics, which are called the discrete systems of the phase synchronization. We shall consider the simplest version when the corresponding mapping is piecewise-linear and the property 6) holds.

Let T be the following mapping of the square Π into itself

$$\begin{cases} x_{n+1} \mp 1 = \lambda_1(x_n \mp 1) \\ y_{n+1} \mp 1 = \lambda_2(y_n \mp 1) \end{cases} \qquad \text{if } y \gtrless kx$$

where $0 < \lambda_1 < 1/2$, $1 < \lambda_2 < 2$, $k \neq 0$, $|k| < 1$. The main difference between this mapping and the mapping which arises in the Lorenz system is that there is no global contracting foliation now, since the curve of discontinuity is not parallel to contracting fibers and one cannot perform the factorization.

R. Lozi considered the following class of linear mappings of the plane into itself $(x, y) \to (1 + y - a|x|, bx)$ and by computer simulations found that there is a strange attractor in the system when $a = 1.7$ and $b = 0.58$ (cf. [Loz]). The introduction of this class of mappings was stimulated by numerous works devoted to the study of the Henon attractor for the following class of mappings of the plane into itself $(x, y) \to (1+y-ax^2, bx)$ (cf. [H]). The experience accumulated by studying ergodic and topological properties of one-dimensional mappings (see Chap. 9, Sect. 2) shows that it is easier to investigate a Lozi mapping than a Henon one. In fact, the existence and hyperbolicity of a Lozi attractor for some region in the space of parameters a and b was soon proved (cf. [Mi1]). The metric properties of a Lozi attractor were investigated in [Le], where an invariant measure was constructed and its uniqueness and stochasticity were proved. Later, in the series of remarkable papers Benedics, Carleson and Young [BC], [BY] proved the existence of a hyperbolic attractor in the Henon map and investigated its ergodic properties (see Chap. 7).

Ya.B. Pesin introduced a class of generalized hyperbolic attractors, which includes Lorenz, Belykh and Lozi attractors, and studied some of their topological and ergodic properties (cf. [Pe6]). In [Pe6], u-Gibbsian measures were constructed as well for these attractors.

The appearance of hyperbolic strange attractors in model dynamical system is not a rule but rather an exception. Numerous computer simulations and rigorous mathematical investigations have shown that, in a generic dissipative dynamical system with a chaotic behavior, there are usually stable periodic trajectories besides nontrivial hyperbolic subsets. Thus a generic situation in dissipative systems is apparently the same as in Hamiltonian systems, and the dynamics on a generic attractor is similar to the dynamics of generic conservative systems where there are stochastic layers and invariant KAM-tori (see Chap. 7, Sect. 2).

Additional References

[Bi] Bialy, M.: Convex billiards and a theorem by E. Hopf. Math. Zeitschrift *214*, 147–154 (1993)

[Bo] Bolotin, S.: Integrable Birkhoff billiards. Moscow Univ. Vestnik *2*, 33–36 (1990)

[BK] Berglund, N. and Kunz, H.: Integrability and ergodicity of classical billiards in a magnetic field. J. Stat. Phys. *83*, 81–126 (1996)

[Bl] Bleher, P. M.: Statistical properties of two-dimensional periodic Lorentz gas with infinite horizon. J. Stat. Phys. *66*, 315–373 (1992)

[BC] Benedicks, M. and Carleson, L.: The dynamics of the Henon map. Ann. Math. *133*, 73–169 (1991)

[BG] Burns, K., Gerber, M.: Real analytic Bernoulli geodesic flows on S^2. Ergod. Theory and Dyn. Syst. *9*, 27–45 (1989)

[BY1] Benedicks, M. and Young, L.-S.: SBR measures for certain Henon maps. Invent. Math. *113*, 541–576 (1993)

[BY2] Benedicks, M. and Young, L.-S.: Decay of correlations for certain Henon maps. Preprint (1996)

[BGW] Bourgain, J., Golse, F., Wennberg, B.: On the distribution of free path length for the periodic Lorentz gas. Comm. Math. Phys. *190*, 491–508 (1998)

[Bu6] Bunimovich, L. A.: Many-dimensional nowhere dispersing billiards with chaotic behavior. Physica D *33*, 58–64 (1988)

[Bu7] Bunimovich, L. A.: A theorem on ergodicity of two-dimensional hyperbolic billiards. Comm. Math. Phys. *130*, 529–621 (1990)

[Bu8] Bunimovich, L. A.: On absolutely focusing mirrors. Springer Lect. Notes Math. *1514*, 62–82 (1992)

[Bu9] Bunimoivch, L. A.: Conditions of stochasticity of 2-dimensional billiards. Chaos *1*, 187–193 (1992)

[Bu10] Bunimovich, L. A.: Continued fractions and geometrical optics. Amer. Math. Soc. Transl-s, Ser. 2 *171*, 45–56 (1995)

[BLPS] Bunimovich, L. A., Liverani, C., Pellegrinotti, A., Sukhov, Yu. M.: Ergodic systems of *n* balls in a billiard table. Comm. Math. Phys. *146*, 357–396 (1992)

[BSp] Bunimovich, L. A., Spohn, H.: Viscosity for a periodic two disk fluid: An existence proof. Comm. Math. Phys. *176*, 661–680 (1996)

[BR1] Bunimovich, L. A., Rehacek, J.: Nowhere dispersing 3D billiards with non-vanishing Lyapunov exponents. Comm. Math. Phys. *189*, 729–758 (1997)

[BR2] Bunimovich, L. A., Rehacek, J.: How many-dimensional stadia look like. Comm. Math. Phys. *197*, 277–301 (1998)

[BR3] Bunimovich, L. A., Rehacek, J.: On the ergodicity of many-dimensional focusing billiards. Ann. Inst. H. Poincaré, *68*, 421–448 (1998)

[BSC1] Bunimovich, L. A., Sinai, Ya. G., Chernov, N. I.: Markov partitions for two-dimensional hyperbolic billiards. Russ. Math. Surv. *45*, 105–152 (1990)

[BSC2] Bunimovich, L. A., Sinai, Ya. G., Chernov, N. I.: Statistical properties of two-dimensional hyperbolic billiards. Russ. Math. Surv. *46*, 47–106 (1991)

[Ch2] Chernov, N. I.: A new proof of Sinai's formula for the calculation of entropy of hyperbolic billiards. Applications to the Lorentz gas and Bunimovich's stadium. Funct. Anal. and Appl. *25*, 50–69 (1991)
[Ch3] Chernov, N. I.: The ergodicity of a Hamiltonian system of two particles in an external field. Physica D *53*, 233–239 (1991)
[Ch4] Chernov, N. I.: On local ergodicity in hyperbolic systems with singularities. Funct. Anal. and Appl. *27*, 60–64 (1993)
[Ch5] Chernov, N. I.: Statistical properties of the periodic Lorentz gas. Multidimensional case. J. Stat. Phys. *74*, 11–53 (1994)
[Ch6] Chernov, N. I.: Limit theorems and Markov approximations for chaotic dynamical systems. Probab. Theory and Related Fields *101*, 321–362 (1995)
[Ch7] Chernov, N. I.: Entropy, Lyapunov exponents and mean free path for billiards. J. Stat. Phys. *88*, 1–29 (1997)
[CdV] Colin de Verdiere, Y.: Ergodicité et fonctions propes du Laplacien. Comm. Math. Phys. *102*, 497–502 (1985)
[CELS] Chernov, N. I., Eyink, G. I., Lebowitz, J. L., and Sinai, Ya. G.: Steady state electrical conduction in the periodic Lorentz gas. Comm. Math. Phys. *154*, 569–601 (1993)
[CH] Chernov, N. I. and Haskell, C.: Nonuniformly hyperbolic K-systems are Bernoulli. Ergod. Theory and Dyn. Syst. *16*, 19–44 (1996)
[CM] Chernov, N. I. and Markarian, R.: Entropy of non-uniformly hyperbolic plane billiards. Bol. Soc. Bras. Mat. *23*, 121–135 (1992)
[Co] Coddington, H.: Treatise on reflection and refraction of light. Simpkin and Marshall, London (1989)
[Do1] Donnay, V. J.: Geodesic flow on the two-sphere. I: Positive measure entropy. Ergod. Theory and Dyn. Syst. *8*, 531–553 (1988)
[Do2] Donnay, V. J.: Geodesic flow on the two-sphere. II: Ergodicity. In: J. C. Alexander (ed.) Dynamical Systems. Lect. Notes *1342*, 112–153, Springer-Verlag, New York (1988)
[Do3] Donnay, V. J.: Using integrability to produce chaos: Billiards with positive entropy. Comm. Math. Phys. *141*, 225–257 (1991)
[DDG] Dumas, H. S., Dumas, L., Golse, F.: Remarks on the notion of mean free path for a periodic array of spherical obstacles. J. Stat. Phys. *87*, 943–950 (1997)
[DL] Donnay, V. J., Liverani, C.: Potentials on the two-torus for which the Hamiltonian flow is ergodic. Comm. Math. Phys. *135*, 267–302 (1990)
[G] Gutkin, E.: Billiards in polygons: survey of recent results. J. Stat. Phys. *87*, 7–26 (1996)
[GK] Gutkin, E. and Katok, A. B.: Caustics for inner and outer billiards. Commun. Math. Phys. *173*, 101–133 (1995)
[Ha] Halpern, B.: Strange billiard tables. Trans. Amer. Math. Soc. 297–305 (1977)
[Hu] Hubacher, A.: Instability of the boundary in the billiard ball problem. Comm. Math. Phys. *108*, 483–488 (1987)
[KMS] Kerckhoff, S., Masur, H., Smillie, J.: Ergodicity of billiard flows and quadratic differentials. Ann. Math. *124*, 293–311 (1986)
[KT] Kozlov, V. V., Treshchev, D. V.: Billiards. A genetic introduction to the dynamics of systems with impacts. Transl-s. of Math. Monographs *98* Amer. Math. Soc (1991)
[KSS1] Kramli, A., Simanyi, N., Szasz, D.: A "transversal" fundamental theorem for semi-dispersing billiards. Comm. Math. Phys. *129*, 535–560 (1990)
[KSS2] Kramli, A., Simanyi, N., Szasz, D.: Three billiard balls on the n-dimensional torus is a K-flow. Ann. Math. *133*, 37–72 (1991)
[KSS3] Kramli, A., Simanyi, N., Szasz, D.: The K-property of four billiard balls. Comm. Math. Phys. *144*, 107–148 (1992)
[LM] Lehtichet, H. E., Miller, B. N.: Numerical study of a billiard in a gravitational field. Physica D *21*, 94–104 (1986)

[LW] Liverani, C. and Wojtkowski, M.: Ergodicity in Hamiltonian systems. Dynamics Reported *4*, 130–202 (1995)

[M1] Markarian, R.: Billiards with Pesin region of measure one. Comm. Math. Phys. *118*, 87–97 (1988)

[M2] Markarian, R.: Nonuniform hyperbolicity, quadratic forms and billiards. Annals de la Faculte des Sciences de Toulouse *IV* (1994)

[M3] Markarian, R.: New ergodic billiards. Exact results. Nonlinearity *6*, 819–841 (1993)

[Ma] Mather, J.: Glancing billiards. Ergod. Th. and Dyn. Syst. *2*, 397–403 (1982)

[SC2] Sinai, Ya. G., Chernov, N. I.: Ergodic properties of certain systems of two-dimensional discs and three-dimensional balls. Russ. Math. Surveys *42*, 181–207 (1987)

[SS] Simanyi, N., Szasz, D.: Hard ball systems are completely hyperbolic. Ann. Math. (to be published)

[Sp] Spohn, H.: Large Scale Dynamics of Interacting Particles. Springer-Verlag, Heidelberg (1991)

[Sz] Szasz, D.: Boltzmann ergodic hypothesis, a conjecture for centuries? Studia Sci. Math. Hung. *31*, 299–322 (1996)

[Ta] Tabachnikov, S.: Billiards, Panoramas et Syntheses. Soc. Math. France (1995)

[V] Vorobets, Ya. B.: Ergodicity of billiards in polygons: explicit examples. Russ. Math. Surveys *51*, 151–152 (1996)

[Wi] Williams, R.: The structure of Lorenz attractors. Lect. Notes Math *645*, 94–112 (1977) Springer-Verlag, New York

[Wo2] Wojtkowski, M.: Linearly stable orbits in 3-dimensional billiards. Comm. Math. Phys. *129*, 319–328 (1990)

[Wo3] Wojtkowski, M.: A system of one dimensional balls with gravity. Comm. Math. Phys. *126*, 505–531 (1990)

[Wo4] Wojtkowski, M.: The system of one dimensional balls in an external field. II. Comm. Math. Phys. *127*, 425–432 (1990)

[Y3] Young, L.-S.: Statistical properties of dynamical systems with some hyperbolicity including certain billiards. Ann. Math. *147*, 585–650 (1998)

[Y4] Young, L.-S.: Recurrence times and rates of mixing. Israel J. Math. *9* (to be published)

[Ze] Zelditch, S.: Uniform distribution of eigenfunctions on compact hyperbolic surfaces. Duke Math. J. *55*, 909–941 (1987)

Chapter 9
Ergodic Theory of One-Dimensional Mappings

M.V. Jakobson

In this chapter, we shall consider one-dimensional mappings. They have been intensely studied during the last few years from the point of view of both dynamical systems and ergodic theory. The phase spaces of these systems are intervals $I \subset \mathbb{R}^1$ and the transformations are real-valued functions determined on I and taking their values in I. We shall study the invariant measures of one-dimensional maps, especially absolutely continuous invariant measures. The topological aspects of the problem will be discussed in one of the later volumes.

§ 1. Expanding Maps

1.1. Definitions, Examples, the Entropy Formula. We start the study of one-dimensional dynamics by considering the map T_2: $[0, 1] \to [0, 1]$ given by

$$T_2(x) = \begin{cases} 2x, & x \in [0, \frac{1}{2}] \\ 2x - 1, & x \in (\frac{1}{2}, 1]. \end{cases}$$

If we denote by $T_2^{-1}(x)$ the preimage of x, then for any measurable set C we have $l(T_2^{-1}C) = l(C)$ which means that the Lebesgue measure $l(dx) = dx$ is invariant under T_2.

Let us denote by $(\Omega_2, S, \mu(\frac{1}{2}, \frac{1}{2}))$ the *Bernoulli shift* S: $(\omega_0, \omega_1, \omega_2, \ldots) \mapsto (\omega_1, \omega_2, \ldots)$ on the set Ω_2 of one-sided sequences with the measure μ on Ω_2 given by $\mu(K_n) = \frac{1}{2^n}$ for any cylindric set $K_n = \{\omega\colon \omega_{i_1} = \omega_{i_1}^0, \ldots, \omega_{i_n} = \omega_{i_n}^0\}$. Consider the measurable partition $\xi = \{A_0 = [0, \frac{1}{2}], A_1 = (\frac{1}{2}, 1]\}$ and let the point $\pi(\omega) = \bigcap_{k=0}^{\infty} T_2^{-k} A_{\omega_k}$ correspond to a sequence $\omega = (\omega_0, \omega_1, \ldots, \omega_n, \ldots) \in \Omega_2$. Then the equality $l(\bigcap_{k=0}^{n-1} T_2^{-k} A_{\omega_k}) = \frac{1}{2^n}$ shows that π is an isomorphism (mod 0) between noninvertible transformations of measure spaces (endomorphisms) $([0, 1], T_2, dl) \overset{\pi}{\approx} (\Omega_2, S, \mu(\frac{1}{2}, \frac{1}{2}))$. Similarly the map T_n: $x \mapsto nx$ (mod 1), $2 \leqslant n \in \mathbb{N}$ is isomorphic (mod 0) to the one-sided shift on the alphabet of n symbols.

Now let T: $[0, 1] \to [0, 1]$ be a map, such that the derivative $|dT/dx|$ is not constant (Fig. 17) but satisfies the *expanding property*

$$|dT/dx| \geqslant C_1 > 1.$$

For any expanding map the distance between the trajectories of near-by initial points grows exponentially. This provides the analogy between such

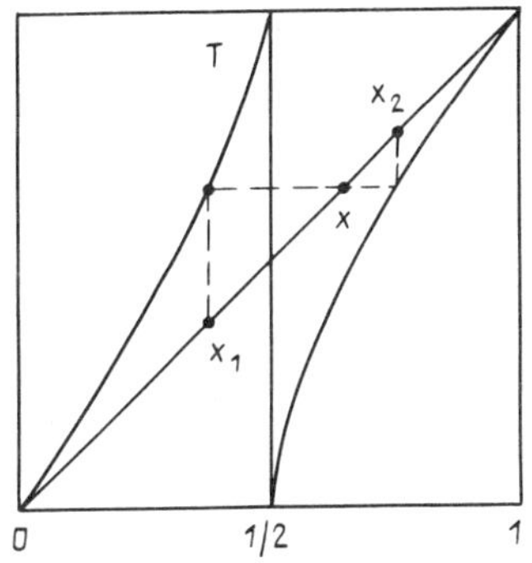

Fig. 17

maps and hyperbolic systems considered in Chap. 7, 8. In particular, expanding maps generally admit absolutely continuous invariant measures.

In this section, we shall consider piecewise monotone transformations T with a finite or countable number of maximal intervals (laps) on which T is monotone. The following argument due to Ya.G. Sinai [Si5] shows that an expanding map $T\colon [0,1] \to [0,1]$ with a finite number of laps which is C^2 on each lap and maps it onto $[0,1]$ admits an invariant measure $\mu(dx) = h(x)dx$ where $h(x)$ is continuous and strictly positive.

The map T induces in L^1_{dl} the so-called *Perron-Frobenius operator*

$$\mathscr{L}\varphi(x) = \sum_{y \in T^{-1}(x)} \varphi(y)/|T'(y)|.$$

Suppose for definiteness that T is monotone increasing from 0 to 1 on the intervals $[0, \frac{1}{2}]$ and $[\frac{1}{2}, 1]$. Then

$$\mathscr{L}\varphi(x) = \frac{\varphi(x_1)}{T'(x_1)} + \frac{\varphi(x_2)}{T'(x_2)}$$

where $\{x_1 \in [0, \frac{1}{2}],\ x_2 \in [\frac{1}{2}, 1]\} = T^{-1}(x)$ (see Fig. 17). If φ is the density of an invariant measure, then $\mathscr{L}\varphi = \varphi$. We shall find some φ satisfying this equation in the set $\mathbb{A}_c = \{\varphi(x)\}$ such that, for any $x, y \in [0, 1]$

$$\exp(-cd(x,y)) \leqslant \varphi(x)/\varphi(y) \leqslant \exp(cd(x,y))$$

where $c > 0$ is some constant. We have

$$\frac{\mathscr{L}\varphi(x)}{\mathscr{L}\varphi(y)} \leqslant \max\left\{\frac{\varphi(x_1)}{\varphi(y_1)}\cdot\frac{T'(y_1)}{T'(x_1)};\ \frac{\varphi(x_2)}{\varphi(y_2)}\cdot\frac{T'(y_2)}{T'(x_2)}\right\}$$

Since $\varphi \in \mathbb{A}_c$, we obtain

$$\frac{\varphi(x_i)}{\varphi(y_i)}\cdot\frac{T'(y_i)}{T'(x_i)} \leqslant \exp[c\cdot d(x_i, y_i) + \ln T'(y_i) - \ln T'(x_i)]$$

Now we use $|\ln T'(y_i) - \ln T'(x_i)| = \left|\dfrac{T''(\theta)}{T'(\theta)}\right| . d(x_i, y_i)$ where $\theta \in [x_i, y_i]$ and the expanding property $d(x_i, y_i) \leqslant c_1^{-1} d(x, y)$. If we denote by c_2 the quantity $\max_{z \in [0,1]} |T''(z)/T'(z)|$, then we have

$$\frac{\mathscr{L}\varphi(x)}{\mathscr{L}\varphi(y)} < \exp[d(x, y)(c_1^{-1} \cdot c + c_1^{-1} \cdot c_2)].$$

If c is sufficiently large for $c_1^{-1}(c + c_2) < c$, then

$$\mathscr{L}\mathbb{A}_c \subset \mathbb{A}_c$$

and the Shauder-Tychonoff theorem implies the existence of a fixed point $h = \mathscr{L}h \in \mathbb{A}_c$ which is the density of an absolutely continuous invariant measure. □

A traditional object of ergodic theory is the distribution of fractional parts for various functions. Namely, for a monotone function $g(x)$ defined on $(0, 1)$, we consider the transformation

$$Tx = \{g(x)\}.$$

The question is whether or not T admits an invariant measure μ equivalent to the Lebesgue measure and what are the ergodic properties of μ which define ergodic properties of the corresponding fractional parts.

A classical example due to Gauss is $g(x) = 1/x$. In this case the invariant measure is given by the explicit formula

$$\mu(X) = \frac{1}{\ln 2} \int_X \frac{dx}{1 + x}.$$

A. Rényi and V.A. Rokhlin considered a wide class of examples (cf [RO]). Their results concern the mappings T satisfying the following conditions. Suppose that T has a finite or a countable set of points of discontinuity $\{a_i\}$ (these are points where $g(x)$ takes integer values). The points a_i partition I into intervals $\Delta_i = (a_{i-1}, a_i)$ which are referred to as intervals of the first rank. Let this partition be denoted by $\xi^{(1)}$ and let $\xi^{(n)} = \bigvee_{i=0}^{n-1} T^{-i}\, \xi^{(1)}$ be the partition of $[0, 1]$ into the intervals $\Delta_{i_1 i_2 \ldots i_n} = \Delta_{i_1} \cap T^{-1}\Delta_{i_2} \cap \cdots \cap T^{-(n-1)}\Delta_{i_n}$ of rank n. Then T^n is uniquely defined on $\Delta_{i_1 i_2 \ldots i_n}$. We assume that:

(i) There exists $k_0 \in \mathbb{N}$ such that T^{k_0} is an expanding map.

(ii) There exists $c_1 > 0$ such that for any n and for any y, z belonging to $\Delta_{i_1 i_2 \ldots i_n}$, $c_1^{-1} \leqslant \dfrac{dT^n}{dx}(y) \bigg/ \dfrac{dT^n}{dx}(z) \leqslant c_1$ holds.

(iii) $T\Delta_i \supset (0, 1)$ for any interval of rank 1.

Notice that (iii) is a Markov condition of special type. If (iii) holds, then we have $T^n \Delta_{i_1 i_2 \ldots i_n} \supset (0, 1)$ for intervals of any rank. Condition (ii) which concerns intervals of any rank is much more cumbersome.

If for any i $T|\Delta_i$ is a C^2-map, then we may substitute (ii) by another condition which concerns only the intervals of the first rank

(ii)′

$$\sup_i \sup_{z\in\Delta_i} \left|\frac{d^2T(z)}{dT(z)}\right| |\Delta_i| \leqslant c_2.$$

If T satisfies (i), (ii)′ and (iii) then (ii)′ holds for $T^n|\Delta_{i_1 i_2 \dots i_n}$ on any interval of any rank (cf [J1]). Thus (i), (ii)′ and (iii) imply (ii). For T satisfying these conditions, the above arguments prove the existence of an absolutely continuous measure μ invariant under T^{k_0}. Then $\mu' = \sum_{i=0}^{k_0-1} T^{i*}\mu = \rho(x)\,dx$ is T-invariant and the density $\rho(x)$ is continuous and strictly positive.

V.A. Rokhlin proved in [Ro] that (T, μ') is an exact endomorphism (see Chapter 1, Section 4 for definition) and the following *entropy formula* holds:

$$h_{\mu'}(T) = \int_0^1 \log|dT/dx|\mu'(dx).$$

We shall now show that the entropy formula follows from (i)–(iii). Condition (i) implies that $\xi^{(1)}$ is a one-sided generator. Therefore using Shannon-McMillan-Breiman theorem, we have $-1/n \log \mu'(\Delta_{i_1 \dots i_n}(x)) \to h_{\mu'}(T)$ for μ' almost all x. By the intermediate value theorem we have

$$l(\Delta_{i_1 \dots i_n}(x)) = 1\Big/\left|\frac{dT^n(\bar{x})}{dx}\right|, \quad \bar{x} \in \Delta_{i_1 \dots i_n}(x).$$

Then (ii) and the properties of μ' give for some $c_2, c_3 > 0$

$$c_3 \leqslant \mu'(\Delta_{i_1 \dots i_n}(x))\Big/\left|\frac{dT^n(x)}{dx}\right|^{-1} \leqslant c_2$$

Since by the Birkhoff theorem $\frac{1}{n} \log\left|\frac{dT^n}{dx}\right| = \frac{1}{n}\sum_{i=0}^{n-1} \log\left|\frac{dT}{dx}(T^i(x))\right| \to \int_0^1 \log\left|\frac{dT}{dx}\right| \cdot$ $\mu'(dx)$ for μ' almost all x, the entropy formula is proved. □

In particular for $Tx = \{1/x\}$, we obtain

$$h_\mu(T) = \frac{\pi^2}{6} \frac{1}{(\ln 2)^2}.$$

1.2. Walters Theorem. In [W2], Walters studied the ergodic properties of mappings which expand distances. The examples considered above satisfy the conditions of the Walters theorem.

Let $\bar{X}$ be a compact metric space with the metric denoted by d, $X \subset \bar{X}$ an open dense subset of X, X_0—an open dense subset of X and let $T\colon X_0 \to X$ be a continuous surjective map with the following properties.

I. There exists an $\varepsilon_0 > 0$ such that $\forall x \in X$ the set $T^{-1}(B_{\varepsilon_0}(x) \cap X)$ is a disjoint union of open subsets (components) $A_i(x) \subset X_0$, such that $T\colon A_i(x) \to B_{\varepsilon_0}(x) \cap X$ is a homeomorphism satisfying $d(Ty, T'y) \geqslant d(y, y')$ for each i.

II. For any $\varepsilon > 0$ there exists $m \in \mathbb{N}$ such that for any $x \in X$ $\{T^{-m}x\}$ is ε-dense in X.

Let ν be a probability Borel measure on $\bar{X}$ non-singular with respect to T (i.e. $\nu(E) = 0$ implies $\nu(TE) = 0$, $\nu(T^{-1}E) = 0$). Suppose that the Radon-Nikodyrn derivative $d\nu T/d\nu$ is continuous on X_0 and the following conditions hold.

$$\text{(a)} \qquad \frac{d\nu T^{-1}}{d\nu}(x) = \sum_{y \in T_x^{-1}} 1 \Big/ \frac{d\nu T}{d\nu}(y) \leqslant K_1 \quad \forall x \in X$$

$$\text{(b)} \qquad \sup_{n \geqslant 1} \sup_{y \in T_x^{-1}} \frac{d\nu T^n}{d\nu}(y') \Big/ \frac{d\nu T^n}{d\nu}(y) \leqslant R_2$$

when $d(x, x') < \varepsilon_0$ and this expression converges to 1, as $d(x, x') \to 0$.

Here we use the fact that for x, $x' \in X$, $d(x, x') < \varepsilon_0$ the property I gives rise to a natural bijection: $y \in T^{-n}x \leftrightarrow y' \in T^{-n}x'$ for y, y' lying in the same component.

The Perron-Frobenius operator is defined for $f \in C(X)$ by

$$\mathscr{L}f(x) = \sum_{y \in T_x^{-1}} \left(1 \Big/ \frac{d\nu T}{d\nu}(y)\right) f(y).$$

It follows from a) and b) that $\mathscr{L}$ extends to a continuous linear operator $z\colon C(\bar{X}) \to C(\bar{X})$.

Theorem 1.1 (cf [W2]).

(1) *There exists $h(x) \in C(\bar{X})$, $h(x) > 0$ such that for any $f \in C(\bar{X})$*

$$\mathscr{L}^n f \rightrightarrows h(x) \cdot \nu(f);$$

(2) *The measure $\mu(dx) = h(x) \cdot \nu(dx)$ is T-invariant and (T, μ) is an exact endomorphism;*

(3) *$\nu \circ T^{-n} \to \mu$ in weak topology;*

(4) *The Variational principle. For any T-invariant Borel probability measure m*

$$0 = H_\mu(\mathscr{B}/T^{-1}\mathscr{B}) - \mu\left(\log\left(\frac{d\nu T}{d\nu}\right)\right) \geqslant H_m(\mathscr{B}/T^{-1}\mathscr{B}) - m\left(\log\frac{d\nu T}{d\nu}\right)\Bigg)$$

and μ is the unique measure with this property.

If there exists a finite or a countable partition ξ which is a one-side generator, then $H_\mu(\mathscr{B}/T^{-1}\mathscr{B}) = h_\mu(T)$ and the left part of (4) gives the entropy formula. Suppose further that any element of ξ is contained in some set $T^{-1}B_{\varepsilon_0}(x)$ for some x and that ξ is a Markov partition satisfying $\mu(\partial\xi) = 0$. Then we have

(5) *The natural extension of (T, μ) is isomorphic to a Bernoulli shift.*

Apart from one-dimensional examples considered above, Theorem 1 gives the existence and describes the ergodic properties of absolutely continuous invariant measures for expanding maps of compact manifolds (cf [W2]).

According to the property 1 of the map T, we have for any $x \in XT^{-1}B_{\varepsilon_0}(x) = \bigcup_i A_i(x)$ where $T_i A_i(x) B_{\varepsilon_0}(x)$ are homeomorphisms for all i. This is a restrictive condition: for example β-transformations $x \mapsto \beta x \pmod 1$ with irrational β do not satisfy it.

Hofbauer and Keller obtained a result similar to Theorem 1 for a wide class of piecewise-monotone transformations of the interval including β-transformations and studied ergodic properties of the corresponding invariant measures (cf [HK]).

For a large class of expanding maps of the interval with infinitely many laps, Rychlik (cf [Ry1]) considered the action of Perron-Frobenius operator on the space of functions with bounded variation. He proved the existence of a spectral gap between the maximal eigenvalue and the remaining part of the spectrum and, as a corollary, proved the exponential decay of correlations.

§2. Absolutely Continuous Invariant Measures for Nonexpanding Maps

2.1. Some Examples. A classical example of a one-dimensional nonexpanding map admitting absolutely continuous invariant measure is

$$F\colon x \mapsto 4x(1-x).$$

It was studied by Ulam and von Neiman (cf [CE]). It is interesting to point out that Fatou is his famous work (1920) on the iteration of rational maps of the Riemann sphere considered the maps of that type and studied their relation with expanding maps (cf [MO]). For $F\colon x \to 4x(1-x)$ there is a change of variable $y = \varphi(x) = \frac{2}{\pi} \arcsin \sqrt{x}$ transforming F into the piecewise linear map

$$T(x) = \begin{cases} 2x, & x \in [0, \frac{1}{2}] \\ 2x - 1, & x \in (\frac{1}{2}, 1]. \end{cases}$$

Since the Lebesgue measure dl is T-invariant, its image $\varphi^{-1*}\, dl = \mu(dx) = \dfrac{dx}{\pi\sqrt{x(1-x)}}$ is F-invariant. Let $\xi = \{A_0 = [0, \frac{1}{2}], A_1 = (\frac{1}{2}, 1]\}$ be the partition of $[0,1]$ into intervals of monotonicity. The correspondence $\omega = (\omega_0 \omega_1 \ldots) \leftrightarrow \pi(\omega) = \bigcap_{k=0}^{\infty} T^{-k} A_{\omega_k}$ generates an isomorphism between the system $([0,1], T, dl)$ and the one-sided Bernoulli shift. Thus the system $\left([0,1], F, \dfrac{dx}{\pi\sqrt{x(1-x)}}\right)$ is also Bernoullian.

Generally the invariant measure is not given by an explicit formula, and it is a more difficult task to study ergodic properties of invariant measures for nonexpanding maps than for expanding ones. The first results in this direction were obtained for unimodal maps, satisfying the following property: some iteration of the critical point coincides with an unstable periodic point (see references in [CE], [J2]). For such mappings it is possible to overcome the difficulties

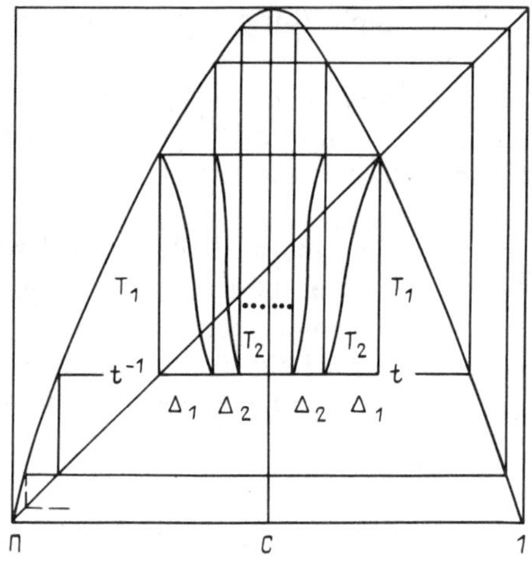

Fig. 18

related to the critical point by considering the induced map (see Chapter 1, Section 4).

Let $F \in C^2([0,1],[0,1])$ be a unimodal map with a nondegenerate critical point c satisfying $F(0) = F(1) = 0$, $F'(0) = A > 1$, $F(c) = 1$. Let $t \in (c,1)$ be the fixed point of F, $t^{-1} \in (0,c)$, the preimage of t. We consider the induced map T on the interval $[t^{-1}, t] = I$ (Fig. 18).

This map has a countable number of laps Δ_i with $T|\Delta_i = T_i = F^{i+1}$, $i = 1, 2, 3, \ldots$, $T_i\colon \Delta_i \to I$. Using the nondegeneracy of c, it is straightforward to check that for some $c_1, c_2, c_3 > 0$, independent of i, the following inequalities hold:

$$c_1 A^{i/2} < \left|\frac{dT_i(x)}{dx}\right| < c_2 A^{i/2} \left|\frac{d^2 T_i(x)}{dx^2}\right| < c_3 A^i.$$

If i is sufficiently large then we obtain

$$\left|\frac{dT_i}{dx}\right| > c > 1.$$

Besides for any i and for any $x, y, z \in \Delta_i$, we have

$$\left|\frac{d^2 T_i(x)}{dx^2}\right| \Bigg/ \left|\frac{dT_i}{dx}(y)\right| \cdot \left|\frac{dT_i}{dx}(z)\right| < c'.$$

If some iteration of T is expanding, then T satisfies the conditions of Theorem 1.1 and we conclude that an absolutely continuous T-invariant measure ν exists and obtain the ergodic properties of ν. The measure ν uniquely defines an F-invariant ergodic measure μ absolutely continuous with respect to the Lebesgue measure on $[0, 1]$. The density of μ is separated from zero and is continuous everywhere except the orbit of the critical point (which in our case consists of two points 1 and 0), where it has singularities of the type $1/\sqrt{x}$. The above conditions are satisfied for unimodal *maps with negative Schwarzian derivative*

$$SF = \frac{F'''}{F'} - \frac{3}{2}\left(\frac{F''}{F'}\right)^2 < 0.$$

The topological structure of these maps is well known (cf [CE]). The induced map method alllows us to prove the existence of an absolutely continuous invariant measures for different kinds of maps with negative Schwarzian derivatives. Namely this is the case when some iterate of the critical point falls into an unstable orbit or when it falls into an invariant unstable Cantor set (cf [J1], [Mi2]). K. Ziemian generalized the results of [HK] to this case (cf [Z]).

The book of de Melo and van Strien (cf [MS)) is an excellent introduction to both the topological and ergodic aspects of one-dimensional dynamics. In particular, it contains a detailed study of maps with negative Schwarzian derivative.

2.2. Intermittency of Stochastic and Stable Systems. When considering several one-parameter families of one-dimensional maps $x \to f_\lambda(x)$, we see different types of behavior. For the maps of the first type the trajectories of almost every point (with respect to dx) converge to a stable periodic orbit, and the nonwandering set of f_λ consists of a finite number of such orbits and of an invariant unstable Cantor set. These maps are structurally stable. We shall call such maps *hyperbolic*.

The maps of the second type display stochastic behavior on a set of positive Lebesgue measure. As pointed out above this happens for a unimodal map with negative Schwarzian derivative when some iterate of the critical point falls into a periodic unstable orbit or into an invariant unstable Cantor set. In such cases the critical point is not recurrent: $c \notin \omega(c)$. These maps do not present all the possibilities for stochastic dynamics, it seems quite probable that the corresponding parameter values form a set of measure zero, although the whole set of parameter values corresponding to the stochastic dynamics has positive measure (cf [J2]).

Let F: $[0, 1] \to [0, 1]$, $F(0) = F(1) = 0$, $F'(0) \neq 0$ be a unimodal C^2-map with a nondegenerate critical point. Consider a one parameter family of piecewise smooth mappings

$$T_\lambda: x \to \lambda \cdot F(x) (\mathrm{mod}\ 1), \qquad \lambda > 1.$$

T_λ has a finite number of monotone branches depending on λ and a middle parabolic branch (Fig. 19). If $\lambda \cdot F(c) \in \mathbb{N}$, then the middle branch bifurcates into two monotone branches and a new middle branch is born. We shall use λ_n to denote the corresponding parameter values and we shall use mes to denote the Lebesgue measure on the axis of parameter.

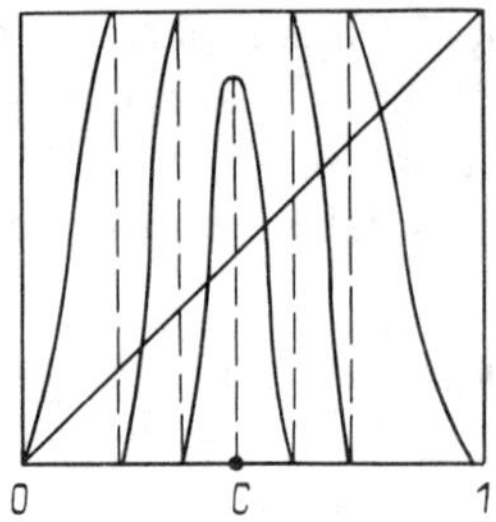

Fig. 19

Theorem 2.1 (cf [J2]). *For any $\varepsilon > 0$ there exists $\Lambda(\varepsilon)$ such that for $\lambda_n > \Lambda(\varepsilon)$ there is a set of positive measure $M = \{\lambda \in [\lambda_n, \lambda_{n+1}]$: T_λ admits an ergodic absolutely continuous invariant measure with positive entropy$\}$ and mes $M > (\lambda_{n+1} - \lambda_n)(1 - \varepsilon)$.*

We construct the set M as a complement to an open set $V = \bigcup_{n=0}^{\infty} V_n$, where V_n consists of λ such that $T_\lambda^{k_n}(c)$ falls into an open interval containing c. The presence of a large parameter implies that the lengths of intervals removed at the n-th step decrease very quickly and mes $V \to 0$, as $\lambda \to \infty$.

Theorem 2.1 may be generalized to the mappings with several extrema, see [J4]. For a smooth one-parameter family of maps such as $x \mapsto ax(1-x)$, $a \in [0, 4]$ there is no large parameter. In order to reduce the situation to the preceding one, we construct the induced map as described above. Then for a close to 4, we obtain a piecewise smooth map T_a similar to the one depicted in Fig. 19. The limit map T (Fig. 18) has a countable number of laps and is expanding. This is essential in proving the following results (cf [J2]).

Theorem 2.2. *Let F be a map of $[0,1]$, $F(0) = F(1) = 0$, C^2-close to $x \to x(1-x)$. Define λ_0 by $\lambda_0 \cdot F(c) = 1$. Then the Lebesgue measure of the set $M = \{\lambda \in (0, \lambda_0] | F_\lambda: x \to \lambda \cdot F(x)$ has an ergodic absolutely continuous invariant measure of positive entropy$\}$ is positive, and λ_0 is a density point of M.*

A similar theorem holds for the family of smooth unimodal maps $x \to a \cdot F(x)$, where $F(x)$ is a map with negative Schwarzian derivative.

Several results of that kind were obtained for families of unimodal maps by J. Guckenheimer [Gu], M. Benedicks and L. Carleson [BC], P. Collet and J.P. Eckmann (cf [CE]), M. Rychlik [Ry1]. For other generalizations see [J4] and [TTY].

In the opposite direction, the following theorem was proved by G. Swiatek and J. Graczyk (cf [GS1]).

Theorem 2.3. *For the family $x \to \lambda x(1-x)$ the set of parameters λ corresponding to hyperbolic maps is dense in* $[0, 4]$.

G. Swiatek circulated the first version of the proof of Theorem 2.3 in 1992. More transparent arguments due to both authors were given in [GS2] in 1993. They also proved the uniform growth of certain complex moduli.

Another proof of Theorem 2.3 was given in 1995 by M. Lyubich (published in [Lyu5]), who also extended it to some examples of complex quadratic polynomials.

The following questions concern the problem of intermittency between structurally stable periodic systems and stochastic systems.

1. Is the set of hyperbolic systems dense in C^r-topology?

It was proved in ([J6]) for $r = 1$. In the case when the maps are immersions, R. Mañé ([M]) proved it for any r. For unimodal maps and any r, it was recently proved by O. Kozlovski (cf [Ko]).

2. Does the union of hyperbolic and stochastic systems form a set of full measure in the parameter space?

For real quadratic family, the question was recently answered positively by M. Lyubich (cf [Lyu1]).

An important tool in the proof of Theorem 2.3 is Sullivan's method of quasiconformal deformations (see [Su1]). Based on this method, the strategy is to prove that topological equivalence between certain maps is actually quasiconform (quasisymmetric in real case). For families of unimodal maps with negative Schwarzian derivative topological equivalence implies quasisymmetric equivalence for many stochastic maps. Such maps have positive measure in the parameter space (see [JS]).

One class of maps, where topological equivalence implies strong additional properties, are maps satisfying the Collet-Eckmann condition: derivatives grow exponentially fast along the orbit of every critical value. Recently T. Nowicki and F. Przyticki [NP] proved that Collet-Eckmann condition is topologically invariant in the space of unimodal maps with negative Schwarzian derivative.

2.3. Ergodic Properties of Absolutely Continuous Invariant Measures. If μ is an absolutely continuous f-invariant measure of positive entropy, then f exhibits strong stochastic properties with respect to μ. The most general results in this direction, due to F. Ledrappier, are summarized in Theorems 2.3–2.6 below.

We consider a map $f: [0,1] \to [0,1]$ with a finite number of points of discontinuity $0 = b_0 < b_1 < \cdots < b_m < b_{m+1} = 1$ and we suggest that for $j \in [0, m]$ the restriction $f|[b_j, b_{j+1}]$ satisfies the following conditions.

C_1. f is a $C^{1+\varepsilon}$ map, i.e. f' satisfies the Hölder condition of order $\varepsilon > 0$.

C_2. f has a finite number of critical points.

C_3. There are positive numbers k_i^-, k_i^+, such that

$$\log \frac{|f'(x)|}{|x - a_i|^{k_i - (+)}}$$

is bounded in a left (right) neighborhood of a_i.

Theorem 2.3 (cf [L1]). *Let f satisfy C_1–C_3 and let μ be an ergodic absolutely continuous f-invariant measure of positive entropy. If μ is ergodic for f^k for all $k > 0$, then the natural extension of f is isomorphic to a Bernoulli shift. In any case there exists $k_0 \geqslant 1$ such that the natural extension of f^{k_0} is Bernoulli on every ergodic component. Furthermore, the Rokhlin formula holds*

$$h_\mu(f) = \int \log|f'|\, d\mu.$$

In order to prove Theorem 2.3 the local unstable manifolds are constructed, and the absolute continuity of the unstable foliation is proved with methods similar to the case of hyperbolic attractors considered in Chaps. 7, 8.

According to Chap. 7, we classify f as a nonuniformely completely hyperbolic map. The size of a local unstable manifold at some point of the natural extension may become small: it is related to the iterates of critical points and of points of discontinuity (similar to the systems considered in Chap. 8).

The similarity becomes quite clear if we establish an isomorphism between the natural extension of a one-dimensional map and the action of some multidimensional map on an attractor with one dimensional unstable manifolds. A well known example of this type corresponding to a one-dimensional map with discontinuities, is the Lorenz attractor (cf [R] and Chap. 8). An example, corresponding to a one-dimensional map with critical points, is the so-called twisted horseshoe (cf [J3]).

Let us denote by Q a partition of I into intervals whose end points are the critical points and the points of discontinuity. Let $Q_i = [\alpha_i, \alpha_{i+1}]$ be the elements of Q. Then the points of the natural extension (Y, f, μ) of the endomorphism (I, f, μ) may be identified with $y = (x; z) = (x; z_1, z_2, \ldots)$, where $x \in I$ and z_k is the index of the element Q_{z_k} containing x_k (see Chap. 1, Sect. 4.6 for definition). We denote by Z the set of sequences z and by $\Pi : y = (x; z) \mapsto x$ the projection on the first coordinate.

Theorem 2.4 (cf [L1]). *Suppose that f is a map satisfying $C_1 - C_3$ and μ is an absolutely continuous f-invariant measure, and let χ be a number satisfying $0 < \chi < \int \log|f'|\, d\mu$. Then there exist measurable functions α, β, γ, $\tilde{\gamma}$ on Y and a constant D such that*

(i) *μ-almost everywhere on Y holds $\alpha > 0$, $1 < \beta < \infty$, $0 < \gamma < \tilde{\gamma} < \infty$*

(ii) *If $y = (x; z)$ and $t < \alpha(y)$ then $y_t = ((x + t); z) \in Y$*

(iii) $|\Pi(\tilde{f}^{-n} y) - \Pi(\tilde{f}^{-n} y_t)| \leqslant |t| \cdot \beta(y) \exp(-\chi \cdot n)$

(iv) *for any n*

$$1/D \leqslant \frac{f'(\Pi(\tilde{f}^{-n} y))}{f'(\Pi(\tilde{f}^{-n} y_t))} \leqslant D$$

(v) $\gamma(y) \leqslant \prod_{n=1}^{\infty} \frac{f'(\Pi(\tilde{f}^{-n}y))}{f'(\Pi(\tilde{f}^{-n}y_t))} = \Delta(y, y_t) \leqslant \tilde{\gamma}(y)$.

Theorem 2.4 asserts in particular that classes of equivalence relation on Y-defined by the projection on Z are open subsets of the intervals. These classes define a measurable partition ξ. Conditional measures of $\tilde{\mu}$ with respect to ξ are described in the following theorem.

Theorem 2.5 (cf [L1]). *Let f, μ be as in Theorem 2.4, and let $q(y,\cdot)$ denote the conditional measures of μ with respect to ξ. Then for $\tilde{\mu}$-almost every y, the measure $q(y, \Pi^{-1}, \cdot)$ on I is absolutely continuous with respect to the Lebesgue measure. There exists a partition $\eta > \xi$ such that $f^{-1}\eta > \eta$, $\bigvee_{k=0}^{\infty} f^{-k}\eta = \varepsilon$ and*

$$q(y, B) = \frac{\int_{B\cap\eta(y)} \Delta(y, y')\,dy'}{\int_{\eta(y)} \Delta(y, y')\,dy'},$$

where $\Delta(y, y')$ is defined in (v) *of Theorem* 2.4.

When considering individual trajectories it is natural to call a point satisfying the following conditions *regular*:

R_1) the sequence of measures $\frac{1}{n}\sum_{i=0}^{n-1} \delta_{f^i_{(x)}}$ weakly converges towards an ergodic measure μ_x

R_2) the sequence of numbers $\frac{1}{n} log|df^n/dx|$ converges to

$$\lambda_x = \int \log|f'|\,d\mu_x.$$

A regular point is called *positive regular* if $\lambda_x > 0$.

Theorem 2.6 (cf [L1]). *Let $f: I \to I$ be a smooth map satisfying $C_1 - C_3$. Then the set of positive regular points has positive Lebesgue measure if and only if there is an absolutely continuous invariant measure with positive entropy.*

The "if" of this theorem may be considered as a one-dimensional version of the Sinai-Ruelle-Bowen theorem about limit measures on attractors (see Chap. 7).

The exponential decay of correlations for a positive measure set of parameter values in quadratic family was proved by L.-S. Young in [Y3].

§ 3. Feigenbaum Universality Law

3.1. The Phenomenon of Universality. Several one-parameter families of differential equations depending on some external parameter μ demonstrate a cascade of successive *period doubling bifurcations* of stable periodic orbits when μ is varied. The Lorenz system, nonlinear electrical circuits, and the Galerkin approximations of the Navier-Stokes equations all examples of this phenomena.

Let γ be some stable periodic orbit continuously varying with μ. For some μ_0 an eigenvalue $\lambda(\mu)$ of the Poincaré map along γ takes the value $\lambda(\mu_0) = -1$. It is a generic property that for $\mu > \mu_0$ the orbit γ becomes unstable and a new stable periodic orbit γ' is born which coincides for $\mu = \mu_0$ with γ passed twice. If we denote by $\lambda'(\mu)$ the eigenvalue of the Poincaré map along γ', then $\lambda'(\mu_0) = (\lambda(\mu_0))^2 = 1$. The orbit γ' varies with μ and in turn for some $\mu_i > \mu_0$ becomes unstable, so that $\lambda'(\mu_1) = -1$ and a new periodic orbit is born whose period is twice the period of γ', and so on. The parameter values μ_i corresponding to successive bifurcations converge to some limit μ_∞. When $\mu_i \to \mu_\infty$, the form of the bifurcating orbits becomes more and more complicated and they converge to an invariant set whose structure does not depend on the family under consideration.

Several one-parameter families of discrete transformations, in particular of one-dimensional maps exhibit a similar phenomenon. Moreover, it follows from the theory of kneading, by Milnor and Thurston, that for any one-parameter family of smooth unimodal maps $F_t\colon [0,1] \to [0,1]$, $t \in [0,1]$ continuously depending on t in C^1-topology and satisfying $F_0(x) \equiv 0$, $\max_{x\in[0,1]} F_1(x) = 1$ there is an infinite number of different period doubling series.

Let us consider for example the family of quadratic polynomials $G_\mu\colon x \mapsto 1 - \mu x^2$, $x \in [-1,1]$, $\mu \in [0,2]$. The value $\mu_0 = 0.75$ corresponds to the first period doubling: the fixed point $x_0(0.75) = 2/3$ bifurcates into a cycle of period 2 $(x_1(\mu), x_2(\mu))$. The subsequent bifurcation values μ_n corresponding to the birth of 2^n-cycles are $\mu_1 = 1{,}25$, $\mu_2 = 1{,}3681$, $\mu_3 = 1{,}3940, \ldots$. The sequence μ_n converges to $\mu_\infty = 1.40155\ldots$ G_{μ_∞} has orbits of period 2^n for any n and no other periodic orbits. The subsequent quotients $\dfrac{\mu_n - \mu_{n-1}}{\mu_{n+1} - \mu_n} = \delta_n$ take values: $\delta_1 = 4.23$, $\delta_2 = 4.55$, $\delta_3 = 4.65$, $\delta_4 = 4.664$, $\delta_5 = 4.668$, $\delta_6 = 4.669, \ldots$. The values δ_n tend to the limit $\delta = 4.669\ldots$. The universality law states that δ_n computed for different families converge to δ which does not depend either on the dimension of the phase space or on the family. In other words

$$|\mu_n - \mu_\infty| \sim c \cdot \delta^{-n}$$

where c depends on the family and δ is universal. The sizes of some domains in the phase space are also characterized by universal scaling. For example if we consider a bifurcation parameter value μ_n such that for a 2^n-cycle $\{x_1, \ldots, x_{2^n}\}$ $\dfrac{dF^{2^n}}{dx} = \prod_{i=1}^{2^n} \dfrac{dF}{dx}(x_i) = -1$ holds, and if we let $x^{(n)}$ denote the point of this cycle the closest to zero, then we have $x^{(n)} \sim c' \cdot \lambda^n$, where $\lambda = -0.3995\ldots$ is universal i.e. independent of the family.

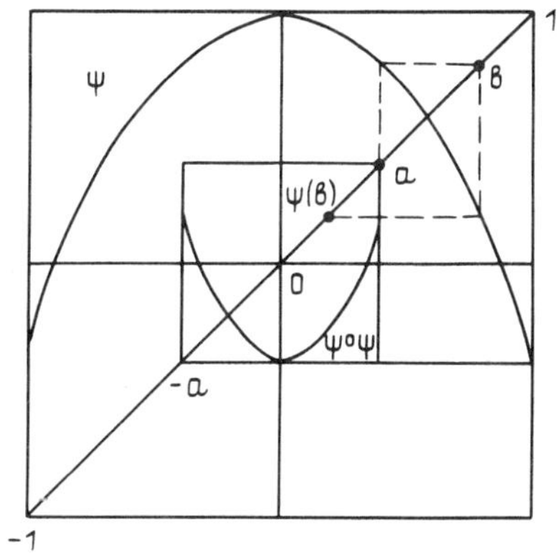

Fig. 20

3.2. Doubling Transformation. The universal behavior in nonlinear systems was discovered by M. Feigenbaum in 1978 and stimulated a number of theoretical and experimental works (see [CE] for references). In order to explain the Feigenbaum universality law let us consider the so-called *doubling transformation* which consists in taking the second iterate of the map with simultaneous rescaling. Namely, let $\mathfrak{U}_1$ be the set of even unimodal C^1-mappings ψ: $[-1,1] \to [-1,1]$ satisfying the following conditions (see Fig. 20):

1) $\psi'(0) = 0$; $\psi(0) = 1$;
2) $\psi(1) = -a < 0$;
3) $b = \psi(a) > a$; $\psi(b) = \psi^2(a) < a$.

For $\psi \in \mathfrak{U}_1$ we define

$$\mathcal{T}\psi(x) = -\frac{1}{a}\psi \circ \psi(-ax).$$

The nonlinear operator $\mathcal{T}$ is said to be the doubling transformation. The following theorems state the main properties of $\mathcal{T}$.

Theorem 3.1. *The doubling transformation has an isolated fixed point which is an even analytic function*
$\Phi(x) = 1 - 1.52763x^2 + 0.104815x^4 - 0.0267057x^6 + \cdots$. *The universal constant* $\lambda = -0.3993\ldots$ *coincides with* $\Phi(1)$.

We denote by $\mathfrak{H}$ the Banach space of even analytic functions $\psi(z)$ limited in some neighborhood of the interval $[-1,1]$, and real on the real axis. Let $\mathfrak{H}_0$ be a subspace of $\mathfrak{H}$ given by $\mathfrak{H}_0 = \{\psi(z)\colon \psi(0) = \psi'(0) = 0\}$ and let $\mathfrak{H}_1 = \mathfrak{H}_0 + 1$.

Theorem 3.2. *There exists a neighborhood* $\mathcal{U}_\Phi$ *of* Φ *in* $\mathfrak{H}_1$ *such that the doubling transformation* $\mathcal{T}$ *is a* C^∞*-map from* $\mathcal{U}_\Phi$ *into* $\mathfrak{H}_1$. $D\mathcal{T}_\Phi$ *is a hyperbolic operator with a single unstable eigenvalue* $\delta = 4.6692\ldots$ *and a stable subspace of codimension one.*

We denote by $\Sigma_0 \subset \mathfrak{U}_1$ the bifurcation surface consisting of ψ satisfying the following conditions: $\psi'(x_0) = -1$, $(\psi \circ \psi)'''(x_0) < 0$, where $x_0 = x_0(\psi)$ is the single fixed point of ψ lying in $[0,1]$. *In consequence* of Theorem 3.2 a local unstable one-dimensional manifold $W^u_{loc}(\Phi)$ and a local stable manifold $W^s_{loc}(\Phi)$ of codimension 1 are defined.

Theorem 3.3. *A local unstable manifold may be extended to the global one $W^u(\Phi)$ which has a point of transversal intersection with Σ_0. $W^u(\Phi)$ consists of maps with the negative Schwarzian derivative.*

The proofs of Theorems 3.1–3.3 known to date use computer estimates (see [VSK] for references). Similar analytic results were obtained by Collet, Eckmann and Lanford for a class of maps of the form $f(|x|^{1+\varepsilon})$, where f is analytic and ε is sufficiently small (cf [CE]).

A new approach using complex variable methods was developed by D. Sullivan [Su2] and C. Mcmullen [Mc].

In a neighborhood U_Φ of the fixed point, we obtain the following picture (Fig. 21). The hyperbolic structure of $\mathcal{T}$ implies that codimension one surfaces $\mathcal{T}^{-n}\Sigma_0$ are near to parallel to the stable manifold $W^s(\Phi)$. Consider a curve F_μ corresponding to a one parameter family of maps. If this curve has a point of transversal intersection with $W^s(\Phi)$, then it is the same for $\mathcal{T}^{-n}\Sigma_0$ with sufficiently large n. The parameter values μ_n corresponding to $F_{\mu_n} \in \mathcal{T}^{-n}\Sigma_0$ are the values of doubling bifurcations $2^n \to 2^{n+1}$, and the map $F_{\mu_\infty} \in W^s(\Phi)$ corresponds to the value $\mu_\infty = \lim_{n\to\infty} \mu_n$. Since the unstable eigenvalue of $D\mathcal{T}_\Phi$ equals δ, the distances between $\mathcal{T}^{-n}\Sigma_0$ and $W^s(\Phi)$ decreases as δ^{-n}. Thus

$$|\mu_\infty - \mu_n| \sim c \cdot \delta^{-n}$$

which implies the universal scaling of parameter.

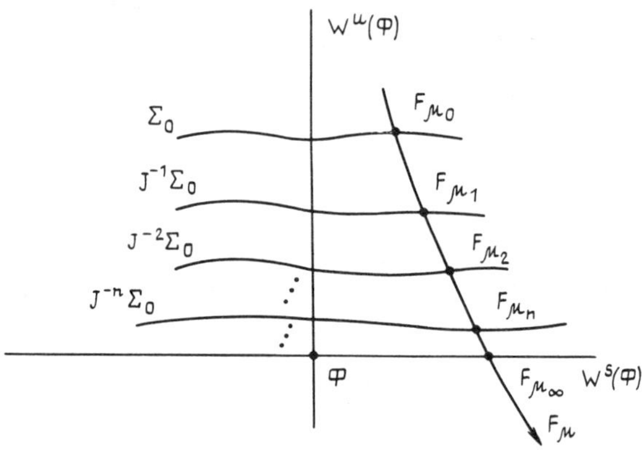

Fig. 21

Now let $x_{(n)}$ be the closest to zero 2^n-periodic point of F_{μ_n}. Since $\mathcal{T}$ is hyperbolic, the map $\mathcal{T}^n F_{\mu_n}$, which belongs to the surface Σ_0, lies in a small neighborhood of $W^u\Phi \cap \Sigma_0 = \Sigma_*$. Let us denote by x_0 the fixed point of the map $\mathcal{T}^n F_{\mu_n}$. We have $x_{(n)} = x_0 \prod_{i=1}^n (-a_i)$ where $(-a_i)$ are the rescaling constants for $\mathcal{T}^i F_{\mu_n}$. Almost all points $\mathcal{T}^k F_{\mu_n}$, $k \in [1, n]$, lie in a small neighborhood of the fixed point Φ and the corresponding a_i are exponentially close to $a(\Phi) = 0.3995\ldots$. This explains the universal scaling in the phase space.

If we want to prove that a given family F_μ admits universal scaling and if an experimental accuracy of the relation $|\mu_n - \mu_\infty| \sim c \cdot \delta^{-n}$ does not satisfy us, then we must verify that F_μ is in the domain of the definition of $\mathcal{T}$ and that it intersects $W^s(\Phi)$ transversely. This is the case for the family of quadratic polynomials.

E.B. Vul and K.M. Khanin proposed the following method of constructing $W^u(\Phi)$ (cf [VSK]). One considers the space of one-parameter families F_μ and defines an operator T acting like the doubling transformation on F_μ and like some rescaling on the parameter. If a family F_μ lies in a small neighborhood of $W^u(\Phi)$ (which is also considered as a one-parameter family), then the iterations $T^n F_\mu$ converge, with an exponential rate, to $W^u(\Phi)$ which is an attracting fixed point of T.

3.3. Neighborhood of the Fixed Point. The following description of a C^3-neighborhood of the fixed point Φ seems to be quite probable. The stable manifold $W^s(\Phi)$ divides this neighborhood in two parts (Fig. 22). There are two fundamental domains of $\mathcal{T}^{-1}$ in $\mathcal{U}_\Phi$: the first one denoted by $\mathcal{U}$ lies above $W^s(\Phi)$, and the second denoted by $\mathcal{U}'$ lies below $W^s(\Phi)$. The boundaries of $\mathcal{U}$ are: $\Sigma = \{f : f^2(0) = 0\}$ and $\mathcal{T}^{-1}\Sigma$ (Fig. 23).

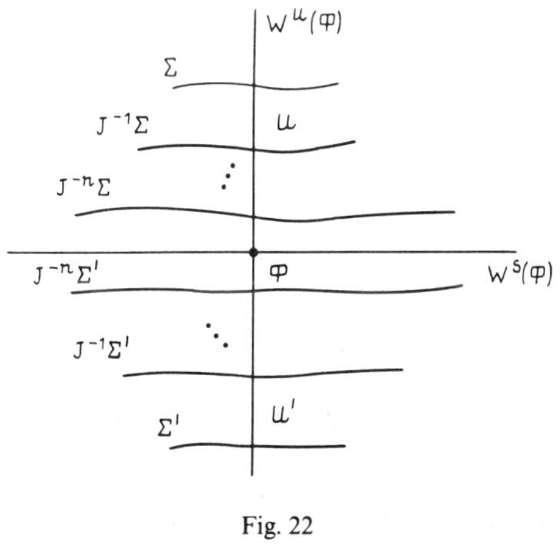

Fig. 22

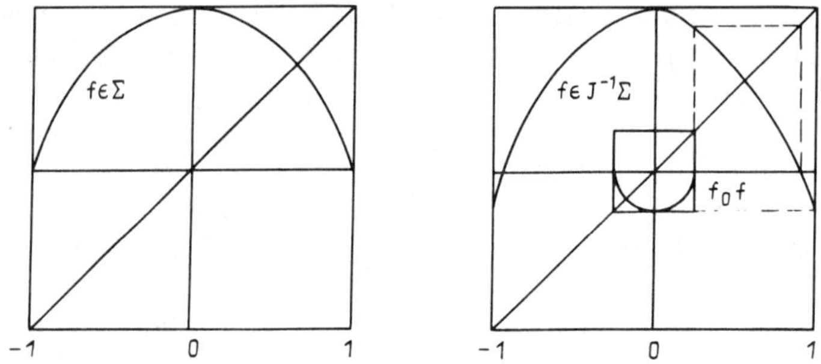

Fig. 23

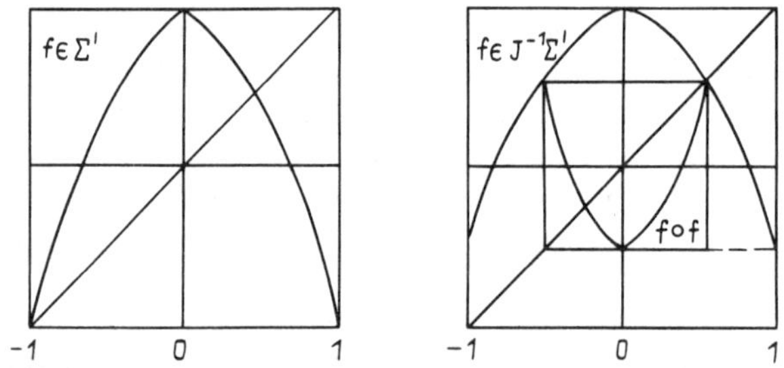

Fig. 24

The boundaries of $\mathscr{U}'$ are: $\Sigma' = \{f: f^2(0) = -1\}$ and $\mathscr{T}^{-1}\Sigma'$ (Fig. 24).

$W^s(\Phi)$ separates the mappings with a simple structure (finite non-wandering set and zero topological entropy) from the mappings with complicated structure (infinite non-wandering set and positive topological entropy).

If $\psi \in \mathscr{T}^{-n}\mathscr{U}$, then $\Omega(f)$ is a union of a stable (or neutral if $\psi \in \mathscr{T}^{-(n+1)}\Sigma_0$) orbit of period 2^{n+1} (or 2^{n+2}) and unstable periodic orbits of periods 2^k, $k < n + 1$ (or $k < n + 2$).

If $\psi \leqslant \mathscr{T}^{-n}\mathscr{U}'$, $n \geqslant 1$, then the nonwandering set $\Omega(\psi)$ is contained in the union of 2^n disjoint intervals (with the exception of a finite number of 2^k-cycles, $k < n$). Any map $\psi \in \mathscr{U}'$ has some periodic orbit of odd period, while for $\psi \in \mathscr{T}^{-n}\mathscr{U}'$ the period of any orbit divides 2^n (apart from 2^k-cycles mentioned above).

There are stochastic maps below and infinitely close to $W^s(\Phi)$. In order to see this, consider a curve $F_\mu \subset \mathscr{U}_\Phi$ which transversely intersects $W^s(\Phi)$ and let $\gamma^{(n)} = \{F_\mu, \mu \in [\mu_1^{(n)}, \mu_2^{(n)}]\}$ denote $F_\mu \cap \mathscr{T}^{-n}\mathscr{U}'$. The maps $G_\mu = \mathscr{T}^n F_\mu \in \mathscr{T}^n\gamma^{(n)} \subset \mathscr{U}'$ have negative Schwarzian derivatives since they are close to $W^u(\Phi)$ (recall that $W^u(\Phi)$ consists of maps satisfying $S\Phi < 0$). Since the curve $\mathscr{T}^n\gamma^{(n)}$ is transversal to Σ', the results of Section 2 are applicable. They imply that mes$\{\mu$: G_μ admits an absolutely continuous invariant measure$\}$ is positive. Coming back to F_μ, we obtain a positive measure set $M_n = \{\mu$: $F_\mu \in \gamma^{(n)}$ has absolutely continuous invariant measure with the support inside 2^n disjoint cyclically permutted intervals}. For this measure, the restriction of $F_\mu^{2^n}$ on the ergodic component is Bernoulli.

Moreover the relative measure of stochastic maps on any arc $\gamma^{(n)}$ is bounded from below. There exists $c > 0$ depending only on $\mathscr{U}_\Phi$ such that

$$\operatorname{mes} M_n > c|\mu_1^{(n)} - \mu_2^{(n)}|$$

(cf [J5]).

Similar arguments are valid for any family of maps (not necessary close to Φ) which transversely intersects $W^s(\Phi)$. In particular, it holds for the family of quadratic polynomials.

3.4. Properties of Maps Belonging to the Stable Manifold of Φ. The stable manifold $W^s(\Phi)$ separates mappings with simple dynamics from those with complex one. Any map $\psi \in W^s(\Phi) \cap \mathscr{U}_\Phi$ satisfies two specific conditions.

(i) ψ has 2^n—periodic points for any n, and has no points of other periods.

(ii) $S\psi < 0$.

The following theorem proved independently by M. Misiurewicz and by Yu.S. Barkowski and G.M. Levin describes the properties of unimodal maps ψ satisfying (i), (ii) (cf [Mi3], [VSK]).

Theorem 3.4. $\Omega(\psi) = \sigma \cup \operatorname{Per}\psi$, *where* $\operatorname{Per}\psi$ *is the set of periodic points, and* σ *is a closed* ψ*-invariant Cantor set with the following properties*:

a) $\psi\colon \sigma \to \sigma$ *is a minimal homeomorphism*;

b) $\lim_{n\to\infty} \operatorname{dist}(\psi^n x, \sigma) = 0\ \forall x \notin \bigcup_{k=0}^{\infty} \psi^{-k}(\operatorname{Per}\psi)$;

c) $\forall x \in [-1, 1] \exists t(x) = \lim_{n\to\infty} \psi^{2^n}(x)$, *and* $t(x) = x$ *for* $x \in \Omega(\psi)$;

d) *The unique* ψ*-invariant non atomic measure* μ *is supported by* σ; *the system* (σ, ψ, μ) *is isomorphic to the unit translation on the group of 2-adic integers which is called the "adding machine".*

This isomorphism is a result of the following construction. There exists a system of intervals $\Delta_i^{(n)}$, $n \geqslant 1$, $i = 0, 1, \ldots, 2^n - 1$, satisfying: $\psi\Delta_i^{(n)} = \Delta_{i+1}^{(n)}$ for $i < 2^n - 1$ and $\psi\Delta_{2^n-1}^{(n)} \subset \Delta_0^{(n)}$; $\Delta_i^{(n)} \cap \Delta_j^{(n)} = \varnothing$ for $i \neq j$; $\Delta_k^{(n-1)} \supset \Delta_k^{(n)} \cup \Delta_{k+2^{n-1}}^{(n)}$; $\operatorname{diam}_{n\to\infty} \Delta_k^{(n)} \to 0$. Then

$$\sigma = \bigcap_{n\geqslant 1} \bigcup_{i=0}^{2^n-1} \Delta_i^{(n)}.$$

The spectrum of the dynamical system (σ, ψ, μ) consists of the binary rationals. The corresponding eigenfunctions are

$$e^{(n)}(x) = \exp(2\pi i l 2^{-n}) \text{ for } x \in \Delta_l^{(n)}$$

and

$$e_r^{(n)}(x) = (e^{(n)}(x))^{2r+1}, \qquad r = 0, 1, \ldots, 2^{n-1} - 1.$$

The fixed point equation $\Phi(x) = \lambda^{-1}\Phi \circ \Phi(\lambda x)$ implies, besides a)–d) of Theorem 3.4, some additional properties of $\Phi(x)$. The following results were obtained in [VSK].

Theorem 3.5. *Let $\varphi(x)$ be a C^1-function on $[-1,1]$. Consider the spectral measure of φ: $\rho_\varphi(\omega) = \sum_{n=1}^{\infty} \sum_{r=0}^{2^{n-1}-1} \rho_r^{(n)} \delta\left(\omega - \frac{2r+1}{2^n}\right)$. Then*

$$\rho_r^{(n)} \leqslant \frac{1}{2} \max_{x \in [-1,1]} |\varphi'(x)| \cdot \max_{0 \leqslant l \leqslant 2^{n-1}-1} |\Delta_l^{(n-1)}|.$$

Theorem 3.6. *There exist $\gamma < 0$, $\beta_0 > 0$ such that for x constituting some subset of σ of full μ-measure*

$$\lim_{n \to \infty} \frac{1}{n} \ln |\Delta_i^{(n)}(x)| = \gamma$$

$$\lim_{n \to \infty} \frac{1}{n} \ln \left(\sum_{i=0}^{2^n-1} |\Delta_i^{(n)}|^{\beta_0} \right) = 0 \qquad \text{holds.}$$

Here β_0 is the Hausdorff dimension of σ and

$$\beta_0 > -\frac{\ln 2}{\gamma}$$

which signifies that the Hausdorff dimension is reached on the set of zero μ-measure. A similar result has been obtained by F. Ledrappier and M. Misiurewicz ([LM]).

For later developments in renormalization theory see ([Su2]) and ([Mc]).

§ 4. Rational Endomorphisms of the Riemann Sphere

4.1. The Julia Set and Its Complement. The dynamics of rational endomorphisms $z \mapsto R(z) = \dfrac{a_0 + a_1 z + \cdots + a_n z^n}{b_0 + b_1 + \cdots + b_n z^n}$ of the Riemann sphere was studied in the fundamental memoirs of Julia and Fatou at the beginning of the century (see [Mo], the surveys of P. Blanchard ([B1]) and M. Lyubich ([Lyu4]), and the recent books by L. Carleson and T. Gamelin ([CG]) and A. Beardon ([Bea]). This is an area of active research. Many important, new methods have been developed and many strong results obtained since the first edition of this Volume appeared in 1985. We will not cover all of these advances. But rather, we mention only a few results which are relevant to our discussion.

Similarly to one-dimensional real maps, rational endomorphisms may have a stable or a stochastic asymptotic behavior. In order to describe possible types of dynamics, we shall use the following classification of periodic orbits (cycles) of $R(z)$. A cycle $\alpha = (\alpha_0, \alpha_1, \ldots, \alpha_{m-1})$ is *attracting* if $|(R^m)'(\alpha)| = |\prod_{i=0}^{m-1} R'(\alpha_i) < 1$, *rational neutral* if $(R^m)'(\alpha) = \exp 2\pi i r$, $r \in \mathbb{Q}$, *irrational neutral* if $(R^m)'(\alpha) = \exp 2\pi i \theta$, θ-irrational and *repelling* if $|(R^m)'(\alpha)| > 1$.

The points with irregular aperiodic behavior are inside the so-called *Julia set* $J(R)$ which coincides with the closure of repelling periodic orbits. $J(R)$ is a perfect set invariant under R and under R^{-1}. There exist endomorphisms R satisfying $J(R) = \bar{\mathbb{C}}$ (cf [Mo]) and moreover the measure of such R in the parameter space is positive (see [Re2]). Nevertheless a generally accepted hypothesis is that the set of such endomorphisms is not generic, i.e. it is a countable union of nowhere dense sets.

Any component of the open set $\Delta(R) = \bar{\mathbb{C}} \backslash J(R)$ consists of points with the same asymptotic behavior. Different types of dynamics for $z \in \Delta(R)$ were studied by Fatou and Julia. A final result was obtained by D. Sullivan (cf [Su1]).

Theorem 4.1. *Let $\mathscr{D}$ be a component of $\Delta(R)$ Then*

(i) *there exists k_0 such that the domain $\mathscr{D}_1 = R^{k_0}\mathscr{D}$ is periodic, i.e. $\mathscr{D}_1 = R^m\mathscr{D}_1$ for some m and for $|k - l| < m$ $R^k\mathscr{D}_1 \cap R^l\mathscr{D}_1 = \emptyset$,*

(ii) *the number of periodic components is finite,*

(iii) *the only possible types of dynamics on any periodic component $\mathscr{D}$ are the following:*

a) *for any $z \in \mathscr{D}$ the trajectory $\{R^n(z)\}$ converges to an attracting cycle $\alpha = (\alpha_0, \alpha_1, \ldots, \alpha_{m-1})$ where $\alpha_i \in R^i\mathscr{D}$;*

b) *for any $z \in \mathscr{D}$ the trajectory $\{R^n(z)\}$ converges to a rational neutral cycle $\alpha = (\alpha_0, \alpha_1, \ldots, \alpha_{m-1})$ where α_i lies on the boundary of $R^i\mathscr{D}$;*

c) *$\mathscr{D}$ contains a point of some irrational neutral cycle $\alpha = (\alpha_0, \alpha_1, \ldots, \alpha_{m-1})$ and $R^m|\mathscr{D}$ is topologically conjugate to an irrational rotation of a disc;*

d) *$R^m|\mathscr{D}$ is topologically conjugate to an irrational rotation of a ring.*

Let us consider for example the map $z \to R(z) = \frac{1}{2}(z + z^3)$ studied in detail by Julia. The structure of $J(R)$ and of components of $\Delta(R)$ is illustrated in Fig. 25. $\Delta(R)$ is composed of the basin of attraction to infinity, denoted by $\mathscr{D}_\infty$, of two basins of attraction to the fixed points 1 and -1 denoted respectively by $\mathscr{D}_1$ and by $\mathscr{D}_{-1}$ and of an infinite number of preimages $\bigcup_{n=1}^{\infty} R^{-n}(\mathscr{D}_{\pm 1})$. For R as well

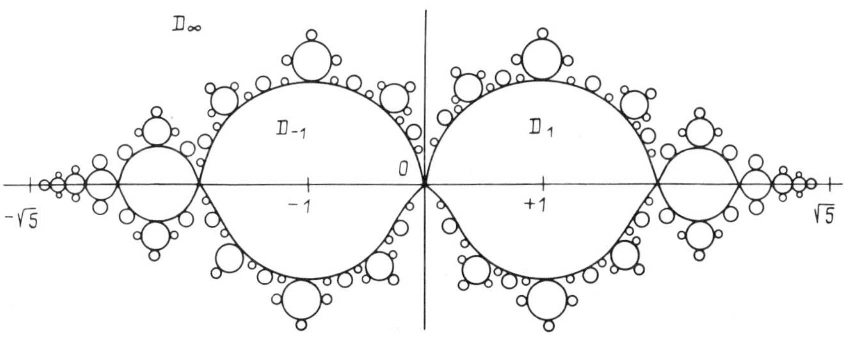

Fig. 25

as for any other polynomial, the set $J(R)$ coincides with the boundary of $\mathcal{D}_\infty$, which is connected and invariant under $R(z)$ and under $R^{-1}(z)$.

4.2. The Stability Properties of Rational Endomorphisms. We have a good understanding of the dynamics of rational maps satisfying the following hyperbolicity condition:

there exist $c > 0$ and $l_0 \in \mathbb{N}$ such that for any $z \in J(R)$

$$|(R^{l_0})'(z)| > c.$$

In this case $\Delta(R)$ is non empty and any component of $\Delta(R)$ eventually falls in some periodic component of type a). The following condition is necessary and sufficient for hyperbolicity: the iterates of any critical point of the map R converge to some attracting cycle.

Rational endomorphisms with the hyperbolic Julia set are structurally stable. They admits Markov partitions any may be studied using the method of symbolic dynamics (cf [J5]).

The following problem formulated by Fatou is still unsolved: is the set of hyperbolic rational (polynomial) endomorphisms dense in the space of all rational (polynomial) endomorphisms? In particular the question remains unsolved for the family of quadratic polynomials $z \mapsto z^2 + a$. For real quadratic family see Theorem 2.3 above.

For polynomials this conjecture is implied by the following: for a dense set of polynomials the Lebesgue measure of the Julia set is zero (cf [MSS]).

Although the density of hyperbolic maps is unknown, the following J-stability theorem holds as proved by Mañé, Sad, Sullivan (cf [MSS]), and in a weaker form by Levin (cf [Lev]) and by Lyubich (cf [Lyu3]).

Theorem 4.2. *For any family $R_\omega(z)$ of rational endomorphisms holomorphicly depending on a parameter $\omega \in U \subset \mathbb{C}^k$ the set $S = \{\omega\colon R_\omega(z)$ is J-stable$\}$ is open and dense in U. If R_1 and R_2 are in the same component of S then $R_1|J(R_1)$ is quasi-conformally conjugate to $R_2|J(R_2)$.*

For quadratic polynomials $f_c' : z \mapsto z^2 + c$, the *Mandelbrot set* M is the set of parameter values c such that the critical orbit $f_c^n(0)_{n=0}^\infty$ is bounded.

The famous problem is whether the Mandelbrot set is locally connected.

J.-C. Yoccoz proved that Mandelbrot set is locally connected at every point corresponding to the finitely renormalizable quadratic polynomial (see [Mi]).

Let ∂M be the boundary of the Mandelbrot set. M. Shishikura (cf [Shi]) proved that $H_{\dim}(\partial M) = 2$ and parameter values c such that $H_{\dim}(J(f_c)) = 2$ are typical in ∂M in topological sense.

The local connectivity of Julia sets was studied for many polynomials. There are examples of quadratic polynomials, which Julia sets are not locally connected, see [CG]. We mention two recent results in that area.

G. Levin and S. van Strien proved that Julia sets are locally connected for real polynomials $z^n + c$ for all n, see [LS].

J. Graczyk and S. Smirnov proved that Julia sets for polynomials satisfying Collet-Eckmann condition are locally connected, see [GSm].

4.3. Ergodic and Dimensional Properties of Julia Sets. The investigation of ergodic properties of rational endomorphisms i.e. of invariant measures supported by Julia sets originates from an article of Brolin who proved that for a polynomial map $z \mapsto P(z)$ the preimages of any $z\{P^{-n}(z)\}_{n=0}^{\infty}$ (with the possible exception of one exclusive point z_0) are asymptotically equidistributed with respect to some measure μ on $J(P)$.

Lyubich in [Lyu2] generalized the result of Brolin for rational maps. The corresponding measure μ turns out to be the unique measure of maximal entropy $h\mu(R) = h(R) = \log \deg R$.

D. Ruelle in [Ru2] studied Gibbs measures for endomorphisms satisfying the hyperbolicity condition and their relation to the Hausdorff dimension. He proved that $H_{\dim}(J(R))$ is a real analytic function of R. In particular, if $R(z) = z^d + \varepsilon$ where $|\varepsilon|$ is small, then $H_{\dim}(J(R)) = 1 + (|\varepsilon|^2/4)\log d +$ higher order terms in ε.

A. Manning proved in [Ma2] that for an open set of polynomials of degree d (including in particular $z^d + \varepsilon$) the Hausdorff dimension of the measure μ of maximal entropy (which is defined as $\inf\{H_{\dim}(Y)\colon \mu(Y) = 1\}$) equals 1 (see [Pr2] for generalization of this result).

Let $D_{\dim}(f)$ (*dynamical dimension*) be the supremum of $H_{\dim}(\mu)$ over invariant probabilistic measures with positive Lyapunov exponents. It was conjectured that for any rational map R holds $H_{\dim}(J(R)) = D_{\dim}(R)$.

M. Denker and M. Urbanski proved it for Julia sets which do not contain critical points, see [DU]. F. Przytycki proved it for polynomials satisfying Collet-Eckmann condition (see [Pr3]).

Returning to the ergodic properties of rational maps with respect to the Lebesgue measure, we notice that M. Rees (cf [Re2]) proved that there exists a set of positive measure in the parameter space of R such that $J(R) = \bar{C}$ and R has an invariant ergodic measure equivalent to the Lebesgue measure.

Thus in the parameter space of rational endomorphisms of the Riemann sphere as well as for the maps of the interval, there exist two sets with opposite properties. The first one is an open and presumably dense set consisting of structurally stable endomorphisms with hyperbolic nonwandering sets. Under the action of such a map, the iterates of almost every (with respect to the Lebesgue measure) point z converge toward a finite number of attracting cycles. The second set is a set of positive measure consisting of endomorphisms which are ergodic with respect to the Lebesgue measure. It is natural to ask whether or not the union of structurally stable hyperbolic maps and of stochastic maps, which have Sinai-Ruelle-Bowen measure is a set of full measure in the parameter space.

Bibliography

The main results and methods of KAM theory are discussed in the books [Ar1], [Ar2], [Mos]. The latest advances in this area are related to the works of Aubry [AD], Parseval [P], Mather [Mat2].

The main results concerning topological and measure theoretical aspects of hyperbolic theory in different stages of its development are discussed in the surveys [KSS], [PS1] and in the books [N], which presents topological aspects of the theory, and [CSF] where measure theoretical aspects are considered. They also contain various examples, a voluminous bibliography, and other applications.

The theory of Anosov systems preserving Liouville measure is discussed in the monograph [An1] which is the first systematic and fundamental research in hyperbolic theory. There is also an exposition of general results concerning Anosov systems in the book [An2] and in the review article [AS]. A theory of hyperbolic sets (topological properties, various examples) and some related problems (the A-system and so on) are treated in the book [N] (see also [K1] where a full proof of the theorem about the families of ε-trajectories is given). Symbolic dynamics for Anosov systems (Markov partitions, equilibrium states, measures of maximal entropy) is constructed in [Si6] (see also [Si2], [Si3]); a generalization to hyperbolic sets is given in several articles of Bowen (c.f. [Bo2]); some further generalizations may be found in [AJ], which also contains a short review of topological Markov chains theory. The foundations of the theory of UPH-systems are contained in [BP]. NCH-systems are introduced in [Pe2] where their local properties and ergodic properties with respect to Liouville measure are considered (see also [K3]). A generalization to Sinai measures is given in [L3].

The main results concerning topological and ergodic properties of geodesic flows with hyperbolic behavior of trajectories (on manifolds without conjugate points, or without focal points, or on manifolds with negative curvature) and their relation to Riemannian geometry and classical mechanics are given in the survey [Pe3]. In this survey some other dynamical systems of geometric origin are considered (frame flows and horocycle flows) and a voluminous bibliography is given.

Proofs of general results about billiard systems may be found in the monograph [CSF], which also contains a detailed explanation of the relation between some models of classical mechanics and these systems. Ergodic properties of dispersed billiards are considered in [Si4] where in particular the ergodicity of a system of two disks on two-torus is proved. Some aspects of ergodic theory for dispersed and semidispersed billiards (in particularly for the gas of firm spheres) are given in [Si8]. A detailed exposition of the theory of billiards in two-dimensional convex domains and its relation to Dirichlet problem is contained in [La2]. In the survey [PS1], general stochastic properties of dynamical systems are discussed, in particular the properties of stochastic attractors. Geometrical and topological properties of attractors of the Lorenz type are considered in [ABS], their ergodic properties in [BS1] and [Bu4].

In the theory of non-invertible one-dimensional maps, an important role was played by the work [Ro], where ergodic properties of an endomorphism were related to those of its natural extension. The properties of Gibbs measures for different classes of expanding maps are considered in [W2], [HK] and [Ry1]. For more recent developments see the book [MS]. One-dimensional maps played a crucial role in the development of the renormalization group method for dynamical systems, see [CE], [VSK], [Su2], [Mc]. Recently there has been an explosion of interest in ergodic theory of holomorphic dynamical systems on the Riemann sphere. Part of results concerning this theme may be found in [Bl], [Lyu4], [D], ([CG]) and [Bea].

For the convenience of the reader, references to reviews in Zentralblatt für Mathematik (Zbl.), compiled using the MATH database, and Jahrbuch über die Fortschritte der Mathematik (Jrb.) have, as far as possible, been included in this bibliography.

[AW] Adler, R.L., Weiss, B.: Entropy be a complete metric invariant for automorphisms of the torus. Proc. Natl. Acad. Sci. USA 57, 1573–1576 (1967). Zbl. 177.80

[ABS1] Afraimovich, V.S., Bykov, V.V., Shilnikov, L.P.: On the origin and structure of the Lorenz attractor. Dokl. Akad. Nauk SSSR 234, 336–339 (1977) [Russian]. Zbl. 451.76052. English transl.: Sov. Phys., Dokl. 22, 253–255 (1977)

[ABS2] Afraimovich, V.S., Bykov. V.V., Shilnikov, L.P.: On attracting structurally unstable limit sets of Lorenz type attractor. Tr. Mosk. Mat. O.-va 44, 150–212 (1982) [Russian]. Zbl. 506.58023. English transl.: Trans. Mosc. Math. Soc. 1983, No. 2, 153–216 (1983)

[AP] Afraimovich, V.S., Pesin, Ya.B.: An estimate of Hausdorff dimension of a basic set in a neighborhood of homoclinic trajectory. Usp. Mat. Nauk 39, No. 2, 135–136 (1984) [Russian]. Zbl. 551.58026. English transl.: Russ. Math. Surv. 39, No. 2, 137–138 (1984)

[A1] Alekseev, V.M.: Perron sets and topological Markov chains. Usp. Mat. Nauk 24, 227–228 (1969) [Russian]. Zbl. 201.566

[A2] Alekseev, V.M.: Quasirandom oscillations and qualitative questions in celestial mechanics. In: Ninth Summer School., Izd. Inst. Mat. Akad. Nauk Ukr. SSR. Kiev. 212–341 (1972) [Russian]. Zbl. 502.70016. English transl.: Transl., II. Ser., Am. Math. Soc. 116, 97–169 (1981)

[AJ] Alekseev, V.M., Jakobson, M.V.: Symbolic dynamics and hyperbolic dynamical systems, Phys. Rep. 75, 287–325 (1981)

[An1] Anosov, D.V.: Geodesic flows on closed Riemann manifolds with negative curvature, Tr. Mat. Inst. Steklov 90, 210 p. (1967) [Russian]. Zbl. 163.436. English transl.: Proc. Steklov Inst. Math. 90, (1967), 235 p.

[An2] Anosov, D.V.: On a certain class of invariant sets of smooth dynamical systems. In: Proc. 5th Intern. Conf. on non-linear oscillations, pp. 39–45, 2, (1970) [Russian]. Zbl. 243.34085 English transl.: Russ. Math. Surveys 22, No. 5, 103–167 (1967)

[AS] Anosov, D.V., Sinai, Ya.G.: Some smooth ergodic systems, Usp. Mat. Nauk 22, No.5, 107–172 (1967) [Russian]. Zbl. 177.420

[Ar1] Arnold, V.I.: Mathematical methods of classical mechanics. Nauka, Moscow 1974 [Russian]. Zbl. 386.70001. English transl.: Springer-Verlag: New York-Heidelberg-Berlin, 462 p. (1978)

[Ar2] Arnold, V.I.: Some additional topics of the theory of ordinary differential equations. Nauka, Moscow 1978 [Russian]

[AD] Aubry, S., Le Daeron, P.V.: The discrete Frenkel-Kontorova model and its extension. Physica D., 8, 381–422 (1983)

[Be] Belykh, V.N.: Models of discrete systems of phase synchronization. Ch. 10 In: Systems of phase synchronization (ed. by V.V. Shakhgildyan and L.N. Belyustina), pp. 161–176. Radio and Communication, Moscow 1982 [Russian]

[Bea] Beardon, A.F.: Iteration of rational functions. Complex analytic dynamical systems. Graduate Texts in Mathematics, 132. Springer-Verlag, New York, 1991. xvi+280 pp.

[BC] Benedicks, M., Carleson, L.: On iterations of $1 - ax^2$ on $(-1, 1)$, Ann. Math. II. Ser. 122, 1–25 (1985). Zbl. 597.58016

[BeS] Benettin, G., Strelcyn, J.-M.: Numerical experiments on a billiard. Stochastic transition and entropy. Phys. Rev. A 17, 773–786 (1978)

[Bil] Billingsley, P.: Ergodic theory and information. John Wiley and Sons, Inc., New York-London-Sydney; 1965. Zbl. 141.167

[Bi] Birkhoff, G.D.: Dynamical systems. New York, Am. Math. Soc., Colloq. Publ. 9 1927. Jrb. 53.732

[Bl] Blanchard, P.: Complex analytic dynamics on the Riemann sphere. Bull. Am. Math. Soc., New Ser. 11, 85–141 (1984). Zbl. 558.58017

[BKM] Boldrighini, C., Keane, M., Marchetti, F.: Billiards in polygons. Ann. Probab. 6, 532–540 (1978). Zbl. 377.28014

[Bo1] Bowen, R.: Periodic orbits for hyperbolic flows. Am. J. Math. 94, 1–30 (1972). Zbl. 254.58005

[Bo2] Bowen R.: Methods of symbolic dynamics, collected papers, Mir, Moscow 1979 [Russian]

[BFK] Brin, M., Feldman, J., Katok, A.: Bernoulli diffeorphisms and group extensions of dynamical systems with non-zero characteristic exponents. Ann. Math., II. Ser. 113, 159–179 (1981). Zbl. 477.58021

[BP] Brin, M.I., Pesin, Ya.B.: Partially hyperbolic dynamical Systems, lzv. Akad. Nauk SSSR, Ser. Mat. 38, 170–212 (1974) [Russian]. Zbl. 304.58017

[Bu1] Bunimovich, L.A.: On billiards close to dispersing. Mat. Sb., Nov. Ser. 94 (136), 49–73 (1974) [Russian]. Zbl. 309.58018. English transl.: Math. USSR, Sb. 95, 49–73 (1974).

[Bu2] Bunimovich, L.A.: On the ergodic properties of nowhere dispersing billiards. Commun. Math. Phys. 65, 295–312 (1979). Zbl. 421.58017

[Bu3] Bunimovich, L.A.: Some new advancements in the physical applications of ergodic theory. In: Ergodic theory and Related Topics (ed. by Michel, H.), pp. 27–33, Berlin, Akademie-Verlag 1982. Zbl. 523.58025

[Bu4] Bunimovich, L.A.: Statistical properties of Lorenz attractors. In: Nonlinear Dynamics and Turbulence (ed. by GI. Barenblatt et al.), pp. 71–92, Boston-London-Melbourne: Pitman 1983. Zbl. 578.58025

[Bu5] Bunimovich, L.A.: Decay of correlations in dynamical systems with chaotic behavior. J. Exp. Theor. Phys. 89, 1452–1471 (1985) [Russian].

[Bu6] Bunimovich, L.A.: On stochastic dynamics of rays in resonators. Radiofizika 28, 1601–1602 (1985) [Russian].

[BS1] Bunimovich, L.A., Sinai, Ya.G.: On a fundamental theorem in the theory of dispersing billiards. Mat. Sb.. Nov. Ser. 90 (132), 415–431 (1973) [Russian]. Zbl. 252.58006. English transl.: Math. USSR, Sb. 19, 407–423 (1974)

[BS2] Bunimovich, L.A., Sinai, Ya.G.: Markov partitions for dispersed billiards. Commun. Math. Phys. 78, 247–280 (1980), Erratum, ibid, 107, 357–358 (1986). Zbl. 453.60098

[BS3] Bunimovich, L.A., Sinai, Ya.G.: Stochasticity of attractor in the Lorenz model. In Nonlinear waves (ed. by A.V. Gaponov-Grekhon), pp. 212–226, Moscow, Nauka 1980 [Russian].

[BS4] Bunimovich, L.A., Sinai, Ya.G.: Statistical properties of Lorentz gas with periodic configuration of scatterers. Commun. Math. Phys. 78, 479–497 (1981). Zbl. 459.60099

[BK] Burns, K., Katok, A.: Manifolds with non positive curvature. Ergodic Theory Dyni. Syst. 5, 307–317 (1985). Zbl. 572.58019

[C] Chentsova, N.I.: A Natural invariant measure on Smale's horseshoe, Dokl. Akad. Nauk SSSR 256, 294–298 (1981) [Russian]. Zbl. 475.58013. English transl.: Sov. Math., Dokl. 23, 87–91 (1981)

[Ch] Chernov, N.J.: Construction of transversal fibres in multidimensional semidispersing billiards. Funkt. Anal. Prilozh. 16, No. 4, 35–46 (1982) [Russian]; English transl.: Funct. Anal. Appl. 16, 270–280 (1983). Zbl. 552.28015

[Chi] Chirikov, B.V.: An universal instability of many-dimensional oscillator systems. Phys. Repts. 53, 263–319 (1979)

[CE] Collet, P., Eckmann, J.-P.: Iterated maps on the interval as dynamical systcms. Basel-Boston-Stuttgart Birkhäuser 1980. 248 p. Zbl. 458.58002

[CG] Carleson, L., Gamelin, T.W.: Complex dynamics. Universitext: Tracts in Mathematics. Springer-Verlag, New York, 1993. x+175 pp.

[CSF] Cornfeld, I.P., Sinai, Ya.G., Fomin, S.V.: Ergodic theory, Moscow, Nauka 1980, 383 p. [Russian]. English transl.: Springer 1982, 486 p. Zbl. 493.28007

[D] Douady, A.: Disques de Siegel et anneax de Herman. Sémm, Bourbaki 677 (1987)

[DH] Donady, A. Hubbard, J.: On the dynamics of polynomial-like mappings. Ann. Scient. Ec. Norm. Sup. 18 (1985), 287–343.

[DU] Denker, M., Urbanski, M.: On Sullivan's conformal measures for rational maps of the Riemann sphere. Nonlinearity 4 (1991) 365–384.

[EW] Eckmann, J.-P., Wittwer, P.: A complete proof of the Feigenbaum conjectures, preprint (1986)

[FOY] Farmer, J.D., Ott, E., Yorke, J.A.: The dimension of chaotic attractors. Phys. 7D, 153–180 (1983). Zbl. 561.58032

[Gac] Gachok, P.V.: Construction of models of dynamical systems with a strange attractor. Dokl. Akad. Nauk SSSR 274, 1292–1294 (1984) [Russian]. Zbl. 596.34028. English transl.: Sov. Math. Dokl. 29, 123–125 (1984)

[G] Gallavotti, G., Ornstein, D.: Billiards and Bernoulli schemes. Commun. Math. Phys. 38, 83–101(1974). Zbl. 313.58012

[Ga] Galperin, G.A.: Non-periodic and not everywhere dense billiard trajectories in convex polygons and polyhedrons. Commun. Math. Phys. 91, 187–211 (1983). Zbl. 529.70001

[GGV] Gelfand, I.M., Graev, M.I., Vilenkin, I.Ya.: Integral geometry and some related aspects of representation theory, Moscow 1962, 656 p. [Russian]. Zbl. 115.167

[GSm] Graczyk, J., Smirnov, S. Collet, Eckmann and Holder, Preprint (1995), to appear in Invent. Math.

[GS1] Graczyk, J., Swiatek, G.: Generic hyperbolicity in the logistic family. Ann. of Math. (2) 146 (1997), no. 1, 1–52.

[GS2] Graczyk, J., Swiatek, G.: Induced expansion for quadratic polynomials. Ann. Scient. Ec. Norm. Sup. 29 (1996), 399–482.

[GU] Guckenheimer, J.: Renormalization of one dimensional mappings and strange attractors, Contemp. Math. 58, No. 3, 143–160 (1987). Zbl. 622.58030

[H] Henon, M.: A two-dimensional mapping with a strange attractor. Commun. Math. Phys. 50, 69–76 (1976). Zbl. 576.58018

[HK] Hofbauer, F., Keller G.: Equilibrium states for piecewise monotonic transformations. Ergod. Theory and Dyn. Syst. 2, 23–43 (1982). Zbl. 499.28012

[J1] Jakobson, M.V.: Markov partitions for rational endomorphisms of the Riemann sphere, In: Multicomponent random systems, (R.L. Dobrushin, Ya.G. Sinai eds.) Advances in probability and related topics 6, 381–396 (1980). Zbl. 439.58021

[J2] Jakobson, M.V.: Absolutely continuous invariant measures for one-parameter families of one-dimensional maps. Commun. Math. Phys. 81, 39–88 (1981). Zbl. 497.58017

[J3] Jakobson, M.V.: Invariant measures of some one-dimensional attractors. Ergodic Theory Dyn., Syst. 2, 317–337 (1982). Zbl. 521.58039

[J4] Jakobson, M.V.: Families of one-dimensional maps and near-by diffeomorphisms. In: Proceedings of ICM-86, Berkeley 1987.

[J5] Jakobson, M.V.: Universal behavior and stochasticity for one-dimensional dynamical systems. In: Proceedings of the 1st world congress of the Bernoulli society: VNU Sc. Press. Utrecht 1, 85–90, 1987

[J6] Jakobson, M.V.: On Smooth mappings of the circle into itself. (English) Math. USSR Sbornik. v. 14 (1971), no. 2, 161–185.

[JS] Jakobson, M., Swiatek, G.: Metric properties of non-renormalizable S-unimodal maps: II. Quasisymmetric conjugacy classes. Ergod. Th. and Dynam. Sys. 15 (1995), 871–938.

[K1] Katok, A.B.: Local properties of hyperbolic sets, Appendix 1 to Nitecki Z., Differentiable dynamics, Moscow, Mir, (translated from English), pp. 214–232, 1975 [Russian]. Zbl. 246.58012

[K2] Katok, A.: Bernoulli diffeomorphisms on surfaces. Ann. Math., II. Ser. 110, 529–547 (1979). Zbl. 435.58021

[K3] Katok, A.: Lyapunov exponents, entropy and periodic orbits for diffeomorphisms, Publ. Math., Inst. Hautes Ethol. Sci. 51, 137–173 (1980). Zbl. 445.58015

[K4] Katok, A.: Some remarks on Birkhoff and Mather twist map theorems. Ergodic Theory Dyn. Syst. 2, 185–194 (1982). Zbl. 521.58048

[KSS] Katok, A.B., Sinai, Ya.G., Stepin, A.M.: Theory of dynamical systems and general transformation groups with invariant measure, Itogi Nauki Tekhn. Ser. Math. Anal. 13, 129–262, 1975 [Russian]. Zbl. 399.28011. English transl.: J. Sov. Math. 7, 974–1065 (1977)

[KS] Katok, A., Strelcyn, J-M.: Invariant Manifolds, Entropy and Billiards; Smooth Maps with Singularities. Lect. Notes Math. 1222, Berlin-Heidelberg, Springer 1986

[Ko] O.S. Kozlovski, O.S. Structural stability in one-dimensional dynamics. PhD thesis. (1998), Univ. of Amsterdam.

[Kr] Krylov, N.S.: Works on the foundations of statistical physics. Moscow: Akad. Nauk SSSR Publ., 1950 [Russian]

[L1] Ledrappier, F.: Some properties of absolutely continuous invariant measures on an interval. Ergodic Theory Dyn. Syst. 1, 77–93 (1981). Zbl. 487.28015

[L2] Ledrappier, F.: Some relations between dimension and Lyapunov exponents. Commun. Math. Phys. 81, 229–238 (1981). Zbl. 486.58021

[L3] Ledrappier, F.: Propriétés ergodiques des mesures de Sinai, C.R. Acad. Sd., Paris, Se'r. I, 294, 593–595 (1982). Zbl. 513.58030

[LM] Ledrappier, F., Misiurewicz, M.: Dimension of invariant measures for maps with exponent zero. Ergod. Th. Dyn. Syst. 5, 595–610 (1985). Zbl. 608.28008

[LS] Levin, G., van Strien, S.: Local connectivity of the Julia set of real polynomials. Annals of Math. 147 (1998), 471–541.

[LY] Ledrappier, F., Young, L.-S.: The metric entropy of diffeomorphism, I, II, Ann. Math., II. Ser. 122, 509–539, 540–574 (1985). Zbl. 605.58028

[Lanf] Lanford III, O.E.: Computer-assisted proofs in analysis. Proceedings of ICM-86 (1987)

[La1] Lazutkin, V.F.: The existence of caustics for a billiard problem in a convex domain. Izv. Akad. Nauk SSSR, Ser. Mat. 37, 186–216 (1983) [Russian]. Zbl. 256.52001. English transl.: Math. USSR, Izv. 7, 185–214 (1974)

[La2] Lazutkin, V.F.: Convex billiards and eigenfunctions of Laplace operator. Leningrad Univ. Publ. 1981 [Russian]. Zbl. 532.58031

[Le] Levy, Y.: Ergodic properties of the Lozi mappings. Commun. Math. Phys. 93, 461–481 (1984). Zbl. 553.58019

[Lev] Levin, G.M.: Irregular values of the parameter of a family of polynomial mappings. (English) Russian Math. Surveys 36 (1981), no. 6 189–190.

[Lor] Lorenz, E.N.: Deterministic nonperiodic flow. J. Atmos. Sci. 20, 130–141 (1963)

[Loz] Lozi, R.: Un attracteur etrange du type attracteur de Henon. J. Phys., Paris 39, 9–10 (1978)

[Lyu1] Lyubich, M.: Almost every real quadratic map is either regular or stochastic. Preprint SUNY StonyBrook, 1997.

[Lyu2] Lyubich, M.Yu.: On the measure of maximal entropy for rational endomorphism of Riemann sphere, Funkt. Anal. Prilozh. 16, No. 4, 78–79 (1982) [Russian]. Zbl. 525.28021. English transl.: Funct. Anal. Appl. 16, 309–311 (1983)

[Lyu3] Lyubich, M.Yu.: The study of stability for dynamics of rational maps, In: Teor. Funkts., Funkt. Anal. Prilozh. 42, 72–91 (1984) [Russian]. Zbl. 572.30023

[Lyu4] Lyubich, M.Yu.: Dynamics of rational maps: topological picture. Usp. Mat. Nauk 41, No. 4 (250) 35–95 (1986) [Russian]. Zbl. 619.30033. English transl.: Russ. Math. Surv. 41, No. 4, 43–117 (1987)

[Lyu5] Lyubich, M.: Dynamics of complex polynomials. Acta Math. 178 (1997), 185–297.

[M] Marie, R.: Hyperbolicity, sinks and measure in one dimensional dynamics, Commun. Math. Phys. 100, 495–524 (1985). Zbl. 583.58016

[MS] de Melo, W., van Strien, S.: One-dimensional dynamics. Ergebnisse der Mathematik und ihrer Grenzgebiete (3) [Results in Mathematics and Related Areas (3)], 25. Springer-Verlag, Berlin, 1993. xiv+605 pp.

[MSS] Marie, R.: Sad P., Sullivan D.: On the dynamics of rational maps. Ann. Sci. E.N.S. 16, 193–217 (1982). Zbl. 524.58025

[Ma1] Manning, A.: A relation between Lyapunov exponents, Hausdorff dimension and entropy. Ergodic Theory Dyn. Systems 1, 451–459 (1981). Zbl. 478.58011

[Ma2] Manning, A.: The dimension of the maximal measure for a polynomial map., Ann. Math., II. Ser. 119, 425–430 (1984). Zbl. 551.30021

[Mar] Margulis, G.A.: On some measures related to C-flows on compact manifolds, Funkt. Anal. Prilozh. 4, No. 1, 62–76 (1970) [Russian]. Zbl. 245.58003. English transl.: Funct. Anal. Appl. 4, 55–67 (1970)

[Mat1] Mather, J.: Characterization of Anosov diffeomorphisms. Indagaiones Math. 30, 479–483 (1968). Zbl. 165.570

[Mat2] Mather, J.: Concavity of the Lagrangian for quasiperiodic orbits. Comment Math. Helv. 57, 356–376 (1982). Zbl. 508.58037

[Mc] McMullen, C.: Complex dynamics and renormalization. Annals of Mathematics Studies, 135. Princeton University Press, Princeton, NJ, 1994. x+214 pp.

[Mil] Milnor, J.: The Yoccoz theorem on local connectivity of Julia sets, preprint, SUNY StonyBrook, 1991.

[Mi1] Misiurewicz, M.: Strange attractors for the Lozi mappings. In: Nonlinear dynamics (ed. by R.G. Helleman), pp. 348–358, New York, Ann. N.Y., Acad. Sd, 357, 1980. Zbl. 473.58016

[Mi2] Misiurewicz, M.: Absolutely continuous measures for certain maps of an interval. Pubi. Math., Inst. Hautes Etud. Sci. 53, 17–51 (1981). Zbl. 477.58020

[Mi3] Misiurewicz, M.: Structure of mappings of an interval with zero entropy, Pubi. Math., Inst. Hautes Etud. Sci. 53, 5–13 (1981). Zbl. 477.58030

[Mo] Montel, P.: Leçons sur les families normales de fonctions analytiques et leurs applications. Paris, Gauthier-Villars 1927. Jrb. 53.303

[Mor] Mon, H.: Fractal dimension of chaotic flows autonomous dissipative systems. Progr. Theor. Phys. 63, 1044–1047 (1980)

[Mos] Moser, J.: Lectures on Hamiltonian systems. New York, Courant Inst. Math. Sci. 1968

[N] Nitecki, Z.: Differentiable dynamics. MIT Press, London, 1971, 282 p. Zbl. 246.58012

[NP] Nowicki, T., Przytycki, F.: Topological invariance of Collet-Eckmann property for S-unimodal maps. Fundamenta Mathematicae 155 (1998), 33–43.

[OS] Osserman, R., Sarnak, P.: A new curvature invariant and entropy of geodesic flows, Invent. Math. 77, 455–462 (1984). Zbl. 536.53048

[P] Parseval, I.C.: Variational principles for invariant tori and cantori. In: Symp. on Nonlinear Dyn. and Beam-Beam Interactions, Amer. Inst. Phys. Conf. Proc., pp. 310–320 (1980)

[Pe1] Pesin, Ya.B.: An example of non ergodic flow with non zero characteristic exponents., Funkt. Anal. Prilozh. 8, 81–82 (1974) [Russian]. Zbl. 305.58012. English transi.: Funct. Anal. Appl. 8, 263–264 (1975)

[Pe2] Pesin, Ya.B.: Characteristic Lyapunov exponents and smooth ergodic theory, Usp. Mat. Nauk 32, No. 4 (196), 55–112 (1977) [Russian]. Zbl. 359.58010. English transl.: Russ. Math. Surv. 32, No. 4, 55–114 (1977)

[Pe3] Pesin, Ya.B.: Geodesic flows with hyperbolic behavior of the trajectories and objects, connected with them, Usp. Mat. Nauk 36, No. 4, 3–51 (1981) [Russian]. Zbl. 482.58002. English transl.: Russ. Math. Surv. 36, No. 4, 1–59 (1981)

[Pe4] Pesin, Ya.B.: On the notion of the dimension with respect to a dynamical system. Ergodic Theory Dyn. Syst. 4, 405–420 (1984). Zbl. 616.58038

[Pe5] Pesin, Ya.B.: The generalization of Caratheodory's construction for dimensional characteristics of dynamical systems. In: Statistical physics and dynamical systems, Progr. Phys. 10, pp. 191–202, Birkhäuser 1985

[Pe6] Pesin, Ya.B.: Ergodic properties and dimensional-like characteristics of strange attractors closed to hyperbolic one. Proc. of ICM-86, Berkeley 1987

[PP] Pesin, Ya.B., Pitskel, B.S.: Topological pressure and variational principle for noncompact sets, Funkt. Anal. Prilozh. 18, 50–63 (1984) [Russian]. Zbl. 567.54027. English transl.: Funct. Anal. Appl. 18, 307–318 (1984)

[PS1] Pesin, Ya.B., Sinai, Ya.G.: Hyperbolicity and stochasticity of dynamical systems. In: Mathematical Physics Reviews, Gordon and Breach Press, Harwood Acad. Publ., USA, 2, pp. 53–115, 1981. Zbl. 561.58038

[PS2] Pesin, Ya.B., Sinai, Ya.G.: Gibbs measures for partially hyperbolic attractors. Ergodic Theory Dyn. Syst. 2, 417–438 (1982). Zbl. 519.58035

[Pl] Plykin, R.V.: About hyperbolic attractors of diffeomorphisms, Usp. Mat. Nauk 35, No. 3, 94–104 (1980) [Russian]. Zbl. 445.58016. English transl.: Russ. Math. Surv. 35, No. 3, 109–121 (1980)

[Pla] Plante, J.F.: Anosov flows. Am. J. Math. 94, 729–754 (1972). Zbl. 257.58007

[Pr1] Przytycki, F.: Ergodicity of toral linked twist mappings. Ann. Sci. Ec. Norm. Super., IV. Ser. 16, 345–354 (1983). Zbl. 531.58031

[Pr2] Przytycki, F.: Hausdorff dimension of harmonic measure on the boundary of an attractive bassin for a holomorphic map, Invent. Math. 80, 161–179 (1985). Zbl. 569.58024

[Pr3] Przytycki, F.: Iteration of holomorphic Collet-Eckmann maps: Conformal and invariant measures. Trans. AMS 350 (1998) no. 2, 717–742.

[R] Rand, D.: The topological classification of Lorenz attractors. Math. Proc. Camb. Philos. Soc. 83, 451–460 (1978). Zbl. 375.58015

[Re1] Rees, M.: Ergodic rational maps with dense critical point forward orbit. Ergodic Theory and Dyn. Syst. 4, 311–322 (1984). Zbl. 553.58008

[Re2] Rees, M.: Positive measure sets of ergodic rational maps. Ann. Sc. Ec. Norm. Super. 19, 383–407 (1986). Zbl. 611.58038

[Ro] Rokhlin, V.A.: Exact endomorphisms of Lebesgue space, Izv. Akad. Nauk SSSR, Ser. Mat. 25, 499–530 (1961) [Russian]. Zbl. 107.330

[Ry1] Rychlik, M.: Another proof of Jakobson's theorem and related results. Ergod. Th. and Dynam. Sys. 8 (1988), 83–109.

[Ry2] Rychlik, M.: Bounded variation and invariant measures. Studia Math. 76 (1983), no. 1, 69–80.

[RT] Ruelle, D., Takens, F On the nature of turbulence. Commun. Math. Phys. 20, 167–192 (1971). Zbl. 223.76041

[S] Series, C.: The infinite word problem and limit sets in Fuchsian groups. Ergodic Theory Dyn. Syst. 1, 337–360 (1981). Zbl. 483.30029

[Sh] Shnirelman, A.I.: Statistical properties of eigenfunctions. In: Trans. of Dilijan Math. School. Erevan, pp. 267–278 (1974) [Russian]

[Shi] Shishikura, M.: The boundary of the Mandelbrot set has dimension two. Asterisque (1994), no. 222, 389–405.

[Si1] Sinai, Ya.G.: Geodesic flows on compact surfaces of negative curvature. Dokl. Akad. Nauk SSSR 136, 549–552 (1961) [Russian]. Zbl.133.110. English transl.: Sov. Math., Dokl. 2, 106–109 (1961)

[Si2] Sinai, Ya.G.: Markov partitions and C-diffeomorphisms. Funkt. Anal. Prilozh. 2, No. 1, 64–89 (1968) [Russian]. Zbl. 182.550. English transl.: Funct. Anal. Appl. 2, 61–82 (1968)

[Si3] Sinai, Ya.G.: Construction of Markov partitions. Funkt. Anal. Prilozh. 2, No. 3, 70–80 (1968) [Russian]. Zbl. 194.226. English transl.: Funct. Anal. Appl. 2, 245–253 (1968)

[Si4] Sinai, Ya.G.: Dynamical systems with elastic reflections. Ergodic properties of dispersing billiards. Usp. Mat. Nauk 25, No. 2 (152), 141–192 (1970) [Russian]. Zbl. 252.58005. English transl.: Russ. Math. Surv. 25, No. 2, 137–189 (1970)

[Si5] Sinai, Ya.G.: Some rigorous results on decay of correlations. Supplement to the book: G.M. Zaslavskij, Statistical irreversibility in nonlinear systems, pp. 124–139, Moscow, Nauka 1970 [Russianj

[Si6] Sinai, Ya.G.: Gibbs measures in ergodic theory. Usp. Mat. Nauk 27, No. 4 (166), 21–64, (1972) [Russian]. Zbl. 246.28008. English transi.: Russ. Math. Surv. 27, No. 4, 21–69 (1973)

[Si7] Sinai. Ya.G.: Ergodic properties of Lorentz gas. Funkt. Anal. Prilozh. 13, No. 3, 46–59, (1979) [Russian]. Zbl. 414.28015. English transl.: Funct. Anal. Appl. 13, 192–202 (1980)

[Si8] Sinai, Ya.G.: Development of Krylov ideas. An addendum to the book: N.S. Krylov, Works on the foundations of statistical physics, pp. 239–281, Princeton, Princeton Univ. Press 1979

[SC] Sinai, Ya.G., Chernov, N.I.: Entropy of hard spheres gas with respect to the group of space-time translations. Tr. Semin. Im. I.6. Petrovskogo 8, 218–238, 1982 [Russian]. Zbl. 575.28014

[SV] Sinai, Ya.G., Vul, E.B.: Hyperbolicity conditions for the Lorenz model. Phys. D. 2, 3–7 (1981)

[Su1] Sullivan, D.: Quasi conformal homeomorphisms and dynamics, I. Solution of the Fatou-Julia problem on wandering domains, Ann. Math., II. Ser. 122, 401–418 (1985). Zbl. 589.30022

[Su2] Sullivan, D.: Bounds, quadratic differentials, and renormalization conjectures. American Mathematical Society centennial publications, Vol. II (Providence, RI, 1988), 417–466, Amer. Math. Soc., Providence, RI, 1992.

[T] Theory of solitons. Inverse scattering method. (ed. by S.P. Novikov), Moscow, Nauka 1980 [Russian]. Zbl. 598.35003

[TTY] Thieullen, Ph., Tresser, C., Young, L.S.: Positive Lyapunov exponent for generic one-parameter families of unimodal maps. J. Anal. Math. 64 (1994), 121–172.

[VSK] Vul, E.B., Sinai, Ya.G., Khanin, K.M.: Feigenbaum universality and thermodynamic formalism. Usp. Mat. Nauk, 39, No. 3 (237), 3–37 (1984) [Russian]. Zbl. 561.58033. English transl.: Russ. Math. Surv. 39, No. 3, 1 40(1984)

[W1] Walters, P.: A variational principle for the pressure of continuous transformations. Am. J. Math. 97, 937–971 (1975). Zbl. 318.28007

[W2] Walters, P.: Invariant measures and equilibrium states for some mappings which expand distances. Trans. Am. Math. Soc. 236, 127–153 (1978). Zbl. 375.28009

[Wo] Wojtkowski, M.: Principles for the design of billiards with nonvanishing Lyapunov exponent. Commun. Math. Phys. 105, 391–414 (1986). Zbl. 602.58029

[Y1] Young, L.-S.: Capacity of attractors. Ergodic Theory Dyn. Syst. 1, 381–388 (1981). Zbl. 501.58028

[Y2] Young, L.-S.: Dimension, entropy and Lyapunov exponents. Ergodic Theory Dyn. Syst. 2, 109–124 (1982). Zbl. 523.58024

[Y3] Young, L.S.: Decay of correlations for certain quadratic maps. Comm. Math. Phys. 146 (1992), no. 1, 123–138.

[Z] Ziemian, K.: Almost sure invariance principle for some maps of an interval, Ergodic Theory Dyn. Syst. 5, 625–640 (1985). Zbl. 604.60031

III. Dynamical Systems on Homogeneous Spaces

Contents

Chapter 10
Dynamical Systems on Homogeneous Spaces

S.G. Dani

§ 1. Introduction

In this Chapter we describe the ergodic and dynamical properties of a special class of systems, namely flows arising from translations and affine automorphisms of homogeneous spaces of Lie groups, and their applications in the areas of Diophantine approximation and geometry.

Let G be a Lie group and C be a closed subgroup of G. We denote by G/C the homogeneous space (quotient space) consisting of all cosets of the form gC, $g \in G$. We consider the space equipped with the quotient topology (and the Borel structure arising from it). It is then a locally compact second countable topological space. Also, it has the structure of a differentiable manifold; this will however be involved in our discussion only occasionally. Each $g \in G$ determines a homeomorphism (in fact a diffeomorphism) of G/C, called the *translation* by g and denoted by T_g, defined by $T_g(xC) = gxC$, for all $x \in G$. Also, if α is a (bicontinuous) automorphism of G such that $\alpha(C) = C$ then we get a homeomorphism of G/C, denoted by $\overline{\alpha}$, given by $\overline{\alpha}(xC) = \alpha(x)C$ for all $x \in G$; we refer to a homeomorphism arising in this way as an *automorphism* of G/C. A homeomorphism of the form $T_g \circ \overline{\alpha}$ will be called an *affine automorphism* of G/C.

1.1. Measures on homogeneous spaces. By a 'measure' on a locally compact space, unless specified otherwise, we shall mean a locally finite Borel measure (namely a Borel measure such that all compact subsets are assigned finite measure). The space of all probability measures on a locally compact space X will be denoted by $\mathscr{P}(X)$ and will be considered equipped with the weak* topology, namely the coarsest topology such that for every bounded continuous function f on X the map $\mu \mapsto \int f \, d\mu$ is continuous. Given locally compact spaces X and Y and a continuous map $\theta : X \to Y$, for a finite measure μ on X we denote by $\theta(\mu)$ the measure on Y defined by $\theta(\mu)(E) = \mu(\theta^{-1}(E))$ for all Borel subsets E of Y.

Now let G be a Lie group and C be a closed subgroup of G. Since G/C is a differentiable manifold it admits a measure whose restriction to each coordinate chart is equivalent to the Lebesgue measure on the chart (two measures are considered equivalent if the collections of their sets of measure zero coincide); the equivalence class of such a measure is unique and is called the Lebesgue-measure class. Each measure from this class is quasi-invariant under every affine automorphism of G/C; that is, affine automorphisms carry sets of measure zero to sets of measure zero. By a theorem of G.W. Mackey (cf. [Mac]) any measure on G/C which is quasi-invariant under all translations belongs to the Lebesgue-measure class; in other words, the Lebesgue-measure class is the unique quasi-invariant measure class for the G-action on G/C.

A homogeneous space G/C as above may in general not admit a G-invariant measure (namely a measure invariant under the G-action by translations); for example, the action of the general linear group $G = GL(n, \mathbb{R})$, $n \geq 2$, on the $(n-1)$-dimensional projective space, arising as the quotient of the natural action of G on $\mathbb{R}^n - \{0\}$, does not admit any invariant measure (note that the projective space is a homogeneous space of $GL(n, \mathbb{R})$). The following proposition describes some conditions which ensure existence of a G-invariant measure on G/C in terms of the modular homomorphisms of G and C; we recall that the modular homomorphism Δ_H of a locally compact group H is the homomorphism into $\mathbb{R}^+$ (positive real numbers) defined by the condition $\lambda(Eh) = \Delta_H(h)\lambda(E)$ for all Borel subsets E of H and $h \in H$, where λ is a (left) Haar measure on H.

Proposition 1.1 (cf. [Na], Ch. III). *Let G be a Lie group and C be a closed subgroup of G. Let $\Delta_G : G \to \mathbb{R}^+$ and $\Delta_C : C \to \mathbb{R}^+$ be the modular homomorphisms of G and C respectively. Let $\rho : G \to \mathbb{R}^+$ be a continuous homomorphism. Then there exists a measure m on G/C such that $m(T_g E) = \rho(g)m(E)$, for all $g \in G$ and Borel subsets E of G/C, if and only if $\Delta_C(x) = \rho(x)\Delta_G(x)$ for all $x \in C$. Moreover, any two such measures are multiples of each other. In particular this implies the following:*

i) if C is unimodular and G/C is compact then G/C admits a unique G-invariant probability measure;

ii) any finite G-invariant measure on G/C is invariant under every automorphism of G/C.

For a homogeneous space G/C as above, a measure on G/C which is invariant under all translations will be called a *Haar measure* on G/C. It is easy to see that the Haar measure (unique up to scalar multiples) belongs to the Lebesgue-measure class.

For the most part of this Chapter we will be concerned only with homogeneous spaces with finite Haar measure. Such a homogeneous space has only finitely many connected components. For a Lie group G we denote by G^0 the connected component of the identity in G. It is a connected Lie group and acts transitively on each connected component of any homogeneous space of G. In the sequel we shall often restrict to connected Lie groups. Many of the results however can be extended to disconnected Lie groups, using these observations.

Given a Lie group G and a closed subgroup C of G, we call G/C (or, strictly speaking the pair (G, C)) a *locally faithful* homogeneous space if C contains no nontrivial connected normal subgroup of G, and we say that it is *faithful* if C contains no nontrivial normal subgroup of G. Most questions about homogeneous spaces and flows on them can be reduced to locally faithful homogeneous spaces in a routine manner and furthermore they can be reduced to faithful homogeneous spaces when the hypothesis does not involve the ambient group being simply connected.

For a Lie group G we shall denote by $\mathcal{F}(G)$ the class of all closed subgroups C of G such that G/C admits a finite Haar measure and by $\mathcal{F}_0(G)$ its subclass consisting of those $C \in \mathcal{F}(G)$ for which G/C is locally faithful.

Let G be a Lie group and $C \in \mathcal{F}(G)$. Let m be a (finite) Haar measure on G/C. We recall that the action of G on G/C induces a (continuous) unitary representation of G over the Hilbert space $L^2(G/C)$ of all square-integrable functions on G/C, with respect to m, defined by $U_g(f)(x) = f(g^{-1}x)$ for all $g \in G$, $f \in L^2(G/C)$ and $x \in G/C$. The representation plays an important role in the analysis of the spectra of the unitary operators associated to affine automorphisms of G/C and in turn in the study of ergodic properties of the systems; the unitary operator associated to a measure-preserving automorphism τ of G/C is the operator on $L^2(G/C)$ defined by $U_\tau(f)(x) = f(\tau(x))$ for all $f \in L^2(G/C)$ and $x \in G/C$ (see Ch. 2, and [CFS], Part III, for details).

Let G be a Lie group. A discrete subgroup Γ of G is called a *lattice* in G if G/Γ admits a finite Haar measure. If Γ is a discrete subgroup such that G/Γ is compact then by Proposition 1.1 it is a lattice in G; such a lattice is called a *uniform lattice.* If G is a connected Lie group with no noncompact simple Lie group as a factor (in particular, if G is solvable) all lattices in G are uniform. However in general there exist lattices which are not uniform; see below.

Two lattices Γ_1 and Γ_2 in a Lie group G are said to be *commensurable* with each other if $\Gamma_1 \cap \Gamma_2$ is of finite index in both Γ_1 and Γ_2.

1.2. Examples of lattices. In this subsection we shall describe some examples of lattices, introducing along the way some notation which will be used in the sequel.

1. We denote by $\mathbb{R}^n$ the n-dimensional real vector space equipped with the usual Euclidean topology, realised also as the space of n-rowed column vectors with real entries (here, and in the sequel where it is not specified otherwise, n denotes an arbitrary natural number). We denote by $\mathbb{Z}^n$ the subgroup of $\mathbb{R}^n$ consisting of all columns with integral entries. Clearly $\mathbb{Z}^n$ is a lattice in $\mathbb{R}^n$. Furthermore, every lattice in $\mathbb{R}^n$ is of the form $A(\mathbb{Z}^n)$ where A is an automorphism of $\mathbb{R}^n$, namely a nonsingular linear transformation.

2. We denote by $SL(n, \mathbb{R})$ the Lie group of $n \times n$ real matrices with determinant 1 and by $SL(n, \mathbb{Z})$ the subgroup consisting of all its elements with integral entries. Then $SL(n, \mathbb{Z})$ is a lattice in $SL(n, \mathbb{R})$ (see [Rag], Ch. X). It is a nonuniform lattice. The homogeneous space $SL(n, \mathbb{R})/SL(n, \mathbb{Z})$ can be realised as the space, say $\mathcal{M}$, of all lattices in $\mathbb{R}^n$ which have a fundamental parallelopiped of unit volume. The intrinsic topology on the space of lattices (see [Ca2], Ch. V, and [Rag], Ch. I) coincides with the homogeneous space topology on $SL(n, \mathbb{R})/SL(n, \mathbb{Z})$. We recall that by Mahler's criterion (see [Ca2], Ch. V, and [Rag], Ch. X) a sequence $\{\Lambda_i\}$ of lattices in $\mathcal{M}$ has a convergent subsequence if and only if there does not exist any sequence $\{x_i\}$ in $\mathbb{R}^n - \{0\}$ such that $x_i \in \Lambda_i$ for all i and $x_i \to 0$ as $i \to \infty$. The realisation of $\mathcal{M}$ as $SL(n, \mathbb{R})/SL(n, \mathbb{Z})$ and Mahler's criterion play an important role in the applications to problems in Diophantine approximation, which will be discussed in § 9 (this Chapter).

3. Various subgroups of $SL(n, \mathbb{R})$ also provide interesting examples of connected Lie groups with lattices. For example, let N be the subgroup consisting of all upper triangular $n \times n$ matrices with 1's on the diagonal. Then $\Gamma = N \cap (SL(n, \mathbb{Z}))$ is a lattice in N. This provides a class of examples of lattices in nilpotent Lie groups; see § 2.2 for results on flows on homogeneous spaces of nilpotent Lie groups.

4. We shall also recall here some more examples of nilpotent Lie groups with lattices. Let V be a finite-dimensional real vector space. Let G denote the vector space $V \oplus \wedge^2 V$, where $\wedge^2 V$ is the second exterior power of V as a vector space. We define on G a product by setting $(v_1, w_1) \cdot (v_2, w_2) = (v_1 + v_2, w_1 + w_2 + \frac{1}{2} v_1 \wedge v_2)$ for all $v_1, v_2 \in V$ and $w_1, w_2 \in \wedge^2 V$. Then G is a nilpotent Lie group with respect to the product. It can be seen that if V is identified as $\mathbb{R}^n$ then $\Gamma = \{(p, \frac{1}{2}q) \mid p \in \mathbb{Z}^n, q \in \wedge^2 \mathbb{Z}^n\}$ is a lattice in G ($\wedge^2 \mathbb{Z}^n$ denotes the subgroup of $\wedge^2 V$ generated by exterior products of integral vectors). The group G constructed here is the free 2-step simply connected Lie group over V. Similarly one can construct free k-step nilpotent Lie groups and lattices in them (see [AS], for instance).

5. Another interesting class of lattices arises from geometry. Let S be a surface of constant negative curvature and finite Riemannian area. Then the unit tangent bundle of S can be realised as G/Γ, where $G = PSL(2, \mathbb{R}) = SL(2, \mathbb{R})/\{\pm I\}$ (I being the identity matrix) and Γ is a lattice in G. Furthermore the geodesic flow associated to S (defined on its unit tangent bundle) can be realised as the action on G/Γ of the image of the diagonal one-parameter subgroup $\{\mathrm{diag}\,(e^t, e^{-t})\}$ in G (see [GF], [Gh], [Man2]). The classical horocycle flow which plays an important role in the study of the geodesic flow, is given by the action of the one-parameter subgroup consisting of the image in G of strictly upper triangular matrices.

6. More generally if M is a Riemannian manifold of constant negative curvature and finite Riemannian volume, then the unit tangent bundle of M

can be realised as $K\backslash G/\Gamma$, where G is the Lie group of isometries of the universal covering space of M, K is a compact subgroup of G and Γ is a lattice in G, and there is a one-parameter subgroup $\{g_t\}$ of G such that each g_t commutes with all elements of K and the $\{g_t\}$-action on $K\backslash G/\Gamma$ (defined by $Kx\Gamma \mapsto Kg_tx\Gamma$ for all $x \in G$ and $t \in \mathbb{R}$) coincides with the geodesic flow. A similar realisation also holds for geodesic flows on the unit tangent bundles of the irreducible Riemannian locally symmetric spaces of noncompact type and rank 1 (see [Mau]). The study of dynamics of flows on G/Γ induced by the one-parameter subgroups of G has interesting applications to the geometry of M. We shall discuss some of these in the following sections at appropriate places.

7. Now let G be a real algebraic subgroup of $SL(n, \mathbb{R})$ defined over the rationals, namely a subgroup of the form $\{g = (g_{ij}) \in SL(n, \mathbb{R}) \mid P(g_{ij}) = 0\}$, where P is a polynomial in the n^2 variables corresponding to the matrix entries, with rational coefficients. For instance, the subgroup N as in Example 3 above is a real algebraic subgroup (we choose $P(g_{ij}) = \Sigma_{i>j} g_{ij}^2 + \Sigma_i (g_{ii} - 1)^2$). Such a subgroup may in general not be connected but necessarily has only finitely many connected components (see [Wh], for instance). Now suppose that the unipotent matrices contained in G together with a compact subgroup of G generate a subgroup of finite index in G (we recall here that a matrix is said to be unipotent if it has no eigenvalues, even complex, other than 1). A theorem of Borel and Harish-Chandra (cf. [BH], [B1]) then asserts that the subgroup $\Gamma = G \cap (SL(n, \mathbb{Z}))$ is a lattice in G.

Also, for any subgroup G' of finite index in G (e.g. G^0), $\Gamma' = \Gamma \cap G'$ is a lattice in G'; any lattice in G' commensurable with Γ' is called an arithmetic lattice associated with the given rational structure. We note also that if M is a compact normal subgroup of G' then $\Gamma' M/M$ is a lattice in G'/M. This yields a larger class of lattices in Lie groups. Lattices arising in this way, as also those commensurable with them, are called *arithmetic* lattices.

A theorem of Margulis asserts in particular that if G is a simple noncompact connected Lie group with trivial center and $\mathbb{R}$-rank (namely, maximum dimension of a subgroup whose adjoint action on the Lie algebra of G is diagonalisable over $\mathbb{R}$) at least 2 then any lattice in G is arithmetic (see [Marg4], [Zi3]).

Remark 1.1. We note that there also exist locally faithful homogeneous spaces G/C, with G a connected Lie group and C a nondiscrete closed subgroup. A class of examples of such homogeneous spaces can be constructed as follows: Let $G = K \cdot V$, semidirect product, where $V = \mathbb{R}^n$ and K is a compact group of (linear) automorphisms of V. Let L be a closed subgroup of K and M be a closed L-invariant subgroup of V such that V/M is compact and M^0 does not contain any nonzero K-invariant subspace. Then $C = LM \in \mathscr{F}_0(G)$ and it is nondiscrete whenever L or M is of positive dimension.

1.3. Ergodicity and its consequences. Let (X, μ) be a measure space and G be a group of measurable transformations of X such that μ is quasi-invariant under the action of each element of G; in such a case G is said to be a group of *nonsingular transformations* of (X, μ). A measurable subset E of X is said to be essentially invariant under the G-action if $\mu(E \Delta g^{-1}E) = 0$ for all $g \in G$. The G-action is said to be *ergodic* if for every essentially invariant measurable subset E of X, either $\mu(E) = 0$ or $\mu(X - E) = 0$. We shall now briefly recall certain consequences of ergodicity of an action on a topological space, which will be involved crucially in the sequel. Firstly, the following is straightforward to verify.

Proposition 1.2. *Let X be a second countable Hausdorff topological space and H be a group acting on X. Let μ be a measure on X which is quasi-invariant and ergodic under the H-action. Suppose also that $\mu(\Omega) > 0$ for every nonempty open subset Ω of X. Let S be a subsemigroup of H generating H. If either μ is a finite H-invariant measure or $S = H$ then there exists a Borel subset Y of X such that $\mu(X - Y) = 0$ and for every $y \in Y$ the orbit Sy is dense in X.*

In particular, for an action on a homogeneous space by a group of affine automorphisms, ergodicity implies that almost all the orbits are dense. However, as we shall see later, it is rare for *all* orbits to be dense.

A considerably stronger assertion than Proposition 1.2 arises, under similar conditions, from the Birkhoff-Khintchin ergodic theorem, when H is either a cyclic group or the real line (a similar result holds also for other groups for which an analogous ergodic theorem is available, but we shall not concern ourselves with it here).

A sequence $\{x_i\}$ in a topological space X is said to be *uniformly distributed* with respect to a probability measure μ on X if for every bounded continuous function f on $\operatorname{supp} \mu$ (the support of μ)

$$\frac{1}{k}\sigma_{i=0}^{k-1} f(x_i) \longrightarrow \int_X f \, d\mu,$$

as $k \to \infty$. Similarly a curve $\{x_t\}_{t \geq 0}$ in X is said to be uniformly distributed with respect to a probability measure μ on X if

$$\frac{1}{T}\int_0^T f(x_t)\, dt \longrightarrow \int_X f \, d\mu,$$

as $T \to \infty$, for all bounded continuous functions f on $\operatorname{supp} \mu$. The ergodic theorem readily implies the following; a similar assertion also holds for one-parameter flows.

Theorem 1.1 (cf. [DGS], § 5). *Let X be a locally compact second countable Hausdorff topological space and let τ be a homeomorphism of X. Let μ be an ergodic τ-invariant probability measure on X. Then there exists a Borel subset Y of X such that $\mu(X - Y) = 0$ and for every $y \in Y$, the sequence $\{\tau^i(y)\}$ is uniformly distributed with respect to μ.*

Let X be a compact second countable Hausdorff topological space and τ be a homeomorphism of X. We know that $\mathscr{P}(X)$ is a compact Hausdorff metrizable space (see § 1.1 for notation and definition of the topology). For $x \in X$ and $k \geq 1$ let σ_k be the probability measure evenly distributed on the orbit-segment $\{x, \tau(x), \tau^2(x), \ldots, \tau^{k-1}(x)\}$. Any accumulation point (limit point) of the sequence $\{\sigma_k\}$ in $\mathscr{P}(X)$ will be called a *limit distribution* of the orbit of x under τ. The orbit of a point $x \in X$ is uniformly distributed with respect to a probability measure μ if and only if $x \in \operatorname{supp} \mu$ and μ is the unique limit distribution of the orbit (see [DGS], § 5). A point $x \in X$ is said to be *quasi-regular* if the sequence $\{\tau^i(x)\}$ is uniformly distributed with respect to some probability measure μ on X. The set, say $Q(\tau)$, of quasi-regular points is a Borel set and $\mu(Q(\tau)) = 1$ for all τ-invariant probability measures on X (see [DGS], § 4). We note however that in general $Q(\tau)$ is a proper subset of X.

If X is a locally compact noncompact second countable Hausdorff space and τ is a homeomorphism of X then for $x \in X$ a probability measure μ on X will be called a limit distribution if it is a limit distribution on the one-point compactification of X (when the homeomorphism and the probability measure are extended to the compactification in the natural way).

1.4. Isomorphisms and factors of affine automorphisms. Let G_1 and G_2 be two Lie groups and C_1 and C_2 be closed subgroups of G_1 and G_2 respectively such that G_1/C_1 and G_2/C_2 have finite Haar measures, say m_1 and m_2 respectively. Let τ_1 and τ_2 be affine automorphisms of G_1/C_1 and G_2/C_2 respectively. We shall say that τ_1 and τ_2 are *topologically isomorphic* if there exists a homeomorphism $\theta : G_1/C_1 \to G_2/C_2$ such that $\theta(m_1) = m_2$ and $\theta \circ \tau_1 = \tau_2 \circ \theta$; a homeomorphism θ satisfying the conditions will be called a topological isomorphism of τ_1 and τ_2. If two affine automorphisms are topologically isomorphic then their ergodic as well as topological dynamical properties coincide.

A map $\psi : G_1/C_1 \to G_2/C_2$ is called an *affine map* if there exists a continuous homomorphism $\eta : G_1 \to G_2$, such that $\psi(gC_1) = \eta(g)\psi(C_1)$ for all $g \in G_1$, and ψ is called an *affine isomorphism* if, in addition, η is a topological isomorphism of the topological groups. Two affine automorphisms τ_1 and τ_2 as above are said to be *affinely isomorphic* if they are topologically isomorphic via an affine isomorphism of the homogeneous spaces. We note that topologically isomorphic affine automorphisms need not be affinely isomorphic in general; we shall see some natural examples of this in § 3 (see Remark 3.1).

Remark 1.2. Let $\tau = T_g \circ \overline{\alpha}$ be an affine automorphism of a homogeneous space G/C, in the notation as before. If τ has a fixed point, namely if there exists $x \in G/C$ such that $\tau(x) = x$, then τ is affinely isomorphic to the automorphism $\overline{\alpha}$ of G/C.

Let $\tau = T_g \circ \overline{\alpha}$ be an affine automorphism of a homogeneous space G/C, where G is a Lie group, C is a closed subgroup, $g \in G$ and α is an automorphism of G such that $\alpha(C) = C$. Let C' be an α-invariant (namely such that $\alpha(C') = C'$) closed subgroup of G containing C. Then we get an affine automorphism τ' of G/C' defined by $\tau'(gC') = \tau(g)C'$ for all $g \in G$. We call τ' the *factor* of τ on G/C'.

§ 2. Affine automorphisms of tori and nilmanifolds

In this section we discuss ergodic and dynamical properties of affine automorphisms of tori and nilmanifolds. In the description of ergodic properties the underlying measure will be implicitly assumed to be the normalised Haar measure.

2.1. Ergodic properties; the case of tori. Any automorphism of the n-dimensional torus $\mathbb{T}^n = \mathbb{R}^n/\mathbb{Z}^n$, $n \geq 1$, is given by an $\alpha \in GL(n, \mathbb{Z})$ (an integral $n \times n$ matrix with determinant ± 1), by $\overline{\alpha}(v + \mathbb{Z}^n) = \alpha(v) + \mathbb{Z}^n$, for all $v \in \mathbb{R}^n$, α being viewed as a linear transformation of $\mathbb{R}^n$, via the standard basis (see [Wal5]).

Theorem 2.1 (cf. [Br], [Wal5]). *An affine automorphism $\tau = T_a \circ \overline{\alpha}$ of $\mathbb{T}^n$, where $a \in \mathbb{T}^n$ and $\alpha \in GL(n, \mathbb{Z})$, is ergodic if and only if the following holds: no root of unity other than 1 is an eigenvalue of α and the subgroup generated by a and $\{x^{-1}\overline{\alpha}(x) \mid x \in \mathbb{T}^n\}$ is dense in $\mathbb{T}^n$. In particular, an automorphism $\overline{\alpha}$, $\alpha \in GL(n, \mathbb{Z})$, is ergodic if and only if no eigenvalue of α is a root of unity.*

We note here that $\mathbb{T}^n$, $n \geq 1$, are the only Lie groups admitting ergodic affine automorphisms, with respect to the Haar measure (which is always quasi-invariant; see § 1.1); see [D11] and [D12]; see also Remark 2.9. It may also be mentioned that more generally, a locally compact group admitting an affine automorphism which is ergodic with respect to the Haar measure, is necessarily compact (see Dateyama and Kasuga [DK]).

We next recall a criterion for ergodicity of translation flows in terms of 'rationality'. We shall say that a vector $v \in \mathbb{R}^n$ is in *general position* if it is not contained in any proper rational subspace of $\mathbb{R}^n$ or, equivalently, if the coordinates of v are linearly independent over the rationals.

Theorem 2.2. *Let $v \in \mathbb{R}^n$. Then the flow induced by $\{tv \mid t \in \mathbb{R}\}$ on $\mathbb{T}^n = \mathbb{R}^n/\mathbb{Z}^n$ is ergodic if and only if v is in general position.*

The reader is also referred to [AA], [CFS] and [Wal5] for other similar results on ergodicity of translation actions.

We note also the following:

Remark 2.1. If $\alpha \in GL(n, \mathbb{Z})$ is such that 1 is not an eigenvalue of α then for any $a \in \mathbb{T}^n$ the affine automorphism $\tau = T_a \circ \overline{\alpha}$ has a fixed point and hence, by Remark 1.2, it is affinely isomorphic to $\overline{\alpha}$.

Remark 2.2. For an affine automorphism $\tau = T_a \circ \overline{\alpha}$ of $\mathbb{T}^n$, where $a \in \mathbb{T}^n$ and α is a unipotent integral matrix, there is a unique minimal $\overline{\alpha}$-invariant proper toral subgroup A such that the factor of τ on $\mathbb{T}^n/A$ is a translation; furthermore, τ is ergodic if and only if the factor on $\mathbb{T}^n/A$ is ergodic.

We now describe the spectra of the unitary operators associated to the affine automorphisms as above. We realise $\mathbb{Z}^n$ as the dual group of $\mathbb{T}^n$ in the usual way and for an automorphism $\overline{\alpha}$ of $\mathbb{T}^n$ denote by $\widehat{\overline{\alpha}}$ the dual automorphism of $\mathbb{Z}^n$. For each $p \in \mathbb{Z}^n$ let χ_p denote the corresponding character, viewed as a function on $\mathbb{T}^n$.

Theorem 2.3 (cf. [Br], [Wal5]). *Let $\tau = T_a \circ \overline{\alpha}$ be an affine automorphism of $\mathbb{T}^n$, where $a \in \mathbb{T}^n$ and $\alpha \in GL(n, \mathbb{Z})$. Let $\mathcal{H}$ denote the Hilbert space $L^2(\mathbb{T}^n)$ and $U : \mathcal{H} \to \mathcal{H}$ be the unitary operator associated to τ. Let $\mathcal{E}$ be the smallest closed subspace of $\mathcal{H}$ containing the set*

$$S = \{\chi_p \in \mathcal{H} \mid p \in \mathbb{Z}^n \text{ such that } \widehat{\overline{\alpha}}^r(p) = p \text{ for some } r \geq 1\}.$$

Then the eigenfunctions of U are all contained in $\mathcal{E}$ and span a dense subspace of $\mathcal{E}$. Furthermore, U has Lebesgue spectrum of infinite multiplicity on the orthocomplement of $\mathcal{E}$ in $\mathcal{H}$, unless $\mathcal{E} = \mathcal{H}$.

Corollary 2.1. *Let $\tau = T_a \circ \overline{\alpha}$ be an affine automorphism of $\mathbb{T}^n$, where $a \in \mathbb{T}^n$ and $\alpha \in GL(n, \mathbb{Z})$. Then τ is weak mixing if and only if no root of unity is an eigenvalue of α. In particular any ergodic automorphism of $\mathbb{T}^n$ is weak mixing. When τ is weak mixing, it is also (strong) mixing and the associated unitary operator has Lebesgue spectrum of infinite multiplicity on the orthocomplement of constant functions.*

The following stronger ergodic property is known for weak mixing affine automorphisms.

Theorem 2.4 (Katznelson [Ka], see also Aoki [Ao]). *Any weak mixing affine automorphism of $\mathbb{T}^n$ is isomorphic to a Bernoulli shift.*

The entropy of an affine automorphism $\tau = T_a \circ \overline{\alpha}$ of $\mathbb{T}^n$, where $a \in \mathbb{T}^n$ and $\alpha \in GL(n, \mathbb{Z})$, is given by $\Sigma_{i=1}^k \log |\lambda_i|$, where $\lambda_1, \lambda_2, \ldots, \lambda_k$, $1 \leq k \leq n$, are all the (complex) eigenvalues of α with absolute value greater than 1 (see [Br], [Wal5]); as may be expected the value of the entropy does not depend on the translating element a as above. We note also that since the eigenvalues are algebraic numbers, only countably many numbers occur as entropies of affine automorphisms of tori.

Remark 2.3. Let $\tau = T_a \circ \overline{\alpha}$ be an affine automorphism of $\mathbb{T}^n$, where $a \in \mathbb{T}^n$ and $\alpha \in GL(n, \mathbb{Z})$. Let V (respectively W) be the largest α-invariant subspace of $\mathbb{R}^n$ such that no eigenvalue of the restrictions of α to V (resp. W) is a root of unity (resp. equals 1). Then V and W are rational subspaces (defined by rational equations) and hence $(V + \mathbb{Z}^n)/\mathbb{Z}^n$ and $(W + \mathbb{Z}^n)/\mathbb{Z}^n$ are toral (closed connected) subgroups and, furthermore, they are invariant under $\overline{\alpha}$. We shall call the factors of τ on $\mathbb{R}^n/(V + \mathbb{Z}^n)$ and $\mathbb{R}^n/(W + \mathbb{Z}^n)$ respectively the *maximal quasi-unipotent* and *maximal unipotent* factors. If τ is ergodic the two factors are the same.

We note the following result concerning quasi-discrete spectrum (see Ch. 2, for definition).

Theorem 2.5 (cf. [P1], [Wal1]). *Let $\tau = T_a \circ \overline{\alpha}$ be an ergodic affine automorphism of $\mathbb{T}^n$, where $a \in \mathbb{T}^n$ and $\alpha \in GL(n, \mathbb{Z})$. Then τ has quasi-discrete spectrum if and only if α is unipotent. Moreover, the maximal unipotent factor (see Remark 2.3) of τ is the largest measurable factor of τ with quasi-discrete spectrum.*

By a theorem of Hahn and Parry [HP] the first part of the theorem implies the following:

Corollary 2.2. *Let $\tau = T_a \circ \overline{\alpha}$ be an ergodic affine automorphism of $\mathbb{T}^n$, with $a \in \mathbb{T}^n$ and α a nontrivial unipotent integral matrix. Then τ is not embeddable in a measurable one-parameter flow; that is, there does not exist a measurable, measure-preserving flow $\{\phi_t\}$ on $\mathbb{T}^n$ such that $\phi_1 = \tau$.*

2.2. Ergodic properties on nilmanifolds. We next consider affine automorphisms of nilmanifolds, namely homogeneous spaces of nilpotent Lie groups. These include in particular the affine automorphisms of tori considered earlier. A simple non-toral example of a compact nilmanifold is given by H/Γ, where H is the 3-dimensional Heisenberg group, consisting of all 3×3 upper triangular unipotent matrices with real entries, and Γ is the subgroup consisting of all integral matrices in H (see also Examples 3 and 4 in § 1.2).

Let G be a connected nilpotent Lie group and let C be a closed subgroup such that G/C admits a finite Haar measure. Then G/C is compact (see [Rag], Ch. II). Moreover, the connected component C^0 of the identity in C is a normal subgroup of G (see [Rag], Corollary 2 of Theorem 2.3). Then G/C can be viewed as the quotient of the Lie group G/C^0 by the discrete subgroup C/C^0. In view of this, for our purposes it suffices to consider nilmanifolds G/C with C discrete, namely a (uniform) lattice. By passing to the universal covering group of G we may also assume in general that G is simply connected.

Now let G be a connected nilpotent Lie group and Γ be a lattice in G. Let $[G, G]$ be the commutator subgroup of G. Then $[G, G]$ is a closed normal subgroup of G and $[G, G] \cap \Gamma$ is a lattice in $[G, G]$; in fact even $[\Gamma, \Gamma]$ is a lattice in $[G, G]$ (cf. [Rag], Theorem 2.1). In particular it follows that $[G, G]\Gamma$ is a closed subgroup of G. The quotient space $G/[G, G]\Gamma$ is a torus, which we shall call the *maximal torus quotient* of G/Γ. Any affine automorphism τ of G/Γ factors to $G/[G, G]\Gamma$ (see § 1.4). Interestingly, many ergodic properties of affine automorphisms of G/Γ depend essentially only on this factor.

Theorem 2.6 (Parry [P2]). *Let G be a connected nilpotent Lie group and Γ be a lattice in G. Let τ be an affine automorphism of G/Γ and let $\overline{\tau}$ be the factor of τ on the maximal torus quotient $G/[G, G]\Gamma$. Then*

i) τ is ergodic if and only if $\overline{\tau}$ is ergodic;

ii) if $\overline{\tau}$ is weak mixing then τ is a K-automorphism (see Ch. 3, for definition).

A simple proof of (i) for the case of automorphisms may be found in Parry [P3]. For assertion (ii) alternative proofs may be found in [Th] and [D1].

Example 2.1. Let $G = V \oplus \wedge^2 V$, where $V = \mathbb{R}^n$ and $\Gamma = \{(p, \frac{1}{2}q) \mid p \in \mathbb{Z}^n,\ q \in \wedge^2\mathbb{Z}^n\}$ be as in Example 4 in § 1.2. Let $A \in GL(n, \mathbb{Z})$ and let $\alpha : G \to G$ be defined by $\alpha(v, w) = (Av, (\wedge^2 A)w)$ for all $v \in V$ and $w \in \wedge^2 V$, where $\wedge^2 A$ is the second exterior power transformation of $\wedge^2 V$ corresponding to A. Then α is an automorphism of G such that $\alpha(\Gamma) = \Gamma$. By Theorems 2.6 and 2.1 the automorphism $\overline{\alpha}$ induced by α is ergodic if and only if A has no eigenvalue which is a root of unity, and whenever $\overline{\alpha}$ is ergodic it is a K-automorphism.

In analogy with assertion (i) of Theorem 2.6, a one-parameter flow of translations of a nilmanifold is ergodic if and only if its factor on $G/[G, G]\Gamma$ is ergodic; this is known as Green's criterion (cf. [AGH], Ch. V).

As in the case of tori, ergodicity of one-parameter translation flows on nilmanifolds can also be characterised in terms of 'rationality' considerations. Let G be a connected nilpotent Lie group, $\mathcal{G}$ be the Lie algebra of G and let $\exp : \mathcal{G} \to G$ be the exponential map. Let Γ be a lattice in G. Then in $\mathcal{G}$, viewed as a vector space, the $\mathbb{Z}$-span of $\exp^{-1}(\Gamma)$ (namely the subgroup generated by $\exp^{-1}(\Gamma)$) is a lattice, and hence it determines a rational structure on $\mathcal{G}$ (see [Rag], Theorem 2.12 and the remark following its statement); see also [AGH], Ch. IV and [Mo1], Theorem 2, for other results on the structure of a lattice in a nilpotent Lie group.

Theorem 2.7. *Let G be a connected nilpotent Lie group and Γ be a lattice in G. Let $\mathcal{G}$ be the Lie algebra of G. Then for $\xi \in \mathcal{G}$ the flow on G/Γ induced by the one-parameter subgroup $\{\exp t\xi\}$ is ergodic if and only if ξ is in general position in $\mathcal{G}$, with respect to the rational structure determined by the $\mathbb{Z}$-span of* $\exp^{-1}(\Gamma)$.

Example 2.2. Let N and Γ be as in Example 3 in § 1.2. The commutator subgroup of N is given by $\{g = (g_{ij}) \in N \mid g_{i(i+1)} = 0$ for all $i = 1, \ldots, n-1\}$. Any one-parameter subgroup of N is of the form $\{\exp t\xi\}$ where ξ is a $n \times n$ matrix (x_{ij}) with $x_{ij} = 0$ whenever $j \leq i$, and the flow induced by it is ergodic if and only if $x_{12}, x_{23}, \ldots, x_{(n-1)n}$ are linearly independent over the rationals.

The spectrum of an ergodic affine automorphism of a nilmanifold is described by the following.

Theorem 2.8 (Parry [P4]). *Let G be a connected nilpotent Lie group, Γ be a lattice in G and τ be an ergodic affine automorphism of G/Γ. Let U be the unitary operator on $L^2(G/\Gamma)$ associated to τ and let $\mathcal{D}$ be the closed subspace of $L^2(G/\Gamma)$ spanned by the eigenfunctions of U. Then U has Lebesgue spectrum of infinite multiplicity on the orthocomplement of $\mathcal{D}$ in $L^2(G/\Gamma)$, unless $\mathcal{D} = L^2(G/\Gamma)$.*

Let G and Γ be as above and let $\tau = T_g \circ \overline{\alpha}$ be any affine automorphism of G/Γ, where $g \in G$ and α is an automorphism of G such that $\alpha(\Gamma) = \Gamma$. In analogy with the case of affine automorphisms of tori, the entropy of τ with respect to the normalised Haar measure, is given by $h(\tau) = \Sigma_{\lambda \in \Lambda} \log|\lambda|$, where Λ is the set of all eigenvalues λ of $d\alpha$ such that $|\lambda| > 1$, $d\alpha$ being the derivative of α on the Lie algebra of G (see Parry [P2]). The topological entropy of τ is also given by $h(\tau)$ as above (see Bowen [Bo2]) and moreover, when τ is ergodic, the normalised Haar measure is the only τ-invariant measure with entropy $h(\tau)$ (see Walters [Wal4]); in common parlance, the normalised Haar measure is the unique probability measure of maximal entropy.

It follows from a theorem of D. Lind [Li1] that if G, Γ and $\tau = T_g \circ \overline{\alpha}$ are as above and no eigenvalue of $d\alpha$ is a root of unity then τ is isomorphic to a Bernoulli shift (see also Theorem 4.16, infra, for another sufficient condition for Bernoullicity). However, an affine automorphism can be a K-automorphism without the above condition on the eigenvalues being satisfied; e.g. this happens for $\overline{\alpha}$ as in Example 2.1 if $A \in GL(n, \mathbb{Z})$ is such that for some $\lambda \neq 0$ both λ and λ^{-1} are eigenvalues of A and there are no eigenvalues which are roots of unity. It does not seem to be known whether an affine automorphism of G/Γ is isomorphic to a Bernoulli shift whenever it is a K-automorphism.

Remark 2.4. Let G be a simply connected nonabelian nilpotent Lie group and Γ be a lattice in G. Let $g \in G$ be such that the translation T_g of G/Γ is ergodic. Using embeddability of T_g in a one-parameter flow and Corollary 2.2 it can be seen that T_g is not metrically isomorphic to an affine automorphism of a torus (see also Theorem 10.1 in § 10). The theorem of Hahn and Parry involved in Corollary 2.2 also implies that T_g does not have quasi-discrete spectrum. These translations however have generalised discrete spectrum in the sense of Zimmer [Zi1].

2.3. Unipotent affine automorphisms. For a translation of $\mathbb{T}^n$ the closure of any orbit is a coset of a compact subgroup of $\mathbb{T}^n$ and the orbit is uniformly distributed in the coset (with respect to the measure which is a translate of the normalised Haar measure of the subgroup). When $\mathbb{T}^n$ is viewed as $\mathbb{R}^n/\mathbb{Z}^n$, for the translation by $(a_1, \ldots, a_n)$, where $a_1, \ldots, a_n \in \mathbb{R}$, all orbits are dense and uniformly distributed in the torus (with respect to the Haar measure) whenever the translation is ergodic, namely when $a_1, \ldots, a_n$ and 1 are linearly independent over the rationals (see [Br], [CFS], [Wal5]). In this subsection we shall see similar properties being true for the larger class of 'unipotent' affine automorphisms of nilmanifolds (see below). Before passing over to the general case let us note a result on the uniformity of the uniform distribution, for certain translations of tori.

Theorem 2.9 (Miles and Thomas [MiT]). *Let $a = (a_1, \ldots, a_n) \in \mathbb{R}^n$, be such that $a_1, \ldots, a_n$ are real algebraic numbers and together with 1 form a linearly independent set over the field of rational numbers. Let m denote the normalised Haar measure on $\mathbb{T}^n$. Then there exists a constant d such that the following holds: for any subset Ω of $\mathbb{T}^n$ with positive measure and rectifiable boundary there exists a constant c such that for any $\epsilon \in (0,1)$, $x \in \mathbb{R}^n$ and $k > c/\epsilon^d$ we have*

$$\left| \frac{\#\{0 \le j \le k-1 \mid (ja + x) + \mathbb{Z}^n \in \Omega\}}{k} - m(\Omega) \right| < \epsilon,$$

where, for any set E, $\#E$ stands the cardinality of E.

A similar assertion also holds for almost all $a \in \mathbb{R}^n$ (cf. [MiT]). There are also results on 'how rapidly' the orbits of translations fill the torus, depending on the translating element. We shall not go into the details here, as the conditions involved are rather technical, but refer the reader to Meyer [Me].

It was shown by Furstenberg that for an affine automorphism $T_a \circ \overline{\alpha}$ of $\mathbb{T}^n$, where $a \in \mathbb{T}^n$ and α is a unipotent integral matrix, ergodicity implies unique ergodicity (see below). More generally we have the following theorem, due to W. Parry [P2], for all nilmanifolds.

Let G be a connected nilpotent Lie group and $\mathscr{G}$ be the Lie algebra of G. We call an automorphism α of G *unipotent* if its differential $d\alpha$ is a unipotent linear transformation of $\mathscr{G}$; when G is simply connected the condition is also equivalent to the factor of α on $G/[G, G]$ (which is a vector space) being a unipotent linear transformation. For a lattice Γ in G an affine automorphism $\tau = T_g \circ \overline{\alpha}$ of G/Γ, where $g \in G$ and α is an automorphism of G such that $\alpha(\Gamma) = \Gamma$, is called a *unipotent affine automorphism* if α is unipotent.

Theorem 2.10 (Parry [P2]). *Let G be a connected nilpotent Lie group and Γ be a lattice in G. Let m be the G-invariant probability measure on G/Γ. Let τ be a unipotent affine automorphism of G/Γ. Then the following conditions are equivalent:*

i) τ is ergodic;

ii) τ has a dense orbit;

iii) τ is minimal (equivalently, all orbits of τ are dense);

iv) τ is uniquely ergodic (that is, m is the only τ-invariant probability measure on G/Γ);

v) all orbits of τ are uniformly distributed with respect to m.

Remark 2.5. Any of the properties as in the Theorem 2.10 hold if and only if the corresponding properties holds for the factor on the maximal torus quotient of τ, namely the factor on $G/[G, G]\Gamma$. In turn they hold if and only if the translation factor as in Remark 2.2 has a dense orbit.

Remark 2.6. Let τ be an affine automorphism of $\mathbb{T}^n$. If it is minimal (or satisfies any of the conditions as in Theorem 2.10) then it has quasi-discrete spectrum as a homeomorphism; see Hoare and Parry [HoP1] for details. The analogous statement is however not true for nilmanifolds (cf. [HP]).

Any unipotent affine automorphism τ of a nilmanifold G/Γ can be readily seen to be distal (that is, for $x, y \in G/\Gamma$ there exist a $z \in G/\Gamma$ and a sequence $\{i_k\}$ of integers such that $\tau^{i_k}(x) \to z$ and $\tau^{i_k}(y) \to z$ only if $x = y$). In the case of tori the following result of Parry shows that the maximal unipotent factor of an ergodic affine automorphism is also its maximal distal factor. The analogous statement may also be expected to be true for any compact nilmanifold, but it is not found in literature.

Theorem 2.11 (Parry [P1]). *Let τ be an ergodic affine automorphism of $\mathbb{T}^n$. Let W be the closed subgroup of $\mathbb{T}^n$ invariant under the automorphism part of τ, such that the factor of τ of $\mathbb{T}^n/W$ is the maximal unipotent factor (see Remark 2.3). If $\psi : \mathbb{T}^n \to X$ is a continuous map onto a compact space X, such that the partition of $\mathbb{T}^n$ into fibers of ψ is invariant under τ and the factor of τ on X is distal, then the map ψ is constant on the cosets of W.*

Remark 2.7. The condition in Theorem 2.10 that τ be unipotent is in fact necessary; that is, for G and Γ as in the theorem, if an affine automorphism τ of G/Γ is minimal then it is unipotent.

In particular, if an affine automorphism is not unipotent then not all of its orbits can be uniformly distributed with respect to the Haar measure on G/Γ. We note however that each orbit can be uniformly distributed with respect to some probability measure (namely, all points can be quasi-regular) under the weaker condition of all the eigenvalues being of absolute value 1 (see Theorem 2.12, infra).

2.4. Quasi-unipotent affine automorphisms. Let G be a connected nilpotent Lie group, Γ be a lattice in G and $\tau = T_g \circ \overline{\alpha}$ be an affine automorphism of G/Γ, where $g \in G$ and α is an automorphism of G such that $\alpha(\Gamma) = \Gamma$. If all the eigenvalues of $d\alpha$ are of absolute value 1 then we say that α and τ are *quasi-unipotent*; this holds if and only if all the eigenvalues are roots of unity; further, when G is simply connected, the factor of α on $G/[G, G]$ is a linear automorphism and the preceding condition holds if and only if all the eigenvalues of the factor are roots of unity.

Theorem 2.12. *Let G be a connected nilpotent Lie group and Γ be a lattice in G. Let τ be an affine automorphism of G/Γ. Then the following conditions are equivalent:*

i) τ is quasi-unipotent;

ii) τ has zero entropy;

iii) τ is distal;

iv) all points are quasi-regular for τ.

The equivalence of (i) and (ii) follows from the entropy formula noted earlier and that of (i) and (iii) is straightforward to verify. See [DMu] and [L2] for the equivalence of (i) and (iv).

The equivalence of (i) and (iv) as in Theorem 2.12 implies in particular that if $\tau_1, \ldots, \tau_k$ are quasi-unipotent affine automorphisms of a nilmanifold G/Γ as above, and $f_1, \ldots, f_k$ are continuous functions on G/Γ then the sequence

$$\frac{1}{j}\Sigma_{i=0}^{j-1} f_1(\tau_1^i x) f_2(\tau_2^i x) \ldots f_k(\tau_k^i x)$$

converges as $j \to \infty$, for all $x \in G/\Gamma$. The reader is referred to Lesigne [L1], for some results about such averages, in certain special cases, for bounded measurable functions in the place of the continuous functions $f_1, \ldots, f_k$ as above.

2.5. Closed invariant sets of automorphisms. Let G be a connected nilpotent Lie group and Γ be a lattice in G. Let $\tau = T_g \circ \overline{\alpha}$ be an affine automorphism of G, where $g \in G$ and α is an automorphism of G such that $\alpha(\Gamma) = \Gamma$. Just as in the case of tori (see Remark 2.1) it can be deduced from Remark 1.2 that if 1 is not an eigenvalue of $d\alpha$ (the differential of α on the Lie algebra of G) then τ is affinely isomorphic to $\overline{\alpha}$. Through the rest of this section, except for the last subsection, we shall restrict to automorphisms; the results can be applied to more general affine automorphisms via the preceding observation. We recall in this subsection some natural closed invariant subsets of automorphisms, arising algebraically. More results about invariant subsets are included in the following subsections.

To begin with it may be noted that any automorphism of $\mathbb{T}^n$ has a fixed point, namely the identity of $\mathbb{T}^n$, and a dense set of periodic orbits, namely orbits of points with rational coordinates (or equivalently elements of finite order in $\mathbb{T}^n$). This shows in particular that not all orbits are dense, even for ergodic automorphisms. Also, when $\alpha \in GL(n, \mathbb{Z})$ is not irreducible over the field of rational numbers, there exist proper nontrivial toral subgroups invariant under the automorphism $\overline{\alpha}$ of $\mathbb{T}^n$ induced by α.

If G is a connected nilpotent Lie group, Γ is a lattice in G and α is an automorphism of G such that $\alpha(\Gamma) = \Gamma$ then for the automorphism $\overline{\alpha}$ of G/Γ, $[G, G]\Gamma/\Gamma$ is a closed $\overline{\alpha}$-invariant subset. It can also be seen that there is a dense set of points of G/Γ (namely points of the form $(\exp \xi)\Gamma$, where ξ is a rational point for the $\mathbb{Q}$-structure corresponding to Γ - see § 2.2) which are periodic for all automorphisms of G/Γ.

2.6. Dynamics of hyperbolic automorphisms. Let G be a connected nilpotent Lie group and Γ be a lattice in G. Let $\theta = \overline{\alpha}$ be an automorphism of G/Γ, induced by an automorphism α of G such that $\alpha(\Gamma) = \Gamma$. We call θ a *hyperbolic automorphism* if no eigenvalue of the derivative $d\alpha$ of α (viewed as a linear transformation of the Lie algebra of G) is of absolute value 1; in the case of tori the condition is the same as α having no eigenvalue of absolute value 1. A hyperbolic automorphism is an Anosov diffeomorphism (see Ch. 8); in view of this we shall call such an automorphism an *Anosov automorphism.* While tori of all dimensions admit Anosov automorphisms the same does not hold for a general compact nilmanifold (see § 5.1 for a discussion on this aspect).

A rich theory covering various aspects of dynamical behaviour is available for Anosov diffeomorphisms and more generally for what are called axiom A diffeomorphisms (cf. Ch. 8, [KH], [Ma3], [Bo6] and [Bo8]). We shall not go into the general theory here but content ourselves by describing some results arising from the study, concerning mainly invariant sets and measures. We mention in particular that there is a considerable body of work on the growth of the number of periodic orbits of hyperbolic automorphisms, the associated zeta functions and their significance to the dynamics of the systems, that we will not go into; the reader is referred to Parry and Pollicott [PP] for a survey of the topic (see also Mañé [Ma3] and the recent paper of Degli Eposti and Isola [DI]).

Anosov automorphisms admit Markov partitions (see [DGS]). By a result of Keane and Smorodinsky (see [KeS]) this implies that $\overline{\alpha}$ is finitarily Bernoulli. Another interesting consequence is that the automorphisms have exponential rate of mixing; that is, the correlation functions $\int f(\tau^i x)\varphi(x)\mathrm{d}m(x)$ converge to zero exponentially fast, for Hölder continuous functions f and φ with $\int f\mathrm{d}m(x) = 0$, where m is the Haar measure (see [Bo6], p. 38).

Representation of the dynamical systems in terms of symbolic systems also yields interesting information about orbit closures and closed invariant subsets. The following is one of the results arrived at from such a representation.

Theorem 2.13 (Bowen [Bo1]). *Any minimal (nonempty) closed invariant subset of a hyperbolic automorphism of a compact nilmanifold is* 0*-dimensional (in particular it is totally disconnected).*

It can also be seen from the construction in [Bo1] that there exist minimal closed invariant sets other than periodic orbits; furthermore, the union of such minimal subsets is dense in the manifold.

There also exist, in general, compact invariant subsets of positive dimension other than those arising algebraically, as noted in § 2.5. There has been considerable work on this aspect, inspired by a question of S. Smale as to whether a hyperbolic toral automorphism admits one-dimensional compact invariant sets. A related aspect is to understand closures of orbits of paths, in the place of points. In the 2-dimensional case it was shown by Smale that any proper nonempty compact invariant set of a hyperbolic automorphism is 0-dimensional and in particular the orbit of any nonconstant path is dense. For a hyperbolic toral automorphism of $\mathbb{T}^n$, $n \geq 3$, there are no $(n-1)$-dimensional compact invariant sets (see Hirsch [Hi]) but there exist compact invariant sets of all dimensions between 1 and $n-2$ (see Przytycki [Pr]). For an Anosov automorphism on a compact m-dimensional nilmanifold, for any $k \leq (m-2)$ there exist closed invariant subsets of some dimension d such that $k \leq d \leq \min\{k+s-1, k+u-1, (k+m-1)/2\}$, where s and u denote the dimensions of the stable and unstable submanifolds for the diffeomorphisms (equivalently, the numbers of eigenvalues of the derivative, counting multiplicities, of absolute value less than 1 and greater than 1, respectively) see [Pr]; in a statement added in proof, in [Pr], it is asserted that the author was able to prove subsequently that, as in the case of the toral automorphisms, Anosov automorphisms of nilmanifolds also admit closed invariant sets of all dimensions up to $m - 2$, m being the dimension of the manifold.

Remark 2.8. The preceding discussion indicates that for an Anosov automorphism there must exist a 'large' set of points whose orbits are not dense. In fact, by a result of Urbanski [Ur2] the set of such points has Hausdorff dimension (see Ch. 8, § 6 for definition) equal to the dimension of the manifold (see Theorem 2.23 below for a similar result).

There are however not many smooth invariant sets. In fact we have the following:

Theorem 2.14 (Zeghib [Ze1]). *Let G be a connected nilpotent Lie group and Γ be a lattice in G. Let θ be a hyperbolic automorphism of G/Γ. Let M be a connected immersed C^1 submanifold with finite Riemannian volume, such that $\theta(M) = M$. Then the closure $\overline{M}$ of M in G/Γ is an orbit Hx, where H is a closed subgroup of G and $x \in G/\Gamma$. Furthermore, M is open in $\overline{M}$.*

Zeghib [Ze1] also describes the invariant sets without the connectedness assumption. Further, a similar assertion is made under the weaker assumption of M being rectifiable, in the place of an immersed smooth manifold. We shall however not go into the details of these here.

We note that if H as in the conclusion of Theorem 2.14 is a proper subgroup (namely if M is not dense) then the projection of $\overline{M}$ on the maximal torus quotient is a coset of a proper toral subgroup (cf. Theorem 2.7). It follows that there exist only countably many maximal proper θ-invariant smooth subsets, for θ as above, in contrast to the situation with regard to arbitrary closed invariant sets as in Remark 2.8.

2.7. More on invariant sets of hyperbolic toral automorphisms. We already noted in § 2.6 various results on invariant sets of hyperbolic automorphisms, applicable in particular to hyperbolic toral automorphisms. Here we describe some further results for the case of $\mathbb{T}^n$, especially $n \geq 3$.

We shall say that an automorphism θ of $\mathbb{T}^n$, $n \geq 2$, is *strongly irreducible* if for any $k \geq 1$, $\mathbb{T}^n$ is the only θ^k-invariant subgroup of positive dimension. For $\theta = \overline{\alpha}$, where $\alpha \in GL(n, \mathbb{Z})$, this holds if and only if α^k is irreducible over $\mathbb{Q}$ for all $k \geq 1$.

We first consider the orbit of a path, namely the smallest invariant set containing the given path, under a hyperbolic toral automorphism. It turns out that regularity assumptions on the path play a role in the study of orbit closures, as seen from the following results.

Theorem 2.15 (Irwin [Ir]). *Let $\theta = \overline{\alpha}$ be a hyperbolic automorphism of $\mathbb{T}^n$, $n \geq 3$, where $\alpha \in GL(n, \mathbb{Z})$. Let $\lambda_1, \cdots, \lambda_n$ be the (possibly complex) eigenvalues of α and let δ_0 be the maximum of the ratios of the form* $\log|\lambda_i|/\log|\lambda_j|$, *where $i \neq j$ and $|\lambda_i| \leq |\lambda_j|$. Then for $0 < \delta < 1$ there exists a δ-Hölder path with 1-dimensional orbit-closure under θ, if and only if $\delta \leq \delta_0$.*

For rectifiable paths we have the following.

Theorem 2.16 (Mañé [Ma2], see also Zeghib [Ze1]). *Let $\theta = \overline{\alpha}$ be a hyperbolic automorphism of $\mathbb{T}^n$, $n \geq 2$, where $\alpha \in GL(n, \mathbb{Z})$. Let $\gamma : [0, 1] \to \mathbb{T}^n$ be a nonconstant rectifiable path. Then the closure of the θ-orbit of $\gamma([0, 1])$ contains a coset of a toral subgroup of positive dimension which is invariant under θ^k for some $k \geq 1$. In particular if θ is strongly irreducible then the θ-orbit of $\gamma([0, 1])$ is dense in $\mathbb{T}^n$ and, furthermore, there exists a $t \in (0, 1)$ such that the θ-orbit of $\gamma(t)$ is dense in $\mathbb{T}^n$.*

We also recall the following result, which applies in particular to orbits of continuous paths.

Theorem 2.17 (Hancock [Ha]). *Let $\theta = \overline{\alpha}$ be a hyperbolic automorphism of $\mathbb{T}^n$, where $\alpha \in GL(n, \mathbb{Z})$. Let p be the number of eigenvalues of α with absolute value greater than 1 and let $1 \leq k < \min\{p, n-p\}$. Let $\mathcal{M}$ be the set of maps $\sigma : [0, 1]^k \to \mathbb{T}^n$ such that $\sigma([0, 1]^k)$ is contained in a proper closed θ-invariant subset of $\mathbb{T}^n$. Then $\mathcal{M}$ is a dense subset of the space of all maps of $[0, 1]^k$ into $\mathbb{T}^n$, in the topology of uniform convergence.*

Urbanski [Ur1] studied orbits of continua under hyperbolic automorphisms of $\mathbb{T}^n$. His results, which are rather too technical to recall in full generality, show in particular the following.

Theorem 2.18 (Urbanski, [Ur1]). *Let $\theta = \overline{\alpha}$ be a hyperbolic automorphism of $\mathbb{T}^n$, $n \geq 3$, where $\alpha \in GL(n, \mathbb{Z})$, and let δ_0 be the number as in Theorem 2.15. Let K be a continuum of capacity less than $2 - \delta_0$. Then the closure of the orbit $\cup_{-\infty}^{\infty} \theta^i(K)$ of K in $\mathbb{T}^n$ contains a coset of a closed subgroup of dimension at least 2, invariant under θ^k for some $k \geq 1$. On the other hand, for any $\epsilon > 0$ there exist continua of capacity less than $2 - \delta_0 + \epsilon$, contained in one-dimensional compact θ-invariant subsets.*

Given a homeomorphism θ of a compact space X, a compact θ-invariant subset C of X is said to be *isolated* if it has a neighbourhood Ω such that $\cap_{-\infty}^{\infty} \theta^i(\Omega) = C$. It is known that if θ is a hyperbolic automorphism of $\mathbb{T}^n$, $n \geq 3$, then every isolated compact connected θ-invariant set is a coset of a compact subgroup of $\mathbb{T}^n$ (cf. [Ma1]). This in particular yields that every compact θ-invariant C^1 submanifold of $\mathbb{T}^n$ is a finite union of cosets of toral subgroups (see the proof of Theorem A in [Ma1]; cf. also Theorem 2.14).

It is known that a topological surface in $\mathbb{T}^n$ which is invariant under a hyperbolic toral automorphism cannot be homeomorphic to a sphere or a projective plane or a Klein bottle (see [Hi]); validity of the analogous statement is not known however for surfaces of higher genus. A. Fathi [Fa] has constructed a class of examples of compact subsets of $\mathbb{T}^n$, $n \geq 3$, invariant under a hyperbolic toral automorphism of $\mathbb{T}^n$, which are images of surfaces of higher genus under certain finite-to-one maps.

2.8. Distribution of orbits of hyperbolic automorphisms. We next discuss the distribution of orbits of hyperbolic automorphisms of nilmanifolds as in § 2.6. Just as there are points whose orbit closures have a complicated structure, there are also points whose orbits are distributed in the manifold in a complicated way, quite at variance with the behaviour of the typical orbits, which are uniformly distributed (see Theorem 1.1). We note the following result, which is a consequence of the fact that hyperbolic automorphisms of nilmanifolds have the specification property (cf. [DGS], §§ 21 and 23); see § 1.3 for the terminology involved.

Theorem 2.19. *Let M be a compact nilmanifold and let θ be a hyperbolic automorphism of M. Then the following conditions hold:*

i) given any nonempty compact connected subset $\mathscr{C}$ of $\mathscr{P}(M)$ (see § 1.1 for notation and the topology) consisting of θ-invariant measures, there exists $x \in M$ such that $\mathscr{C}$ is precisely the set of limit distributions of the orbit of x under θ; moreover, the set of points x for which this holds is dense in M.

ii) there is a residual (dense G_δ) set of points $x \in M$ for which every probability measure is a limit distribution of the orbit of x.

iii) the set of quasi-regular points of θ is a set of first category.

For an automorphism θ of $\mathbb{T}^n$ let $\mathscr{G}(\theta)$ denote the set of points $x \in \mathbb{T}^n$ such that the orbit of x under θ is uniformly distributed with respect to the normalised Haar measure on $\mathbb{T}^n$; we shall also refer to these points as *generic* points. Recall also that $Q(\theta)$ denotes the set of quasi-regular points of θ (see § 1.3). We note the following.

Theorem 2.20 (cf. [Si]). *Let θ and ψ be two hyperbolic automorphisms of $\mathbb{T}^2$. Then the following statements hold:*

i) $\mathscr{G}(\theta) = \mathscr{G}(\psi)$ if and only if θ and ψ are rationally dependent (that is, there exist integers k and l such that $\theta^k = \psi^l$).

ii) $Q(\theta) = Q(\psi)$ if and only if $\theta = \psi$.

For any $\mathbb{T}^n$, $n \geq 2$, the sets of generic points coincide for any two ergodic automorphisms which are rationally dependent (see [Ci] for a general result in this respect). It is conjectured in [Sch3] that the converse is also true in all dimensions; it is proved there that the corresponding statement holds for endomorphisms for which the eigenvalues are all of absolute value greater than 1 (so they are not automorphisms).

2.9. Dynamics of ergodic toral automorphisms. Consider an ergodic automorphism θ of $\mathbb{T}^n$. As recalled earlier, each point with rational coordinates (when $\mathbb{T}^n$ is viewed as $\mathbb{R}^n/\mathbb{Z}^n$) is a periodic point of θ. Each periodic orbit carries a unique θ-invariant probability measure; we call an invariant probability measure on $\mathbb{T}^n$ which is supported on a periodic orbit a *periodic orbit measure.* Also, for any positive integer k let $\mu_k(\theta)$ be the probability measure equally distributed on all points of the set $\{x \in \mathbb{T}^n \mid \theta^k(x) = x\}$; the latter is a closed subgroup for any k and, furthermore, it is finite in view of Theorem 2.1 and the ergodicity of θ. The following results throw light on the distribution of the periodic points in $\mathbb{T}^n$.

Theorem 2.21 (Lind [Li2], Marcus [Marc2], Waddington [Wa]). *Let θ be an ergodic automorphism of $\mathbb{T}^n$. Then*

i) the periodic orbit measures of θ are dense in the space of all θ-invariant measures on $\mathbb{T}^n$;

ii) as $k \to \infty$ the measures $\mu_k(\theta)$ converge to the normalised Haar measure on $\mathbb{T}^n$; furthermore, for any nontrivial character χ on $\mathbb{T}^n$, $\int \chi \, \mathrm{d}\mu_k = 0$ for all large k.

The reader is also referred to Veech [V3] for a result similar to assertion (i) for invariant pseudomeasures. We note that the second assertion means that points with (nonreduced) period p become uniformly distributed in $\mathbb{T}^n$ as p tends to infinity. Waddington [Wa] also gives an asymptotic formula for the number of prime orbits of period at most p, as a function of p, analogous to the prime number theorem.

It may also be worthwhile to note the following result related to the convergence of $\{\mu_k(\theta)\}$ as in Theorem 2.21. A measurable subset E of $\mathbb{T}^n$ is called a *Lipschitz subset* if there exists a constant $c > 0$ such that $m(\{x \in \mathbb{T}^n \mid d(x, E) < \epsilon\}) \le c\epsilon$, for all $\epsilon > 0$, where d denotes the usual distance on $\mathbb{T}^n$ and m the Haar measure. Observe that any Lipschitz set has zero Haar measure.

Theorem 2.22 (Lind [Li2]). *Let θ be an ergodic automorphism of $\mathbb{T}^n$. Then for any Lipschitz subset E of $\mathbb{T}^n$, $\mu_k(\theta)(E) \to 0$, exponentially fast.*

We next discuss results on orbit closures of ergodic toral automorphisms. Even without hyperbolicity there exist strange kinds of invariant sets and orbit closures. The following result shows that there is a large subset of $\mathbb{T}^n$ of points whose orbits are not dense under any semisimple automorphism of $\mathbb{T}^n$; we recall that an automorphism $\overline{\alpha}$ of $\mathbb{T}^n$, corresponding to $\alpha \in GL(n, \mathbb{Z})$, is said to be *semisimple* if α is diagonalisable over the field of complex numbers.

Theorem 2.23 (cf. [D17]). *Let E be the subset of $\mathbb{T}^n$ consisting of all $x \in \mathbb{T}^n$ such that for any semisimple automorphism θ of $\mathbb{T}^n$ the closure of the orbit $\{\theta^i(x) \mid i \in \mathbb{Z}\}$ does not contain any point of $\mathbb{T}^n$ with rational coordinates (or equivalently a periodic point). Then the following conditions are satisfied:*

1. *for any nonempty open subset Ω of $\mathbb{T}^n$, $E \cap \Omega$ is of Hausdorff dimension n;*
2. *if $\{f_i\}$ is a sequence of bi-Lipschitz maps of $\mathbb{T}^n$ onto itself with common Lipschitz constants then $\bigcap_i f_i^{-1}(E)$ is of Hausdorff dimension n.*

The set E as above is an α-winning set of the (α, β)-game introduced by W.M. Schmidt [Schl], for all $\alpha \in (0, \frac{1}{2})$. Assertions (1) and (2) as in the conclusion of the theorem are consequences of this fact (see [D17] and [D19] for details; see also § 7.7 for some properties of α-winning sets and other results involving the game).

We note also the following result of D. Lind [Li3] on the return times of orbits of ergodic automorphisms; it is not known whether every ergodic automorphism of $\mathbb{T}^n$ is finitarily Bernoulli, and the property as in the following theorem is noted in Lind [Li3] to be a necessary condition for that to hold.

Theorem 2.24 (cf. Lind [Li3]). *Let θ be an ergodic automorphism of $\mathbb{T}^n$. Let Ω be a nonempty open subset of $\mathbb{T}^n$ and for $x \in \Omega$ let $r(x)$ be the first return time to Ω (namely the smallest positive integer k such that $\theta^k(x) \in \Omega$). Then $m(\{x \in \Omega \mid r(x) = k\})$ tends to zero exponentially fast, as $k \to \infty$.*

2.10. Actions of groups of affine automorphisms. In the literature on ergodic theory and dynamics of flows on tori and nilmanifolds most works are for the case of a single affine automorphism or a one-parameter flow of translations, as discussed in the preceding subsections. We describe in this subsection some results for actions of more general groups of affine automorphisms. We begin by noting the following criterion for ergodicity of an action (cf. Hoare and Parry [HoP2]).

Theorem 2.25. *Let T be a group of affine automorphisms of $\mathbb{T}^n$. Let A be the group of automorphism parts of elements of T. Let S be the set of all $a \in \mathbb{T}^n$ such that $T_a \circ \theta \in T$ for some $\theta \in A$. Then the T-action on $\mathbb{T}^n$ is ergodic if and only if for the dual action of A on $\mathbb{Z}^n$ every finite orbit consists of a fixed point and the subgroup of $\mathbb{T}^n$ generated by S and $\{x^{-1}\theta(x) \mid x \in \mathbb{T}^n, \theta \in A\}$ is dense in $\mathbb{T}^n$.*

In analogy with the comment following Theorem 2.1 we note here the following:

Remark 2.9. While there do exist noncompact Lie groups with ergodic actions by groups of automorphisms (e.g. $\mathbb{R}^n$, $n \geq 1$, or the Heisenberg group), it may be mentioned that if G is a connected Lie group with no compact central subgroup of positive dimension and A is the group of all automorphisms of G then all A-orbits on G are open in their closures, and consequently the A-action on G is ergodic (with respect to the Haar measure on G) only if it is essentially transitive, in the sense that there is an orbit whose complement in G has zero measure; see [DRj], for a more general result.

For actions of more general groups (not necessarily cyclic or one-parameter group) various mixing properties are defined as follows. Consider an action of a topological group T on a probability space (X, μ), as a group of measure-preserving transformations. It is said to be *weak mixing* if any finite-dimensional subspace of $L^2(X, \mu)$ which is invariant under the action induced by the T-action on X, contains only constant functions. The action is said to be *mixing of order* l (or l-*mixing*) if given any l sequences $\{g_k^{(1)}\}, \ldots, \{g_k^{(l)}\}$ in T such that for all $1 \leq i < j \leq l$, $(g_k^{(i)})^{-1} g_k^{(j)} \to \infty$ as $k \to \infty$ (namely the sequence escapes every compact subset in T) and any measurable sets $E_1, \ldots, E_l$ we have

$$\lim_{k\to\infty} \mu(\cap_{i=1}^{l} g_k^{(i)} E_i) = \prod_{i=1}^{l} \mu(E_i).$$

Mixing of order 2 is simply called *mixing*. It may be seen that these definitions coincide with those of the corresponding classical notions (see [CFS], Ch. I) for the case of measure-preserving automorphisms and one-parameter flows.

The following is straightforward to verify using arguments as in the one-variable case.

Theorem 2.26. *Let T be a group of affine automorphisms of $\mathbb{T}^n$, $n \geq 2$. Let A be the group consisting of automorphism parts of elements of T and let $p : T \to A$ be the canonical projection of T onto A. The T-action on $\mathbb{T}^n$ is weak mixing if and only if $\{0\}$ is the only finite orbit in $\mathbb{Z}^n$ under the dual action of A on $\mathbb{Z}^n$. The action is mixing, with respect to the discrete topology on T, if and only if the kernel of p is finite and for the A-action on $\mathbb{Z}^n$ the stability subgroup of every nonzero element is finite.*

We next recall the following result for higher order mixing; we state it in its original generality, though here we are considering only toral automorphisms.

Theorem 2.27 (Schmidt and Ward [SW]). *Let G be a compact connected abelian group. Then any mixing $\mathbb{Z}^n$-action on G by group automorphisms is mixing of all orders.*

We noted earlier that for an affine automorphism of a compact nilmanifold the normalised Haar measure is the unique invariant probability measure of maximal entropy (see § 2.2). A similar result is known for certain actions of $\mathbb{Z}^k$ on $\mathbb{T}^n$ by affine automorphisms, which we now recall. It may be mentioned that the result was inspired by a question of Furstenberg for endomorphisms of the circle, which is still open, and a significant partial result on the problem, obtained by D. Rudolph (see Katok and Spatzier [KS2] for details).

An action of $\mathbb{Z}^k$, $k \geq 2$, on $\mathbb{T}^n$ is called a *standard action* if the following conditions are satisfied: i) every nontrivial element of $\mathbb{Z}_+^k$ acts ergodically and ii) no finite cover of the action splits as a cartesian product of two actions.

Theorem 2.28 (Katok and Spatzier [KS2]). *Consider a standard action of $\mathbb{Z}^k$, $k \geq 2$, on $\mathbb{T}^n$. Let λ be an invariant probability measure on $\mathbb{T}^n$ such that the action is weak mixing. Suppose that for the action of some element of $\mathbb{Z}^k$ the entropy with respect to λ is positive. Then λ is the Haar measure on a subtorus of $\mathbb{T}^n$. In particular, if there is no proper nontrivial subtorus invariant under the action then λ is the Haar measure on $\mathbb{T}^n$.*

§ 3. Group-induced translation flows; special cases

In this section we discuss ergodic properties of flows on homogeneous spaces of certain special classes of Lie groups, namely solvable and semisimple groups; we also discuss the possibilities for flows on low-dimensional homogeneous spaces, in terms of ergodic behaviour. The underlying measure for considerations of ergodic properties will be understood to be the normalised Haar measure. The general case will be considered in the next section.

3.1. Flows on solvmanifolds. A homogeneous space G/C, where G is a solvable Lie group and C is a closed subgroup of G, is called a solvmanifold. A solvmanifold G/C admits a finite Haar measure if and only if it is compact (see [Rag], Theorem 3.1). A simple class of examples of connected solvable Lie groups is provided by semidirect product groups of the form $G = A \cdot N$, where N is a connected nilpotent Lie group and A is a connected abelian Lie group acting on N (continuously) as a group of automorphisms. Such a group G admits a lattice only if elements of A acting as automorphisms of N preserve the Haar measure on N. If A and N have lattices Δ and Λ respectively such that Δ normalises Λ (that is Λ is invariant under the automorphisms corresponding to each $\delta \in \Delta$) then $\Delta\Lambda$ is a lattice in G. (See also Example 3.1, which is a further specialised case of this construction).

The structure of a general solvmanifold is considerably more complicated than that of a nilmanifold and the study of affine automorphisms and translation flows on it is quite intricate. For a detailed exposition of the theory the reader is referred to Auslander [Au], Brezin and Moore [BM] and Starkov [St3]. Roughly speaking the theory relates the general situation to the simple case of semidirect products as above. We shall discuss here only the results which can be described without going into certain rather technical constructions involved in the theory.

Let G be a connected solvable Lie group. Then G contains a unique maximal closed connected normal nilpotent Lie subgroup, called the *nilradical* of G. By a theorem of G.D. Mostow, if N is the nilradical of G and $C \in \mathcal{F}_0(G)$

(notation as introduced in § 1.1) then $N/(N \cap C)$ is compact (see [Rag], Theorem 3.3).

It turns out that for solvmanifolds also a factor can be described, which plays a role similar to that of the maximal torus quotient in the case of nilmanifolds.

A solvable Lie group is called *Euclidean* if it can be expressed as $A \cdot V$, semidirect product, where V is a vector group (topologically isomorphic to $\mathbb{R}^n$ for some n) and A is a connected abelian Lie group, acting via a representation $\rho : A \to GL(V)$, such that $\rho(A)$ is compact. A compact solvmanifold is called *Euclidean* if it can be expressed as a homogeneous space of a Euclidean solvable Lie group. A compact solvmanifold is Euclidean if and only if it has a finite cover which is homeomorphic to a torus. Any compact solvmanifold admits a unique maximal Euclidean quotient; that is, given a solvable Lie group G and a closed subgroup C of G such that G/C is compact there exists a unique minimal closed subgroup L containing C, such that G/L is Euclidean; if M is the largest closed normal subgroup of G contained in L (namely, the intersection of all conjugates of L) then G/M is a Euclidean solvable Lie group; see [BM] for details on these and other properties of Euclidean solvmanifolds.

Example 3.1. Let $V = \mathbb{R}^n$ and $\Lambda = \mathbb{Z}^n$. Let $\alpha \in SL(n, \mathbb{Z})$ be an element which can be embedded in a one-parameter subgroup $\phi = \{\phi_t\}$ of $SL(n, \mathbb{R})$, as $\alpha = \phi_1$. Let $G = \phi \cdot V$, semidirect product (where ϕ is viewed as a one-parameter subgroup of automorphisms of V). Then Λ, realised as a subgroup of G, is normalised by α and the subgroup Γ generated by α and Λ is a lattice in G. The homogeneous space G/Γ is Euclidean if and only if α is of finite order. In general the subgroup L corresponding to the maximal Euclidean quotient of G/Γ is given by $L = \Gamma W$, where W is the smallest α-invariant subspace of V such that the factor of α on V/W is of finite order. (See Remark 3.2 for more about this example).

Theorem 3.1 (Brezin and Moore [BM]). *Let G be a connected solvable Lie group and $C \in \mathscr{F}(G)$ (notation as in § 1.1). Let L be the closed subgroup of G containing C such that G/L is the maximal Euclidean factor of G/C. Then the flow on G/C induced by a one-parameter subgroup of G is ergodic if and only if its factor on G/L is ergodic.*

When a flow as in the above theorem is ergodic the associated unitary one-parameter group has Lebesgue spectrum of infinite multiplicity on the orthocomplement of $L^2(G/L)$, unless $L = G$ (see Theorem 4.9).

If G is a connected solvable Lie group such that for every $g \in G$ and every eigenvalue λ of $\operatorname{Ad} g$ either $\lambda = 1$ or $|\lambda| \neq 1$ then it follows that any Euclidean factor group of G is abelian. Thus in this case L as in Theorem 3.1 is precisely $[G, G]C$. For a simply connected solvable Lie group G the above condition on eigenvalues is equivalent to the exponential map of G being surjective (cf. [DH]); when the latter condition holds G is called an *exponential Lie group*. If G is a simply connected exponential solvable Lie group and M is the smallest closed connected normal subgroup of G such that G/M is nilpotent, then for any (uniform) lattice Γ in G, $M\Gamma$ is closed (cf. [AGH], Ch. VII). Hence Theorem 3.1 implies that as in the case of nilpotent Lie groups (see Theorem 2.6(i)) the behaviour of flows on G/Γ is determined by the maximal torus quotient. Altogether the following holds in this case.

Theorem 3.2 (cf.[AGH], Ch. VII, and [Ste1]). *Let G be a simply connected exponential solvable Lie group of positive dimension and Γ be a lattice in G. Then $[G, G]\Gamma$ is closed and for any one-parameter subgroup ϕ of G the flow induced by ϕ on G/Γ is ergodic if and only if its factor flow on $G/[G, G]\Gamma$ is ergodic. When the flow is ergodic every eigenfunction of the flow on G/Γ factors through $G/[G, G]\Gamma$ and on the orthocomplement of $L^2(G/[G, G]\Gamma)$ the associated unitary one-parameter group has Lebesgue spectrum of infinite multiplicity.*

We shall next describe conditions for ergodicity of flows on Euclidean solvmanifolds. Let G be a simply connected Euclidean Lie group. Then G has the form $G = A \cdot V$, semidirect product, where V and A are vector groups, V is a normal subgroup of G and the conjugation action of A on V is given by a representation $\rho : A \to GL(V)$, such that $\rho(A)$ is compact. Let $W = A \oplus V$, the direct sum of the vector groups. Then we have the canonical correspondence between G and W given by $a \cdot v \leftrightarrow (a, v)$ for all $a \in A$ and $v \in V$. Let $M = \ker \rho$. Suppose that C is a cocompact subgroup of G contained in MV. Then the set in W corresponding to C is a subgroup of W and furthermore the correspondence induces a canonical bijection of G/C onto W/C; we shall denote it by θ. We note that θ is a measure-preserving homeomorphism of the homogeneous space G/C onto the torus W/C. Also, if $\{a_t\}$ is a one-parameter subgroup contained in A then θ is a topological isomorphism (see § 1.4 for definition) between the flow induced by $\{a_t\}$ on G/C and the flow induced by it on W/C.

Theorem 3.3 (cf. [BM], § 5). *Let $G = A \cdot V$ be a simply connected Euclidean Lie group as above and let $C \in \mathcal{F}_0(G)$. Suppose that G/C admits an ergodic flow induced by a one-parameter subgroup of G. Then, in the notation as above, C is contained in MV and if $\{g_t\}$ is a one-parameter subgroup of G and $\{a_t\}$ is its image under the canonical projection onto A then, the flow induced by $\{g_t\}$ on G/C is topologically isomorphic to the flow induced by $\{a_t\}$ on W/C.*

Consequently, G/C admits an ergodic flow induced by a one-parameter subgroup of G if and only if G/C is homeomorphic to a torus and the subgroup Z of W corresponding to centraliser of A in G (under the correspondence as above) acts ergodically on W/C.

Remark 3.1. Theorem 3.3 provides a general class of examples of topologically isomorphic translations of homogeneous spaces which are not affinely isomorphic.

We recall that a connected Lie group G is said to be of *class* $\mathscr{R}$ if for all $g \in G$ all eigenvalues of $\operatorname{Ad} g$ are of absolute value 1.

Theorem 3.4 (Auslander [Au], Part II). *Let G be a connected solvable Lie group of class $\mathscr{R}$ and $C \in \mathscr{F}(G)$. Let ϕ be a one-parameter subgroup of G and Φ be the flow on G/C induced by ϕ. Then the following holds:*

i) Φ is a distal flow;

ii) Φ is ergodic if and only if it is minimal, and in that case G/C is homeomorphic to a nilmanifold and Φ is topologically isomorphic to a nilflow.

It may be observed that assertion (ii) generalises a similar statement in Theorem 3.3 for Euclidean solvmanifolds. The following theorem shows that the converse of assertion (i) holds for ergodic flows:

Theorem 3.5 (cf. [Au], Part II). *Let G be a connected solvable Lie group and $C \in \mathscr{F}_0(G)$. If there exists a one-parameter subgroup ϕ of G such that the flow induced by ϕ on G/C is ergodic and distal then G is of class $\mathscr{R}$. Similarly, if there exists a one-parameter subgroup whose action on G/C is minimal then G is of class $\mathscr{R}$.*

Remark 3.2. Let $G = \phi \cdot V$, A and Γ be as in Example 3.1. We note that $U = (A - I)(V)$, where I is the identity matrix, is a closed normal subgroup of G such that $U\Gamma$ is closed. Further, G/U is a vector group and $U\Gamma/U$ is a lattice in G/U. Let ψ be any one-parameter subgroup of G. Then Theorems 3.1 and 3.3 show that the flow on G/Γ induced by ψ is ergodic if and only if A has no eigenvalue which is a root of unity other than 1, and the image of ψ in G/U acts ergodically on the torus $(G/U)/(U\Gamma/U)$. The group G is of class $\mathscr{R}$ if and only if all eigenvalues of A are roots of unity. Thus if the flow on G/Γ induced by ψ is ergodic and either distal or minimal then Theorem 3.5 together with the above implies that A is unipotent.

3.2. Homogeneous spaces of semisimple groups. We now consider homogeneous spaces of semisimple Lie groups. The special linear group $SL(n, \mathbb{R})$, the symplectic group $Sp(n, \mathbb{R})$, the special orthogonal groups associated to nondegenerate quadratic forms, are examples of semisimple Lie groups; in fact they are simple Lie groups. Any connected semisimple Lie group G can be expressed as $G = G_1 G_2 \cdots G_k$, where $G_1, \ldots, G_k$, $k \geq 1$, are all the simple normal connected Lie subgroups of G; we note also that for $i \neq j$ the elements of G_i commute with the elements of G_j and that the intersection $G_i \cap G_j$ is a discrete subgroup contained in the center of G. It follows from Weyl's theorem (cf. [Va], Theorem 4.11.6 for instance) that a connected semisimple Lie group has no (nontrivial) compact factors if and only if no G_i as above is compact.

A subgroup H of a connected semisimple Lie group G (without compact factors) is said to be *totally noncompact* if for every proper closed normal subgroup M of G the closure of HM/M in G/M is noncompact.

Now let G be a connected semisimple Lie group with no compact factors and let $C \in \mathscr{F}(G)$. By Borel's density theorem (see [Rag]; see also [D9] and [D12] for more general versions of Borel's density theorem) C^0 is a normal subgroup of G. For the purpose of our exposition we may therefore assume, without loss of generality, C to be a lattice in G. Borel's density theorem also implies that for any lattice Γ in G, $\Gamma \cap Z$ is a subgroup of finite index in Z, where Z is the center of G (see [Rag], Corollary 5.17).

A lattice Λ in a semisimple Lie group L is said to be *irreducible* if for every closed normal subgroup F of positive dimension $F\Lambda$ is dense in L. Let G be a connected semisimple Lie group without compact factors. Then any lattice Γ in G can be 'decomposed' as a product of irreducible lattices in the following sense: there exist closed connected normal subgroups $G_1, \dots, G_r$, for some $r \geq 1$, such that the following conditions are satisfied: i) $G = G_1 G_2 \cdots G_r$ and the pairwise intersections of G_i's are discrete and ii) $G_i \cap \Gamma$ is an irreducible lattice in G_i for each $i = 1, \dots, r$ (see [Rag], Theorem 5.22); if Γ_0 is the product of $G_i \cap \Gamma$, $i = 1, \dots, r$, then it follows that Γ_0 is a lattice in G and hence, in particular, Γ_0 is of finite index in Γ. This shows that G/Γ is finitely covered by the product homogeneous space $\Pi\ (G_i/(G_i \cap \Gamma))$. The subgroups G_i as above are uniquely determined by conditions (i) and (ii). For each $i = 1, \dots, r$ let G_i' be the product of G_j, $j \neq i$. Then $\Gamma G_i'/G_i'$ is an irreducible lattice in G/G_i'. We call $G/G_i'\Gamma$, $i = 1, \dots, r$, the *irreducible quotients* of G/Γ.

Theorem 3.6 (Moore [Mo2]). *Let G be a connected semisimple Lie group with no compact factors and Γ be a lattice in G. Let Z be the center of G. Let F be a subgroup of G and consider the action of F on G/Γ by translations. Under the assumption that Γ is an irreducible lattice, the action is ergodic if and only if FZ/Z is not contained in a compact subgroup of G/Z. In the general case, the F-action is ergodic if and only if its factor on each irreducible quotient of G/Γ is ergodic or, equivalently, if and only if for each $i = 1, \dots, r$ the closure of $FG_i'Z/G_i'Z$ is a noncompact subgroup of $G/G_i'Z$, in the notation as above. In particular, if F is totally noncompact then the action of F on G/Γ is ergodic.*

We mention in particular the following case of the theorem:

Corollary 3.1. *Let G be a noncompact connected simple Lie group with finite center, e.g. $SL(n, \mathbb{R})$, and Γ be a lattice in G. Then the action of a subgroup F of G on G/Γ is ergodic if and only if F is not contained in a compact subgroup of G.*

Any ergodic translation of a finite-volume homogeneous space of a semisimple Lie group is mixing (see [Mo2]); furthermore the unitary operator associated to it has Lebesgue spectrum of infinite multiplicity on the orthocomplement of the constants (see Theorem 4.9; see also [Ste2]). The following theorem yields a generalisation of the mixing property (see Corollary 3.2 below) to translation actions by subgroups of the ambient group; (see § 2.10 for definition of mixing for group actions).

Theorem 3.7 (Howe and Moore [HM], Zimmer [Zi3]). *Let G be a connected semisimple Lie group with finite center and no compact factors. Let $\pi : G \to \mathcal{U}(\mathcal{H})$ be a (continuous) unitary representation of G over a separable Hilbert space $\mathcal{H}$. Suppose that no nonzero element of $\mathcal{H}$ is fixed by $\pi(g)$ for all g in a nondiscrete normal subgroup of G. Then for any $\alpha, \beta \in \mathcal{H}$ the function φ on G defined by $\varphi(g) = \langle \pi(g)(\alpha), \beta \rangle$, for all $g \in G$, vanishes at infinity (that is, for all $\epsilon > 0$ the set $\{g \in G \mid |\varphi(g)| \geq \epsilon\}$ is compact).*

Remark 3.3. In the context of the above result on the 'matrix coefficients' of unitary representations it may also be mentioned here that for any representation of a connected simple noncompact Lie group G over a reflexive Banach space the matrix coefficients extend to the one-point compactification of G; this is proved via the study of weakly almost periodic functions (see Veech [V2] for details, including a more general result and applications to ergodic theory).

Corollary 3.2. *Let G be a connected semisimple Lie group with finite center and no compact factors. Let Γ be a lattice in G and H be a closed noncompact subgroup of G. Suppose that for every proper closed connected normal subgroup M of G such that $M\Gamma$ is closed, $H \cap M$ is compact. Then the action of H on G/Γ is mixing. In particular, if either Γ is an irreducible lattice or H is a totally noncompact one-parameter subgroup then the H-action on G/Γ is mixing.*

It may observed that the condition on the intersection of H with closed connected normal subgroups, as above, is also a necessary condition for the conclusion of the corollary to hold.

The following consequence of Corollary 3.2 generalises a theorem of W.M. Schmidt (cf. [Sch4], Theorem 2).

Corollary 3.3. *Let $G = SL(n, \mathbb{R})$, $\Gamma = SL(n, \mathbb{Z})$ and $\{g_i\}$ be an unbounded sequence in G. Then for almost all $g \in G$ (with respect to the Haar measure) the sequence $\{g_i g\Gamma\}$ is dense in G/Γ. Consequently, if $\mathcal{M}$ is the space of all lattices in $\mathbb{R}^n$ with a fundamental parallelopiped of unit volume, as in Example 2 in § 1.2, then for almost all $\Lambda \in \mathcal{M}$ the sequence of lattices $\{g_i \Lambda\}$ is dense in $\mathcal{M}$.*

3.3. Flows on low-dimensional homogeneous spaces. In this section we shall briefly discuss the ergodic properties of flows on finite-volume homogeneous spaces of small dimensions; we shall denote the dimension of a Lie group G by $\dim G$.

We first consider homogeneous spaces of the form G/Γ, where G is a simply connected Lie group and Γ is a lattice in G. If $\dim G = 2$ then G is topologically isomorphic to $\mathbb{R}^2$ and G/Γ is a torus, the flows on which were discussed in § 2.1.

Now let G be a 3-dimensional simply connected Lie group. Then G is either solvable or semisimple; this follows from the fact that semisimple Lie groups have dimension at least 3. Let us consider the two cases separately:

Case i) Suppose G is solvable. Then G is a semidirect product of $\mathbb{R}$ and $\mathbb{R}^2$ with respect to an action of $\mathbb{R}$ on $\mathbb{R}^2$ as a group of (linear) automorphisms. Since by assumption G contains a lattice Γ it follows (cf. Proposition 1.1) that the conjugation action of $\mathbb{R}$ on $\mathbb{R}^2$ is given by a one-parameter subgroup, say $\phi = \{\phi_t\}$, contained in $SL(2, \mathbb{R})$; thus $G = \phi \cdot V$, where $V = \mathbb{R}^2$. If ϕ is the trivial one-parameter subgroup, then $G = \mathbb{R}^3$ and the flows in question are translation flows on $\mathbb{T}^3$, as in § 2.1. Now suppose that ϕ is nontrivial. Let A be a nontrivial element of ϕ. As $A \in SL(2, \mathbb{R})$, it is either semisimple or unipotent. If it is unipotent then (ϕ consists unipotent elements and) G is topologically isomorphic to the Heisenberg group, namely the group of upper triangular unipotent 3×3 matrices (as in Example 3 in § 1.2 with $n = 3$). In this case G is nilpotent and the results of § 2.2 apply. In particular, we have a criterion for ergodicity of the flows induced by one-parameter subgroups of G, as seen in Example 2.2.

Now suppose that A is semisimpe. Then V is the nilradical of G and by Mostow's theorem (see [Rag], Theorem 3.3) $V \cap \Gamma$ is a lattice in G. Also, A is either an hyperbolic or an elliptic element. Suppose A is hyperbolic. Then G is an exponential group (see § 3.1). Furthermore, after some modifications we may assume that $A \in SL(2, \mathbb{Z})$ and Γ is the subgroup generated by A and $\mathbb{Z}^2$; namely G and Γ are as in Example 3.1. Since V is also the commutator subgroup of G, it follows that the flow on G/Γ induced by a one-parameter subgroup ψ is ergodic if and only if $G = \psi V$ (cf. Theorem 3.2) or, equivalently, if and only if ψ is not contained in V.

Finally, if A is an elliptic element then G is a Euclidean group, and in this case a criterion for ergodicity is given by Theorem 3.3. We see that G/Γ admits an ergodic flow induced by a one-paramenter subgroup of G only if Γ is contained in ZV, where Z is the subgroup of ϕ acting trivially on V (namely the center of G) and when the condition holds the flow induced by a one-parameter subgroup ψ is ergodic if and only if ψ is not contained in V and the projection of Γ on V is dense in V (or equivalently $\phi\Gamma$ is dense in G). When ergodic the flow is topologically isomorphic to a translation flow on a three-dimensional torus.

Case ii) Next suppose that G is semisimple; since it is 3-dimensional it is in fact a simple Lie group. If it is compact then Γ is finite and in this case no one-parameter subgroup of G acts ergodically on G/Γ. Suppose next that it is noncompact. Then, being simply connected, it is the universal covering group of $SL(2, \mathbb{R})$. Let the Lie algebra of G be identified as the Lie algebra $\mathscr{M}_0$ of 2×2 real matrices with trace 0. Then any one-parameter subgroup of G is conjugate to $\{\exp t\xi\}$, where ξ is either the diagonal matrix diag $(1, -1)$ or a nilpotent matrix or a matrix with purely imaginary nonzero eigenvalues. In the last case the image of $\{\exp t\xi\}$ under the adjoint homomorphism of G is compact and hence the flow induced by $\{\exp t\xi\}$ is nonergodic (cf. Theorem 3.6). In the other two cases the induced flows are ergodic and mixing (cf. Theorem 3.6, Corollary 3.1 and Corollary 3.2). It may be seen that these examples include the geodesic and horocycle flows associated to surfaces of constant negative curvature (cf. Example 5 of § 1.2) and the general example as above is essentially one of these, upto finite coverings.

We next consider homogeneous spaces G/Γ where G is a simply connected Lie group of dimension 4 and Γ is a lattice in G. A detailed analysis as in the case of three-dimensional groups, though possible, would be too cumbersome. We content ourselves making some observations which may be helpful to a reader interested in the case. In this case either G is solvable or the radical of G is one-dimensional. Suppose first that G is solvable. Let N be the nilradical of G. We note that if N is abelian then the conjugation action of G on N is locally faithful. Using this, one can deduce from Mostow's theorem (see [Rag], Theorem 3.3) that N is of dimension at least 3. Then G is either nilpotent or of the form $\phi \cdot N$, where ϕ is a one-parameter subgroup of G. If the latter holds, considering the possibilities for the conjugation action of ϕ on N one can further conclude that G is either exponential or of class $\mathscr{R}$. Ergodicity of the flows on G/Γ can therefore be analysed using Theorems 3.2 and 3.3.

Now suppose that the radical, say R, is one-dimensional. Then G is reductive and hence a direct product of $[G, G]$ and R (see [Va], Theorem 3.16.3). Now, $[G, G]$ is a 3-dimensional semisimple Lie group and hence it is a simple Lie group. Suppose that $[G, G]$ is compact. By Auslander's theorem (see [Rag], Theorem 8.24) $\overline{R\Gamma}^0/R$ is solvable and hence it is of dimension 1. It follows that in this case there is no one-parameter subgroup of G inducing an ergodic flow on G/Γ. Now suppose that $[G, G]$ is a noncompact simple Lie group. We note that $[G, G]\Gamma$ is either closed or dense. In the former case G/Γ is finitely covered by the homogeneous space $(R/R\cap\Gamma)\times([G, G]/[G, G]\cap\Gamma)$. Since for homogeneous spaces of semisimple Lie groups a flow induced by a one-parameter subgroup is mixing whenever it is ergodic (see Corollary 3.2) it follows that for a one-parameter subgroup ψ of G the flow on G/Γ induced by ψ is ergodic if and only if $\psi[G, G] = G$ and $\operatorname{Ad}\psi$ is noncompact (as $[G, G]$ is 3-dimensional the latter is a closed subgroup of the adjoint group). In the case when $[G, G]\Gamma$ is dense, applying Theorem 3.7 to $[G, G]$ one can see that the flow on G/Γ induced by a one-parameter subgroup ψ is ergodic if and only if $\operatorname{Ad}\psi$ is noncompact.

Finally let us consider homogeneous spaces G/C of finite Haar measure, where G is a connected Lie group and $C \in \mathscr{F}_0(G)$, but not a lattice. It turns out that there does not exist such a homogeneous space with dimension 1 (cf. [Ho]). A simple 2-dimensional example can be given by taking G to be a compact simple 3-dimensional Lie group and C a closed one-parameter subgroup. A more general class of low-dimensional examples can be described using the construction as in Remark 1.1; in fact from the construction one can get, for any $k \geq 2$, Lie groups G of any given dimension $d \geq k+1$ with nondiscrete $C \in \mathscr{F}_0(G)$ such that G/C is of dimension k. On the other hand if G is a simply connected Lie group and $C \in \mathscr{F}_0(G)$ is nondiscrete and such that G/C is of dimension at most 3 then it can be shown that G has a closed connected normal abelian subgroup V such that G/V is either compact or abelian; in the former case G is a semidirect product of V with a compact group K acting on V, and it can be shown further that if ϕ is a one-parameter subgroup of G acting ergodically on G/C then there exists a closed connected solvable subgroup H of G containing ϕ and acting transitively on G/C. Thus for purposes of ergodic theory and dynamics in this context, it suffices to consider flows on solvmanifolds. A criterion for ergodicity of these flows is given by Theorem 3.3.

§ 4. Ergodic properties of flows on general homogeneous spaces

We now consider flows on homogeneous spaces of a more general Lie group. For a Lie group G, $\mathscr{F}(G)$ and $\mathscr{F}_0(G)$ will denote the classes of subgroups of G as in § 1.1; as before, the discussion of ergodic properties will be with respect to the normalised Haar measure, and we omit explicit mention of the latter from the individual statements. The first subsection mainly concerns the Mautner phenomenon, which we describe for any affine automorphism of a homogeneous space. In the later subsections we shall restrict to translations and actions of subgroups by translation, as has been the practice in literature; we shall sometimes refer to these actions as *group-induced flows*. Some of the techniques and results that we describe can be extended to affine automorphisms, but there are no suitable references available. It may be mentioned in this respect that ergodic theory and dynamics of an affine automorphism τ can be studied by considering the suspension flow of a suitable power of τ, as the suspension can be realised as a translation flow on a homogeneous space. It may also be noted that any affine automorphism τ of a homogeneous space

G/C, where G is a Lie group and C is a closed subgroup of G, is canonically isomorphic to a translation on a homogeneous space G'/C', with G' a semidirect product of a cyclic group with G; however G' is not connected and most of the results that we describe will not be directly applicable to translations of its homogeneous spaces.

4.1. Horospherical subgroups and Mautner phenomenon. Let G be a Lie group and $\sigma = T_g \circ \alpha$ be an affine automorphism of G, where $g \in G$ and α is an automorphism of G. We denote by e the identity element in G. Let D_σ be the automorphism of G defined by $D_\sigma(x) = g\alpha(x)g^{-1}$ for all $x \in G$. The subgroup U_σ^+ defined by

$$U_\sigma^+ = \{x \in G \mid D_\sigma^i(x) \to e \text{ as } i \to \infty\}$$

is called the *horospherical subgroup* or, more specifically, the *contracting horospherical subgroup* associated to σ. The (contracting) horospherical subgroup corresponding to σ^{-1} is called the *expanding horospherical subgroup* associated to σ and will be denoted by U_σ^-. When σ is a translation T_g, where $g \in G$, the horospherical subgroups are said to be associated to g, rather than T_g. The horospherical subgroups are connected Lie subgroups of G. They consist of Ad-unipotent elements (namely $\operatorname{Ad} u$ is unipotent for every u in such a subgroup) and are therefore nilpotent Lie subgroups. They need not be closed in general (e.g. when σ is a hyperbolic toral automorphism). However if G is a Lie group with discrete center then the horospherical subgroup associated to any affine automorphism is closed.

Let σ be an affine automorphism of a Lie group G and let M_σ be the smallest closed subgroup of G containing the horospherical subgroups U_σ^+ and U_σ^-; we call this the *Mautner subgroup* or the *unstable subgroup* associated to σ. It can be shown that M_σ is a normal subgroup of G^0, so in particular it is a normal subgroup when G is connected (see [D2], for instance). Given an affine automorphism $\tau = T_g \circ \overline{\alpha}$ of a homogeneous space G/C, where G is a Lie group, C is a closed subgroup of G, $g \in G$ and α is an automorphism of G such that $\alpha(C) = C$, by the Mautner subgroup associated to τ we mean the Mautner subgroup associated to the affine automorphism $\sigma = T_g \circ \alpha$ of G; it will be denoted by M_τ.

Example 4.1. Let $G = SL(n, \mathbb{R})$ and $g \in G$. Then g has a Jordan decomposition as $g = su$, where s is a semisimple element (diagonalisable over the complex numbers) and u is a unipotent element commuting with s. The horospherical subgroup associated to g is the same as the one associated to s. For $s \in G$ which is diagonalisable over the complex numbers, there exist subspaces $V_1, \ldots, V_k$ of $\mathbb{R}^n$ and $0 < \lambda_1 < \cdots < \lambda_k$ such that $\mathbb{R}^n = V_1 \oplus \cdots \oplus V_k$, each V_j is s-invariant and all eigenvalues of the restriction of s to V_j are of absolute value λ_j. The (contracting) horospherical subgroup associated to s then consists of all $g \in G$ such that $g(v) - v \in V_1 + \cdots + V_{j-1}$ ($\{0\}$ if

$j = 1$) for all $v \in V_j$, $j = 1, \ldots, k$. In particular if s is the diagonal matrix diag $(d_1, \ldots, d_n)$ with $0 < d_1 < \cdots < d_n$, then the group of all upper triangular unipotent matrices is the contracting horospherical subgroup associated to s.

The following simple lemma is very useful in concluding ergodicity of affine automorphisms in many cases. The first assertion in the lemma is generally known as the Mautner phenomenon and was first noted in [AG]. The second was formulated in [Marg3].

Lemma 4.1. *Let H be a topological group and $\pi : H \to \mathcal{U}(\mathcal{H})$ be a (continuous) unitary representation of H over a Hilbert space $\mathcal{H}$. Then we have the following:*

i) if s and t are two elements of H such that $s^i t s^{-i}$ converges to the identity as $i \to \infty$ then any eigenvector of $\pi(s)$ is fixed by $\pi(t)$.

ii) if F is a subgroup of H and $t \in H$ is such that $t \in \overline{F \Omega F}$ for every neighbourhood Ω of the identity in H, then every vector in $\mathcal{H}$ which is fixed by $\pi(s)$ for all $s \in F$ is also fixed by $\pi(t)$.

The Mautner phenomenon immediately yields the following.

Theorem 4.1. *Let G be a connected Lie group and $C \in \mathcal{F}(G)$. Let $\mathcal{H} = L^2(G/C)$ and let $x \mapsto U_x$ be the unitary representation of G over $\mathcal{H}$ induced by the G-action on G/C (see § 1.1). Let $\tau = T_g \circ \overline{\alpha}$ be an affine automorphism of G/C and let U_τ be the unitary operator corresponding to τ. Let M_τ be the Mautner subgroup associated to τ. Then we have the following:*

i) if $f \in \mathcal{H}$ and $U_\tau(f) = f$ then $U_x(f) = f$ for all $x \in M_\tau$.

ii) τ is ergodic if and only if the factor of τ on $G/\overline{M_\tau C}$ is ergodic.

The theorem reduces the question of checking ergodicity of an affine automorphism to the special case when the associated Mautner subgroup is trivial; for $\tau = T_g \circ \overline{\alpha}$ the latter condition is equivalent to all eigenvalues of $(\operatorname{Ad} g) \circ d\alpha$ (which is a linear transformation of the Lie algebra of G) being of absolute value 1. The general results in this respect for translation actions by subgroups will be discussed in the following subsections. We note here that in the case when $G = SL(2, \mathbb{R})$ and $\{u_t\}$ is a unipotent one-parameter subgroup of G, it can be deduced from Lemma 4.1(ii) that for any lattice Γ in G the flow induced by $\{u_t\}$ on G/Γ (namely the horocycle flow; see Example 5 in § 1.2) is ergodic. More generally, the assertion of Theorem 3.6 for the case of horospherical subgroups can be readily deduced from Lemma 4.1(ii).

We conclude this subsection with the following remark concerning 'stable' affine automorphisms.

Remark 4.1. Let G be a connected Lie group, $C \in \mathcal{F}_0(G)$ and let $\tau = T_g \circ \overline{\alpha}$ be an affine automorphism of G/C, where $g \in G$ and α is an automorphism of G such that $\alpha(C) = C$. We say that τ is *stably ergodic* if there exists a neighbourhood Ω of g in G such that for all $g' \in \Omega$, $\tau' = T_{g'} \circ \overline{\alpha}$ (the 'perturbed' affine automorphism) is ergodic. A result of Brezin and Shub [BS] shows that if τ is stably ergodic then $\overline{M_\tau C} = G$, where M_τ is the Mautner subgroup associated to τ (the 'admissibility' condition assumed in [BS] holds for all C; see the comments following Theorem 4.3, infra). It is conjectured that conversely if $\overline{M_\tau C} = G$ then τ is stably ergodic; this is shown to hold in the cases when G is either nilpotent or semisimple [BS].

4.2. Ergodicity of one-parameter flows. We now proceed to discuss a more general criterion for ergodicity. For this purpose we need some more terminology. Given a Lie group L with Lie algebra $\mathscr{L}$, a subgroup F of L is said to be *Ad-compact* if the subgroup $\{\mathrm{Ad}\, x : \mathscr{L} \to \mathscr{L} \mid x \in F\}$ is contained in a compact subgroup of $GL(\mathscr{L})$. Now let G be a connected Lie group. Then for any subgroup F of G there exists a unique minimal closed normal subgroup M'_F such that FM'_F/M'_F is Ad-compact as a subgroup of G/M'_F (see Moore [Mo3]); we call M'_F the *Mautner-Moore subgroup* associated to F. Clearly M'_F contains the Mautner subgroups associated to the translations T_x, for all $x \in F$.

Remark 4.2. Let G be a connected Lie group and F be a subgroup of G. Suppose that F is generated by the elements $x \in F$ such that all eigenvalues of $\mathrm{Ad}\, x$ (over the Lie algebra of G) are real and positive. Then M'_F is the smallest closed normal subgroup of G such that FM'_F/M'_F is contained in the center of G/M'_F. We note that when this happens, FM'_F is a normal subgroup of G and hence $FM'_F C$ is a subgroup of G.

Theorem 4.2 (Moore [Mo3]). *Let G be a connected Lie group, $\{g_t\}$ be a one-parameter subgroup of G and let M' be the Mautner-Moore subgroup corresponding to $\{g_t\}$. Let $\pi : G \to \mathscr{U}(\mathscr{H})$ be a unitary representation of G over a Hilbert space $\mathscr{H}$. Then the infinitesimal generator of $\{\pi(g_t)\}$ has absolutely continuous spectrum on the orthocomplement of the subspace $\{\xi \mid \pi(x)(\xi) = \xi$ for all $x \in M'\}$. In particular, if $\mathscr{V}$ is a finite-dimensional subspace of $\mathscr{H}$ invariant under $\{\pi(g_t)\}$ then $\pi(x)(v) = v$ for all $x \in M'$ and $v \in \mathscr{V}$.*

This yields in particular the following criteria for ergodicity of group-induced flows.

Corollary 4.1. *Let G be a connected Lie group and $C \in \mathscr{F}(G)$. Let $\phi = \{g_t\}$ be a one-parameter subgroup of G and M' be the associated Mautner-Moore subgroup. Then the flow induced by ϕ on G/C is ergodic if and only if the flow induced by it on the quotient $G/\overline{M'C}$ is ergodic.*

Corollary 4.1 and Remark 4.2 together imply the following:

Corollary 4.2. *Let the notation be as in Corollary 4.1. Suppose further that all eigenvalues of* $\operatorname{Ad} g_t$, $t \in \mathbb{R}$, *are real. Then the flow induced by* ϕ *on* G/C *is ergodic if only if* $\phi M'C$ *is dense in* G.

Ergodicity of flows on a homogeneous space G/C of a general Lie group can also be characterised in terms of factors on homogeneous spaces of semisimple and Euclidean groups (see § 3.1 for definition).

Theorem 4.3 (Brezin and Moore [BM]). *Let* G *be a connected Lie group and* $C \in \mathscr{F}(G)$. *Let* R *be the solvable radical of* G *and* L *be the smallest closed normal subgroup of* G *such that* G/L *is Euclidean. Let* $\phi = \{g_t\}$ *be a one-parameter subgroup of* G. *Then the flow induced by* ϕ *on* G/C *is ergodic if and only if the flows induced by* ϕ *on* $G/\overline{RC}$ *and* $G/\overline{LC}$ *are ergodic.*

The assertion in Theorem 4.3 was proved in [BM] under an additional condition of 'admissibility' of C, which has subsequently been shown to hold for all $C \in \mathscr{F}_0(G)$ (see Wu [Wu], Starkov [St1] and Witte [Wi1]; a more general result in this respect may be found in Zimmer [Zi4]); note that in proving the theorem we may assume that $C \in \mathscr{F}_0(G)$.

We observe that in particular a flow as in Theorem 4.3 is ergodic if and only if its factors on $G/\overline{RC}$ and $G/\overline{SC}$ are ergodic, where R (as before) is the radical of G and S is the smallest closed normal subgroup of G such that G/S is a solvable Lie group; namely S is the terminal subgroup in the derived series defined by $G_1 = \overline{[G, G]}$ and $G_{i+1} = \overline{[G_i, G_i]}$ for all $i \geq 1$ (see [D5]). A simple proof of this weaker criterion may be found in Starkov [St2].

We noted earlier (see Theorem 3.4) that a homogeneous space with finite Haar measure may not admit an ergodic group-induced one-parameter flow. The following sufficient condition for existence of such a flow on a homogeneous space can be deduced from Theorem 4.3.

Corollary 4.3. *Let* G *be a Lie group and* $C \in \mathscr{F}(G)$. *Suppose that* G *is generated by one-parameter subgroups* $\{g_t\}$ *such that* $\operatorname{Ad} g_t$ *has only real eigenvalues for all* $t \in \mathbb{R}$. *Then there exists a one-parameter subgroup* $\{g_t\}$ *acting ergodically on* G/C; *furthermore* $\{g_t\}$ *can be chosen such that* $\operatorname{Ad} g_t$ *has only real eigenvalues for all* $t \in \mathbb{R}$. *If* G *is generated by Ad-unipotent one-parameter subgroups then there exists an Ad-unipotent one-parameter subgroup acting ergodically on* G/C.

It may also be worthwhile to recall here that if G is a Lie group as in Corollary 4.3 then given any ergodic measure-preserving action of G on a probability space (X, μ) there exists a $g \in G$ whose action on (X, μ) is ergodic (see Gutschera [Gu]).

4.3. Invariant functions and ergodic decomposition. We first note the following (cf. [BM], Theorem 6.1).

Theorem 4.4. *Let G be a Lie group and $C \in \mathcal{F}(G)$. Let $\phi = \{g_t\}$ be a one-parameter subgroup of G. If the action of ϕ on G/C is not ergodic then there exists a nonconstant C^∞ function on G/C which is ϕ-invariant.*

Theorem 4.4 in particular implies the following.

Remark 4.3. Let G be a Lie group and Γ_1 and Γ_2 be two lattices commensurable with each other (see § 1.1 for definition). Let ϕ be a one-parameter subgroup of G. Then Theorem 4.4 implies in particular that the ϕ-action on G/Γ_1 is ergodic if and only if the ϕ-action on G/Γ_2 is ergodic.

Theorem 4.4 and Proposition 1.2 together show that ergodicity and topological transitivity are equivalent for group-induced flows; that is,

Corollary 4.4. *Let G, C and ϕ be as in Theorem 4.4. Then the flow induced by ϕ on G/C is ergodic if and only if it has a dense orbit.*

In the notation as in Theorem 4.4 if all the eigenvalues of $\operatorname{Ad} g_t$, $t \in \mathbb{R}$, are real then one can further conclude that there exists a proper closed subgroup C' of G containing C such that the factor of ϕ on G/C' is trivial (that is, all points are fixed). The conclusion does not hold in general; for instance if G is a compact simple Lie group, T is a maximal toral subgroup of G and ϕ is a dense one-parameter subgroup of T then the above conclusion does not hold for the flow induced by ϕ on G/T.

It may be recalled that any measure-preserving flow can be decomposed into ergodic components (see [KH], [Ma3]). For a group-induced flow on a homogeneous space there exists an ergodic decomposition into sets which are 'almost' homogeneous, as shown by the following result.

Theorem 4.5 (Starkov [St4]). *Let G be a connected Lie group and let $C \in \mathcal{F}(G)$. Let $\phi = \{g_t\}$ be a one-parameter subgroup of G and let M be the Mautner subgroup associated to the translation by g_1 (or any g_t, $t \neq 0$). Let R be the radical of G and let I be the smallest closed normal subgroup containing M such that $\phi RI/RI$ is contained in an Ad-compact subgroup of G/RI. Then G/C can be partitioned into sets of the form $\overline{\phi IgC}/C$, $g \in G$, each of which is a closed submanifold of G/C, finitely covered by a homogeneous space, and admits a smooth ergodic ϕ-invariant probability measure.*

It may be observed that the subgroup I as in Theorem 4.5 is contained in the Mautner-Moore subgroup, say M', corresponding to ϕ. Maximal subsets of the form $\overline{\phi M'gC}/C$, $g \in G$, also form a partition of G/C into submanifolds with smooth invariant probability measures, which may not all be ergodic. Actually, for almost all $g \in G$, $\overline{\phi M'gC}/C$ and $\overline{\phi IgC}/C$ coincide [St4].

The homogeneous covers of the orbit closures as in Theorem 4.5 can also be described in the following unified manner.

Theorem 4.6 (Starkov [St4]). *Let G be a simply connected Lie group and $C \in \mathscr{F}(G)$. Let $\phi = \{g_t\}$ be a one-parameter subgroup of G and I be the subgroup as in Theorem 4.5. Then there exist a Lie group G^* containing G as a closed normal subgroup such that G^*/G is a torus, a connected Lie subgroup H of G^* and a one-parameter subgroup ψ of H such that the following holds: for almost all $g \in G$ the homogeneous space $H/(H \cap C)$ finitely covers $\overline{\phi I g C}/C$ and the ϕ-action on $\overline{\phi I g C}/C$ is the image under the covering, of the action of $g^{-1}\psi g$ on $H/(H \cap C)$. Furthermore, $(H \cap C)^0$ is a normal subgroup of H.*

In particular, if ϕ acts ergodically on G/C then there exists a connected Lie group F, a lattice Γ in F, a one-parameter subgroup $\{f_t\}$ of F and a covering map $\theta : F/\Gamma \to G/C$ with finite fiber, such that $\theta(f_t x) = g_t\theta(x)$ for all $x \in F/\Gamma$ and $t \in \mathbb{R}$.

4.4. Actions of subgroups. Using Theorem 4.2 and certain arguments as in § 7 of [Mo2] one can deduce the following result on weak mixing of group-induced flows (see § 2.10 for definition of weak mixing for group actions).

Theorem 4.7. *Let G be a connected Lie group, $C \in \mathscr{F}(G)$ and π be the representation of G over $L^2(G/C)$, induced by the action of G on G/C. Let F be a subgroup of G and M'_F be the Mautner-Moore subgroup associated to F. If $\mathscr{V}$ is a finite-dimensional subspace of $L^2(G/C)$ invariant under $\pi(f)$ for all $f \in F$ then $\pi(x)(v) = v$, for all $x \in M'_F$ and $v \in \mathscr{V}$. In particular, if $M'_F C$ is dense in G then the F-action on G/C is weak mixing.*

Together with Remark 4.2 this yields the following generalisation of Corollary 4.2.

Corollary 4.5. *Let the notation be as in Theorem 4.7. Let F be a subgroup of G which is generated by its elements whose adjoint action on the Lie algebra of G has only positive real eigenvalues. Let M'_F be the Mautner-Moore subgroup associated to F. Then the F-action is ergodic if and only if $FM'_F C$ is dense in G.*

It would also be worthwhile to note here the following result on the structure of homogeneous spaces admitting an ergodic action by a subgroup satisfying the condition as in the above Corollary; it can be proved using the version of Borel's density theorem as in [D9] (see Witte [Wi1], § 4, for an idea of the proof).

Theorem 4.8. *Let G be a connected Lie group and $C \in \mathscr{F}(G)$. Suppose that there exists a subgroup F of G which is generated by its elements whose adjoint action on the Lie algebra of G has only positive real eigenvalues, acting ergodically on G/C. Then any closed connected subgroup of G normalised by C is normal in G. In particular, if $C \in \mathscr{F}_0(G)$ then C is a lattice in G.*

The natural analogue of Theorem 4.3 also holds for actions of subgroups F as above. We note in particular the following consequence of this.

Corollary 4.6. *Let G be a connected Lie group such that $[G, G] = G$ and let H be a closed subgroup of G such that HR/R is a totally noncompact subgroup (see § 3.2 for definition) of G/R, where R is the radical of G. Then for any $C \in \mathcal{F}(G)$ the H-action on G/C is ergodic; moreover, it is also weak mixing.*

4.5. Duality. Let G be a connected Lie group, Γ be a lattice in G and H be a closed subgroup of G. Consider the Γ-action on G/H. We recall that G/H admits a measure which is quasi-invariant under the G-action and in particular we get an action of Γ on G/H, as a group of nonsingular transformations (see § 1.1 for terminology). We note the following.

Proposition 4.1. *Let the notation be as above. Then the Γ-action on G/H is ergodic if and only if the H-action on G/Γ is ergodic.*

It is also interesting to note that ergodicity of either of the actions as above is equivalent to ergodicity of the G-action on $G/H \times G/\Gamma$, defined componentwise (see Zimmer [Zi2]); the cartesian product is endowed with the product measure, which can be seen to be quasi-invariant under the action of G. The reader is referred to Zimmer [Zi2] for some general results on the ergodicity of restrictions of ergodic actions of Lie groups to lattices in them, involving a general version of the above duality principle.

Corollary 4.6 and Proposition 4.1 together imply the following; other results of § 4 can also be similarly reinterpreted using the duality principle.

Corollary 4.7. *Let G be a connected Lie group such that $[G, G] = G$ and let H be a closed subgroup of G such that HR/R is totally noncompact in G/R, where R is the radical of G. The for any lattice Γ in G the Γ-action on G/H is ergodic.*

The following result involves restriction to Γ of certain G-actions which are not transitive as in the above corollary, but are essentially transitive (namely, they admit a G-orbit whose complement has zero measure); the result can be deduced from Corollary 3.1.

Corollary 4.8. *Let Γ be a lattice in $SL(n, \mathbb{R})$. Then the action of Γ on $\mathbb{R}^n$ is ergodic. More generally, if $W = \mathbb{R}^n \times \cdots \times \mathbb{R}^n$, a cartesian product of p copies of $\mathbb{R}^n$, where $1 \leq p \leq (n-1)$, then the componentwise action of Γ on W is ergodic with respect to the Lebesgue measure on $W = \mathbb{R}^{pn}$.*

Similarly, it can be seen that if Γ is a lattice in $Sp(2n, \mathbb{R})$, the group of $2n \times 2n$ symplectic matrices, and $W = \mathbb{R}^{2n} \times \cdots \times \mathbb{R}^{2n}$, the cartesian product of n copies of $\mathbb{R}^{2n}$, then the Γ-action on W is ergodic with respect to the Lebesgue measure.

4.6. Spectrum and mixing of group-induced flows. For an ergodic group-induced one-parameter flow the spectrum of the associated unitary one-parameter group is described by the following (see [BM], Theorem 6.2).

Theorem 4.9 (Brezin and Moore [BM]). *Let G be a connected Lie group and $C \in \mathscr{F}(G)$. Let $\{g_t\}$ be a one-parameter subgroup of G acting ergodically on G/C. Let $\{U_t\}$ be the one-parameter group of unitary operators on $L^2(G/C)$ corresponding to the flow induced by $\{g_t\}$ on G/C. Let L be the closed subgroup of G containing C, such that G/L is the maximal Euclidean factor of G/C. Then the infinitesimal generator of $\{U_t\}$ has discrete spectrum on $L^2(G/L)$ and Lebesgue spectrum of infinite multiplicity on the orthocomplement of $L^2(G/L)$, unless $L = G$.*

Theorems 4.2 and 4.9 yield the following criteria for mixing.

Corollary 4.9. *Let G be a connected Lie group and $C \in \mathscr{F}(G)$. Let $\{g_t\}$ be a one-parameter subgroup of G and let Φ be the flow induced by $\{g_t\}$ on G/C. Suppose that Φ is ergodic. Let M' be the Mautner-Moore subgroup associated to $\{g_t\}$. Let S and L be the smallest closed normal subgroups of G such that G/S is solvable and G/L is Euclidean, respectively. Then we have the following:*

i) if Φ is weak mixing then $M'C$, SC, and LC are dense in G and

ii) if any of $M'C$, SC or LC is dense in G then Φ is mixing.

In particular, Φ is mixing whenever it is weak mixing.

The theorem, together with Corollary 4.6 implies in particular that if G is a connected Lie group such that $[G, G] = G$ and ϕ is a one-parameter subgroup such that $\phi R/R$ is totally noncompact, where R is the radical of G, then for any $C \in \mathscr{F}(G)$ the flow induced by ϕ on G/C is mixing.

We next describe some results on the 'rate of mixing' of some group-induced one-parameter flows.

Theorem 4.10 (Kleinbock and Margulis [KM1]). *Let G be a connected semi-simple Lie group with finite center, Γ be a lattice in G and m be the normalised Haar measure on G/Γ. Let G/Γ be equipped with the metric induced by a Riemannian metric on G which is invariant under the G-action by translations on the right and the action of a maximal compact subgroup by translations on the left. Let $\{g_t\}$ be a one-parameter subgroup of G. Suppose that for any simple normal closed subgroup G' of G the restriction of any $Ad\, g_t$, $t \neq 0$, to the Lie subalgebra of G' has eigenvalues of absolute value other than 1. Let f and φ be two Hölder continuous square-integrable functions on G/Γ and suppose that $\int f\, \mathrm{d}m = 0$. Then there exist positive constants a and b such that for all $t \geq 0$*

$$\int f(g_t x)\varphi(x)\mathrm{d}m(x) \leq ae^{-bt}.$$

The results in [KM1] contain further details about the constants a and b, concerning their dependence on the functions f and φ, the one-parameter subgroup $\{g_t\}$ etc. There are also similar, but conditional, results for one-parameter subgroups not satisfying the condition in Theorem 4.10, on the eigenvalues of $\operatorname{Ad} g_t$, $t \neq 0$.

Similar results were proved earlier in Moore [Mo4] and Ratner [R5] for $G = SL(2, \mathbb{R})$ and in Pollicott [Pol] for $SL(2, \mathbb{C})$. Further, in [Mo4] and [R5] the decay of correlation functions for flows induced by unipotent one-parameter subgroups (corresponding to the horocycle flows) are also considered.

Theorem 4.11 (Moore [Mo4], Ratner [R5]). *Let $G = SL(2, \mathbb{R})$, Γ be a lattice in G and m be the normalised Haar measure on G/Γ. Let $\{a_t\}$ be the one-parameter group of diagonal matrices in G, $\{u_t\}$ be the one-parameter group of upper triangular unipotent matrices and $K = \{k_t\}$ be a compact one-parameter subgroup of G. Let f and φ be two square-integrable functions which are Hölder continuous along the orbits of K (in the variable t). Suppose also that $\int f \mathrm{d}m = 0$. Then there exist positive constants a, b, c, d such that for all $t \geq 1$,*

$$\int f(a_t x)\varphi(x)\mathrm{d}m(x) \leq a e^{-bt}$$

and

$$\int f(u_t x)\varphi(x)\mathrm{d}m(x) \leq c t^{-d}.$$

The results in the above mentioned papers ([KM1], [Mo4], [R5] and [Pol]) extend also to correlation functions corresponding to unitary representations other than those on the spaces $L^2(G/\Gamma)$, involved in the above discussion. There are also other results on exponential decay of the correlation functions, for analytic functions on the homogeneous spaces, the references to which may be found in the above mentioned papers.

4.7. Mixing of higher orders. In this subsection we recall results on higher order mixing (see § 2.10 for definition) of group-induced flows. We begin by recalling the following general result of S. Mozes.

Theorem 4.12 (Mozes [Moz1]). *Let G be a connected Lie group with finite center, such that the adjoint representation is a proper homomorphism. Consider a weakly measurable measure-preserving action of G on a Lebesgue probability space (X, μ). If the action is mixing then it is mixing of all orders.*

Together with Corollary 3.2 this implies the following.

Corollary 4.10. *Let G be a noncompact connected semisimple Lie group with finite center and Γ be an irreducible lattice in G. Then the G-action on G/Γ is mixing of all orders. In particular for any one-parameter subgroup $\{g_t\}$ which is not contained in a compact subgroup of G, the flow induced by $\{g_t\}$ on G/Γ is mixing of all orders.*

Recall that for group-induced flows weak mixing implies mixing (see Corollary 4.9). The following result strengthens the assertion further, to higher order mixing; its proof is based on Ratner's classification of invariant measures of Ad-unipotent flows (see § 6.1).

Theorem 4.13 (Starkov [St7]). *Let G be a connected Lie group and $C \in \mathscr{F}(G)$. Let $\{g_t\}$ be a one-parameter subgroup of G. If the flow induced by $\{g_t\}$ on G/Γ is mixing then it is mixing of all orders.*

The theorem shows in particular that the horocycle flow associated to a surface of constant negative curvature and finite area is mixing of all orders; this result (and also a generalisation) was proved earlier by B. Marcus, who had conjectured the statement as in Theorem 4.13 (cf. Marcus [Marc1] and Starkov [St7] for details).

4.8. Entropy. It may be recalled that a smooth dynamical system on a compact manifold has finite entropy (see [AA], § 12, and [KH] §§3.2 and 4.5). For an affine automorphism this is true even if the homogeneous space is noncompact, but has finite Haar measure. More specifically, the following is known about the value of the entropy.

Theorem 4.14 (cf. Bowen [Bo2], [Co] and [D5], Appendix). *Let G be a connected Lie group and $C \in \mathscr{F}(G)$. Let $\tau = T_g \circ \overline{\alpha}$ be an affine automorphism of G/C, where $g \in G$ and α is an automorphism of G such that $\alpha(C) = C$. Let τ_* be the automorphism of the Lie algebra of G, defined by $\tau_* = \mathrm{Ad}\, g \circ d\alpha$. Let $\lambda_1, \lambda_2, \ldots, \lambda_k$ be the distinct eigenvalues (possibly complex) of τ_*, with absolute value greater than 1 and let $m_1, m_2, \ldots, m_k$ be their respective multiplicities. Let $h(\tau)$ be the entropy of τ with respect to the normalised Haar measure and $h_{\mathrm{top}}(\tau)$ be the topological entropy. Then we have the following:*

i) $h(\tau) \leq \Sigma_{i=1}^{k} m_i \log |\lambda_i|$.

ii) If G/C is compact then $h(\tau) = h_{\mathrm{top}}(\tau) \leq \Sigma_{i=1}^{k} m_i \log |\lambda_i|$.

iii) If C is a uniform lattice in G then $h(\tau) = h_{\mathrm{top}}(\tau) = \Sigma_{i=1}^{k} m_i \log |\lambda_i|$.

Theorem 4.14 shows in particular that if all the eigenvalues of τ_* are of absolute value 1 then the entropy of τ is zero. Hence we see that there exist translations and one-parameter flows with zero entropy which are mixing of all orders (see Corollary 3.2 and Theorem 4.13). Thus, for flows on homogeneous spaces of finite Haar measure, though weak mixing implies mixing and also mixing of all orders, it does not imply K-mixing.

Let G be a Lie group and Γ be a uniform lattice in G. Let $g \in G$ be such that all eigenvalues of $\mathrm{Ad}\, g$ are of absolute value 1. Then for the translation T_g of G/Γ by g, the sequential entropy along the sequence $\{2^k\}$ is finite (see [D4]). It is proved to be positive for a class of translations by Ad-unipotent elements, when G is semisimple [D4]; it may be expected to be positive whenever the unipotent Jordan component of $\mathrm{Ad}\, g$ is nontrivial.

4.9. K-mixing, Bernoullicity. The following result shows that for affine automorphisms as in our discussion the maximal homogeneous factor of zero entropy is also the Pinsker factor, namely the maximal metric factor with zero entropy (see [Wal5]).

Theorem 4.15 (cf. [Co] and [D1]). *Let G be a connected Lie group and $C \in \mathscr{F}(G)$. Let τ be an affine automorphism of G/C. Let M_τ be the Mautner subgroup associated to τ. The partition of G/C into $\{x\overline{M_\tau C}/C\}_{x \in G}$ is the Pinsker partition for τ. In particular, τ is a K-automorphism if and only if $M_\tau C$ is dense in G.*

Remark 4.4. The following analogue of the Mautner phenomenon holds for K-mixing: Let G be a Lie group and $\sigma = T_g \circ \alpha$ be an affine automorphism of G, where $g \in G$ and α is an automorphisms of G. Let D_σ be the automorphism of G defined as in § 4.1, corresponding to σ. Consider a measurable measure-preserving action of G on a probability space (X, μ). If T is a measure-preserving transformation of X such that $T(ax) = D_\sigma(a)T(x)$ for almost all $x \in X$ and all $a \in G$ then the Pinsker partition of T is coarser than the ergodic decomposition under the action of M_σ on X, where M_σ is the Mautner subgroup associated to σ; in particular, if the M_σ-action is ergodic then τ is a K-automorphism (cf. [D1] for details).

It may be recalled that a (measurable) flow is K-mixing whenever the action of every nontrivial element is a K-automorphism (see [CFS], Ch. VIII). A notion generalising the K-property in the cyclic and one-parameter group cases was introduced in [D3] for actions of groups of the form $\mathbb{R}^a \times \mathbb{Z}^b$, ($a$ and b being nonnegative integers) and it was shown to hold for actions of certain subgroups of semisimple Lie groups, on homogeneous spaces of the latter, with finite Haar measure. We shall however not go into the details here.

Theorem 4.16 (cf. [D2]). *Let G be a connected Lie group and Γ be a lattice in G. Let $\tau = T_g \circ \overline{\alpha}$ be an affine automorphism of G/Γ, where $g \in G$ and α is an automorphism of G such that $\alpha(\Gamma) = \Gamma$. Let $\mathscr{G}$ be the Lie algebra of G and τ_* be the automorphism of $\mathscr{G}$ defined by $\tau_* = \mathrm{Ad}\, g \circ d\alpha$. Let $\mathscr{H}$ be the largest τ_*-invariant subspace of $\mathscr{G}$ on which all eigenvalues of τ_* are of absolute value 1. Suppose that τ is a K-automorphism of G/Γ and that the restriction of τ_* to $\mathscr{H}$ is a semisimple linear transformation. Then τ is isomorphic to a Bernoulli shift.*

The condition of semisimplicity of the restriction of τ_* as in the theorem is perhaps not necessary, but this is not established. Thus there is no known example of a translation flow with K-property which is not Bernoullian. However, Furman and Weiss have exhibited (see [FW]) a class of skew product flows involving translations of homogeneous spaces of $SL(2, \mathbb{R})$ and $SO(n, 1)$, $n \geq 3$, which have the K-property but are not Bernoullian.

The case of the theorem with $G = SL(2, \mathbb{R})$ and $\tau = T_g$, where g is a diagonal matrix with positive entries other than 1, means that the geodesic flow associated to surfaces of constant negative curvature and finite area is isomorphic to a Bernoulli flow; this result was first proved, for compact surfaces, by D. Ornstein and B. Weiss [OW]. Theorem 4.16 also implies the corresponding statement for manifolds of all dimensions, when applied to the special orthogonal groups $G = SO(n, 1)$, $n \geq 2$.

§ 5. Group-induced flows with hyperbolic structure

Hyperbolic systems constitute an important class of dynamical systems for which a rich theory is available, dealing with various aspects of their behaviour (cf. Ch. 8 and [Ma3]). Systems on homogeneous spaces satisfying the hyperbolicity conditions have often played an important role, both by serving as useful models and also sometimes in classification of systems satisfying certain conditions. This section is devoted to a discussion of such systems, concerning mainly construction and identification of such systems; their dynamical properties are discussed in the respective sections.

5.1. Anosov automorphisms. Hyperbolic automorphisms of $\mathbb{T}^n$, $n \geq 2$, (see § 2.6) are the simplest examples of Anosov diffeomorphisms. More generally one may look for affine automorphisms of homogeneous spaces with the Anosov property. It turns out that for a general homogeneous space G/C existence of such an affine automorphism is a rather restrictive condition. A simple Lie group-theoretic argument shows the following.

Proposition 5.1. *Let G be a connected Lie group and C be a closed subgroup such that G/C is a locally faithful homogeneous space (see § 1.1) with finite Haar measure. If G/C admits an affine automorphism which is an Anosov diffeomorphism then G is a nilpotent Lie group.*

If θ is an automorphism of a nilmanifold N/Γ, where N is a nilpotent Lie group and Γ is a lattice in N, then it can be seen that $T_g \circ \theta$ is an Anosov diffeomorphism for $g \in N$ if and only if θ is an Anosov diffeomorphism. Also, for any $g \in N$, $T_g \circ \theta$ has a fixed point and hence (by Remark 1.2) it is affinely isomorphic to θ. We need therefore describe only automorphisms with the Anosov property; we shall call these Anosov automorphisms. Note that for N and Γ as above and an automorphism α of N leaving Γ invariant, $\overline{\alpha}$ is an Anosov automorphism of N/Γ if and only if α is hyperbolic, namely all eigenvalues of the differential $d\alpha$ are of absolute value different from 1.

Not all compact nilmanifolds admit Anosov automorphisms. For instance if the center of the nilpotent Lie group N is one-dimensional (e.g. if N is the Heisenberg group) then N does not admit hyperbolic automorphisms. There are also other conditions which preclude existence of Anosov automorphisms on nilmanifolds (see Auslander and Scheuneman [AS]). The first non-toral example of an Anosov automorphism of a compact nilmanifold was described by S. Smale (see [Sm], § I.3; the example is attributed to A. Borel). L. Auslander and J. Scheuneman [AS] gave another class of examples, which includes in particular the following.

Example 5.1. Let $N = V \oplus \wedge^2 V$ be a Lie group as in Example 4 in § 1.2, with $V = \mathbb{R}^n$ and Γ be the lattice $\{(p, \frac{1}{2}q) \mid p \in \mathbb{Z}^n, q \in \wedge^2\mathbb{Z}^n\}$. Let $A \in GL(n, \mathbb{Z})$ be a matrix with no eigenvalues of absolute value 1 and α be the corresponding automorphism of N, as introduced in the example. Let W_1 be the subspace of $\wedge^2 V$ spanned by the generalised eigenspaces of $\wedge^2 A$ corresponding to all the eigenvalues of absolute value 1 (equivalently the subspace spanned by the vectors of the form $v_1 \wedge v_2$, with $v_1 \in V_\lambda$ and $v_2 \in V_{\lambda^{-1}}$ for some $\lambda > 0$ where, for $\mu = \lambda$ or λ^{-1}, V_μ denotes the sum of all generalised eigenspaces in V corresponding eigenvalues of absolute value μ). Let W be the smallest subspace of $\wedge^2 V$ such that $W_1 \subseteq W$ and $W \cap \wedge^2\mathbb{Z}^n$ is a lattice in W. Then W is $\wedge^2 A$-invariant. Also W is a normal subgroup of N and ΓW is a closed subgroup. Let $N' = N/W$, $\Gamma' = \Gamma W/W$ and α' be the factor of α on N'. Then it is easy to see that $\alpha'(\Gamma') = \Gamma'$ and the automorphism $\overline{\alpha}'$ of N'/Γ' induced by α' is an Anosov automorphism.

A somewhat more general construction of Anosov automorphisms is described in [D6], via the following result.

Theorem 5.1. *Let N be a simply connected nilpotent Lie group and let Γ be a lattice in N. Let G be the group of all Lie automorphisms of N, viewed canonically as a subgroup of $GL(\mathcal{N})$, where $\mathcal{N}$ is the Lie algebra of N. Let G_0 be the subgroup of G generated by all unipotent linear transformations of $\mathcal{N}$ contained in G. If G_0 contains a hyperbolic element then there exists a hyperbolic automorphism α of N such that $\alpha(\Gamma) = \Gamma$, inducing an Anosov automorphism of N/Γ.*

A larger class of Anosov diffeomorphisms can be obtained by considering factors of Anosov automorphisms. Let N be a simply connected nilpotent Lie group and Γ be a (uniform) lattice in N. Suppose that there exists a hyperbolic automorphism α of N such that $\alpha(\Gamma) = \Gamma$, so we have an Anosov automorphism of N/Γ. Suppose also that there exists a finite group F of affine automorphisms of N with the following properties: i) F normalises Γ, when the latter is viewed as the group of right translations of N, ii) if τ is a nontrivial element of the subgroup generated by F and Γ then τ does not fix any element of N and iii) α normalises F (in the group of affine automorphisms). Now if Δ is the subgroup generated by F and Γ then the quotient space $M = \Delta\backslash N$ under the action of Δ (which is the same as the quotient space $F\backslash(N/\Gamma)$) is a smooth manifold finitely covered by N/Γ. Furthermore, in view of the last condition α factors to a diffeomorphism of

M, which is evidently an Anosov diffeomorphism. A quotient space as above is called an *infra-nilmanifold* and an Anosov diffeomorphism arising as above is called an infra-nilmanifold Anosov automorphism. It is conjectured that any Anosov diffeomorphism of a compact manifold is topologically conjugate to an infra-nilmanifold Anosov automorphism. The conjecture is known to hold under the assumption that the underlying manifold is an infra-nilmanifold (see Manning [Man1]); in particular, any Anosov diffeomorphism of a torus is topologically conjugate to a hyperbolic automorphism of the torus (see [Man1]). It is also known that if M is a compact manifold and φ is an Anosov diffeomorphism such that either the expanding or the contracting foliation is one-dimensional then M is a torus and φ is topologically conjugate to a hyperbolic toral automorphism (see Franks [Fr] and Newhouse [Ne]). It may also be noted here that a statement analogous to the above mentioned conjecture for Anosov diffeomorphisms, holds in the case of expanding maps (see Franks [Fr], Gromov [Gr]).

In the special case when N is taken to be $\mathbb{R}^n$, the corresponding infra-nilmanifolds are flat manifolds. A characterisation of flat manifolds admitting Anosov automorphisms may be found in Porteous [Por].

5.2. Affine automorphisms with a hyperbolic fixed point. In the theory of smooth dynamical systems Anosov diffeomorphisms have been generalised to axiom A diffeomorphisms (see Ch.7). It turns out that in the case of homogeneous and infra-homogeneous spaces the latter arise only in a limited way. In this respect the following class of systems have been analysed in [D8]. Let G be a connected Lie group, H and K be closed subgroups of G such that K is compact and G/H is either compact or has finite Haar measure. Let α be an automorphism of G such that $\alpha(H) = H$ and $\alpha(K) = K$ and let $\overline{\alpha}$ be the homeomorphism of $K\backslash G/H$ defined by $\overline{\alpha}(KgH) = K\alpha(g)H$ for all $g \in G$. Suppose that the identity double coset is a hyperbolic fixed point of $\overline{\alpha}$; this is meaningful even if $K\backslash G/H$ is not a manifold, since in the smooth case the condition corresponds to the condition that the largest $d\alpha$-invariant subspace of the Lie algebra of G on which all eigenvalues of $d\alpha$ are of absolute value 1 be contained in the sum, as vector subspaces, of the Lie subalgebras associated to K and H. It is shown that if G/H has finite Haar measure then up to a natural 'affine' equivalence the system is an infra-nilmanifold Anosov automorphism. When G/H is compact (but not necessarily with a finite Haar measure) then it has a factor with zero topological entropy such that the fiber over the fixed point at the identity double coset in the factor, is an infra-nilmanifold Anosov automorphism (see [D8] for details).

We note that if $G = GL(n, \mathbb{R})$, H is the isotropy subgroup of a point under the action of G on the $(n-1)$-dimensional projective space, K is the trivial subgroup, and α is the automorphism given by inner conjugation by a diagonal matrix g in G whose diagonal entries have distinct absolute values, then it can be seen, by realising $\overline{\alpha}$ as a projective transformation, that the nonwandering set of $\overline{\alpha}$ as above is finite and consists of hyperbolic fixed points. It is also possible to construct similar automorphisms on certain homogeneous spaces of more general connected Lie groups (not admitting a finite Haar measure); see [D8] for details. These examples are at the other extreme than the Anosov automorphisms, in the classification recalled above.

5.3. Anosov flows. The geodesic flows associated to manifolds of constant negative curvature, which can be realised as actions of one-parameter subgroups on infra-homogeneous spaces (double coset spaces) of Lie groups (see Example 6 in § 1.2) are the prototypes for the class of Anosov flows (cf. Ch. 8 and [An]).

A general construction of Anosov flows on infra-homogeneous spaces can be formulated as follows. Let G be a Lie group and let Γ be a lattice in G. Suppose that there exists a one-parameter subgroup $A = \{a_t\}$ of G such that for $t \neq 0$, $\mathrm{Ad}\, a_t$ has no eigenvalue λ other than 1 with $|\lambda| = 1$ and the connected component of the centraliser of A in G has the form AK_0, where K_0 is a compact subgroup of G. Let K be a compact subgroup of G containing K_0, and normalised by a_t for all $t \in \mathbb{R}$. Now let $M = K\backslash G/\Gamma$ and $\{\phi_t\}$ be the flow on M defined by $\phi_t(Kg\Gamma) = Ka_t g\Gamma$ for all $g \in G$ and $t \in \mathbb{R}$. Then $\{\phi_t\}$ can be verified to be an Anosov flow (see Tomter [To]). An Anosov flow arising in this way, which we shall call a group-induced Anosov flow, has a finite invariant measure with full support (positive for all nonempty open subsets) and a dense set of periodic points. It is conjectured that any Anosov flow is topologically conjugate to a group-induced Anosov flow (see [To]). The reader is referred to [R1] for some work in this direction.

Given a Lie group G and a lattice Γ in G there need not exist a one-parameter subgroup A satisfying the conditions as above. In particular, the conditions force that if R is the radical of G then the semisimple group G/R has $\mathbb{R}$-rank at most 1.

The Anosov flows as above fall broadly in two classes as seen from the following results of Tomter [To].

Theorem 5.2. *Let G be a connected solvable Lie group and Γ be a lattice in G. Let A be a one-parameter subgroup of G inducing an Anosov flow on an infra-homogeneous space $K\backslash G/\Gamma$ as above. Then the Anosov flow is a suspension of an Anosov automorphism on a nilmanifold. The flow is ergodic but not weak mixing.*

Suspensions of Anosov automorphisms on infra-nilmanifolds (see § 5.1) can be realised in the above general form, with G a Lie group which is not necessarily connected, but has finitely many connected components.

If G is a semisimple Lie group of $\mathbb{R}$-rank 1 then any nontrivial one-parameter subgroup whose adjoint action is diagonalisable over $\mathbb{R}$ satisfies the condition on the one-parameter subgroup A as in the above construction. The Anosov flows arising from these turn out, up to finite coverings, to be precisely the geodesic flows on the unit tangent bundles of the irreducible Riemannian locally symmetric spaces of noncompact type and rank 1.

Theorem 5.3 (Tomter [To]). *Let G be a connected Lie group, Γ be a lattice in G and A be a one-parameter subgroup of G inducing an Anosov flow on an infra-homogeneous space $K\backslash G/\Gamma$ as above. Suppose that G is not solvable. Let R be the radical of G. Then R is nilpotent and G/R is of $\mathbb{R}$-rank 1. Furthermore, A is contained in a semisimple Levi subgroup of G and its adjoint action is diagonalisable over $\mathbb{R}$. The induced Anosov flow has Lebesgue spectrum on the orthocomplement of constant functions. Moreover, it is isomorphic to a Bernoulli flow.*

In recent years Anosov actions of higher dimensional groups have also been studied and considerable work is done concerning such actions on homogeneous spaces, especially with regard to their rigidity. The reader is referred to [KS1], [Hu], [Qi] and other references in these papers, for pursuing the theme. We shall not go into the details here.

§ 6. Invariant measures of group-induced flows

In this section we discuss invariant measures of translation actions on homogeneous spaces, by subgroups of the ambient group. There is a parallel between the case of affine automorphisms of nilmanifolds seen earlier (in § 2) and the case at hand. The condition of the adjoint transformations associated to the translating elements being unipotent ensures a considerable degree of 'regularity' in the dynamical behaviour.

6.1. Invariant measures of Ad-unipotent flows. A remarkable general result on classification of invariant measures of certain flows on homogeneous spaces was proved in 1990 by Marina Ratner. Theorem 6.1 below is a strengthened version, proved recently by Nimish Shah.

We recall that an element u (respectively a subgroup U) of a Lie group G is said to be *Ad-unipotent* if $\operatorname{Ad} u$ is unipotent (resp. $\operatorname{Ad} u$ is unipotent for all $u \in U$), as a linear transformation of the Lie algebra of G. A group-induced flow arising from an Ad-unipotent subgroup will be called an Ad-unipotent flow. The reader is cautioned that if G is a Lie group and H is a Lie subgroup of G then for $h \in H$ the condition of h being Ad-unipotent in H may not coincide with h being Ad-unipotent in G; we shall however consistently mean Ad-unipotent as an element of the ambient group G.

If G is a Lie group and W and C are closed subgroups of G, in general it is possible to have proper closed subgroups H containing W such that HC/C admits a finite H-invariant measure and, furthermore, it can be ergodic as a W-invariant measure. For example consider the case of $G = SL(n, \mathbb{R})$

and $\Gamma = SL(n, \mathbb{Z})$. Let N be the subgroup of G consisting of all upper triangular unipotent matrices. Then $N \cap \Gamma$ is a lattice in N and $N/(N \cap \Gamma)$ is a compact nilmanifold. In N we have one-parameter subgroups ϕ whose actions on $N/(N \cap \Gamma)$ are ergodic (see Theorem 2.7). Any such subgroup also acts ergodically on G/Γ (see Corollary 3.1). Furthermore, for $n \geq 3$ there also exist other closed subgroups H, containing N as a proper subgroup, for which $H\Gamma/\Gamma$ admits a finite H-invariant measure and the action of ϕ on $H\Gamma/\Gamma$ is ergodic. A classification of ϕ-invariant measures on G/Γ has to take into account all such measures.

Given a Lie group G and a closed subgroup C of G, we call a measure μ on G/C *homogeneous* if there exist a closed subgroup H of G and an element $x \in G/C$ such that μ is H-invariant and $\operatorname{supp} \mu = Hx$; it may be noted that such a measure is uniquely determined up to a scalar multiple, by the subgroup H and the orbit Hx.

Given a Lie group G with Lie algebra $\mathscr{G}$ and subgroups H and L such that $H \subseteq L$, we shall say that $\operatorname{Ad} H$ is *Zariski-dense* in $\operatorname{Ad} L$ if every real algebraic subgroup of $GL(\mathscr{G})$ containing $\operatorname{Ad} H$ contains $\operatorname{Ad} L$; by a real algebraic subgroup of $GL(\mathscr{G})$ we mean the group of real points of a complex algebraic subgroup of $GL(\mathscr{G} \otimes \mathbb{C})$ defined over $\mathbb{R}$; for subgroups contained in the special linear group $SL(\mathscr{G})$ this coincides with the notion as in Example 7 in § 1.2, when $SL(\mathscr{G})$ is realised as $SL(n, \mathbb{R})$ via a basis of $\mathscr{G}$, n being the dimension of G.

Theorem 6.1 (Ratner [R8], Shah [Sh5]). *Let G be a Lie group and C be a closed subgroup of G. Let W be a closed subgroup of G and U be the set of all Ad-unipotent elements in W. Suppose that the subgroup generated by $\{Ad\, u \mid u \in U\}$ is Zariski-dense in $Ad\, W$. Then any finite ergodic W-invariant measure on G/C is homogeneous.*

The theorem was proved by Ratner [R8] under the additional conditions that W^0 be generated by the Ad-unipotent elements (of G) contained in it, W/W^0 be finitely generated and each coset of W^0 in W contain an Ad-unipotent element (the middle condition is not needed for Ad-unipotent W); this includes in particular the cases of Theorem 6.1 when W is connected or cyclic. A reader interested in a sketch of her proof is referred to [R11]. The general theorem as above is deduced from her results (see [Sh5]). It may be emphasized that G/C is not assumed to have finite Haar measure; on the other hand the conclusion in the theorem is only for finite invariant measures. Another proof of the theorem for the case of algebraic groups, involving the same overall theme but differing in various details, may be found in Margulis and Tomanov [MT1]. A proof for the case of horocycle flows, namely when $G = SL(2, \mathbb{R})$, is presented by Ratner in [R10] (see also [Gh]).

It may also be mentioned here that similar results on classification of invariant measures have been obtained recently for flows on homogeneous spaces of algebraic groups over local fields, and also products of such groups (see

Ratner [R13], [R14] and Margulis and Tomanov [MT1], [MT2]). We shall not go into the details.

For the case when $G = SL(2, \mathbb{R})$ and W is the one-parameter group of upper triangular unipotent matrices Theorem 6.1 was proved by Furstenberg for uniform lattices [Fu] and the present author for all lattices (see [D10]). Various authors including Veech [V1], Bowen [Bo7], Ellis and Perrizo [EP] and the present author [D10] had proved further results in the direction of the theorem, for actions of horospherical subgroups (see [D21], [D22], [Marg3] and [R12] for details). Various partial results were also proved by M. Ratner, before the general result mentioned above, including the cases of Theorem 6.1 when G is a cartesian product of copies of $SL(2, \mathbb{R})$, C is a product of lattices in the copies and W is a unipotent one-parameter subgroup (cf. [R4]), or G is solvable, C is discrete and W is connected (cf. [R6]) or G is semisimple, C is a uniform lattice and W is either cyclic or a one-parameter subgroup (cf. [R7]), and a general result on invariant measures of horospherical subgroups (see [R6], Theorem 4).

Remark 6.1. The case of Theorem 6.1 for connected subgroups W, which was proved by Ratner some years ago, has been fundamental to many subsequent developments in ergodic theory and dynamics of group-induced flows and various areas of applications (the general version is very recent). In particular the result is involved in the proofs of several theorems in the present and the next sections, and through them in various applications discussed in § 9. The reader is also referred to [D22], [R12] and [R14] for certain applications of Ratner's work that are beyond the scope of this chapter.

Remark 6.2. In view of ergodic decomposition (see [KH], [Ma3]) to classify all finite invariant measures it suffices to describe all ergodic invariant measures. Theorem 6.1 reduces the task to determining all closed subgroups, containing the given subgroup W as in the theorem, and their orbits with finite invariant measures which are ergodic for the action of W. This may be dealt with using some arithmetical information about W. Since supports of these measures are orbit closures of W, information about the latter may also be used. Various results on orbit closures in special situations are described in §§ 7.2 and 7.3 below.

It may be observed, following the notation as in Theorem 6.1, that if W acts ergodically on G/C and μ is a finite W-invariant measure such that $\mu(Hx) = 0$ for all proper closed subgroups H and $x \in G/C$ such that $W \subseteq H$ and Hx admits a finite H-invariant measure then, by ergodic decomposition, μ is G-invariant. We shall now see a variation of this, which serves as a useful characterisation of the Haar measure on G/C among the W-invariant measures, involving a single condition; (for convenience we shall restrict to lattices; in view of Theorem 4.8 this involves no loss of generality).

Let G be a connected Lie group and Γ be a lattice in G. We denote by $\mathscr{H}$ the class of all proper closed subgroups H of G such that $H \cap \Gamma$ is a lattice in H and $\operatorname{Ad}(H \cap \Gamma)$ is Zariski-dense in $\operatorname{Ad} H$. Then $\mathscr{H}$ is countable (see [DM3], Proposition 2.1). Furthermore, if H is a closed subgroup of G such that $H \cap \Gamma$ is a lattice in H and there exists an Ad-unipotent one-parameter subgroup U of G contained in H and acting ergodically on $H/(H \cap \Gamma)$ then $H \in \mathscr{H}$ (see [Sh1], Corollary 2.13). A direct proof of countability of the class of subgroups satisfying the latter condition was given earlier by Ratner [R8].

For any closed subgroups H and U of a Lie group G we denote by $X(H, U)$ the set $\{g \in G \mid Ug \subseteq gH\}$; then $X(H, U)$ is a real analytic subset of G and, furthermore, if G is a real algebraic group and U is connected then $X(H, U)$ is an algebraic subvariety of G.

Theorem 6.2 (cf. [DM3]). *Let G be a connected Lie group, Γ be a lattice in G and $\mathscr{H}$ be the class of subgroups of G as defined above. Let U be a closed connected subgroup of G generated by the Ad-unipotent one-parameter subgroups contained in it. Suppose that U acts ergodically on G/Γ (with respect to the Haar measure). Then a finite U-invariant measure on G/Γ is G-invariant if and only if $\mu(X(H, U)\Gamma/\Gamma) = 0$ for all $H \in \mathscr{H}$ or, equivalently, $\mu(\cup_{H\in\mathscr{H}} X(H, U)\Gamma/\Gamma) = 0$.*

In certain contexts, apart from knowing the individual invariant measures, it is also of interest to know about their totality, and especially the topology on the collection of the invariant measures (see § 1.1 for the definition of the topology). We now recall some results in this respect.

Theorem 6.3 (Mozes and Shah [MS]). *Let G be a Lie group and Γ be a discrete subgroup of G. Let $\mathscr{M}$ be the set of all probability measures μ on G/Γ for which there exists an Ad-unipotent one-parameter subgroup $U = \{u_t\}$ of G such that μ is invariant and ergodic with respect to the flow induced by U. Then $\mathscr{M}$ is a closed subset in the space of probability measures on G/Γ.*

It may be noted that in view of Theorem 6.1 all the probability measures in $\mathscr{M}$ as above are homogeneous measures; not all homogeneous measures may be in $\mathscr{M}$ however. We note also that the homogeneous measure on a closed orbit Hx of a connected subgroup H belongs to $\mathscr{M}$ (namely it is ergodic for an Ad-unipotent one-parameter subgroup) if H is generated by the Ad-unipotent one-parameter subgroups contained in it (see Corollary 4.3).

6.2. Invariant measures and epimorphic subgroups. In this subsection we discuss some results on invariant measures for actions of certain subgroups which are not necessarily generated by Ad-unipotent elements.

Theorem 6.4 (cf. Mozes [Moz2]). *Let G be a Lie group and F and H be closed connected subgroups of G such that $F \subseteq H$ and the following conditions are satisfied:*

i) every connected Lie subgroup U of G which is normalised by F and has a Haar measure invariant under all automorphisms induced by conjugation by elements of F, is normalised by H;

ii) F is generated by its elements whose adjoint actions on the Lie algebra of G have only real eigenvalues;

iii) there is no proper closed normal subgroup of H containing all Ad-unipotent one-parameter subgroups of F.

Then for any discrete subgroup Γ of G every finite F-invariant measure on G/Γ is H-invariant.

A subgroup F of a real algebraic group H is said to be *epimorphic* in H if for every algebraic representation of H over a vector space V, every vector in V which is fixed by F is also fixed by H; given subgroups F and H of a real algebraic group L such that $F \subseteq H$, we say that F is epimorphic in H if F is epimorphic in the Zariski-closure of H in L. It was proved by Mozes [Moz2] using Theorem 6.4 that if F and H are two real algebraic subgroups of a real algebraic group G, with H generated by unipotent elements and F contained and epimorphic in H, then for any discrete subgroup Γ of G any finite F-invariant measure on G/Γ is H-invariant. Shah and Weiss have now proved the following result which in particular implies the theorem of Mozes (see [ShW]).

Theorem 6.5 (Shah and Weiss [ShW]). *Let G be a Lie group and Γ be a lattice in G. Let H be a subgroup of G, which is generated by Ad-unipotent one-parameter subgroups. Let F be a connected Lie subgroup of H such that $\mathrm{Ad}\, F$ is epimorphic in $\mathrm{Ad}\, H$. Then any finite F-invariant measure on G/Γ is $F[H, H]$-invariant. Furthermore, if H intersects the center of G in a discrete subgroup then any finite F-invariant measure on G/Γ is H-invariant; in particular, if it is also ergodic then it is homogeneous.*

We recall here that if $G = SL(n, \mathbb{R})$ and P is the subgroup consisting of all elements leaving invariant a given flag, then P^0 is epimorphic in G; hence by Theorem 6.5, for any lattice Γ in G, the normalised Haar measure is the only P^0-invariant probability measure on G/Γ. In particular this applies to the subgroup consisting of all upper triangular matrices in G, with positive diagonal entries. More generally, if G is a semisimple Lie group with no nontrivial compact factors, $\phi = \{g_t\}$ is a totally noncompact (see § 3.2 for definition) one-parameter subgroup and U is the horospherical subgroup (see § 4.1 for definition) associated to g_1 then (U is normalised by ϕ and) ϕU is an epimorphic subgroup of G (cf. Shah [Sh4]); in particular the parabolic subgroups of G and their connected components of identity are epimorphic subgroups of G.

By a result of Bien and Borel any simple noncompact real algebraic group G contains a 3-dimensional solvable subgroup H which is epimorphic in G;

by Theorem 6.5 the H-action on G/Γ is uniquely ergodic (has a unique invariant probability measure) for any lattice Γ in G (see Mozes [Moz2]).

6.3. Invariant measures of actions of diagonalisable groups. Relatively little is known about the invariant measures (as also about orbit closures) for actions of diagonalisable subgroups, though the problem is of considerable interest. Some of the results that may be mentioned in this connection concern the group-induced Anosov flows as in § 5.3. In this respect first we recall a result on equidistribution of periodic orbits of these flows.

For a periodic orbit P of a flow Φ let μ_P be the (unique) Φ-invariant probability measure supported on P and for a set S which is a finite union of periodic orbits let μ_S be the average of the measures corresponding to the periodic orbits contained in S.

Theorem 6.6 (Bowen [Bo3]). *Let Φ be a group-induced Anosov flow on an infra-homogeneous space $K\backslash G/\Gamma$ (notation as in § 5.3). Let m be the image on $K\backslash G/\Gamma$ of the normalised Haar measure on G/Γ. For any $t > 0$ and $\epsilon > 0$ let $S(t,\epsilon)$ be the set of points on periodic orbits of Φ with (minimal) period in the interval $[t-\epsilon, t+\epsilon]$. Then for all small $\epsilon > 0$, $\mu_{S(t,\epsilon)}$ converges to m, as $t \to \infty$.*

A general result of Bowen [Bo5] for basic sets of axiom A flows implies in particular that for a group-induced Anosov flow on a compact homogeneous space the image of the Haar measure is the unique invariant measure of maximal entropy.

The reader is also referred to Katok and Spatzier [KS2] for a study of measures on G/Γ, where G is a semisimple Lie group with no compact factors and Γ is an irreducible lattice in G, invariant under the action of the connected component of the identity in a split Cartan subgroup (maximal diagonalisable subgroup over the reals), through a local analysis in terms of what are called Lyapunov spaces. Under certain local geometrical conditions that we shall not go into here, it is concluded that an ergodic invariant measure which has positive entropy with respect to the action of some element is a homogeneous measure.

6.4. A weak recurrence property and infinite invariant measures. While in the earlier subsections we considered finite invariant measures for various actions, there are also results for infinite (locally finite) invariant measures. These arise from an interesting property of orbits of unipotent one-parameter subgroups:

Theorem 6.7 (cf. [D15]). *Let G be a connected Lie group and Γ be a lattice in G. Let a compact subset F of G/Γ and an $\epsilon > 0$ be given. Then there exists a compact subset K of G/Γ such that for any Ad-unipotent one-parameter* *subgroup* $\{u_t\}$ *of* G, $x \in G/\Gamma$ *and* $T \geq 0$ *such that* $u_s x \in F$ *for some* $s \in [0, T]$ *we have*

$$l(\{t \in [0, T] \mid u_t x \in K\}) \geq (1 - \epsilon)T;$$

(we denote by l *the Lebesgue measure on* $\mathbb{R}$*). Moreover,* K *can be chosen so that for any Ad-unipotent one-parameter subgroup* $\{u_t\}$ *and* $g \in G$ *such that* $\{g^{-1}u_t g\}$ *is not contained in a proper closed subgroup* H *intersecting* Γ *in a lattice, the above inequality holds for* $x = g\Gamma$ *for all large* T.

In particular, in the notation as in Theorem 6.7, for any Ad-unipotent one-parameter subgroup $\{u_t\}$ of G and $x \in G/\Gamma$ there exists a compact subset K of G/Γ and a sequence $\{t_i\}$ tending to infinity such that $u_{t_i}x \in K$ for all i; in other words, the trajectory 'does not go to infinity'. This 'qualitative' recurrence property was first proved by G.A. Margulis for $G = SL(n, \mathbb{R})$ and $\Gamma = SL(n, \mathbb{Z})$, and used in proving one of his arithmeticity theorems for lattices in semisimple Lie groups.

Recall that the homogeneous space $SL(n, \mathbb{R})/SL(n, \mathbb{Z})$ is canonically equivalent to the space of unimodular lattices in $\mathbb{R}^n$ (see Example 2 in § 1.2). Via the correspondence, Theorem 6.7 translates into a statement about minimal lengths of nonzero vectors in lattices. The following result provides quantitative information in this respect.

Theorem 6.8 (Kleinbock and Margulis [KM2]). *For any lattice* Λ *in* $\mathbb{R}^n$, $n \geq 2$, *there exists a constant* $\rho > 0$ *such that for any unipotent one-parameter subgroup* $\{u_t\}$ *of* $SL(n, \mathbb{R})$, $T > 0$ *and* $\epsilon \in (0, \rho)$,

$$l(\{t \in [0, T] \mid u_t(\Lambda) \cap B(\epsilon) \neq \{0\}\}) \leq 2n^3 6^n (n^2 + 1)^{1/n^2} (\epsilon/\rho)^{1/n^2} T,$$

where $B(\epsilon)$ *denotes the open ball of radius* ϵ *centered at* 0.

Similar quantitative estimates can be deduced from the theorem for a more general arithmetic lattice, via an embedding of the homogeneous space into $SL(n, \mathbb{R})/SL(n, \mathbb{Z})$ (see [Rag], Proposition 10.15).

Theorem 6.7 implies in particular, by Birkhoff's ergodic theorem, that all ergodic invariant (locally finite) measures of Ad-unipotent flows are finite (we caution the reader that the local finiteness condition cannot be weakened to σ-finiteness; each orbit of the flow supports a σ-finite measure and it is locally finite only when the orbit is closed; the results of [Sc], on infinite invariant measures of irrational rotations of the circle, suggest that there could also be other σ-finite ergodic invariant measures which are not locally finite). More generally we have the following result, which complements, in the case of lattices, the classification of finite invariant measures provided by Theorem 6.1.

Theorem 6.9 (cf. [D15] and [Marg1]). *Let* G *be a connected Lie group and* Γ *be a lattice in* G. *Let* H *be a closed subgroup of* G *which is generated by* *the Ad-unipotent one-parameter subgroups contained in it. Let μ be a (locally finite) H-invariant measure on G/Γ. Then there exists a sequence $\{X_i\}$ of H-invariant Borel subsets of G/Γ such that $\mu(X_i) < \infty$ for all i and $\bigcup X_i = G/\Gamma$. In particular if μ is ergodic with respect to the H-action then it is a finite measure.*

Applying the theorem to $G = SL(n, \mathbb{R})$ and $\Gamma = SL(n, \mathbb{Z})$ one can deduce the theorem of Borel and Harish-Chandra on finiteness of volumes of homogeneous spaces of Lie groups by arithmetic subgroups (see Margulis [Marg1]).

A generalisation of Theorem 6.7 is proved in [EMS2], where the authors consider actions of higher-dimensional subgroups, which may also contain non-unipotent elements (the results are local in nature, when the latter is allowed). The results are applied in particular to show that under certain conditions the set of all translates of a homogeneous probability measure (as in § 6.7 infra) is relatively compact.

6.5. Distribution of orbits and polynomial trajectories. Using her theorem on classification of invariant measures and (a variation of) Theorem 6.7 Ratner deduced that any orbit of an Ad-unipotent flow is uniformly distributed with respect to a homogeneous measure; her theorem, stated for lattices, readily generalises to the following (see Theorem 4.8, in this respect).

Theorem 6.10 (Ratner [R9]). *Let G be a connected Lie group and C be a closed subgroup of G such that G/C admits a finite G-invariant measure. Let $\{u_t\}$ be an Ad-unipotent one-parameter subgroup of G. Then for any $x \in G/C$ there exists a homogeneous probability measure μ such that the $\{u_t\}$-orbit of x is uniformly distributed with respect to μ.*

Ratner also proved the corresponding assertion for cyclic subgroups generated by Ad-unipotent elements.

When $G = SL(2, \mathbb{R})$ Theorem 6.10 implies that for any lattice Γ in G and unipotent one-parameter subgroup $\{u_t\}$ of G every nonperiodic orbit is uniformly distributed with respect to the normalised Haar measure on G/Γ; this was proved earlier in [DS] (see also [Gh] for a discussion on this case). It means in particular that every nonperiodic horocycle on a surface of constant negative curvature is uniformly distributed on the surface, with respect to the Riemannian area (cf. Example 5 in § 1.2). For the case when G is a simple Lie group of $\mathbb{R}$-rank 1 and C is a uniform lattice Theorem 6.10 was obtained by N.A. Shah in [Sh1], using Ratner's classification of invariant measures of unipotent flows; a geometric application of the result is given in [Sh2] (see also § 7.3, infra).

We say that a point x of a homogeneous space is *generic* for a flow induced by a one-parameter subgroup if its orbit is uniformly distributed with respect to the Haar measure. In view of Theorem 6.10 generic points of an Ad-unipotent flow can be characterised as follows (a similar assertion also holds in the case of cyclic subgroups); see [Sh1] and [DM3] for an idea of the arguments involved.

Corollary 6.1. *Let G be a connected Lie group and Γ be a lattice in G. Let $U = \{u_t\}$ be an Ad-unipotent one-parameter subgroup of G. Then for $x \in G/\Gamma$ the following conditions are equivalent:*

i) x is generic;

ii) there does not exist any proper closed subgroup H containing U such that Hx has a finite H-invariant measure;

iii) there does not exist any proper closed subgroup H containing U such that Hx is closed.

The classification of invariant measures of Ad-unipotent flows has also been applied to study the distribution of trajectories of polynomial curves and, more generally, images of regular algebraic maps. If $\mathbf{L}$ is a (complex) algebraic group defined over $\mathbb{R}$ and L is the corresponding real algebraic group, namely the group of real points of $\mathbf{L}$, then a map $f : \mathbb{R}^k \to L$ is said to be *regular algebraic* if it is the restriction of a regular map of $\mathbb{C}^k$ into $\mathbf{L}$.

Theorem 6.11 (Shah [Sh3]). *Let G be a real algebraic group and H and C be closed subgroups of G such that $C \subseteq H$ and H/C admits a finite H-invariant measure. Let $\Theta : \mathbb{R}^k \to G$ be a regular algebraic map such that $\Theta(0) = e$, the identity, and $\Theta(\mathbb{R}^k) \subseteq H$. Then there exists a closed subgroup F of H such that FC is closed, FC/C admits a F-invariant probability measure μ, and for any bounded continuous function f on H/C*

$$\frac{1}{\lambda(B_r)} \int_{B_r} f(\Theta(v)C)\mathrm{d}\lambda(v) \to \int_{H/C} f \mathrm{d}\mu \quad \text{as } r \to \infty,$$

where B_r denotes the ball in $\mathbb{R}^k$ with radius r and center at the origin, and λ is the Lebesgue measure on $\mathbb{R}^k$.

The result is applied to deduce an analogue of Theorem 6.10 for distributions of orbits of higher-dimensional nilpotent Lie groups (see Shah [Sh3]).

6.6. A uniform version of uniform distribution. In certain problems of Diophantine approximation it is of interest to have results on the distribution of trajectories, which are uniform over sets of initial points as well as over classes of one-parameter subgroups inducing the flows. We now describe a result from [DM3] in that direction; a quantitative version of Oppenheim's conjecture obtained by application of the theorem will be described in § 9. Here the class of Ad-unipotent one-parameter subgroups of G is considered equipped with the topology of pointwise convergence, as maps from $\mathbb{R}$ to G.

By Theorem 6.2 the set of generic points for the action of a one-parameter subgroup $U = \{u_t\}$ consists of the complement of $\bigcup_{H \in \mathscr{H}} X(H, U)\Gamma/\Gamma$. The following result shows that for a given bounded continuous function, the time averages over large time are close to the space average, uniformly over compact subsets in the complement of a certain set (of exceptional points) which is a union of some *compact* subsets from *finitely* many $X(H, U)\Gamma/\Gamma$.

Theorem 6.12 (cf. [DM3]). *Let G be a connected Lie group, Γ be a lattice in G and m be the normalised Haar measure on G/Γ. Let $\mathscr{U}$ be a compact set of Ad-unipotent one-parameter subgroups of G. Let a bounded continuous function φ on G/Γ, a compact subset K of G/Γ and an $\epsilon > 0$ be given. Then there exist finitely many subgroups $H_1, \ldots, H_k \in \mathscr{H}$ and a compact subset C of G such that the following holds: For any $U = \{u_t\} \in \mathscr{U}$ and any compact subset F of $K - \cup_{i=1}^{k} (C \cap X(H_i, U))\Gamma/\Gamma$ there exists a $T_0 \geq 0$ such that for all $x \in F$ and $T > T_0$,*

$$\left| \frac{1}{T} \int_0^T \varphi(u_t x)\,\mathrm{d}t - \int_{G/\Gamma} \varphi\,\mathrm{d}m \right| < \epsilon.$$

This includes the following result which was proved earlier (cf. [R12], [D21] and [D22] for details).

Theorem 6.13. *Let G, Γ and m be as in Theorem 6.12. Let $\{u_t^{(i)}\}$ be a sequence of Ad-unipotent one-parameter subgroups of G converging to a one-parameter subgroup $\{u_t\}$, namely such that $u_t^{(i)} \to u_t$ for all $t \in \mathbb{R}$. Let $\{x_i\}$ be a sequence of points in G/Γ converging to a generic point x. Let $\{T_i\}$ be a sequence in $\mathbb{R}^+$, tending to infinity. Then for any bounded continuous function φ on G/Γ we have*

$$\frac{1}{T_i} \int_0^{T_i} \varphi(u_t^{(i)} x_i)\,\mathrm{d}t \to \int_{G/\Gamma} \varphi\,\mathrm{d}m, \quad \text{as } i \to \infty.$$

Remark 6.3. Theorem 6.12 includes also the following result which has been in the folklore: Let $G = SL(2, \mathbb{R})$ and Γ be a nonuniform lattice in G. Let $U = \{u_t\}$ be the upper triangular unipotent one-parameter subgroup of G. Let μ be the U-invariant probability measure supported on a periodic orbit of U (such orbits exist; see § 7.2). Let $\{a_i\}$ be a sequence of diagonal matrices such that for all $t \neq 0$, $a_i u_t a_i^{-1} \to \infty$ as $i \to \infty$, namely $a_i = \operatorname{diag}(\lambda_i, \lambda_i^{-1})$, with $|\lambda_i| \to \infty$. Then $a_i\mu \to m$, the normalised Haar measure on G/Γ, as $i \to \infty$. The reader is referred to Sarnak [Sa] for a quantitative result on the rate of the convergence; it is noted in [Sa] that by a connection observed by D. Zagier a certain improved estimate in the case of $\Gamma = SL(2, \mathbb{Z})$ would prove the Riemann hypothesis. We also refer the reader to Verjovsky (see [Ve]) for some results about the rate of convergence of the sequence $a_i\mu(E)$ (notation as above) for certain sets E.

Remark 6.4. Uniform distribution can also be studied with respect to averages over other sequences, in the place of the time segments of one-parameter flows as in the above discussion. The reader is in particular referred to Auslander and Brezin [AB] for a study of uniform distribution on solvmanifolds, from a different angle.

6.7. Distribution of translates of closed orbits. We next recall the following result which describes a condition under which a sequence of translates of a fixed homogeneous measure converges to the Haar measure on G/Γ; we shall see in § 9 an interesting application of the result to a Diophantine problem.

Theorem 6.14 (Eskin, Mozes and Shah [EMS1]). *Let G be a reductive real algebraic group defined over $\mathbb{Q}$ and let Γ be an arithmetic lattice with respect to the $\mathbb{Q}$-structure. Let H be a reductive real algebraic $\mathbb{Q}$-subgroup of G, not contained in any $\mathbb{Q}$-parabolic subgroup of G. Let $Z(H)$ denote the centraliser of H in G. Let μ be the H^0-invariant probability measure on $H^0\Gamma/\Gamma$ and m be the G^0-invariant probability measure on $G^0\Gamma/\Gamma$. Let $\{g_i\}$ be a sequence in G^0. Then at least one of the following conditions is satisfied:*

i) $\{g_i\mu\}$ converges to m as $i \to \infty$;

ii) there exist a reductive real algebraic $\mathbb{Q}$-subgroup L of G such that $H^0 \subseteq (L \cap G)^0 \neq G^0$ and a compact subset C of G such that infinitely many g_i's are contained in $CL(Z(H) \cap \Gamma)$.

In particular, if G is semisimple, H is a maximal proper connected reductive subgroup of G^0 and $\{g_i H\}$ has no convergent subsequence in G/H then $\{g_i\mu\}$ converges to m as $i \to \infty$.

In a related paper [EMS2] it is also shown that for a measure μ as in Theorem 6.14 the set $\{g\mu \mid g \in G\}$ of all translates of μ is relatively compact in the space of probability measures on G/Γ. For the case when G/H is an affine symmetric space, viz. when H is the set of fixed points of an involution on G, a result as in Theorem 6.14 was proved in [EM], using the mixing property of the G-action on G/Γ for irreducible lattices Γ (cf. Corollary 3.2).

§ 7. Orbit closures of group-induced flows

We now describe results on orbit closures of the translation actions.

7.1. Homogeneity of orbit closures. Theorem 6.10 implies in particular that orbit closures of Ad-unipotent one-parameter subgroups are homogeneous subsets. This was conjectured by M.S. Raghunathan, who observed also that proving it would yield the conjecture of Oppenheim on values of nondegenerate indefinite quadratic forms at integral points (see § 9.4). Ratner also deduced from her results a conjecture of Margulis on the homogeneity of orbit closures of subgroups generated by Ad-unipotent one-parameter subgroups and, more generally, proved Theorem 7.1 below under the (additional) conditions as in her result on classification of invariant measures (see the remarks following Theorem 6.1); see [R12] and [D22] for some historical details. The following generalisation was obtained recently by N.A. Shah [Sh5], via the strengthened result on invariant measures (see Theorem 6.1).

Theorem 7.1 (Ratner [R9], Shah [Sh5]). *Let G be a Lie group and C be a closed subgroup of G such that G/C has finite Haar measure. Let W be a closed subgroup of G containing a set U of Ad-unipotent elements such that the subgroup generated by $\{\mathrm{Ad}\, u \mid u \in U\}$ is Zariski-dense in $\mathrm{Ad}\, W$. Then for any $x \in G/C$, there exists a closed subgroup F such that $\overline{Wx} = Fx$ and F^0x admits a finite F^0-invariant measure. In particular if W is connected then $\overline{Wx}$ is a homogeneous subset with finite Haar measure.*

It is conjectured in Shah [Sh5] that $\overline{Wx} = Fx$ as above has only finitely many connected components, or equivalently that Fx admits a finite F-invariant measure.

Ratner also proved homogeneity of orbit closures, as in the above theorem, for subgroups W which are generated by Ad-unipotent one-parameter subgroups together with certain (non Ad-unipotent) subgroups which are 'diagonal' for some (Ad-unipotent) one-parameter subgroups in W; these include parabolic subgroups in the case of semisimple Lie groups G (see [R11] and [R14] for details). The study of orbit closures has also been carried out for actions of epimorphic subgroups (see § 6.2 for definition).

Theorem 7.2 (Shah and Weiss [ShW]). *Let G be a connected Lie group with Lie algebra $\mathcal{G}$ and let Γ be a lattice in G. Let F and H be connected Lie subgroups of G such that $F \subseteq H$ and $\mathrm{Ad}\, F$ and $\mathrm{Ad}\, H$ are real algebraic subgroups of $GL(\mathcal{G})$. Suppose that $\mathrm{Ad}\, F$ is an epimorphic subgroup of $\mathrm{Ad}\, H$. Then any F-invariant closed subset of G/Γ is $F[H, H]$-invariant. Furthermore if H intersects the center of G in a discrete subgroup then any F-invariant closed subset of G/Γ is H-invariant. Also, if H is generated by Ad-unipotent one-parameter subgroups then so is $F[H, H]$ and the closure of any F-orbit is a homogeneous subset of G/Γ with finite Haar measure.*

A Lie subgroup H of a real algebraic group G is said to be *almost algebraic* if it is an open subgroup (with respect to the Lie group topology) of a real algebraic subgroup of G. For almost algebraic subgroups we have the following variation of Theorem 7.2, with a stronger conclusion (see Shah and Weiss [ShW]).

Theorem 7.3. *Let G be a real algebraic group and F and H, $F \subseteq H$, be connected almost algebraic subgroups of G. Suppose that F is epimorphic* *in H. Then for any lattice Γ in G any closed F-invariant subset of G/Γ is H-invariant.*

In the special case when G is a real algebraic group defined over $\mathbb{Q}$, $H = G$ and Γ an arithmetic lattice associated to the arithmetic structure, the theorem was proved by Weiss [We]; it is pointed out in [We], with an example due to M.S. Raghunathan, that the conclusion that any F-invariant closed subset is G-invariant need not hold in general, if F is not assumed to be almost algebraic.

Shah and Weiss [ShW] also give a characterisation of subgroups in a class of algebraic groups G for which all orbit closures in G/Γ are 'almost homogeneous' for any lattice Γ in G. Given a Lie group G and a lattice Γ in G, a subset is said to be *almost homogeneous* if it is of the form KS, where K is a compact subgroup of G and S is a homogeneous subset with finite Haar measure. Given a real algebraic group L, a real algebraic subgroup F of L is said to be an *observable* subgroup of L if every algebraic linear representation of F extends to an algebraic representation of L. If G is a Zariski-connected real algebraic group (namely such that it has no proper real algebraic subgroup of finite index) and F is a Zariski-connected real algebraic subgroup of G there exists a smallest observable real algebraic subgroup E of G containing F; E is called the *observable envelope* of F in G. We recall also that a real algebraic group is said to be $\mathbb{R}$ split if its $\mathbb{R}$-rank coincides with the rank over $\mathbb{C}$. We have the following:

Theorem 7.4 (Shah and Weiss [ShW]). *Let G be a Zariski-connected $\mathbb{R}$-split real algebraic group (e.g. $G = SL(n, \mathbb{R})$). Let F be a Zariski-connected algebraic subgroup of G and E be the observable envelope of F in G. Then for all lattices Γ in G all orbit closures of F on G/Γ are almost homogeneous subsets if and only if the subgroup of E generated by unipotent elements is cocompact in E.*

For a one-parameter flow induced by an Ad-semisimple one-parameter subgroup, in general the orbit closures are not homogeneous (cf. Proposition 7.2 below). There are, however, certain special cases where closures of smooth manifolds invariant under such flows have to be homogeneous or 'semi-homogeneous', as we shall see in § 7.4 below.

7.2. Orbit closures of horospherical subgroups. Let G be a Lie group, Γ be a lattice in G and W be a closed subgroup generated by Ad-unipotent one-parameter subgroups. Let $x \in G/\Gamma$ and let F be the closed subgroup, as in Theorem 7.1, such that $\overline{Wx} = Fx$. Then F is the smallest closed subgroup containing W, such that Fx is closed and admits a finite F-invariant measure. Thus the problem of determining the orbit closures is reduced to a group-theoretic problem. In certain situations dynamical considerations lead to some more information on F and especially on the question whether $F = G$, namely whether the W-orbit is dense. We now note some results in this respect.

Theorem 7.5. *Let G be a connected Lie group and Γ be a uniform lattice in G. Let $g \in G$ and W be the horospherical subgroup associated to g (see § 4.1). Let M be the Mautner subgroup associated to g and $H = \overline{M\Gamma}$. Then we have the following:*

i) for any $x \in G$, $\overline{Wx\Gamma} = xH\Gamma = (xH^0x^{-1})x\Gamma$; in particular, for the action of W on G/Γ distinct orbit-closures are disjoint, and so if there exists one dense orbit then all orbits are dense (see [St6]*).*

ii) if the translation of G/Γ by g is weak mixing then all orbits of W are dense (see [Bo7], [D2], [EP], [R6], [V1]*).*

For a nonuniform lattice of course neither of the above statements is true, as can be seen from the example of $G = SL(n, \mathbb{R})$, $\Gamma = SL(n, \mathbb{Z})$ and g a diagonal matrix which is not a scalar multiple of the identity. For the general case we note the following.

Theorem 7.6. *Let G be a Lie group and Γ be a lattice in G. Let R be the radical of G. Let $g \in G$ and W be the horospherical subgroup of G associated to g. Suppose that the W-action on G/Γ is ergodic. Then we have the following:*

i) if $\mathrm{Ad}\, g$ is semisimple (diagonalisable over $\mathbb{C}$) and $x \in G/\Gamma$ is such that for some compact subset K of G/Γ, $g^i x \in WK$ for infinitely many positive integers i then the W-orbit of x is dense in G/Γ (cf. [D16]*);*

ii) if G is a semisimple Lie group then for $x = y\Gamma \in G/\Gamma$, where $y \in G$, Wx is dense in G/Γ if and only if there does not exist a (proper) parabolic subgroup P of G such that $W \subseteq P$ and the unipotent radical of P intersects $y\Gamma y^{-1}$ in a lattice (cf. [D16]*);*

iii) for $x \in G/\Gamma$ the W-orbit of x is dense in G/Γ if and only if the W-orbit of $x\overline{RC}$ is dense in $G/\overline{RC}$ (see [D18]*).*

From the proofs of assertions (ii) and (iii) one can also read off the following.

Theorem 7.7. *Let G, Γ, R and W be as in Theorem 7.6. Then the set of all $x \in G/\Gamma$ for which Wx is not dense in G/Γ is contained in a union of finitely many orbits of (finitely many) subgroups Q such that Q contains R and Q/R is a parabolic subgroup of (the semisimple group) G/R.*

The finitely many orbits in the conclusion are related to the 'cusps' of the homogeneous space G/Γ and the structure at infinity. In the case when $G = SL(2, \mathbb{R})$, the upper triangular unipotent one-parameter subgroup is a horospherical subgroup (with respect to a diagonal matrix) and the assertions in the above theorems reduce to the classical results of G.A. Hedlund [He] on horocycle flows corresponding to surfaces of constant negative curvature and finite area (see Example 5 in § 1.2). The horocycle flows are minimal if the surface is compact. In the case of a noncompact surface of finite area each orbit is either periodic or dense and the periodic orbits are contained in finitely many cylinders, which are orbits of the parabolic subgroup consisting of all upper triangular matrices in G. Similar assertions hold also for horospherical flows associated to higher-dimensional manifolds of constant negative curvature and finite Riemannian volume, the Lie groups involved being the groups of isometries of the hyperbolic spaces.

The cylinders as above containing the periodic orbits of a horocycle flow are dense. The following theorem provides a general version of this fact.

Theorem 7.8 (Shah [Sh4]). *Let G be a connected Lie group and Γ be a lattice in G. Let S be a connected semisimple subgroup of G. Let $a \in S$ be such that the restriction of $\mathrm{Ad}\,a$ to the Lie subalgebra of S is a semisimple linear transformation. Let U be the horospherical subgroup corresponding to a. Suppose that $U \cap S$ is not contained in any proper closed normal subgroup of S. Then for any $x \in G/\Gamma$, $\overline{\bigcup_{i=0}^{\infty} a^i(S \cap U)x} = \overline{Sx}$. In particular if the S-action on G/Γ is minimal then so is the action of the subgroup generated by a and $S \cap U$.*

The theorem shows in particular that if G is a semisimple Lie group with no compact factors and Γ is a lattice in G then all orbits of parabolic subgroups of G are dense in G/Γ.

7.3. Orbits of reductive subgroups. The following result on orbits of special orthogonal groups is related to the Oppenheim conjecture (see [BP]; see also [DM2] and [D20] for proofs in the case of $n = 3$).

Theorem 7.9. *Let $G = SL(n, \mathbb{R})$ and $\Gamma = SL(n, \mathbb{Z})$ for some $n \geq 3$. Let Q be a nondegenerate indefinite quadratic form on $\mathbb{R}^n$ and let $H \subseteq G$ be the special orthogonal group associated to Q, namely $H = \{g \in G \mid Q(gv) = Q(v)$ for all $v \in \mathbb{R}^n\}$. Then for $g \in G$ the H-orbit of $g\Gamma$ is either closed or dense in G/Γ. Furthermore, it is closed if and only if the quadratic form Q^g defined by $Q^g(v) = Q(gv)$ for all $v \in \mathbb{R}^n$, is a scalar multiple of a form with rational coefficients.*

The proof of the first assertion depends on Ratner's theorem and the maximality of the special orthogonal groups. In fact, the following stronger assertion can be deduced from Theorems 6.3 and 6.9 for orbits of maximal subgroups generated by Ad-unipotent one-parameter subgroups; for the case when $G = SL(3, \mathbb{R})$ and H is the special orthogonal group of a quadratic form as in Theorem 7.9, the result was noted in [Marg2].

Theorem 7.10. *Let G be a connected Lie group and Γ be a lattice in G. Let H be a subgroup generated by Ad-unipotent one-parameter subgroups and* *suppose that it is a maximal connected Lie subgroup in G. Then any proper closed H-invariant subset of G/Γ is a finite union of closed H-orbits.*

The condition on W as in Theorem 7.1 is satisfied in particular if W is a connected semisimple Lie group without compact factors. Using this together with a characterisation of totally geodesic submanifolds of locally symmetric spaces the following is deduced by Payne [Pa].

Theorem 7.11. *Let M be a locally symmetric space of noncompact type, with finite Riemannian volume. Let $\widetilde{M}$ be the universal covering manifold of M and $p : \widetilde{M} \to M$ be the covering map. Let $\widetilde{M}$ be equipped with the metric such that p is a local isometry. Let N be a locally symmetric space of noncompact type and let $\phi : N \to M$ be a totally geodesic immersion of N into M. Then there exists a closed connected subgroup H of the group of isometries of $\widetilde{M}$ such that the closure of $\phi(N)$ in M is an immersed submanifold of M of the form $p(\widetilde{x}H)$, where $\widetilde{x} \in \widetilde{M}$. Furthermore, if the rank of N equals the rank of M then $\overline{\phi(N)}$ is a totally geodesic submanifold with finite Riemannian volume.*

In particular, for M as in Theorem 7.11, totally geodesic proper immersed submanifolds of maximal dimension and rank equal to that of M, are either closed or dense. This was noted earlier in Shah [Sh2] in the case of compact manifolds of constant negative curvature. A similar assertion is also proved in [Sh2] for orhthonormal frame bundles.

7.4. Orbit closures of one-parameter flows. We next consider orbits of one-parameter subgroups. For a one-parameter subgroup $\phi = \{g_t\}$ of a Lie group G let $Q(\phi)$ denote the connected Lie subgroup of G whose associated Lie subalgebra is the largest $\mathrm{Ad}\,\phi$-invariant subspace on which all eigenvalues of $\mathrm{Ad}\,g_t$, $t \in \mathbb{R}$, are of absolute value 1.

Theorem 7.12 (Starkov [St8]). *Let G be a Lie group and let $C \in \mathscr{F}(G)$ (notation as in § 1.1). Let $\phi = \{g_t\}$ be a one-parameter subgroup of G. Let $x \in G/C$ and suppose that there exist a compact subset K of $Q(\phi)$ and a sequence $\{t_i\}$ in $\mathbb{R}$, tending to infinity, such that $g_{t_i}x \in Kx$ for all i. Then $\overline{\phi x}$ is a smooth submanifold of G/C with a finite ϕ-invariant measure of full support. Furthermore, if all eigenvalues of $\mathrm{Ad}\,g_t$, $t \in \mathbb{R}$, are real, then there exists a closed subgroup F such that $\overline{\phi x} = Fx$ and Fx has a finite F-invariant measure.*

We note that if $\{g_t\}$ as in the theorem is an Ad-unipotent one-parameter subgroup then the condition in the hypothesis is satisfied in view of Theorem 6.7 and thus Theorem 7.12 generalises the case of Theorem 7.1 for one-parameter subgroups.

An element g of a connected Lie group G is said to be *quasi-unipotent* if all eigenvalues of $\mathrm{Ad}\,g$ are of absolute value 1 and a subgroup of G is said to be quasi-unipotent if each of its elements is quasi-unipotent. Theorem 7.12 implies in particular that orbit closures of quasi-unipotent flows are smooth submanifolds (see Proposition 7.2 for a converse). In fact the following also holds:

Proposition 7.1 (Starkov [St4]). *Let G be a connected Lie group and let $\phi = \{g_t\}$ be a quasi-unipotent one-parameter subgroup of G. Then there exist a connected Lie group G^* and an Ad-unipotent one-parameter subgroup ν of G^* such that G can be realised as a closed normal subgroup of G^* so that G^*/G is a torus, and for any $C \in \mathscr{F}(G)$ any orbit-closure of ϕ on G/C is the quotient of a (homogeneous) orbit-closure of ν on G^*/C, under an action of a compact group on G^*/C which commutes with the ν-action and has a single orbit-type.*

The quotients as above are said to be 'compactly covered' by the corresponding homogeneous orbit-closures in the ambient space and the flow on it. The quotients need not in general be homogeneous subsets themselves (see Starkov [St3]).

For one-parameter subgroups which are not quasi-unipotent the orbit closures can be 'bad', as seen from the following.

Proposition 7.2 (cf. [St5]). *Let G be a Lie group and $C \in \mathscr{F}_0(G)$. Let $\phi = \{g_t\}$ be a one-parameter subgroup of G which is not quasi-unipotent. Then there exists a point $x \in G/C$ such that $\overline{\phi x}$ is locally closed but not closed; in particular it is not a smooth submanifold.*

In the case when $G = SL(2, \mathbb{R})$, Γ is a lattice in G and ϕ is the diagonal one-parameter subgroup (namely for the geodesic flow) Furstenberg and Weiss have shown that for any d in the interval $[1, 3]$ there exists an orbit of the ϕ-action on G/Γ, whose closure has Hausdorff dimension d (informal communication from Furstenberg).

Assumption of smoothness however drastically restricts the possibilities for invariant sets, for a large class of flows. In this respect we first recall a result of Zeghib [Ze1] for group-induced Anosov flows (see § 5.3). Given a Lie group G, a lattice Γ in G and a one-parameter subgroup $\phi = \{g_t\}$ we say that a subset E of G/Γ is *semi-homogeneous* relative to ϕ if there exist a closed subgroup H and a $g \in G$ such that $\phi \cap H$ is nontrivial, $Hg\Gamma$ is closed and $E = \phi Hg\Gamma/\Gamma$; we note that such a subset is closed.

Theorem 7.13 (Zeghib [Ze1]). *Let G, Γ, $A = \{a_t\}$ and K be as in § 5.3, so that the A-action on $K\backslash G/\Gamma$ is an Anosov flow. Let M be an A-invariant immersed C^1 submanifold of $K\backslash G/\Gamma$, with finite Riemannian volume. Then $\overline{M}$ is the image in $K\backslash G/\Gamma$ of a finite union of semi-homogeneous subsets of G/Γ relative to A. Moreover, M is open in $\overline{M}$.*

Remark 7.1. If A as above is a subgroup with the property that it is contained in every connected Lie subgroup of G which it intersects nontrivially then any subset which is semi-homogeneous relative to the A-action is in fact homogeneous; this applies in particular to the smooth closed invariant subsets as in the above theorem. The condition is satisfied for instance when $G = SL(2, \mathbb{R})$ and A is the one-parameter subgroup consisting of diagonal matrices and, more generally for any one-parameter subgroup of a simple Lie group of $\mathbb{R}$-rank 1 which is diagonalisable over the reals.

Theorem 7.13 may be compared with Theorem 2.14 for Anosov automorphisms. As in the previous case, an analogous assertion is made in Zeghib [Ze1] for rectifiable sets in the place of immersed submanifolds.

It may be noticed that the Lie group involved in Theorem 7.13 is of $\mathbb{R}$-rank 1. The analogous statement is not to be expected to hold for flows induced by Ad-semisimple one-parameter subgroups on infra-homogeneous spaces of Lie groups of higher rank. We recall here an interesting description of smooth submanifolds invariant under Ad-semisimple one-parameter subgroups on homogeneous spaces of higher rank groups, though it does not seem to be directly helpful in understanding closures of orbits. Let G be a Lie group, Γ be a discrete subgroup of G, K be a compact subgroup of G and ϕ be a one-parameter subgroup of G centralising K. Let M be a smooth submanifold of $K\backslash G/\Gamma$ invariant under the flow induced by ϕ, as the factor of the flow on G/Γ. Under the condition that M admits a finite ϕ-invariant measure equivalent to the Lebesgue measure on each coordinate chart, it is shown that M is the image, up to a 'negligible set', of a countable union of subsets of G of the form SxH, where S is a subset of the centraliser of ϕ, $x \in G$ and H is a closed subgroup of G such that $H \cap \Gamma$ is a lattice in H, a conjugate of ϕ is contained in H and acts ergodically on $H/H \cap \Gamma$; a similar result is also proved more generally for invariant rectifiable subsets. The reader is referred to Zeghib [Ze2] for details. In Zeghib [Ze2] it is also shown that the ergodic components of rectifiable subsets invariant under the flow induced by ϕ on $K\backslash G/\Gamma$ as above, are semi-homogeneous relative to ϕ.

7.5. Dense orbits and minimal sets of flows. Let G be a connected Lie group, Γ be a lattice in G and $\phi = \{g_t\}$ be a one-parameter subgroup of G acting ergodically on G/Γ. If ϕ is Ad-unipotent then by Ratner's theorem (see Theorem 7.1) the orbit closures of the flow are homogeneous subsets with finite Haar measure and any orbit which is not dense is contained in $X(H, \phi)\Gamma/\Gamma$, for some $H \in \mathscr{H}$ (notation as in § 6.1). In particular this means that in this case the set of all points with non-dense orbits is a countable union of lower-dimensional smooth submanifolds of G/Γ. A similar assertion then follows from Proposition 7.1 for quasi-unipotent one-parameter subgroups. The next theorem shows that the situation is quite different however when there is no nontrivial affine factor on which ϕ is quasi-unipotent.

A subset E of a differentiable manifold M is said to be *thick* if for every nonempty open subset Ω of M, $E \cap \Omega$ is of Hausdorff dimension equal to the dimension of M (the latter being also the Hausdorff dimension of M); Hausdorff dimensions are considered with respect to a metric equivalent to the usual metric on coordinate charts; the condition is independent of the specific choice of the metric.

We mention here the following consequence of a result of Kleinbock [Kl1] (see Theorem 7.18).

Theorem 7.14. *Let G be a Lie group and Γ be a lattice in G. Let $\phi = \{g_t\}$ be a one-parameter subgroup of G. Let M be the Mautner subgroup corresponding to g_1 (see § 4.1 for definition) and suppose that $M\Gamma$ is dense in G. Let E be the set of points $x \in G/\Gamma$ such that the orbit ϕx is not dense in G/Γ. If E is nonempty then it is a thick subset of G/Γ.*

Applying the theorem to G and Γ as in the construction of group-induced Anosov flows in § 5.3 and using the fact that the latter always have periodic points we see that the set of points with non-dense orbits is a thick set. The reader is referred to Urbanski [Ur2] for a similar result for general Anosov flows on compact manifolds.

Another question of interest from a topological dynamical point of view is to describe the minimal (nonempty closed invariant) sets of the flows. In this respect let me begin by recalling the following. If N is a compactly generated nilpotent Lie group acting continuously on a locally compact Hausdorff space X and there exists a relatively compact open subset $\Omega \subseteq X$ such that Ω intersects each N-orbit and for every $x \in \Omega$ and $g \in N$ the semi-orbit $\{g^i x \mid i = 0, 1, 2, \dots\}$ has a convergent subsequence in X, then X is compact (see Margulis [Marg5]). Now let G be a Lie group and $C \in \mathcal{F}(G)$. In view of Theorem 6.7 the preceding observation implies that for the action of any compactly generated Ad-unipotent subgroup U of G on G/C, any minimal closed nonempty invariant subset is compact; moreover, by Theorem 7.1 it is a homogeneous subset with finite Haar measure. For a general one-parameter subgroup (not necessarily Ad-unipotent) ϕ of G the above observation implies that every minimal closed invariant subset M is either compact or contains a point x such that either $\{g_t x\}_{t\geq 0}$ or $\{g_{-t}x\}_{t\geq 0}$ diverges (eventually escapes every compact subset of G/C). The compact minimal sets of a general group-induced one-parameter flow exhibit the following dichotomy.

Theorem 7.15 (Starkov [St8]). *Let G be a Lie group and $C \in \mathcal{F}(G)$. Let $\phi = \{g_t\}$ be a one-parameter subgroup of G. Let M be a compact minimal closed nonempty ϕ-invariant subset of G/C. Then M is either a smooth submanifold or it is not locally connected at any point. Further, if M is a smooth submanifold invariant under the flow then the restriction of the ϕ-action to M is compactly covered by an Ad-unipotent flow on a compact homogeneous space with finite Haar measure, and it is uniquely ergodic.*

It may also be mentioned here that it follows from a general result of Bowen [Bo4] that the minimal sets of the Anosov flows as in § 5.3 are all one-dimensional.

7.6. Divergent trajectories of flows. Now let G and C be as before and $\phi = \{g_t\}$ be a one-parameter subgroup of G. Suppose that G/C is not compact. Then in general there can exist orbits $\{g_t x\}$ which diverge in one or both directions (that is, as $t \to \infty$ or as $t \to -\infty$ or both). For example if $G = SL(2, \mathbb{R})$ and $\Gamma = SL(2, \mathbb{Z})$, the orbit of the identity coset Γ under the diagonal one-parameter subgroup $\{\operatorname{diag}(e^t, e^{-t})\}$ may be seen to be divergent in both directions. Geometrically, any noncompact Riemannian manifold of constant negative curvature and finite volume has cusps and there are geodesics which go to the cusp point at infinity; the corresponding trajectory of the geodesic flow, which is a group-induced flow (see Example 6 in § 1.2), is divergent.

Theorem 7.16 (cf. [D13]). *Let G be a semisimple Lie group and Γ be an irreducible lattice in G such that G/Γ is noncompact. Let $\{g_t\}$ be a one-parameter subgroup of G. Then we have the following:*

1. *if $\operatorname{Ad} g_t$ is semisimple (diagonalisable over $\mathbb{C}$) then there exists $x \in G/\Gamma$ such that $\{g_t x \mid t \geq 0\}$ is divergent;*
2. *if G is a real algebraic group defined over $\mathbb{Q}$ such that the $\mathbb{Q}$-rank of G equals the $\mathbb{R}$-rank of G, and Γ is an arithmetic lattice with respect to the $\mathbb{Q}$-structure, then there exists $x \in G/\Gamma$ such that $\{g_t x \mid t \geq 0\}$ is divergent (this holds in particular for $G = SL(n, \mathbb{R})$ and $\Gamma = SL(n, \mathbb{Z})$);*
3. *if G is a simple Lie group of $\mathbb{R}$-rank 1 and $\{\operatorname{Ad} g_t\}$ is diagonalisable over $\mathbb{R}$ then there exists a parabolic subgroup P of G such that the unipotent radical of P intersects Γ in a lattice and the set of x in G/Γ for which $\{g_t x \mid t \geq 0\}$ is divergent is a union of finitely many P-orbits on G/Γ.*

The finitely many P-orbits as in the last assertion in the theorem are in one-one correspondence with the cusps of the corresponding locally symmetric space of rank 1.

For homogeneous spaces of semisimple Lie groups of higher rank the set of points with divergent trajectories is rather complicated in general (see [D13] for details). In the case when $G = SL(n, \mathbb{R})$, $n \geq 2$, $\Gamma = SL(n, \mathbb{Z})$ and $\{g_t^{(p)}\}$, $1 \leq p \leq (n-1)$, is the one-parameter subgroup with $g_t^{(p)}$ as the diagonal matrix in which the first p entries are $e^{-t/p}$ and the last $n-p$ entries are $e^{t/(n-p)}$, for $g \in G$ the trajectory $\{g_t^{(p)} g\Gamma \mid t \geq 0\}$ is divergent if and only if a certain system of $(n-p)$ linear forms in p variables corresponding to g is 'singular', in the sense of J.W.S. Cassels [Ca1] (see [D13] for details; see also § 9.2 for related results). Kleinbock has introduced a notion of a nonsingular $n \times n$ matrix g being '(r, s)-very tight', for $r = (r_1, \ldots, r_p) \in \mathbb{R}^p$, $s = (s_1, \ldots, s_q) \in \mathbb{R}^q$ such that $p + q = n$, $r_i > 0$ and $s_j > 0$ for all i, j, and $\Sigma r_i = \Sigma s_j = 1$, in terms of certain Diophantine conditions similar to those in the definition of singular systems of linear forms, and shown that for G and Γ as above the trajectory $\{a_t g\Gamma \mid t \geq 0\}$ of $g\Gamma$, where $g \in G$ and a_t is the diagonal matrix diag $(e^{r_1 t}, \ldots, e^{r_p t}, e^{-s_1 t}, \ldots, e^{-s_q t})$, is divergent if and only if g is (r, s)-very tight (see [Kl2], [Kl3]).

7.7. Bounded orbits and escapable sets. Let G be a Lie group, Γ be a nonuniform lattice in G and $\phi = \{g_t\}$ be a one-parameter subgroup of G. By a bounded orbit we mean one with compact closure. Such an orbit is not dense and it follows (see § 7.5) that when ϕ is quasi-unipotent the set of points with bounded orbits is contained in certain a countable union of lower-dimensional submanifolds. On the other hand the following results show that for a one-parameter subgroup which is not quasi-unipotent there can exist a thick set (see § 7.5 for definition) of points whose orbits are bounded, and also avoid a given closed invariant set of zero Haar measure.

Theorem 7.17 (Kleinbock and Margulis [KM1]). *Let G be a semisimple Lie group without compact factors and let Γ be an irreducible lattice in G. Let $\phi = \{g_t\}$ be a one-parameter subgroup of G which is not quasi-unipotent. Let S be a closed subset of G/Γ with zero Haar measure such that $g_t S \subseteq S$ for all $t \geq 0$. Let E be the subset consisting of all $x \in G/\Gamma$ for which ϕx is bounded and $\overline{\phi x} \cap S = \emptyset$. Then E is a thick subset of G/Γ.*

More generally, if G is a connected Lie group, Γ is a lattice in G and ϕ is a one-parameter subgroup of G then for a subset S such that $g_t S \subseteq S$ for all $t \geq 0$ the subset E consisting of all $x \in G/\Gamma$ such that ϕx is bounded and $\overline{\phi x} \cap S = \emptyset$ is a thick subset of G/Γ if the following condition is satisfied: there exists a closed normal subgroup N of G such that G/N is a semisimple Lie group without compact factors and $\Gamma N/N$ is an irreducible lattice in G/N, and for every such subgroup N the one-parameter subgroup $\phi N/N$ of G/N is non-quasi-unipotent and $\overline{S\Gamma N/\Gamma N}$ has zero Haar measure in $G/\Gamma N$ (see [KM1]; in [KM1] existence of at least one subgroup N as above is not stated as a condition, but it seems to be required for the argument to hold when S is nonempty; this was pointed out to the author by A.N. Starkov in informal communication). It is conjectured in [KM1] that the condition on the measures of $\overline{S\Gamma N/\Gamma N}$, as above, may be replaced by the (weaker) condition that $\overline{S}$ has zero Haar measure in G/Γ.

In the case when $G = SL(n, \mathbb{R})$, $\Gamma = SL(n, \mathbb{Z})$ and $\{a_t\}$ is a one-parameter group of diagonal matrices, the points $g\Gamma$ in G/Γ for which the trajectory $\{a_t g\Gamma \mid t \geq 0\}$ is bounded in G/Γ are characterised by Kleinbock (see [Kl2], [Kl3]) in terms of a Diophantine condition (see § 9.2 for some details); see below for some stronger results known in the case of the one-parameter subgroups $\{g_t^{(p)}\}$ of diagonal matrices as in § 7.6.

When the conclusion as in Theorem 7.17 holds $S \cup \{\infty\}$ is said to be *escapable* for the flow induced by ϕ on the one-point compactification of G/Γ (see [KM1]). A notion called *horospherical escapability* of a set, which is free from considerations of Hausdorff dimension and is stronger than escapability in the above sense, is introduced in Kleinbock [Kl1]. We shall not go into the details of the notion, but recall the following result proved with the help of it.

Theorem 7.18 (Kleinbock [Kl1]). *Let G be a Lie group and Γ be a lattice in G. Let $g \in G$ and U be the expanding horospherical subgroup associated to g. Let M be a compact C^1 submanifold of G/Γ such that for any $x \in M$ the tangent space to Ux at x is not contained in the tangent space to M. Then M is escapable for the translations of G/Γ by $\{g^i \mid i = 1, 2, \ldots\}$.*

A stronger property than thickness is known in certain cases for the set of points on bounded orbits, in terms of the (α, β) game introduced by W.M. Schmidt, where $\alpha, \beta \in (0, 1)$. The reader is referred to Schmidt [Sch1] for details of the game. Here we content ourselves by recalling that if E is an 'α-winning' subset of a smooth manifold M, for some $\alpha > 0$, then E is not only a thick subset itself, but $\cap_{i=0}^{\infty} f_i^{-1}(E)$ is also a thick subset of M for any sequence $\{f_i\}$ of diffeomorphisms of M onto itself; a similar assertion also holds for any sequence of functions $\{f_i\}$ of biLipschitz maps of M onto itself, with common biLipschitz constants (see [D14] and [D19]).

Schmidt [Sch2] introduced a notion of a system of q linear forms in p variables being badly approximable, where $p, q \geq 1$, generalising the classical notion of badly approximable numbers, and showed that the set of badly approximable systems form an α-winning subset in the space of all systems of q linear forms in p variables, for all $\alpha \leq \frac{1}{2}$. It was shown in [D13] that for the action of the one-parameter subgroup $\{g_t^{(p)}\}$ $(1 \leq p < n)$ of diagonal matrices as in § 7.6 on $SL(n, \mathbb{R})/\Gamma$, where $\Gamma = SL(n, \mathbb{Z})$, the trajectory $\{g_t^{(p)} g\Gamma \mid t \geq 0\}$ of $g\Gamma$ is bounded if and only if a certain system of $(n-p)$ linear forms in p variables corresponding to g is badly approximable; it follows that the set of $g\Gamma$ for which the trajectory is bounded is an α-winning set in $SL(n, \mathbb{R})/\Gamma$, for all $\alpha \leq \frac{1}{2}$ (cf. [D14], [D19]).

A similar result also holds for flows induced by diagonalisable one-parameter subgroups on homogeneous spaces of Lie groups of $\mathbb{R}$-rank 1. We recall the result in a geometrical form, which is more convenient to express.

Theorem 7.19 ([AL], [D14]). *Let M be a locally symmetric space of non-compact type, with rank 1 and finite Riemannian volume. Let $p \in M$ and S_p be the unit tangent sphere at p. Let E be the subset of S_p consisting of all unit tangent directions ξ such that the geodesic through p in the direction of ξ is a bounded subset of M. Then we have the following:*

i) for any nonconstant C^1 curve $\sigma : [0, 1] \to S_p$ the set $\{t \in [0, 1] \mid \sigma(t) \in E\}$ is an α-winning subset of $[0, 1]$ for all $\alpha \leq \frac{1}{2}$.

ii) if M has constant negative curvature then E is an α-winning subset of S_p for all $\alpha \leq \frac{1}{2}$.

It may be noted that assertion (i) as above implies the conclusion as in Theorem 7.17 for Lie groups of $\mathbb{R}$-rank 1.

§ 8. Duality and lattice-actions on vector spaces

In this section we discuss applications of the results of the earlier sections to the study of actions of some discrete groups (lattices in certain Lie groups) on vector spaces.

8.1. Duality between orbits. Let G be a connected Lie group, Γ be a lattice in G and let H be a closed subgroup of G. In the earlier sections we considered dynamics of the H-action on G/Γ. On the other hand one can also consider the Γ-action on G/H. There is a simple duality between orbit behaviour of the two actions (see also § 4.5).

Proposition 8.1. *Let G, Γ and H be as above. Then for $g \in G$, the Γ-orbit of gH is dense in G/H if and only if the H-orbit of $g^{-1}\Gamma$ is dense in G/Γ.*

Now let G be as above and consider a linear action of G on a finite-dimensional vector space V. Let $v \in V$ and H be the isotropy (stability) subgroup of v, with respect to the G-action. Proposition 8.1 can be used to study the closure of Γv in $Gv \subseteq V$, using results on the H-action on G/Γ. The following examples of this approach may be worth noting.

Theorem 8.1 (Greenberg; see [AGH], Appendix). *If either $V = \mathbb{R}^n$ and $G = SL(n, \mathbb{R})$ or $V = \mathbb{R}^{2n}$ and $G = Sp(n, \mathbb{R})$ (the symplectic group viewed as a subgroup of $GL(2n, \mathbb{R})$) then for any uniform lattice Γ in G the Γ-orbit of any nonzero vector is dense in V.*

In these cases the isotropy subgroups contain maximal horospherical subgroups acting minimally on G/Γ as above (cf. Theorem 7.5). Theorem 8.1 proves in particular a conjecture of K. Mahler for values of quadratic forms (see Greenberg [AGH], Appendix).

Another example, where the lattice involved is nonuniform, is given by the following:

Theorem 8.2 (see [DR]). *Let $\Gamma = SL(n, \mathbb{Z})$ and $1 \leq p < n$. Let V be the cartesian product of p copies of $\mathbb{R}^n$. Then for $v = (v_1, \ldots, v_p)$ the Γ-orbit* *of v (under the componentwise action) is dense in V if and only if the linear span of $\{v_1, \ldots, v_p\}$ does not contain any nonzero integral vector.*

More generally, if $v = (v_1, \ldots, v_p) \in V$ and W is a subspace such that the integral vectors in the linear span of $\{v_1, \ldots, v_p\} \cup W$ are contained in W and span W, then for the subgroup $\Gamma = \{\gamma \in SL(n, \mathbb{Z}) \mid \gamma(w) = w$ for all $w \in W\}$ the Γ-orbit of v is dense in V.

(It may be mentioned here that in Proposition 4.3 in [DR] the assertion as in the second part of the theorem is made under a weaker hypothesis, which however is inadequate and needs to be replaced by the one given here).

A result analogous to Theorem 8.2 is also proved in [DR] for orbits under $Sp(n, \mathbb{Z})$ on symplectic p-tuples $(v_1, \ldots, v_p)$, where $v_1, \ldots, v_p \in \mathbb{R}^{2n}$ and $p \leq n$; a p-tuple $(v_1, \ldots, v_p)$ is said to be symplectic if $\omega(v_i, v_j) = 0$ for all $i, j = 1, \ldots, p$, where ω is the symplectic form defining the symplectic group.

The reader is also referred to Witte, Yukie and Zierau [WYZ] for explicit conditions on orbits of lattices on certain prehomogeneous vector spaces to be dense, proved using Ratner's theorem on orbit closures (see Theorem 7.1).

We note also the following consequence of Theorem 6.7; the argument is not via Proposition 8.1, but nevertheless depends on consideration of the dual action as above (see [D15] for details).

Theorem 8.3. *Let G be a Lie group and Γ be a lattice in G. Consider a linear action of G on a finite-dimensional vector space V. Let $v \in V$ be such that the following conditions are satisfied:*

i) there exists a sequence $\{g_i\}$ in G such that $g_i v \to 0$, as $i \to \infty$, and

ii) the isotropy subgroup of v contains a connected Ad-unipotent subgroup U which is not contained in any proper closed subgroup H such that $H \cap \Gamma$ is a lattice in H.

Then there exists a sequence $\{\gamma_i\}$ in Γ such that $\gamma_i v \to 0$, as $i \to \infty$.

The theorem implies in particular that if G is a semisimple Lie group with trivial center and Γ is a lattice in G then for any Ad-unipotent element u which is not contained in a proper closed subgroup H of G such that $H \cap \Gamma$ is a lattice in H, the Γ-conjugacy class $\{\gamma u \gamma^{-1} \mid \gamma \in \Gamma\}$ contains the identity element in its closure (see [D15]). For the case of $G = SL(n, \mathbb{R})$ and $\Gamma = SL(n, \mathbb{Z})$, $n \geq 2$, the argument involved in the proof of Theorem 8.3 in [D15] shows also that any unipotent matrix in $SL(n, \mathbb{R})$ which does not leave invariant a proper nonzero rational subspace of $\mathbb{R}^n$, can be conjugated into any neighbourhood of the identity, by elements of $SL(n, \mathbb{Z})$.

8.2. Duality of invariant measures. Let G be a connected unimodular Lie group and C and H be two closed unimodular subgroups of G. We fix two Haar measures λ_C and λ_H on C and H respectively. Then there is a natural duality between H-invariant measures on G/C and C-invariant measures on G/H arising as follows (see [Fu]). To any measure μ on G/C we associate the (unique) measure μ' on G determined by the condition

$$\int_G f(g)\,\mathrm{d}\mu'(g) = \int_{G/C}\int_C f(gx)\,\mathrm{d}\lambda_C(x)\,\mathrm{d}\mu(gC)$$

for all continuous functions f, on G, with compact support (recall that by a measure we mean a locally finite measure). We note that μ' is invariant under the action of C on G by right translations. Clearly μ is a H-invariant measure on G/C if and only if μ' is H-invariant under the action of H on G by left translations. This gives a one-one correspondence between H-invariant measures on G/C and the measures on G which are C-invariant under the action by right translations and H-invariant under the action by left translations. In view of this, and the corresponding statement for C-invariant measures on G/H, the inversion map of G (namely $g \mapsto g^{-1}$) gives a duality between the two classes of measures as above.

In fact the results on classification of invariant measures of horospherical subgroups prior to Ratner's work (see [Fu], [V1], [D10]) were obtained by studying the dual measures. Results of § 6 can now be applied the other way. We note in particular that by Theorems 6.1 and 6.7 we have the following:

Theorem 8.4. *Let G be a connected unimodular Lie group and Γ be a lattice in G. Let H be a closed subgroup generated by Ad-unipotent one-parameter subgroups. Let π be an ergodic Γ-invariant measure on G/H. Then there exist a closed subgroup L containing H and an element $g \in G$ such that ΓgL is closed, π is supported on $\Gamma gL/H$ and the restriction of π to gL/H is gLg^{-1}-invariant.*

Corollary 8.1 (cf. [D7]). *Consider the natural action of $\Gamma = SL(n,\mathbb{Z})$ on $\mathbb{R}^n$. If π is an ergodic Γ-invariant (locally finite) measure, then π is a scalar multiple of either the Lebesgue measure or the counting measure on the (discrete) orbit of a vector of the form tp, where $t \in \mathbb{R}$ and p is an integral vector.*

Recently A. Nogueira has obtained a proof of the above corollary by a different approach (cf. [No1] and [No2]). In [No1] the condition of local finiteness is also relaxed and it is shown that, in fact, upto scalar multiples the Lebesgue measure is the only $SL(n,\mathbb{Z})$-invariant measure which is locally finite at one point (or, equivalently, some nonempty open set has finite measure) and assigns zero measure to the set of all points of the form tp, with $t \in \mathbb{R}$ and p an integral vector. A similar strengthening can be obtained in the more general situation as in Theorem 8.4, using Ratner's theorem on distribution of orbits of unipotent flows (see Theorem 6.10); we shall however not go into the details of this here.

§ 9. Applications to Diophantine approximation

In this section we discuss applications of various results from earlier sections to Diophantine approximation. The reader is also referred to the surveys [Marg3] and [D22] and the recent papers [Sk] and [ST] for other applications of ergodic theory on homogeneous spaces.

9.1. Polynomials in one variable. Using Theorem 2.10 for affine automorphisms of tori a proof can be given for the following classical result due to H. Weyl on the distribution of the fractional parts of values of a polynomial (see [CFS], Ch. 7, § 2).

Theorem 9.1. *Let $P(t) = a_0t^n + a_1t^{n-1} + \cdots + a_n$ be a polynomial of degree $n \geq 1$, with real coefficients. Suppose that at least one of $a_0, \cdots, a_{n-1}$ is irrational. Then the sequence $\langle P(i) \rangle$ of fractional parts of $P(i)$ is uniformly distributed in the interval* $[0, 1]$, *with respect to the Lebesgue measure.*

Using minimality of certain flows on nilmanifolds the following analogue of another theorem of H. Weyl was deduced in [AGH], Ch. VIII; (we recall that a set of integers is said to be *relatively dense* if the difference between the successive integers in the set is bounded above).

Theorem 9.2. *Let $P_i(t) = \Sigma_{j=1}^n a_{ij}t^j$, be n nonzero polynomials with integral coefficients a_{ij}, $i, j = 1, \ldots, n$. Let $\alpha_1, \ldots, \alpha_n$ be real numbers which together with* 1 *form a linearly independent set over $\mathbb{Q}$. Then for any $\epsilon > 0$ and $\theta_1, \ldots, \theta_n \in \mathbb{R}$ there exists a relatively dense set M of integers such that for each $i = 1, \ldots, n$ and $m \in M$, $\alpha_i P_i(m) - \theta_i$ differs from an integer by at most ϵ.*

9.2. Values of linear forms. Theorem 8.2 in particular yields the following result for values of linear forms. We recall that an integral n-tuple $(x_1, \ldots, x_n)$ is said to be *primitive* if no integer exceeding 1 divides x_i for all i.

Theorem 9.3. *Let $\xi_1, \ldots, \xi_p$ and $\eta_1, \ldots, \eta_q$ be linear forms in n real variables, where $1 \leq p + q < n$. Suppose that $\eta_1, \ldots, \eta_q$ are rational forms (forms with rational coefficients) and any linear combination of the form $\Sigma \lambda_i\xi_i + \Sigma \mu_j\eta_j$, with $\lambda_1, \ldots, \lambda_p, \mu_1, \ldots, \mu_q \in \mathbb{R}$, which is a rational form is contained in the span of $\eta_1, \ldots, \eta_q$. Let $b_1, \ldots, b_q \in \mathbb{R}$ be such that there exist $y_1, \ldots, y_n \in \mathbb{Z}$ satisfying $\eta_j(y_1, \ldots, y_n) = b_j$ for $j = 1, \ldots, q$. Then for any $a_1, \ldots, a_p \in \mathbb{R}$ and $\epsilon > 0$ there exists a primitive integral n-tuple $(x_1, \ldots, x_n)$ such that*

$$|\xi_i(x_1, \ldots, x_n) - a_i| < \epsilon \text{ for all } i = 1, \ldots, p$$

and $\eta_j(x_1, \ldots, x_n) = b_j$ for all $j = 1, \ldots q$.

Similarly the analogous theorem for the $Sp(n, \mathbb{Z})$-orbits of n-tuples from $\mathbb{R}^{2n}$ yields a Diophantine result for n-tuples of symplectic linear forms (see [DR]).

In the theory of Diophantine solutions for systems of linear forms there are notions of singular and badly approximable systems (see [Ca1] and [Sch2], respectively). As mentioned earlier the notions turned out to be related to divergence and boundedness, respectively, of certain trajectories on $SL(n, \mathbb{R})/SL(n, \mathbb{Z})$ of a one-parameter subgroup of diagonal matrices (see §§ 7.6, 7.7 and [D13]). The concepts were recently generalised by D. Kleinbock (see [Kl2] and [Kl3]). The notion of badly approximable systems is generalised as follows (for the generalisation of singular systems we refer the reader to [Kl2] and [Kl3]). Let p and q be two positive integers and $n = p + q$. Let $r_1, \ldots, r_p$ and $s_1, \ldots, s_q$ be positive real numbers such that $\Sigma r_i = \Sigma s_j = 1$. Put $r = (r_1, \ldots, r_p) \in \mathbb{R}^p$ and $s = (s_1, \ldots, s_q) \in \mathbb{R}^q$. For $m = p$ or q, $v = (v_1, \ldots, v_m)$ and $t = (t_1, \ldots, t_m)$ in $\mathbb{R}^m$, with $t_k > 0$ for all $k = 1, \ldots, m$, let $\| v \|_t = \max\{|v_1|^{1/t_1}, \ldots, |v_m|^{1/t_m}\}$. A $n \times n$ matrix $g \in SL(n, \mathbb{R})$ with L and M as the $p \times n$ and $q \times n$ matrices formed by the first p and the last q rows of g respectively, is said to be (r, s)-*loose* if

$$\liminf_{x \in \mathbb{Z}^n, M(x) \to \infty} \| L(x) \|_r \| M(x) \|_s > 0.$$

This is related to the notion of badly approximable systems of linear forms via the following: if $r_i = 1/p$ for all $i = 1, \ldots, p$, $s_j = 1/q$ for all $j = 1, \ldots, q$ and $g = (g_{kl})$ is the matrix such that $g_{kl} = \delta_{kl}$ whenever either $k > p$ or $l \leq p$ then g is (r, s)-loose if and only if the system of p linear forms in q variables corresponding to the $p \times q$ matrix $L = (x_{kl})$ with $x_{kl} = g_{k(p+l)}$ for all $k = 1, \ldots, p$ and $l = 1, \ldots, q$ (namely the upper right block in g) is badly approximable in the sense of Schmidt.

Now let $G = SL(n, \mathbb{R})$, $\Gamma = SL(n, \mathbb{Z})$ and $\{a_t\}$ be the one-parameter subgroup $\{\text{diag}\,(e^{r_1 t}, \ldots, e^{r_p t}, e^{-s_1 t}, \ldots, e^{-s_q t})\}$, of diagonal matrices. Then for $g \in G$ the trajectory $\{a_t g\Gamma \mid t \geq 0\}$ is bounded in G/Γ if and only if g is (r, s)-loose (cf. [Kl2], [Kl3]). Together with Theorem 7.17 this implies the following result, generalising the corresponding result of Schmidt for badly approximable systems of linear forms (cf. [Kl2], [Kl3], [D13]).

Theorem 9.4 (Kleinbock [Kl3]). *The set of* (r, s)*-loose matrices is a thick subset of* $SL(n, \mathbb{R})$.

9.3. Diophantine approximation with dependent quantities. A vector $v = (v_1, \ldots, v_n) \in \mathbb{R}^n$ is said to be *very well approximable* (VWA, for short) if for some $\epsilon > 0$ there exist infinitely many $m \in \mathbb{Z} - \{0\}$ such that $\max_{1 \leq i \leq n} |mv_i - x_i|^n \leq |m|^{-1-\epsilon}$ for some $x_1, \ldots, x_n \in \mathbb{Z}$ and it is said to be *very well multiplicative approximable* (VWMA) if for some $\epsilon > 0$ there exist infinitely many $m \in \mathbb{Z} - \{0\}$ such that $\Pi_{i=1}^n |mv_i - x_i| \leq |m|^{-1-\epsilon}$ for some $x_1, \ldots, x_n \in \mathbb{Z}$. If v is VWA then it is necessarily VWMA. A submanifold M

of $\mathbb{R}^n$ is said to be *extremal* if the set of VWA vectors in M is a null subset of M (with respect to the Lebesgue measure class on M) and it is said to be *strongly extremal* if the set of VWMA vectors in M is a null subset. In Kleinbock and Margulis [KM2], via a sharpening of the arguments involved in proving nondivergence of trajectories of Ad-unipotent flows (see § 6.4) the following result is proved; the result verifies a conjecture of Sprindzuk (see [KM2] for the history of the problem).

Theorem 9.5 (Kleinbock and Margulis [KM2]). *Let $f_1, \ldots, f_n$ be real analytic functions, defined on an open subset U of $\mathbb{R}^d$ for some $d \geq 1$, which together with the constant function 1 form a linearly independent set over $\mathbb{R}$. Then $\{(f_1(x), \ldots, f_n(x)) \in \mathbb{R}^n \mid x \in U\}$ is strongly extremal.*

The theorem is in fact proved under a weaker 'infinitesimal' condition on the functions $f_1, \ldots, f_n$, in the place of real analyticity (see [KM2]).

9.4. Values of quadratic forms. Let Q be a (real) quadratic form in n variables, say $Q(x_1, \ldots, x_n) = \Sigma_{i,j=1}^n a_{ij}x_ix_j$, where (a_{ij}) is a symmetric $n \times n$ matrix with entries in $\mathbb{R}$. We recall that Q is said to be nondegenerate if the matrix (a_{ij}) is nonsingular and it is said to be indefinite if there exist $x_1, \ldots, x_n \in \mathbb{R}$, not all 0, such that $Q(x_1, \ldots, x_n) = 0$. It was a longstanding conjecture, due to A. Oppenheim, that if Q is a nondegenerate indefinite quadratic form in $n \geq 3$ variables, which is not a scalar multiple of a form with rational coefficients, then $\{Q(x_1, \ldots, x_n) \mid x_1, \ldots, x_n \in \mathbb{Z}\}$, namely the set of values of Q on integral n-tuples, is dense in $\mathbb{R}$; that is, given a quadratic form Q satisfying the above conditions for every $a \in \mathbb{R}$ and $\epsilon > 0$ there exist $x_1, \ldots, x_n \in \mathbb{Z}$ such that

$$|Q(x_1, \ldots, x_n) - a| < \epsilon.$$

The conjecture was settled in the affirmative by G.A. Margulis; there was a considerable amount of work done on the conjecture, prior to the work of Margulis, by number-theoretic methods and the conjecture was known hold if $n \geq 21$ for a general quadratic form as above and for lower values of n for forms satisfying certain additional conditions (see [Marg6] for historical details). There were subsequent strengthenings of Margulis' result and simplification of proofs due to Margulis and the present author. The reader is referred to [B2], [D22] and [Marg6] for detailed accounts of the developments in the area. Here we shall recall some select results from the area. For convenience we view the quadratic forms as functions on $\mathbb{R}^n$ (realised as the space of column vectors).

We begin with the following consequence of Ratner's theorem on orbit closures (see Theorem 7.1) which compares also the 'angles' along with the values of the quadratic form as in Oppenheim's conjecture; it may be mentioned that Borel and Prasad [BP] discuss also the values of quadratic forms at S-integral points, which we shall not go into.

Theorem 9.6 (cf. [BP]). *Let Q be a nondegenerate indefinite quadratic form on $\mathbb{R}^n$, $n \geq 3$, and let B be the corresponding bilinear form (defined by $B(v, w) = \frac{1}{4}\{Q(v + w) - Q(v - w)\}$ for all $v, w \in \mathbb{R}^n$). Then for any $v_1, \ldots, v_k \in \mathbb{R}^n$, where $k \leq (n - 1)$, and $\epsilon > 0$ there exist $x_1, \ldots, x_k \in \mathbb{Z}^n$ such that*

$$| B(x_i, x_j) - B(v_i, v_j) | < \epsilon \text{ for all } i, j = 1, \ldots, k;$$

moreover, $x_1, \ldots, x_k$ can be chosen to be a part of a basis of $\mathbb{Z}^n$ (and hence primitive, in particular).

We note also the following result which was deduced in [DM1] from the results on orbit closures of a unipotent one-parameter subgroup of $SL(3, \mathbb{R})$ on $SL(3, \mathbb{R})/SL(3, \mathbb{Z})$.

Theorem 9.7. *Let Q be a nondegenerate indefinite quadratic form on $\mathbb{R}^3$ and let L be a linear form on $\mathbb{R}^3$ such that the plane $\{v \in \mathbb{R}^3 \mid L(v) = 0\}$ and the double cone $\{w \in \mathbb{R}^3 \mid Q(w) = 0\}$ intersect tangentially along one line. Suppose that no linear combination of Q and L^2 is a rational quadratic form. Then for any $a, b \in \mathbb{R}$ and any $\epsilon > 0$ there exists a primitive integral triple x such that*

$$|Q(x) - a| < \epsilon \text{ and } |L(x) - b| < \epsilon.$$

It may be mentioned that the conclusion as above does not hold in general if the cone and the plane intersect in two lines (oral communication from Margulis); see [D23] for details.

We next discuss some results on the asymptotics of the number of solutions of quadratic inequalities as in Oppenheim's conjecture, in certain bounded regions. For this purpose we first fix some notation. For positive integers p and q let $\mathcal{O}(p, q)$ denote the set of all nondegenerate quadratic forms on $\mathbb{R}^n$, where $n = p + q$, with signature (p, q) (namely, equivalent to the quadratic form $x_1^2 + \cdots + x_p^2 - x_{p+1}^2 - \cdots - x_n^2$ over $\mathbb{R}$) and discriminant ± 1. We equip $\mathcal{O}(p, q)$ with the topology of pointwise convergence. For a positive continuous function ν on the unit sphere $\{v \in \mathbb{R}^n \mid \| v \| = 1\}$ and $r > 0$ we denote by $\Omega(\nu, r)$ the set $\{v \in \mathbb{R}^n - (0) \mid \| v \| < r\nu(v/\| v \|)\}$; it may be noted that when ν is identically 1, $\Omega(\nu, r)$ is the intersection with $\mathbb{R}^n - (0)$ of the ball of radius r centered at the origin, which is one of the main cases of interest. We would be interested in the asymptotic behaviour of the number of integral points in regions of the form $\{v \in \Omega(\nu, r) \mid a < Q(v) < b\}$, where $Q \in \mathcal{O}(p, q)$ and $a, b \in \mathbb{R}$, as $r \to \infty$. We note that when p or q is at least 3 there exists a constant $\lambda(\nu, Q)$, depending on ν as above and $Q \in \mathcal{O}(p, q)$, such that asymptotically the volumes of the regions are given by the following (see Eskin, Margulis and Mozes [EMM]):

$$\text{vol}\,(\{v \in \Omega(\nu, r) \mid a < Q(v) < b\}) \sim \lambda(\nu, Q)(b - a)r^{n-2}.$$

For a finite set S we denote by $\#S$ the cardinality of S. The following result describes the number of integral solutions in regions as above, in a way which is uniform over certain compact sets of quadratic forms.

Theorem 9.8 (Eskin, Margulis and Mozes [EMM]; see also [DM3]). *Let $\mathcal{K}$ be a compact subset of $\mathcal{Q}(p,q)$ for some positive integers p and q, with $p+q=n$. Let a positive continuous function ν on the unit sphere in $\mathbb{R}^n$ and $a,b \in \mathbb{R}$, with $a<b$ be given. For $Q \in \mathcal{Q}(p,q)$ and $r>0$ let $f_Q(r)$ be the ratio defined by*

$$f_Q(r) = \frac{\#\{x \in \Omega(\nu,r) \cap \mathbb{Z}^n \mid a < Q(x) < b\}}{\operatorname{vol}(\{v \in \Omega(\nu,r) \mid a < Q(v) < b\})}.$$

Then for any $\theta > 0$ there exists a finite subset $\mathcal{F}$ of $\mathcal{K}$ such that each $Q \in \mathcal{F}$ is a scalar multiple of a rational form, and for any compact subset $\mathcal{C}$ of $\mathcal{K} - \mathcal{F}$ there exists a $r_0 > 0$ such that for all $Q \in \mathcal{C}$ and $r \geq r_0$, $f_Q(r) > (1-\theta)$, and furthermore, if $\max\{p,q\} \geq 3$ *then $f_Q(r) < (1+\theta)$. In particular, when $p \geq 3$, for any $Q \in \mathcal{Q}(p,q)$ which is not a multiple of a rational form, asymptotically as $r \to \infty$ we have*

$$\#\{x \in \Omega(\nu,r) \cap \mathbb{Z}^n \mid a < Q(x) < b\} \sim \lambda(\nu,Q)(b-a)r^{n-2},$$

where $\lambda(\nu,Q)$ is the constant as above.

Exclusion of the signature-pairs with $p,q \in \{1,2\}$ in part of the assertion as above is necessary; in fact there exist quadratic forms Q with these signatures for which for any $\epsilon > 0$, there exists a $c > 0$ such that $f_Q(r_i) > c(\log r_i)^{1-\epsilon}$ for a sequence $\{r_i\}$ tending to infinity (see Eskin, Margulis and Mozes [EMM] or Margulis [Marg6]), f_Q being the function defined as in the theorem. Nevertheless, the asymptotic formula for the number of solutions as in Theorem 9.8 holds for *almost all* quadratic forms in $\mathcal{Q}(p,q)$ for $p,q \in \{1,2\}$ (see [EMM], [Marg6]).

The following uniform estimate for values close to 0 was also proved in [DM3], for $n \geq 5$; the condition $n \geq 5$ may be seen to be necessary as there exist nondegenerate indefinite rational quadratic forms in 4 variables which have no nontrivial integral zeros.

Theorem 9.9 (cf. [DM3]). *Let $\mathcal{K}$ be a compact subset of $\mathcal{Q}(p,q)$ for some positive integers p and q, with $n = p+q \geq 5$, and let ν be a positive continuous function on the unit sphere in $\mathbb{R}^n$. Then for any $\epsilon > 0$ there exist $c > 0$ and $r_0 \geq 0$ such that for any $Q \in \mathcal{K}$ and $r \geq r_0$*

$$\#\{x \in \Omega(\nu,r) \cap \mathbb{Z}^n \mid |Q(x)| < \epsilon\} \geq c \operatorname{vol}(\{v \in \Omega(\nu,r) \mid |Q(v)| < \epsilon\}).$$

We also recall here the following upper bounds on the number of solutions of the inequalities as above.

Theorem 9.10 (cf. [EMM], [Marg6]). *Let the notation p, q, $\mathcal{K}$, ν, a and b be as in Theorem 9.8. Then there exists a constant $c > 0$ such that for any $Q \in \mathcal{K}$,*

$$\#\{x \in \Omega(\nu, r) \cap \mathbb{Z}^n \mid a < Q(x) < b\}$$

is bounded by cr^{n-2}, for all $r \geq 1$ if p or q exceeds 2, and by $cr^{n-2}\log r$, for all $r \geq 2$ for $p, q \in \{1, 2\}$.

9.5. Forms of higher degree. For a generic homogeneous form F of degree $d \geq 3$ in $n \geq 3$ variables, by a theorem of Jordan, the subgroup of $SL(n, \mathbb{R})$ leaving it invariant is finite (see [Jo] or [B1], § 6.5). Hence the techniques as in the preceding discussion are not applicable, in general, to forms of higher degree. However, values at integral points have been studied for certain classes of forms of higher degree, using similar arguments. We mention in particular that using Ratner's theorem on orbit closures (see Theorem 7.1) and the theory of prehomogeneous vector spaces Yukie has described families of cubic forms in 5 variables and quartic forms in 4 variables, for (each of) which the values at primitive integral points form a dense subset of $\mathbb{R}$ (see [Y1] and [Y2]). It may also be observed that one can get a class homogeneous cubic forms from Theorem 9.7, for which the values at primitive integral points are dense.

9.6. Integral points on algebraic varieties. We next describe an application to studying the asymptotic pattern of the integral points on certain algebraic subvarieties of vector spaces. Let W be a finite-dimensional real vector space with a $\mathbb{Q}$ structure and let V be a Zariski-closed subvariety of W defined over $\mathbb{Q}$. For $r > 0$ let B_r denote the ball of radius r (with respect to the Euclidean norm on W) centered at the origin in W. There is a general problem of understanding the growth of the number of integral points in $V \cap B_r$, as $r \to \infty$. The following is deduced in [EMS1] from of Theorem 6.14, towards this problem.

Theorem 9.11 (Eskin, Mozes and Shah [EMS1]). *Let the notation be as above and suppose that V is the orbit of a real reductive algebraic group G defined over $\mathbb{Q}$ under an action via a $\mathbb{Q}$-representation of G into $GL(W)$. Let Γ be an arithmetic lattice in G associated with the $\mathbb{Q}$-structure. Let H be the stabiliser of an integral point x_0 in V. Suppose that H is reductive and that there is no real algebraic $\mathbb{Q}$-subgroup L of G such that $H^0 \subseteq L$ and* $\dim H < \dim L < \dim G$. *Then asymptotically, as $r \to \infty$,*

$$\#(V \cap B_r \cap \Gamma x_0) \sim \lambda(\{gH \mid gx_0 \in B_r\}),$$

where λ is a G-invariant measure on G/H.

The G-invariant measure λ involved in the conclusion is made precise in [EMS1], but we shall not go into it here. By a theorem of Borel and Harish-Chandra the integral points on a variety V as above are contained in finitely many orbits of Γ. The result therefore yields the asymptotics of the number of integral points in $V \cap B_r$. The result applies in particular to affine symmetric varieties for which the asymptotics were first obtained by Duke, Rudnik and Sarnak and a simpler proof was given by Eskin and McMullen (see [EM] and [EMS1] for details and references).

There is also a generalisation of Theorem 9.11 given in [EMS1] which does not involve the maximality condition as in the theorem. The validity of the analogous conclusion then involves, of necessity, certain additional conditions; these are of a technical nature and we shall not go into them here. We note however the following application of the general theorem; in this case the variety involved is not affine symmetric.

Corollary 9.1 (Eskin, Mozes and Shah [EMS1]). *Let P be a monic polynomial in one variable, with integral coefficients and degree $n \geq 2$. Suppose that P is irreducible over $\mathbb{Q}$. Let $M_n(\mathbb{Z})$ denote the set of $n \times n$ matrices with integral entries. For any matrix X let $\chi(X)$ denote the characteristic polynomial of X. Then there exists a computable constant $c_P > 0$ such that, asymptotically as $T \to \infty$,*

$$\#\{X = (x_{ij}) \in M_n(\mathbb{Z}) \mid \chi(X) = P \text{ and } \Sigma_{i,j}\, x_{ij}^2 \leq T^2\} \sim c_P T^{n(n-1)/2}.$$

The value of c_P is determined explicitly in [EMS1] for polynomials all whose roots are real. The proof of the theorem can be modified to obtain an analogous estimate (involving an additional factor of $(\log T)^{k-1}$, where k is the number of irreducible factors of P) for any (not necessarily irreducible) polynomial P which has no multiple roots (informal communication from N.A. Shah).

An application of the results on limits of translates of homogeneous measures is also made in [EMS1] to the asymptotics of the number of integral $m \times n$ matrices X of norm at most T, satisfying ${}^tX A X = B$, where A and B are given integral symmetric matrices of orders $m \times m$ and $n \times n$ respectively (this corresponds to the number of ways of 'representing' B by A over $\mathbb{Z}$, when A and B are viewed as quadratic forms).

§ 10. Classification and related questions

In this section we shall discuss results concerning isomorphisms, equivalences etc. of group-induced systems, both in the measure-theoretic and topological contexts.

10.1. Metric isomorphisms and factors. The classical theorem of von Neumann asserts that any ergodic finite-measure-preserving automorphism with discrete spectrum is isomorphic to a translation of a compact abelian group and that two such translations are isomorphic if and only if their unitary spectra coincide. Among the class of ergodic systems on homogeneous spaces of Lie groups, that we have been dealing with, translations of tori are precisely the ones with discrete spectrum. Thus none of them is isomorphic to any group-induced system outside the class and isomorphisms between them are determined by their spectra.

Two ergodic translations may also be distinguished by the discrete part of their spectrum, if it is different. The discrete part arises precisely from the maximal Euclidean factor of the homogeneous space (cf. Theorem 4.9) and if the corresponding factors are non-isomorphic then the translations themselves are non-isomorphic. For weak mixing translations however the spectrum is of Lebesgue type with infinite multiplicity, so no further information can be drawn from the spectrum, with regard to isomorphisms.

We have noted earlier (see § 2) that certain of our systems have quasi-discrete spectrum and generalised discrete spectrum. The information in this regard can be applied to the problem of metric isomorphisms. We shall however not go into the details (see Ch. 2 of this volume, and Zimmer [Zi1]).

By Ornstein's isomorphism theorem (see [CFS], Ch. X) Bernoulli shifts with equal entropy are isomorphic. Also, by Sinai's theorem on factors any Bernoulli shift is a factor of every ergodic transformation with entropy at least as much as that of the shift. We have noted earlier certain sufficient conditions for affine automorphisms of nilmanifolds and translations of general homogeneous spaces of finite volume to be isomorphic to Bernoulli shifts (see §§ 2.1, 2.2 and 4.9). Also, the entropy of an affine automorphism of G/Γ, where G is a Lie group and Γ is a uniform lattice in G, is known (see § 4.8). Thus we see that various translations are isomorphic (and also obtain information on their being factors of other transformations). E.g. the translations of $SL(m,\mathbb{R})/\Gamma_1$ and $SL(n,\mathbb{R})/\Gamma_2$, where Γ_1 and Γ_2 are uniform lattices in $SL(m,\mathbb{R})$ and $SL(n,\mathbb{R})$ respectively, by $\operatorname{diag}(a_1,\ldots,a_m)$ and $\operatorname{diag}(b_1,\ldots,b_n)$, respectively, where $a_1 \geq \cdots \geq a_m > 0$ and $b_1 \geq \cdots \geq b_n > 0$, are isomorphic if and only if $a_1^{m-1}a_2^{m-3}\cdots a_m^{-m+1} = b_1^{n-1}b_2^{n-3}\cdots b_n^{-n+1}$, the logarithms of which are the entropies of the respective translations (independent of the lattices involved).

On the other hand when two translations have unequal entropies they are nonisomorphic. If G is a connected Lie group such that $G/\overline{NZ}$, where N is the nilradical of G and Z is the center of G, is noncompact then for any uniform lattice Γ in G, the homogeneous space G/Γ admits translations with any given value for the entropy and so there are uncountably many isomorphism classes represented by these systems; for affine automorphisms of nilmanifolds however only countably many values are possible for the entropy.

10.2. Metric rigidity. We have seen above that translations of finite-volume homogeneous spaces of the form $SL(n, \mathbb{R})/\Gamma$ by diagonal matrices can be metrically isomorphic for nonisomorphic lattices and even for different values of n (when the spaces are of different dimensions and, in particular, not affinely isomorphic). Systems involving unipotent transformations however turn out to be rigid, as seen from the following results.

Theorem 10.1 (Parry [P5]). *Let N_1 and N_2 be two simply connected nilpotent Lie groups and Γ_1 and Γ_2 be lattices in N_1 and N_2 respectively. Let τ_1 and τ_2 be affine automorphisms of N_1/Γ_1 and N_2/Γ_2 respectively. Suppose that τ_1 is ergodic and τ_2 is unipotent (see § 2.3 for definition). Let $\varphi : N_1/\Gamma_1 \to N_2/\Gamma_2$ be a measurable map of N_1/Γ_1 onto a conull subset of N_2/Γ_2, such that $\varphi \circ \tau_1 = \tau_2 \circ \varphi$ almost everywhere on N_1/Γ_1. Then there exists an affine map ψ of N_1/Γ_1 onto N_2/Γ_2 such that $\phi = \psi$ almost everywhere.*

The reader is also referred in this context to Parry [P6] where the class of ergodic unipotent affine automorphisms is characterised in terms of abstractly defined conditions.

Theorem 10.2 (Ratner, Witte; cf. [Wi1]). *Let G_1 and G_2 be simply connected Lie groups. Let $C_1 \in \mathscr{F}_0(G_1)$ and $C_2 \in \mathscr{F}_0(G_2)$. Let $u_1 \in G_1$ and $u_2 \in G_2$ be Ad-unipotent elements such that the corresponding translations of G_1/C_1 and G_2/C_2 are ergodic. Let $\theta : G_1/C_1 \to G_2/C_2$ be a measure-preserving Borel map such that $\theta \circ T_{u_1} = T_{u_2} \circ \theta$ a.e.. Then there exists an affine map ψ of G_1/C_1 onto G_2/C_2 such that $\phi = \theta$ almost everywhere.*

The same conclusion also holds if, in the place of Ad-unipotent, we assume u_1 and u_2 to be such that the corresponding translations are weak mixing and have zero entropy (see Witte [Wi1]). The conclusion does not hold, in general, without the additional assumption of weak mixing [Wi1]. One can however assert that if the translations are ergodic and have zero entropy then any map θ as in the theorem induces a map of the maximal solvmanifold factors of the homogeneous spaces and is affine on almost all fibers over the solvmanifold quotient (see Starkov [St9]).

Remark 10.1. If φ or θ as in Theorems 10.1 and 10.2 is a metric isomorphism then it coincides almost everywhere with an affine isomorphism of the homogeneous spaces. Thus the systems in the respective cases are affinely isomorphic whenever they are metrically isomorphic. The phenomenon is generally referred to as 'rigidity'.

Remark 10.2. The case of Theorem 10.2 with $G_1 = G_2 = SL(2, \mathbb{R})$ and C_1 and C_2 lattices in G, namely the rigidity of the horocycle flows was proved by Ratner in [R2]; the general case is obtained by developing upon the ideas in her papers [R2] and [R4]. Ratner's rigidity theorem implies in particular that the horocycle flow can not be embedded in a free measure-preserving action of $\mathbb{R}^d$, for $d \geq 2$ (see [R2]); this may be compared with Corollary 2.2. The reader is also referred to [FS], for a generalisation of the rigidity theorem to certain Fuchsian groups, in the place of lattices.

Similarly, for actions of Ad-unipotent subgroups there is also 'rigidity' with regard to measurable (metric) factors; all the measurable factors arise from the affine structure, in the following way. Let G be a Lie group and C be a closed subgroup such that G/C has finite Haar measure. Consider an action of a nilpotent Ad-unipotent subgroup U of G on G/C. Then firstly we have the homogeneous factors on quotients of the form G/D, where D is a closed subgroup containing C. Now consider one such factor G/D and let N be the largest normal subgroup of G contained in D. Let $G' = G/N$ and $D' = D/N$, so G/D is canonically equivalent to G'/D'. Let K be a compact subgroup of the group of affine automorphisms of G'/D' whose action is normalised by the U-action on $G'/D' = G/D$ and let $X = K\backslash G'/D'$. Then we get an action of U on X as a quotient of the action on G'/D', which is a factor of the U-action on G/C. It turns out that if the U-action on G/C is ergodic and there exists no proper closed subgroup of G containing U and acting transitively on G/C (e.g. if G is connected and Γ is a lattice in G), then these are the only possible factors, up to measure-preserving isomorphisms; furthermore the subgroup K as above can be chosen such that its action commutes with the U-action (see Witte [Wi3]; see [R3] for the case of horocycle flows obtained earlier).

10.3. Topological conjugacy. In this subsection we describe results concerning topological conjugacy of affine automorphisms and group-induced flows.

Theorem 10.3 (Walters [Wal2]). *Let $\tau_1 = T_a \circ \overline{\alpha}_1$ be an affine automorphism of $\mathbb{T}^m$ and $\tau_2 = T_b \circ \overline{\alpha}_2$ be an affine automorphism of $\mathbb{T}^n$, where $\alpha_1 \in GL(m, \mathbb{Z})$, $\alpha_2 \in GL(n, \mathbb{Z})$, $a \in \mathbb{T}^m$ and $b \in \mathbb{T}^n$. Suppose there exists a continuous surjective map $\psi : \mathbb{T}^m \to \mathbb{T}^n$ such that $\psi \circ \tau_1 = \tau_2 \circ \psi$. Then the following conditions are equivalent:*

i) there exists a nonaffine continuous surjective map φ of $\mathbb{T}^m$ onto $\mathbb{T}^n$ such that $\varphi \circ \tau_1 = \tau_2 \circ \varphi$;

ii) there exists a nontrivial character γ on $\mathbb{T}^m$ and an eigenvalue λ of α_2 such that $|\lambda| = 1$, $\gamma \circ \alpha_1^p = \gamma$ for some integer $p \geq 1$ and, for all such p, $\gamma(a\alpha_1(a) \cdots \alpha_1^{p-1}(a)) = \lambda^p$.

Further, if $m = n$ and ψ as above is a homeomorphism (and condition (ii) holds) then φ as in condition (i) can be chosen to be a nonaffine homeomorphism of $\mathbb{T}^n$.

A similar assertion also holds for affine automorphisms of nilmanifolds (see Walters [Wal3]), the details of which we shall not go into. The following results which are consequences of the above theorem in the case of tori, hold in the generality of nilmanifolds.

Theorem 10.4 (Walters [Wal3]). *Let N_1 and N_2 be connected nilpotent Lie groups and Γ_1 and Γ_2 be lattices in N_1 and N_2 respectively. Let τ_1 and τ_2 be affine automorphisms of N_1/Γ_1 and N_2/Γ_2 respectively. Suppose that τ_1 is weak mixing. Then any continuous surjective map $\psi : N_1/\Gamma_1 \to N_2/\Gamma_2$ such that $\psi \circ \tau_1 = \tau_2 \circ \psi$ is an affine map.*

In general two affine automorphisms of a nilmanifold N/Γ, where N is a simply connected nilpotent Lie group and Γ is a lattice in N, are affinely conjugate whenever they are topologically conjugate (see [CM]). Theorem 10.4 implies in particular that if $N_1 = N_2 = N$, $\Gamma_1 = \Gamma_2 = \Gamma$ and τ_1 and τ_2 are ergodic automorphisms of N/Γ then any homeomorphism ψ of N/Γ such that $\psi \circ \tau_1 = \tau_2 \circ \psi$ is an affine isomorphism (see [Wal3]; see also [CM]). However, if one of τ_1 or τ_2 is not ergodic, and there exists a homeomorphism ψ such that $\psi \circ \tau_1 = \tau_2 \circ \psi$ then there also exists one such which is nonaffine (see [Wal3]). Analogous assertions are also proved in [Wal3] and [CM] for translations in the place of automorphisms.

Topological conjugacy of one-parameter flows on compact groups, induced by inner conjugation by one-parameter subgroups is studied in [Bh]. Various classes of groups have been identified with the property that the flows induced by conjugation by one-parameter subgroups $\{g_t\}$ and $\{h_t\}$ are topologically conjugate if and only if there exists an automorphism τ of the group such that $\tau(g_t) = h_t$ for all $t \in \mathbb{R}$.

Concerning topological conjugacy of group-induced flows on more general homogeneous spaces the reader is referred to the next subsection on topological equivalence; various results on topological conjugacy can be read off from there and we shall not note them separately.

A description of topological factors, analogous to that of the measurable factors of Ad-unipotent subgroups described in § 10.2, for actions of subgroups generated by Ad-unipotent one-parameter subgroups is given by N.A. Shah [Sh4], which shows in particular the following. Let G be a connected Lie group with trivial center, C be a lattice in G and H be a subgroup of G generated by Ad-unipotent one-parameter subgroups. Let $\phi : G/C \to X$ be a continuous H-equivariant map onto a locally compact Hausdorff space with a H-action. Then there exists a factor of the form $K\backslash G'/D'$ as in § 10.2, with respect to the H-action, such that ϕ factors through the quotient map of G/C onto $K\backslash G'/D'$ and the quotient map of $K\backslash G'/D'$ onto X is injective on a dense open subset of $K\backslash G'/D'$.

10.4. Topological equivalence. We say that two flows Φ and Ψ on a topological space X are *topologically orbit-equivalent* if there exists a homeomorphism θ of X onto itself taking orbits of Φ into orbits of Ψ; further, the flows are said to be *topologically equivalent* if θ can be chosen so that the orientation of the orbits is also preserved by θ; the homeomorphism θ as above is called a topological orbit-equivalence or topological equivalence, respectively. If X is a homogeneous space, a topological orbit-equivalence (resp. topological equivalence) θ is called an *affine orbit-equivalence* (resp. *affine equivalence*) if it is also an affine map, and when such a map exists the flows are said to be affinely orbit-equivalent (resp. affinely equivalent). A homeomorphism of X is called a *self-orbit-equivalence* (resp. *self-equivalence*) of a flow Φ if it is a topological orbit-equivalence (resp. topological equivalence) of Φ with itself, taking each orbit into itself.

It is a classical result that for $v, w \in \mathbb{R}^n$ the flows on $\mathbb{T}^n = \mathbb{R}^n/\mathbb{Z}^n$ induced by the one-parameter subgroups $\{tv \mid t \in \mathbb{R}\}$ and $\{tw \mid t \in \mathbb{R}\}$ are topologically orbit-equivalent (respectively, topologically equivalent) if and only if there exist $\alpha \in GL(n, \mathbb{Z})$ and a real number $r \neq 0$ (respectively $r > 0$) such that $\alpha(v) = rw$. In a more general setting the following is known.

Theorem 10.5 (Benardete [Be], Witte [Wi2]). *Let G and G' be connected Lie groups and Γ and Γ' be lattices in G and G' respectively. Let Φ and Ψ be group-induced flows on G/Γ and G'/Γ' respectively. Suppose that one of the following conditions holds:*

a) G and G' are simply connected solvable Lie groups and all eigenvalues of $Ad\,g$, $g \in G$, and $Ad\,g'$, $g' \in G'$, are real;

b) G and G' are semisimple Lie groups with trivial center and no compact factors, there is no closed normal subgroup H of G isomorphic to $PSL(2, \mathbb{R})$ and such that $H\Gamma$ is closed, and Φ is ergodic.

Then any topological equivalence (respectively, topological orbit-equivalence) between Φ and Ψ is a composite of an affine equivalence (resp., affine orbit equivalence) between Φ and Ψ and a self-equivalence (resp. self-orbit equivalence) of Ψ; in particular topologically equivalent (resp. topologically orbit-equivalent) flows are affinely equivalent (resp. affinely orbit-equivalent).

Benardete's results are about topological equivalences being affine. The assertion for orbit-equivalences follows as a special case of Witte's results, which are proved in the generality of foliations induced by connected Lie subgroups of possibly higher dimensions; the arguments, which are a generalisation of those of Benardete, apply to topological equivalence as well, leading to the result as above. The reader is also referred to a recent paper of Bernstein and Witte [BW] for a more detailed analysis and further results along these lines.

In the case when $G = G' = PSL(2, \mathbb{R})$, which is excluded from the above theorem, a result of Marcus [Marc3] shows that if ϕ and ψ are unipotent one-parameter subgroups of G and if for two uniform torsion-free lattices Γ and Γ' in G the flows Φ and Ψ on G/Γ and G/Γ', induced by ϕ and ψ respectively, are topologically equivalent then they are affinely equivalent; it may be recalled that the flows here are horocycle flows. Such rigidity however does not hold for the geodesic flows or, equivalently, for flows on G/Γ induced by one-parameter subgroups of $G = PSL(2, \mathbb{R})$, which are diagonalisable over $\mathbb{R}$; in fact any two such flows, say induced on G/Γ and G/Γ' where Γ and Γ' are uniform torsion-free lattices, are topologically equivalent whenever Γ and Γ' are isomorphic (cf. [An], page 26); they are affinely equivalent only if Γ' is conjugate (in G) to either Γ or its transpose. A recent result of Y. Mao [Mao] shows however that if they are topologically conjugate then they are affinely isomorphic and a similar assertion also holds for all geodesic flows on compact locally symmetric Riemannian manifolds of negative curvature.

The rigidity assertion as in Theorem 10.5 also does not hold for solvable groups, in general, when the eigenvalues of the adjoint transformations are not assumed to be all real, as was done in case (a) of the theorem. The special case of $G = G' = N \cdot M$, a semidirect product of vector groups M and N, with M acting on N via a representation $\sigma : M \to GL(N)$ such that $\ker \sigma$ is discrete and $\sigma(M)$ is a compact subgroup of $GL(N)$ acting with no nonzero fixed points, is considered in [BD]. It is proved that for a 'generic' lattice (in a geometric sense that we shall not go into here) flows induced by two one-parameter subgroups, at least one of which is Ad-semisimple, are topologically orbit-equivalent only if they are affinely orbit-equivalent; on the other hand, there are lattices Γ among these such that for $v \in M$ the flows on G/Γ induced by the one-parameter subgroups $\{tv\}_{t\in\mathbb{R}}$ and $\{-tv\}_{t\in\mathbb{R}}$ are topologically equivalent but not affinely equivalent.

Acknowledgement. I (the author) would like to thank G.A. Margulis, Marina Ratner, Nimish Shah, Ya.G. Sinai, A.N. Starkov and Dave Witte for helpful comments on preliminary versions of the article, H. Furstenberg, D. Lind and A. Manning for responding to my queries, and P. Gastesi for technical help in the preparation of the manuscript. I would also like to record thanks to my wife Jyotsna Dani for the support received from her, in various ways, in the venture of writing the article.

Bibliography

[An] Anosov, D.V.: Geodesic flows on closed Riemann manifolds with negative curvature, Trudy Mat. Inst. Steklov 90 (1967), (Russian) = Proceedings of the Steklov Inst. of Math. 90 (1969).

[Ao] Aoki, N.: A simple proof of the Bernoullicity of ergodic automorphisms on compact abelian groups, Israel J. Math. 38 (1981), 189-198.

[AL] Aravinda C.S. and Leuzinger, E.: Bounded geodesics in rank-1 locally symmetric spaces, Ergod. Th. Dynam. Sys. 15 (1995), 813-820.

[AA] Arnold V.I and Avez, A.: Ergodic Problems of Classical Mechanics, W.A. Benjamin Inc., Amsterdam, 1968.

[Au] Auslander, L.: An exposition of the structure of solvmanifolds I and II, Bull. Amer. Math. Soc. 79 (1973), 227-261 and 262-285.

[AB] Auslander, L. and Brezin, J.: Uniform distribution on solvmanifolds, Adv. Math. 7 (1971), 111-144.

[AG] Auslander, L. and Green, L.W.: G-induced flows, Amer. J. Math. 88 (1966), 43-60.

[AGH] Auslander, L., Green L. and Hahn, F.: Flows on homogencous spaces, With the assistance of L. Markus and W. Massey, and an appendix by L. Greenberg, Annals of Mathematics Studies, No. 53, Princeton University Press, Princeton, N.J., 1963.

[AS] Auslander, L. and Scheuneman, J.: On certain automorphisms of nilpotent Lie groups, Global Analysis, Proceedings of Symposia in Pure Mathematics XIV, Berkeley (1968), pp. 9-15, Amer. Math. Soc., Providence, 1970.

[Be] Benardete, D.: Topological equivalence of flows on homogeneous spaces, and divergence of one-parameter subgroups of Lie groups, Trans. Amer. Math. Soc. 306 (1988), 499-527.

[BD] Benardete, D. and Dani, S.G.: Topological equivalence and rigidity of flows on certain solvmanifolds, Ergod. Th. and Dynam. Sys. 19 (1999), 559-569.

[BW] Bernstein, H. and Witte, D.: Foliation-preserving maps between solvmanifolds, Preprint, 1997.

[Bh] Bhattacharya, S.: Topological conjugacy of automorphism flows on compact Lie groups, Ergod. Th. Dynam. Sys. (to appear).

[B1] Borel, A.: Introduction aux Groupes Arithmetiques, Hermann, Paris, 1969.

[B2] Borel, A.: Values of indefinite quadratic forms at integral points and flows on spaces of lattices, Bull. Amer. Math. Soc. 32 (1995), 184-204.

[BH] Borel, A. and Harish-Chandra: Arithmetic subgroups of algebraic groups, Ann. Math. 75(1962), 485-535.

[BP] Borel, A. and Prasad, G.: Values of isotropic quadratic forms at S-integral points, Compositio Math. 83 (1992), 347-372.

[Bo1] Bowen, R.: Markov partitions and minimal sets for axiom A diffeomorphisms, Amer. J. Math. 92 (1970), 907 918.

[Bo2] Bowen, R.: Entropy of group endomorphisms and homogeneous spaces, Tran. Amer. Math. Soc. 153 (1971), 401-414; Erratum: Trans. Amer. Math. Soc. 181 (1973), 509-510.

[Bo3] Bowen, R.: The equidistribution of closed geodesics, Amer. J. Math. 94 (1972), 413-423.

[Bo4] Bowen, R.: Symbolic dynamics for hyperbolic flows, Amer. J. Math. 95 (1973), 429-460.

[Bo5] Bowen, R.: Maximizing entropy for a hyperbolic flow, Math. Syst. Th. 7 (1973-74), 300-303.

[Bo6] Bowen, R.: Equilibrium States and the Ergodic Theory of Anosov Diffeomorphisms, Lect. Notes in Math. 470, Springer-Verlag, 1975.

[Bo7] Bowen, R.: Weak mixing and unique ergodicity on homogeneous spaces, Israel J. Math. 23 (1976), 267-273.

[Bo8] Bowen, R.: On Axiom A Diffeomorphisms, Amer. Math. Soc., Providence, 1978.

[BM] Brezin, J. and Moore, C.C.: Flows on homogeneous spaces: a new look, Amer. J. Math. 103 (1981), 571-613.

[BS] Brezin, J. and Shub, M: Stable ergodicity in homogeneous spaces, Bull. Soc. Brasil. Mat. 28 (1997), 197-210.

[Br] Brown, J.R.: Ergodic Theory and Topological Dynamics, Academic Press, 1976.

[Ca1] Cassels, J.W.S. An Introduction to Diophantine Approximation, Cambridge Tracts 45, Cambridge University Press, 1957.

[Ca2] Cassels, J.W.S. An Introduction to the Geometry of Numbers, Springer-Verlag, 1959.

[Ci] Cigler, J.: Some applications of the individual ergodic theorem to problems in number theory, Compositio Math. 16 (1964), 35-43.

[Co] Conze, J.-P.: Entropie des transformations affines et des flots sur les espaces homogènes compacts, Ann. Inst. H. Poincaré, Sect. B., 8 (1972), 67-81.

[CM] Conze, J.-P. and Marcuard, J.-C.: Conjugasion topologique des des automorphismes et des translations ergodiques de nilvariétés, Ann. Inst. H. Poincaré, Sect. B., 6 (1970), 153-157.

[CFS] Cornfeld, I.P., Fomin, S.V. and Sinai, Ya. G.: Ergodic Theory, Springer-Verlag, 1982.

[D1] Dani, S.G.: Kolmogorov automorphisms on homogeneous spaces, Amer. J. Math. 98 (1976), 119-163.

[D2] Dani, S.G.: Bernoullian translations and minimal horospheres on homogeneous spaces, J. Indian Math. Soc., 40 (1976), 245-284.

[D3] Dani, S.G.: Some mixing multiparameter dynamical systems, J. Indian Math. Soc. 41 (1977), 1-15.

[D4] Dani, S.G.: Sequential entropy of generalised horocycles, J. Indian Math. Soc. 41 (1977), 17-25.

[D5] Dani, S.G.: Spectrum of an affine transformation, Duke Math. J. 44 (1977), 129-155.

[D6] Dani, S.G.: Nilmanifolds with Anosov automorphisms, J. Lond. Math. Soc. 18 (1978), 553-559.

[D7] Dani, S.G.: On invariant measures, minimal sets and a lemma of Margulis, Invent. Math. 51 (1979), 239-260.

[D8] Dani, S.G.: On affine automorphisms with a hyperbolic fixed point, Topology 19 (1980), 351-365.

[D9] Dani, S.G.: A simple proof of Borel's density theorem, Math. Zeits. 174 (1980), 81-94.

[D10] Dani, S.G.: Invariant measures and minimal sets of horospherical flows, Invent. Math. 64 (1981), 357-385.

[D11] Dani, S.G.: Dense orbits of affine automorphisms and compactness of groups, J. Lond. Math. Soc. 25 (1982), 241-245.

[D12] Dani, S.G.: On ergodic quasi-invariant measures of group automorphisms, Israel J. Math. 43 (1982), 62-74.

[D13] Dani, S.G.: Divergent trajectories of flows on homogeneous spaces and Diophantine approximation, J. Reine Angew. Math. 359 (1985), 55-89.

[D14] Dani, S.G.: Bounded orbits of flows on homogeneous spaces, Comment. Math. Helv. 61 (1986), 636-660.

[D15] Dani, S.G.: On orbits of unipotent flows on homogeneous spaces - II, Ergod. Th. Dynam. Sys. 6 (1986), 167-182.

[D16] Dani, S.G.: Orbits of horospherical flows, Duke Math. J. 53 (1986), 177-188.

[D17] Dani, S.G.: On orbits of endomorphisms of tori and the Schmidt game, Ergod. Th. and Dynam. Sys. 8 (1988), 523-529.

[D18] Dani, S.G.: Dense orbits of horospherical flows, In: Dynamical Systems and Ergodic Theory, Banach Center Publications 23, pp. 179-195, PWN Scientific Publ., Warsaw, 1989.

[D19] Dani, S.G.: On badly approximate numbers, Schmidt games and bounded orbits of flows, In: Number Theory and Dynamical Systems, (Proceedings of a conf., York, 1987), pp. 69-86, London Math. Soc. Lect. Notes Ser. 134, Cambridge University Press, 1990.

[D20] Dani, S.G.: A proof of Margulis' theorem on values of quadratic forms, independent of the axiom of choice, L'enseign. Math. 40 (1994), 49-58.

[D21] Dani, S.G.: Flows on homogeneous spaces, Proceed. of the Internat. Congress of Mathematicians, pp. 780-789, Zurich-1994, Birkhauser, Basel, 1995.

[D22] Dani, S.G.: Flows on homogeneous spaces; a review, In: Ergodic Theory of $\mathbb{Z}^n$-actions, (Proceedings of the Warwick Symposium, 1993-94), pp. 63-112, London Math. Soc. Lect. Notes Ser. 228, Cambridge University Press, 1996.

[D23] Dani, S.G.: On values of linear and quadratic forms at integral points, Preprint, 1998.

[DM1] Dani, S.G. and Margulis, G.A.: Orbit closures of generic unipotent flows on homogeneous spaces of $SL(3, \mathbb{R})$, Math. Ann. 286 (1990), 101-128.

[DM2] Dani, S.G. and Margulis, G.A.: Values of quadratic forms at integral points: an elementary approach, L'enseign. Math. 36 (1990), 143-174.

[DM3] Dani, S.G. and Margulis, G.A.: Limit distributions of orbits of unipotent flows and values of quadratic forms, Adv. in Soviet Math. (AMS Publ.) 16 (1993), 91-137.

[DMu] Dani, S.G. and Muralidharan, S.: On ergodic averages for affine lattice actions on tori, Monatsh. Math. 96 (1983), 17-28.

[DR] Dani, S.G. and Raghavan, S.: Orbits of Euclidean frames under discrete linear groups, Israel J. Math. 36 (1980), 300-320.

[DRj] Dani, S.G. and Raja, C.R.E.: Asymptotics of measures under group automorphisms and an application to factor sets, In: Lie Groups and Ergodic Theory (Proceedings, Mumbai, 1996), pp. 59-73, Tata Institute of Fundamental Research, Mumbai, and Narosa Publishing House, New Delhi, 1998.

[DS] Dani, S.G. and Smillie J.: Uniform distribution of horocycle orbits for Fuchsian groups, Duke Math. J. 51 (1984), 185-194.

[DK] Dateyama, M. and Kasuga, T.: Ergodic affine maps of locally compact groups, J. Math. Soc. Japan 37 (1985), 363-372.

[DI] Degli Eposti, M. and Isola, S.: Distribution of closed orbits for linear automorphisms of tori, Nonlinearity 8 (1995), 827-842.

[DGS] Denker, M., Grillenberger, C. and Sigmund, K.: Ergodic Theory on Compact Spaces, Lect. Notes in Math. 527, Springer-Verlag, 1976.

[DH] Djokovic D.Z. and Hofmann, K.H.: The surjectivity question for the exponential function of real Lie groups: a status report, J. of Lie Theory 7 (1997), 171-199.

[EP] Ellis R. and Perrizo, W.: Unique ergodicity of flows on homogeneous spaces, Israel J. Math. 29 (1978), 276-284.

[EMM] Eskin, A., Margulis, G.A. and Mozes, S.: Upper bounds and asymptotics in a quantitative version of the Oppenheim conjecture, Ann. Math. 147 (1998), 93-141.

[EM] Eskin, A. and McMullen, C.T.: Mixing, counting, and equidistribution in Lie groups, Duke Math. J. 71 (1993), 181-209.

[EMS1] Eskin, A., Mozes, S. and Shah, N.A.: Unipotent flows and counting lattice points on homogeneous varieties, Ann. Math. 143 (1996), 253-299.

[EMS2] Eskin, A., Mozes, S. and Shah, N.A.: Non-divergence of translates of certain algebraic measures, Geom. and Funct. Anal. 7 (1997), 48-80.

[Fa] Fathi, A.: Some compact invariant sets for hyperbolic linear automorphisms of tori, Ergod. Th. and Dynam. Sys. 8 (1988), 191-204.

[FS] Flaminio, L. and Spatzier, R.J.: Ratner's rigidity theorem for geometrically finite Fuchsian groups, In: Dynamical Systems (Proceedings, College Park, Maryland, 1986-87), pp. 180-195, Lect. Notes in Math. 1342, Springer-Verlag, 1988.

[Fr] Franks, J.: Anosov diffeomorphisms on tori, Trans. Amer. Math. Soc. 145 (1969), 117-124.

[FW] Furman, A. and Weiss, B.: On the ergodic properties of Cartan flows in ergodic actions of $SL(2, \mathbb{R})$ and $SO(n, 1)$, Ergod. Th. and Dynam. Sys. 17 (1997), 1371-1382.

[Fu] Furstenberg, H.: The unique ergodicity of the horocycle flow, In: Recent Advances in Topological Dynamics, pp. 95-115, Lect. Notes in Math. 318, Springer-Verlag, 1973.

[GF] Gelfand I.M. and Fomin, S.V.: Geodesic flows on manifolds of constant negative curvature, Usp. Mat. Nauk 7 (1952), 118-137 (Russian)= Amer. Math. Soc. Transl. 1 (1955), 49-65.

[Gh] Ghys, E.: Dynamique des flots unipotents sur les espaces homogènes, Sem. Bourbaki 1992-93, Exp. 747, Astérisque 206 (1992), 93-136.

[Gr] Gromov, M.: Groups of polynomial growth and expanding maps, Inst. Hautes Études Sci. Publ. Math. 53 (1981), 53-73.

[Gu] Gutschera, K.R.: Ergodic elements for actions of Lie groups, Ergod. Th. and Dynam. Sys. 16 (1996), 703-717.

[HP] Hahn, F.J. and Parry, W.: Some characteristic properties of dynamical systems with quasi-discrete spectra, Math. Syst. Th. 2 (1968), 179-190.

[Ha] Hancock, S.G.: Construction of invariant sets for Anosov diffeomorphisms, J. Lond. Math. Soc. 18 (1978), 339-348.

[He] Hedlund, G.A.: Fuchsian groups and transitive horocycles, Duke Math. J. 2 (1936), 530-542.

[Hi] Hirsch, M.W.: On invariant subsets of hyperbolic sets, In: Essays on Topology and Related Topics, pp. 126-135, Springer-Verlag 1970.

[HoP1] Hoare, H. and Parry, W.: Affine transformations with quasi-discrete spectrum, I and II, J. London Math. Soc. 41 (1966), 88-96 and 529-530.

[HoP2] Hoare, H. and Parry, W.: Semi-groups of affine transformations, Quart. J. Math. Oxford Ser. (2) 17(1966), 106-111.

[Ho] Hofmann, K.H.: Lie algebras with subalgebras of codimension one, Illinois J. Math. 9 (1965), 636-643.

[HM] Howe, R. and Moore, C.C.: Asymptotic properties of unitary representations, J. Funct. Anal. 32 (1979), 72-96.

[Hu] Hurder, S.: A survey of rigidity theory for Anosov actions, In: Differential Topology, Foliations and Group Actions (Proceedings, Rio de Janeiro, 1992), pp. 143-173, Contemp. Math. 161, Amer. Math. Soc., Providence, 1994.

[Ir] Irwin, M.C.: Hölder continuous paths and hyperbolic toral automorphisms, Ergod. Th. and Dynam. Sys. 6 (1986), 241-257.

[Jo] Jordan, C.: Mémoire sur l'équivalence des formes, J. Ecole Polytech. 48 (1880), 112-150; Collected Works III, pp. 421-461, Gauthier-Villars, Paris, 1962.

[KH] Katok, A. and Hasselblatt, B.: Introduction to the Modern Theory of Dynamical Systems, Cambridge University Press, 1995.

[KS1] Katok, A. and Spatzier, R.J.: First cohomology of Anosov actions of higher rank abelian groups and applications to rigidity, Inst. Haute Études Sci. Publ. Math. 79 (1994), 131-156.

[KS2] Katok, A. and Spatzier, R.J.: Invariant measures for higher rank hyperbolic abelian actions, Ergod. Th. and Dynam. Sys. 16 (1996), 751-778.

[Ka] Katznelson, Y.: Ergodic automorphisms of $\mathbb{T}^n$ are Bernoulli shifts, Israel J. Math. 10 (1971), 186-195.

[KeS] Keane, M. and Smorodinsky, M.: The finitary isomorphism theorem for Markov shifts, Bull. Amer. Math. Soc. 1 (1979), 436-438.

[Kl1] Kleinbock, D.Y.: Nondense orbits of flows on homogeneous spaces, Ergod. Th. and Dynam. Sys. 18 (1998), 373-396.

[Kl2] Kleinbock, D.Y.: Flows on homogeneous spaces and Diophantine properties of matrices, Duke Math. J. 95 (1998), 107-124.

[Kl3] Kleinbock, D.Y.: Bounded orbits conjecture and Diophantine approximation, In: Lie Groups and Ergodic Theory (Proceedings, Mumbai, 1996), pp. 119-130, Tata Institute of Fundamental Research, Mumbai, and Narosa Publishing House, New Delhi, 1998.

[KM1] Kleinbock, D.Y. and Margulis, G.A.: Bounded orbits of nonquasiunipotent flows on homogeneous spaces, In: Sinai Moscow Seminar on Dynamical Systems, Amer. Math. Soc. Translations 171 (1996), pp. 141-172, American Mathematical Society, Providence, 1996.

[KM2] Kleinbock, D.Y. and Margulis, G.A.: Flows on homogeneous spaces and Diophantine approximation on manifolds, Ann. Math. 148 (1998), 339-360.

[L1] Lesigne, E.: Théorème ergodiques pour une translation sur une nilvariété, Ergod. Th. and Dynam. Sys. 9 (1989), 115-126.

[L2] Lesigne, E.: Sur une nil-variété, les parties minimal associées à une translation sont uniquement ergodiques, Ergod. Th. and Dynam. Sys. 11 (1991), 379-391.

[Li1] Lind, D.: The structure of skew products with ergodic group automorphisms, Israel J. Math. 28 (1977), 205–248.

[Li2] Lind, D.: Dynamical properties of quasihyperbolic toral automorphisms, Ergod. Th. and Dynam. Sys. 2 (1982), 49-68.

[Li3] Lind, D.: Ergodic group automorphisms are exponentially recurrent, Israel J. Math. 41 (1982), 313-320.

[Mac] Mackey, G.W.: Induced representations of locally compact groups I, Ann. Math. 55 (1952), 101-139.

[Ma1] Mañé, R.: Invariant sets of Anosovs diffeomorphisms, Invent. Math. 46 (1978), 147-152.

[Ma2] Mañé, R.: Orbits of paths under hyperbolic toral automorphisms, Proc. Amer. Math. Soc. 73 (1979), 121-125.

[Ma3] Mañé, R.: Ergodic Theory and Differentiable Dynamics, Springer-Verlag, 1987.

[Man1] Manning, A.: There are no new Anosov diffeomorphisms on tori, Amer. J. Math. 96 (1974), 422-429.

[Man2] Manning, A.: Dynamics of geodesic and horocycle flows on surfaces of constant negative curvature, In: Ergodic Theory, Symbolic Dynamics and Hyperbolic Spaces, Oxford University Press, 1991.

[Mao] Mao, Y.: Conjugate geodesic flows in negatively curved manifolds, Contemporary Math. 173, pp. 225-229, American Mathematical Society, Providence, 1994.

[Marc1] Marcus, B.: The horocycle flow is mixing of all degrees, Invent. Math. 46 (1978), 201-209.

[Marc2] Marcus, B.: A note on periodic points of toral automorphisms, Monatsh. Math. 89 (1980), 121-129.

[Marc3] Marcus, B.: Topological conjugacy of horocycle flows, Amer. J. Math. 105 (1983), 623-632.

[Marg1] Margulis, G.A.: Lie groups and ergodic theory, In: Algebra - Some Current Trends (Proceedings, Varna, 1986), pp. 130-146, Springer-Verlag, 1988.

[Marg2] Margulis, G.A.: Orbits of group actions and values of quadratic forms at integral points, Festschrift in honour of I. Piatetski Shapiro, Israel Math. Conf. Proc. 3 (1990), 127-150.

[Marg3] Margulis, G.A.: Dynamical and ergodic properties of subgroup actions on homogeneous spaces with applications to number theory, In: Proceed. of the Internat. Congress of Math. (Kyoto, 1990), pp. 193-215, The Mathematical Society of Japan, Tokyo, 1991.

[Marg4] Margulis, G.A.: Discrete Subgroups of Semisimple Lie Groups, Springer-Verlag, 1991.

[Marg5] Margulis, G.A.: Compactness of minimal closed invariant sets of actions of unipotent groups, Geometry Dedicata 37 (1991), 1-7.

[Marg6] Margulis, G.A.: Oppenheim conjecture, In: Fields Medalist Lectures, pp. 272-327, World Scientific, 1997.

[MT1] Margulis G.A. and Tomanov, G.M.: Invariant measures for actions of unipotent groups over local fields on homogeneous spaces, Invent. Math. 116 (1994), 347-392.

[MT2] Margulis G.A. and Tomanov, G.M.: Measure rigidity for almost linear groups and it applications, J. d'Analyse Math. 69 (1996), 25-54.

[Mau] Mautner, F.I.: Geodesic flows on symmetric Riemann spaces, Ann. Math. 65 (1957, 416-431.

[Me] Meyer, Y.: Algebraic Numbers and Harmonic Analysis, North Holland Publishing Company, Amsterdam-London, 1972.

[MiT] Miles, G. and Thomas, R.K.: On the polynomial uniformity of translations of the n-torus, In: Studies in Probability and Ergodic Theory, pp. 231-249, Adv. in Math. Suppl. Stud. 2, Academic Press, New York-London, 1978.

[Mo1] Moore, C.C.: Decompositions of unitary representations defined by discrete subgroups of nilpotent groups, Ann. Math. 82 (1965), 146-182.

[Mo2] Moore, C.C.: Ergodicity of flows on homogeneous spaces, Amer. J. Math. 88 (1966), 154-178.

[Mo3] Moore, C.C.: The Mautner phenomenon for general unitary representations, Pacific J. Math. 86 (1980), 155-169.

[Mo4] Moore, C.C.: Exponential decay of correlation coefficients for geodesic flows, In: Group Representations, Ergodic Theory, Operator Algebras and Mathematical Physics (Proceedings of a conf., Berkeley, 1984), pp. 163-181, MSRI Publications No. 6, Springer-Verlag, 1987.

[Moz1] Mozes, S.: Mixing of all orders of Lie group actions, Invent. Math. 107 (1992), 235-241; Erratum: 119 (1995), 399.

[Moz2] Mozes, S.: Epimorphic subgroups and invariant measures, Ergod. Th. and Dynam. Sys. 15 (1995), 1207-1210.

[MS] Mozes, S. and Shah, N.A.: On the space of ergodic invariant measures of unipotent flows, Ergod. Th. and Dynam. Sys. 15 (1995), 149-159.

[Na] Nachbin, L.: The Haar Integral, Van Nostrand Company, Princeton, 1965.

[Ne] Newhouse, S.E.: On codimension one Anosov diffeomorphisms, Amer. J. Math. 92 (1970), 761-770.

[No1] Nogueira, A.: Invariant measures for the action of $SL(2, \mathbb{Z})$ on $\mathbb{R}^2$, Markov Processes and Related Fields 4 (1998), 571-592.

[No2] Nogueira, A.: Relatively prime numbers and invariant measures under the natural action of $SL(n, \mathbb{Z})$ on $\mathbb{R}^n$, Preprint, 1999.

[OW] Ornstein, D. and Weiss, B.: Geodesic flows are Bernoullian, Israel J. Math. 14 (1973), 184-198.

[P1] Parry, W.: On the coincidence of three invariant σ-algebras associated with an affine transformation, Proc. Amer. Math. Soc. 17 (1966), 1297-1302.

[P2] Parry, W.: Ergodic properties of affine transformations and flows on nilmanifolds, Amer. J. Math. 91 (1969), 757-771.

[P3] Parry, W.: Dynamical systems on nilmanifolds, Bull. London Math. Soc. 2 (1970), 37-40.

[P4] Parry, W.: Spectral analysis of G-extensions of dynamical systems, Topology 9 (1970), 217-224.

[P5] Parry, W.: Metric classification of ergodic nilflows and unipotent affines, Amer. J. Math. 93 (1971), 819-828.

[P6] Parry, W.: Dynamical representations in nilmanifolds, Compositio Math. 26 (1973), 159-174.

[PP] Parry W. and Pollicott, M.: Zeta functions and the periodic orbit structure of hyperbolic dynamics, Astérisque No. 187-188 (1990), 1-268.

[Pa] Payne, T.L.: Closures of totally geodesic immersions into locally symmetric spaces of noncompact type, Proc. Amer. Math. Soc. 127 (1999), 829-833

[Pol] Pollicott, M.: Exponential mixing for the geodesic flow on hyperbolic three-manifolds, J. Statist. Phys. 67 (1992), 667-673.

[Por] Porteous, H.L.: Anosov diffeomorphisms of flat manifolds, Topology 11 (1972), 307-315.

[Pr] Przytycki, F.: Construction of invariant sets for Anosov diffeomorphisms and hyperbolic attractors, Studia Math. 68 (1980), 199-213.

[Qi] Qian, N.T.: Anosov automorphisms for nilmanifolds and rigidity of group actions, Ergod. Th. and Dynam. Sys. 15 (1995), 341-359.

[Rag] Raghunathan, M.S.: Discrete Subgroups of Lie Groups, Springer-Verlag, 1972.

[R1] Ratner, M.: On flows of linear elements on compact surfaces of constant negative curvature, Amer. J. Math. 100 (1978), 443-462.

[R2] Ratner, M.: Rigidity of horocycle flows, Ann. Math. 115 (1982), 597-614.

[R3] Ratner, M.: Factors of horocycle flows, Ergod. Th. and Dynam. Sys. 2 (1982), 465-489.

[R4] Ratner, M.: Horocycle flows: joinings and rigidity of products, Ann. Math. 118 (1983), 277-313.

[R5] Ratner, M.: The rate of mixing for geodesic and horocycle flows, Ergod. Th. and Dynam. Sys. 7 (1987), 267-288.

[R6] Ratner, M.: Strict measure rigidity for unipotent subgroups of solvable groups, Invent. Math. 101 (1990), 449-482.

[R7] Ratner, M.: On measure rigidity of unipotent subgroups of semisimple groups, Acta Math. 165 (1990), 229-309.

[R8] Ratner, M.: On Raghunathan's measure conjecture, Ann. Math. 134 (1991), 545-607.

[R9] Ratner, M.: Raghunathan's topological conjecture and distributions of unipotent flows, Duke Math. J. 63 (1991), 235-280.

[R10] Ratner, M.: Raghunathan's conjectures for $SL(2, \mathbb{R})$, Israel J. Math. 80 (1992), 1-31.

[R11] Ratner, M.: Invariant measures and orbit closures for unipotent actions on homogeneous spaces, Geom. and Funct. Anal. 4(1994), 236-257.

[R12] Ratner, M.: Interactions between ergodic theory, Lie groups and number theory, In: Proceed. of the Internat. Congress of Mathematicians, Zurich-1994, pp. 157-182, Birkhauser, Basel, 1995.

[R13] Ratner, M.: Raghunathan's conjectures for cartesian products of real and p-adic Lie groups, Duke Math. J. 77 (1995), 275-382.

[R14] Ratner, M.: On the p-adic and S-arithmetic generalizations of Raghunathan's conjectures, In: Lie Groups and Ergodic Theory (Proceedings, Mumbai, 1996), pp. 167-202, Tata Institute of Fundamental Research, Mumbai, and Narosa Publishing House, New Delhi, 1998.

[Sa] Sarnak, P.: Asymptotic behaviour of periodic orbits of the horocycle flow and Eisenstein series, Comm. Pure Appl. Math. 34 (1981), 719-739.

[Sc] Schmidt, K.: Infinite invariant measures on the circle, Symposia Mathematica, Vol. XXI, pp. 37-43. Academic Press, London, 1977.

[SW] Schmidt K. and Ward, T.: Mixing automorphisms of compact groups and a theorem of Schlickewei, Invent. Math. 111 (1993), 69-76.

[Sch1] Schmidt, W.M.: On badly approximable numbers and certain games, Trans. Amer. Math. Soc. 123 (1966), 178-199.

[Sch2] Schmidt, W.M.: Badly approximable systems of linear forms, J. Number Th. 1 (1969), 139-154.

[Sch3] Schmidt, W.M.: Normalität bezüglich Matrizen, J. Reine Angew. Math. 214/215 (1969), 227-260.

[Sch4] Schmidt, W.M.: Diophantine approximation and certain sequences of lattices, Acta Arith. 18 (1971), 165-178.

[Sh1] Shah, N.A.: Uniformly distributed orbits of certain flows on homogeneous spaces, Math. Ann. 289 (1991), 315-334.

[Sh2] Shah, N.A.: Closures of totally geodesic immersions of constant negative curvature In: Group Theory from a Geometrical Viewpoint (Proceedings, Trieste, 1990), pp. 718-732, World Scientific Publ., River Edge, NJ, 1991.

[Sh3] Shah, N.A.: Limit distributions of polynomial trajectories on homogeneous spaces, Duke Math. J. 75 (1994), 711-732.

[Sh4] Shah, N.A.: Limit distributions of expanding translates of certain orbits on homogeneous spaces, Proc. Indian Acad. Sci. (Math. Sci.) 106 (1996), 105-125.

[Sh5] Shah, N.A.: Invariant measures and orbit closures on homogeneous spaces for actions of subgroups generated by unipotent elements, In: Lie Groups and Ergodic Theory (Proceedings, Mumbai, 1996), pp. 229-271, Tata Institute of Fundamental Research, Mumbai, and Narosa Publishing House, New Delhi, 1998.

[ShW] Shah N.A. and Weiss, B.: On actions of epimorphic subgroups on homogeneous spaces, Ergod. Th. and Dynam. Sys. (to appear).

[Si] Sigmund, K.: Normal and quasiregular points for automorphisms of the torus, Math. Syst. Th. 8 (1974), 251-255.

[Sk] Skriganov, M.M.: Ergodic theory on $SL(n)$, diophantine approximation and anamolies in the lattice point problem, Invent. Math. 132 (1998), 1-72.

[ST] Skriganov, M.M. and Starkov, A.N.: On logarithmically small errors in the lattice point problem, Preprint.

[Sm] Smale, S.: Differentiable dynamical systems, Bull. Amer. Math. Soc. 73 (1967), 747-817.

[St1] Starkov, A.N.: Spaces of finite volume, Vestnik Mosk. Univ., Ser. I Mat. Mech. 5 (1986), 64-66.

[St2] Starkov, A.N.: On a criterion for the ergodicity of G-induced flows, Usp. Mat. Nauk 42 (1987), No. 3, 197-198 (Russian) = Russian Math. Surveys 42 (1987), 233-234.

[St3] Starkov, A.N.: Solvable homogeneous flows, Mat. Sbornik (N.S.) 134 (176)(1987), No. 2, 242-259 (Russian) = Math. USSR Sbornik 62 (1989), No. 1, 243-260.

[St4] Starkov, A.N.: Ergodic decomposition of flows on homogeneous spaces of finite volume, Mat. Sbornik 180 (1989), No. 12, 1614-1633 (Russian) = Math. USSR Sbornik 68 (1991), No.2, 483-502.

[St5] Starkov, A.N.: Structure of orbits of homogeneous flows and Raghunathan's conjecture, Usp. Mat. Nauk 45 (1990), No. 2, 219-220 (Russian) = Russian Math. Surveys 45 (1990), N. 2, 227-228.

[St6] Starkov, A.N.: Horospherical flows on spaces of finite volume, Mat. Sbornik 182 (1991), No. 5, 774-784 (Russian) = Math. USSR Sbornik 73 (1992), No. 1, 161-170.

[St7] Starkov, A.N.: On mixing of higher degrees of homogeneous flows, Dokl. Russ. Akad. Nauk 333 (1993), 28-31.

[St8] Starkov, A.N.: Minimal sets of homogeneous flows, Ergod. Th. and Dynam. Sys. 15 (1995), 361-377.

[St9] Starkov, A.N.: Mutual isomorphisms of translations of a homogeneous flow, Mat. Zametski 58 (1995), No. 1, 98-110 (Russian) = Math. Notes 58 (1995), No. 1-2, 744-751.

[St10] Starkov, A.N.: New progress in the theory of homogeneous flows, Usp. Mat. Nauk 52, No. 4 (1997), 87-192 (Russian) = Russian Math. Surveys 52 (1997), No. 4, 721-818.

[Ste1] Stepin, A.M.: Flows on solvable manifolds, Usp. Mat. Nauk 24 (1969), No. 5, (149), 241-242 (Russian).

[Ste2] Stepin, A.M.: Dynamical Systems on homogeneous spaces of semisimple Lie groups, Izv. Akad. Nauk SSSR Ser. Mat. 37 (1973), 1091-1107 (Russian)= Math. USSR Izv. 7 (1973), 1089-1104.

[Th] Thomas, R.K.: Metric properties of transformations of G-spaces, Trans. Amer. Math. Soc. 160 (1971), 103-117.

[To] Tomter, P.: Anosov flows on infra-homogeneous spaces, Global Analysis, Proceedings of Symposia in Pure Mathematics, Vol. XIV, pp. 299-327, Amer. Math. Soc., Providence, 1970.

[Ur1] Urbanski, M.: On the capacity of a continuum with a nondense orbit under hyperbolic toral automorphism, Studia Math. 81 (1985), 37-51.

[Ur2] Urbanski, M.: The Hausdorff dimension of the set of points with nondense orbit under a hyperbolic dynamical system, Nonlinearity 4 (1991), 385-397.

[Va] Varadarajan, V.S.: Lie Groups, Lie Algebras and their Representations, Prentice Hall Inc., Engelwood Cliffs, 1974.

[V1] Veech, W.A.: Unique ergodicity of horospherical flows, Amer. J. Math. 99 (1977), 827-859.

[V2] Veech, W.A.: Weakly almost periodic functions on semisimple Lie groups, Monatsh. Math. 88 (1979), 55-68.

[V3] Veech, W.A.: Periodic points and invariant pseudomeasures for toral endomorphisms, Ergod. Th. and Dynam. Sys. 6 (1986), 449-473.

[Ve] Verjovsky, A.: Arithmetic, geometry and dynamics in the unit tangent bundle of the modular orbifold, In: Dynamical Systems (Proceedings, Santiago, 1990), pp. 263-298, Pitman Re. Notes Math. Ser. 285, Longman Sci. Tech., Harlow, 1993.

[Wa] Waddington, S.: The prime orbit theorem for quasihyperbolic toral automorphisms, Monatsh. Math. 112 (1991), 235-248.

[Wal1] Walters, P.: On the relationship between zero entropy and quasi-discrete spectrum for affine transformations, Proc. Amer. Math. Soc. 18 (1967), 661-667.

[Wal2] Walters, P.: Topological conjugacy of affine transformations of tori, Trans. Amer. Math. Soc. 131 (1968), 40-50.

[Wal3] Walters, P.: Conjugacy properties of affine transformations of nilmanifolds, Math. Syst. Theory 4 (1970), 327-333.

[Wal4] Walters, P.: Some transformations having a unique measure with maximal entropy, Proc. Lond. Math. Soc. (3) 28 (1974), 500-516.

[Wal5] Walters, P.: An Introduction to Ergodic Theory, Graduate Texts in Mathematics 79, Springer-Verlag, 1982.

[We] Weiss, B.: Finite dimensional representations and subgroup actions on homogeneous spaces, Israel J. Math. 106 (1998), 189-207.

[Wh] Whitney, H.: Elementary structure of real algebraic varieties, Ann. Math. 66 (1957), 545-556.

[Wi1] Witte, D.: Zero-entropy affine maps on homogeneous spaces, Amer. J. Math. 109 (1987), 927-961; Correction: 118 (1996), 1137-1140.

[Wi2] Witte, D.: Topological equivalence of foliations of homogeneous spaces, Trans. Amer. Math. Soc. 317 (1990), 143-166.

[Wi3] Witte, D.: Measurable quotients of unipotent translations on homogeneous spaces, Trans. Amer. Math. Soc. 345 (1994), 577-594; Correction and extension: 349 (1997), 4685-4688.

[WYZ] Witte, D., Yukie, A. and Zierau, R.: Prehomogeneous vector spaces and ergodic theory II, Trans. Amer. Math. Soc. (to appear).

[Wu] Wu, T.S.: Products of subgroups in Lie groups, Illinois J. Math. 29 (1985), 687-695.

[Y1] Yukie, A.: Prehomogeneous vector spaces and ergodic theory III, J. Number Th. 70 (1998), 160-183.

[Y2] Yukie, A.: Ratner's theorem and invariant theory, Preprint.

[Ze1] Zeghib, A.: Sur une notion d'autonomie de systèmes dynamiques, appliquée aux ensembles invariants des flots d'Anosov algébriques, Ergod. Th. and Dynam. Sys. 15 (1995), 175-207.

[Ze2] Zeghib, A.: Ensembles invariants des flots géodésiques des varietés localment symétriques, Ergod. Th. and Dynam. Sys. 15 (1995), 379-412.

[Zi1] Zimmer, R.J.: Ergodic actions with generalized discrete spectrum, Illinois J. Math. 20 (1976), 555–588.

[Zi2] Zimmer, R.J.: Orbit spaces of unitary representations, ergodic theory, and simple Lie groups Ann. Math. 106 (1977), 573–588.

[Zi3] Zimmer, R.J.: Ergodic Theory and Semisimple Groups, Birkhauser, Boston, 1984.

[Zi4] Zimmer, R.J.: Amenable actions and dense subgroups of Lie groups, J. Funct. Anal. 72 (1987), 58-64.

IV. The Dynamics of Billiard Flows in Rational Polygons

Contents

Chapter 11
The Dynamics of Billiard Flows in Rational Polygons of Dynamical Systems

J. Smillie

Billiard systems provide classic examples of simple mechanical systems. Among such systems the simplest are those that model the motion of a single particle in a region P of the plane. The trajectory of a particle in P is defined by requiring that the particle move in a straight line and at constant velocity in the interior of P and, when it hits the boundary of P at a point where the boundary is a smooth curve, the particle should reflect off of the boundary so that the angle of incidence is equal to the angle of reflection. If the trajectory hits a corner then there may be no good physical principal which selects one particular continuation. In this case the continuation of the trajectory is undefined.

In §2 we will explain how billiard trajectories are projections of orbits of a "geodesic flow" or "billiard flow" on a suitably defined tangent bundle. Thus the behavior of these trajectories is tied to the dynamics of the billiard flow.

Trajectories in planar billiard tables exhibit a wide range of behaviors. Two trajectories which are nearby but not parallel tend to diverge. Trajectories which are nearby and parallel remain parallel until they hit the boundary of P. Two features of the boundary may lead to divergence of nearby parallel trajectories: curvature and corners. The article by Bunimovich in this volume discusses the effects of curvature on planar billiard dynamics. If the sides of P are straight segments then curvature does not play a role. In this case nearby parallel trajectories can only diverge if they hit the boundary on opposite sides of a vertex. Regions with straight sides are called *polygonal billiard tables*. In this article we will consider the dynamics of billiard flows on polygonal tables.

The most familiar and best understood example of a planar polygonal billiard table is the square. The analysis of the dynamical properties of the billiard flow for the square dates to 1913 ([KS] see also [FK] and [HW]). The square has a number of properties that distinguish it from the general polygonal table. One key feature of the square is that each trajectory travels in only finitely many directions. This behavior is a characteristic feature of the class of rational billiards, where a polygonal billiard table with connected boundary is *rational* if each vertex angle has angular measure which is a rational multiple of π. In this article we will focus on the dynamics of billiard flows on rational polygons. For a discussion of general polygonal billiards see the survey articles by Tabachnikov [Ta] and Gutkin [Gu3].

There are several reasons for studying rational billiards. The dynamics of rational billiard tables are simple enough so that a theory can be developed yet they are sufficiently complex that many open questions remain. We will see in §6 that the study of rational billiard dynamics leads to the study of certain flows on moduli spaces which are themselves objects of great dynamical interest. Finally the fact that rational billiards are more complicated than "integrable" systems and yet not fully "chaotic" has led physicists to consider them as test cases for questions relating quantum dynamics to classical dynamics [cf BR]. While this leads to interesting question for future mathematical investigation, in this article we will deal exclusively with the "classical dynamics" of billiard tables.

The main questions that we will consider involve the distribution of billiard trajectories. Here are three specific questions.

(1) If we fix a table and fix a an initial direction of travel, what are the possible behaviors of the trajectories and how does this behavior depend on the initial position?
(2) How does the answer to question (1) change as we vary the direction?
(3) How does the answer to question (2) depend on the table?

§1. Two Examples

To give the flavor of rational billiard dynamics we will consider two examples, the square and the "divided rectangle", where the *divided rectangle* is the table obtained from the square by introducing a vertical reflective barrier in the center of the square which divides the square into two rectangles connected by an opening at the bottom (see figure). In fact the divided rectangle gives us a one parameter family of tables to consider, since we can adjust the length of the barrier. The billiard flow for the divided rectangle is easier to analyze than billiard flows on general rational tables, yet it exhibits many of the features seen in billiard flows on typical tables.

In the square, if two trajectories are parallel and start close together then they remain close together for all time, even if they hit the boundary on opposite sides of a vertex. This is an unusual feature for a rational polygon and accounts for some of the special properties of billiard trajectories in the square. It is no longer true in the divided rectangle, as we can see by considering two nearby parallel trajectories one of which hits the central barrier near its tip and one of which does not.

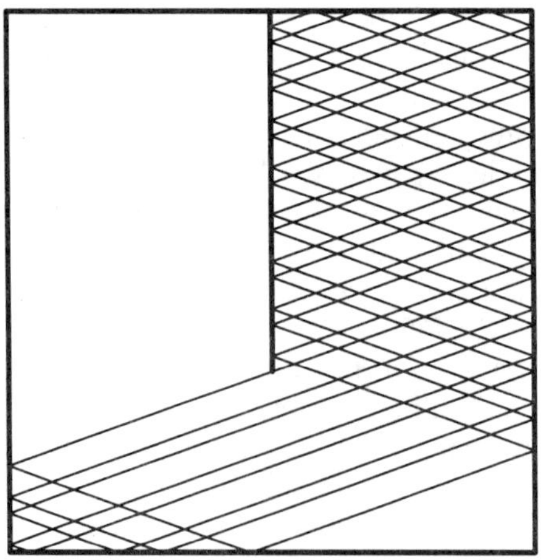

Fig. 1

In both of our examples the trajectories with rational slope have special properties. In the square, a trajectory with rational slope is periodic and the period is independent of the starting point. In the divided rectangle it is still true that all trajectories with rational slope are periodic but the period of the trajectory may depend on the initial point. In the square, each trajectory with irrational slope is dense. In contrast we can find trajectories in the divided rectangle with irrational slope which are not dense in the table (see Fig. 1). The first example of such behavior for polygonal tables was discovered by Galperin [Ga].

A dense orbit "fills up the table" but we can ask more specifically about the *rate* at which it fills up the table. We say that an orbit is *uniformly distributed* if the amount of time that it spends in a region is proportional to the area of the region. A trajectory which is uniformly distributed is necessarily dense. If the direction of the flow is irrational then every orbit in the square is not only dense but uniformly distributed. This is not the case for the divided rectangle; for certain barrier lengths there are orbits with irrational slope which are dense in the table but spend more time on the left of the barrier than on the right. It is not easy to illustrate this behavior with a computer picture but Fig. 2, which shows an ϵ-dense orbit which spends more time in the left half of the table than the right, is meant to be suggestive. At the end of §2 we will explain how the existence of dense but not uniformly distributed trajectories follows from an early theorem of Veech.

We say that a polygonal table has the *dichotomy property* if all non-singular orbits are closed or are uniformly distributed. The divided rectangle does not have this property (at least for most lengths of the gap) and there are reasons to think that this property is quite rare. Nevertheless we will see in §7 that

Fig. 2

there are in fact some interesting examples other than the square where the dichotomy property holds.

§2. Formal Properties of the Billiard Flow

In this section we will relate billiard trajectories to orbits of a billiard flow on an appropriately defined tangent bundle. We will investigate an invariant foliation on this tangent bundle and describe a technique from [ZK] which allows us to reduce questions about billiards to questions about the geodesic flow on certain singular surfaces.

We begin with some basic observations about billiard flows. Let P be a polygon in the plane which, for the moment, we do not assume to be rational. We will describe the construction of a "geodesic flow" or "billiard flow" on the unit tangent bundle to P whose orbits project to billiard trajectories on P which travel at unit speed. Let S^1 be the unit circle in $\mathbf{R}^2$ and let $T(\mathbf{R}^2) = \mathbf{R}^2 \times S^1$ be the unit tangent bundle of $\mathbf{R}^2$. The geodesic flow on $T(\mathbf{R}^2)$ induces a partially defined flow on $P \times S^1 \subset T(\mathbf{R}^2)$ where orbits fail to have continuations when they hit the boundary of P. We would like trajectories to reflect off the boundary, and the simplest way to achieve this is to identify certain inward and outward pointing vectors at points of the boundary of P. If e_i is an edge of P and $\rho_i : S^1 \to S^1$ represents the reflection through e_i then we identify (p, v) with $(p, \rho_i(v))$ for each $p \in e_i$. We define the tangent bundle of P, $T(P)$, to be $P \times S^1/\sim$ where $\sim$ is the equivalence relation generated by identifying (p, v) with $(p, \rho_i(v))$ as above. At the vertex p where the edge e_i meets the edge e_j we identify (p, v) with $(p, \gamma(v))$ for all γ in the group generated by ρ_i and ρ_j. Away from the vertices we can define a billiard flow on $T(P)$ whose trajectories project to billiard trajectories on P. This flow is continuous where it is defined.

The geodesic flow on $\mathbf{R}^2$ has a number of special properties, some of which are reflected in properties of billiard flows. Two tangent vectors (p, v) and (p', v') in $T(\mathbf{R}^2)$ are *parallel* if $v = v'$. The relation of being parallel gives an equivalence relation on $T(\mathbf{R}^2)$ which is preserved by the geodesic flow. The equivalence classes of this relation are copies of $\mathbf{R}^2$ which we can think of as leaves of a "parallel foliation" of $T(\mathbf{R}^2)$. Thus the geodesic flow on $\mathbf{R}^2$ can be decomposed as a family of "directional flows", one for each direction $v \in S^1$. We can view each of these directional flows as a flow on $\mathbf{R}^2$.

There is a related foliation of $T(P)$ which has corresponding properties. The parallel foliation of $T(\mathbf{R}^2)$ induces a foliation of $P \times S^1$. When two parallel billiard trajectories reflect off of the same edge they remain parallel. This implies that the identifications used to create the tangent bundle $T(P)$ in fact preserve the leaves of the parallel foliation. Thus there is a natural induced "parallel" foliation of $T(P)$ where leaves of this new foliation are obtained by gluing together the leaves of the parallel foliation of $P \times S^1$. Two points (p, v) and (p', v') are in the same leaf of this new foliation if there is some sequence of edges $e_{i_1}, \ldots, e_{i_n}$ so that $v' = \rho_{i_1} \circ \ldots \circ \rho_{i_n}(v)$. Let $\Gamma \subset O(2)$ be the group generated by the reflections in sides. The leaves of the foliation correspond to points in the orbit space S^1/Γ.

The properties of the parallel foliation of $T(P)$ depend on the cardinality of Γ. When P is a rational table then Γ is finite. In this case the "leaf space" S^1/Γ is an interval and the leaves of the foliation are closed surfaces. We will assume from now on that the polygon P is rational. Denote the interval S^1/Γ by I. For each $\theta \in S^1$ let M_θ be the surface corresponding to θ. Since the surfaces M_θ are invariant, the billiard flow on $T(P)$ decomposes into a family of *directional flows* on the surfaces M_θ.

The surfaces M_θ are constructed from copies of P which are glued along their edges by isometries. Since each surface M_θ is constructed by gluing together copies of P according to the same pattern all such surfaces can be identified with a single surface $\tilde{P}$. (If θ corresponds to an endpoint of I this is not quite true but this is a minor point which we will ignore.) The surface $\tilde{P}$ appears in [FK] in the case of the square. The general case was considered in [ZK] see also [RB].

Since the surface $\tilde{P}$ is built by gluing together polygons by isometries it has a natural metric space structure (cf [KS]). At points of $\tilde{P}$ corresponding to interior points of P or to edges of P the surface is locally isometric to $\mathbf{R}^2$. In particular there is a natural notion of parallel translation along paths which do not run through the vertices. The behavior at vertices is more complicated. Consider the following situation. Let $p_1 \ldots p_m$ be vertices in polygons $P_1 \ldots P_m$. Let θ_j be the vertex angle at p_j. Glue these polygons together in a cyclic pattern so that all vertices p_j are identified with a single point p. We say that the resulting space has a "cone type singularity" at p and we define the *cone angle at p* to be $\theta = \sum \theta_j$. If the cone angle is equal to 2π then the resulting surface is locally isometric to $\mathbf{R}^2$ at p. We can think of such points as a "removable singularities". Removable singularities arise at points in $\tilde{P}$ corresponding to vertices in P with vertex angles of the form π/n. We call points of $\tilde{P}$ at which the cone angle is not equal to 2π *vertices* of $\tilde{P}$.

When we construct the surface M_θ we are gluing together a finite number of copies of P each with a vector field on it. We perform the gluing so that the vector fields match along the edges of the polygons (though they may not match at the vertices). Since we can identify M_θ with $\tilde{P}$ we can think of these vector fields as vector fields on $\tilde{P}$. The vector fields that arise in this manner are precisely the "parallel vector fields" on $\tilde{P}$. A *parallel* vector field has the property that the vectors at any two points are parallel translates of one another. Thus we can think of the directional flows as the family of (partially defined) flows generated by the collection of parallel vector fields on the single surface $\tilde{P}$.

The fact that $\tilde{P}$ has a parallel vector field implies that the cone angles are multiples of 2π. The vector field can be extended to a point p if and only if the cone angle at p is equal to 2π which is to say that the singularity is removable.

Any two parallel vector fields on $\tilde{P}$ commute. The fact that the billiard flow leaves invariant a decomposition of $T(P)$ into surfaces and on each surface there is a pair of commuting vector fields is reminiscent of the properties of integrable flows on manifolds (cf Chapter 6 §1 of this volume). There is an important distinction between typical billiard flows on $\tilde{P}$ and integrable flows which is related to the existence vertices. In the case of integrable flows the invariant surface is a torus, that is to say a surface of genus one. In the rational billiard case the presence of vertices allows the possibility that the surface can have genus greater than one.

The Gauss-Bonnet theorem shows that the surface $\tilde{P}$ has genus one precisely when all the singular points are removable. In this case the analogy with integrable systems is complete and such polygonal billiard tables are called *integrable*. Richens and Berry [RB] have introduced the term *quasi-integrable* for the more typical case when the genus of $\tilde{P}$ is greater than one.

The list of integrable tables is short. The square is integrable as are rectangles. The only other integrable polygons are the triangle with angles $\pi/4$, $\pi/4$ and $\pi/2$, the equilateral triangle and the triangle with angles $\pi/6$, $\pi/3$ and $\pi/4$. When P is the divided rectangle then $\tilde{P}$ has two vertices each with cone angle 4π. This surface has genus 2 so this example is not integrable. It is nevertheless related to the torus as it can be viewed geometrically as a branched double cover of the torus.

The surfaces $\tilde{P}$ belong to an interesting class of geometric objects called translation surfaces that we will define using a characterization due to Veech. Say that we have a surface M with a specified finite set $\Sigma \subset M$. A *translation structure* on M is given by an atlas of charts in $M - \Sigma$ taking their values in $\mathbf{R}^2$ so that the change of coordinate functions are restrictions of translations of $\mathbf{R}^2$. This atlas of charts induces a Riemannian metric on $M - \Sigma$ and we impose the requirement that M is the metric completion of $M - \Sigma$ with respect to this Riemannian metric. In this case we will say that M is a surface with a translation structure or we will say that M is a *translation surface*. To see that $\tilde{P}$ has a translation structure we let Σ be the set of vertices and we choose a pair of perpendicular parallel vector fields on $\tilde{P}$. Take as a system of charts diffeomorphisms $\phi_j : U_j \to \mathbf{R}^2$ which take these parallel vector fields to the vector fields ∂_x and ∂_y in $\mathbf{R}^2$.

There is a geometric structure closely related to a translation structure where the change of coordinate functions are allowed to have the form $v \mapsto \pm v + c$. These structures are called *admissible $\mathcal{F}$ structures* by Veech or *half-*

integral translation structures in [GJ2] and they arise in the study of quadratic differentials [EG].

The surfaces $\tilde{P}$ provide examples of translation surfaces but there are many translation surfaces that do not arise from polygons via this construction. For any translation structure the geodesic flow decomposes into a family of "directional flows", one for each direction in the unit circle. When the translation structure does come from a rational billiard table this geodesic flow is the same as the billiard flow we have defined. Even when the translation structure does not arise from a billiard table this geodesic flow still constitutes an interesting dynamical systems. Our questions (1), (2) and (3) are still relevant. We can add two questions to our list:

(4) What behaviors are generic for translation structures?
(5) To what extent is the dynamical behavior of billiard tables like the dynamical behavior of translation structures?

§3. The Flow in a Fixed Direction

In this section we survey some results which describe the flow on a translation surface in a fixed direction. A useful technique in studying these flows is to consider the first return map to a transverse interval. Let I denote an interval in the surface M transverse to the flow. Let p be a point in I. The forward trajectory of p will either return to I or hit a vertex. The set of points in I for which forward trajectories hit vertices is finite. These points divide the interval I into subintervals $I_1 \dots I_k$. The restriction of the first return map to one of these intervals is an orientation preserving isometry. In particular the first return map is an interval exchange transformation. (See Chapter 4, Section 2 of this volume.)

Criteria for the minimality of interval exchange transformations lead to criteria for the minimality of directional flows. To describe one such criterion we will introduce some terminology. We will call a geodesic segment which starts at a vertex and ends at a vertex but contains no vertices in its interior a *vertex connection* or when no confusion will arise simply an *edge*. These are sometimes called *saddle connections* in the literature. A vertex connection has a well defined direction.

Theorem 3.1. [ZK], [BKM]. *If there are no vertex connections in direction v then the flow in direction v is minimal.*

If we are given a transverse interval I then there are two ways in which the directional flow could fail to be minimal. First there could be a set of trajectories which never hit I. In this case the union of these trajectories is a manifold with non-empty boundary and the boundary consists of vertex connections. Second it might be the case that all trajectories hit I but that the first return map is not minimal. In this case the Keane criterion for minimality of interval exchange transformations (Ch. 4, Theorem 2.1) shows that there must be a vertex connection. The paper of Arnoux [A1] provides a good reference for the relations between flows on surfaces and interval exchange transformations.

The measure theoretic behavior of interval exchange transformations is more subtle than the topological behavior. Examples of Keane [K] and Keynes and Newton [KN] show that it is possible for an interval exchange transformation to be minimal but not ergodic. On the other hand there are some strong limitations:

Theorem 3.2. [Ka], [V2]. *There is a constant N depending only on the polygon so that for each minimal direction there are at most N ergodic invariant measures for the directional flow.*

If there is only one ergodic component for a directional flow then that flow is *uniquely ergodic*. When the flow is uniquely ergodic all non-singular orbits are uniformly distributed. Of course when the flow is uniquely ergodic it is ergodic with respect to the natural Lebesgue measure.

Criteria for the unique ergodicity of interval exchange transformations lead to criteria for the unique ergodicity of directional flows. The criteria for determining unique ergodicity are more complicated than those for determining minimality. One approach is through a kind of renormalization operator. Renormalization operators occur in many areas of dynamical systems. These operators are maps defined on spaces of dynamical systems. Typically they act by replacing a map by an iterate of the original map restricted to a smaller domain and rescaling the domain.

For an interval exchange transformation defined on an interval I the induced map on a subinterval $I_1 \subset I$ is again an interval exchange transformation so it is natural to consider renormalization operators on the space of interval exchange transformations. Rauzy induction provides one such method. Rauzy induction gives a map $\mathscr{R}$ from the space of interval exchanges to itself and a finite partition of this space. Let α denote an interval exchange transformation. By considering the partition element containing $\mathscr{R}^n(\alpha)$ we assign to α a symbol sequence. In the simplest case when there are only two intervals Rauzy induction corresponds to the continued fraction algorithm. In the general case one can give a criterion for unique ergodicity of α in terms of the symbol sequence of α. (See [Ke], [Ra] of [V2].)

Unique ergodicity is not the only property about the distribution of orbits of an interval exchange transformation that can be deduced from information about Rauzy expansion. If an interval exchange transformation is uniquely ergodic then for any continuous function f on I and any point p the Birkhoff sums along the orbit of p converge to the integral of f. Zorich gives connections between the Lyapunov exponents of the Rauzy induction map and the rate of convergence of Birkhoff sums. See [Z1], [Z2], [Z3] and the joint paper with Kontsevich [KZ].

As an example of these methods we will consider the interval exchange associated with the divided rectangle. Let us start by considering a degenerate case when the gap has length zero and the central barrier, B, completely separates the two halves of the table. In this case $\tilde{P}$ consists of two disjoint tori. The points which map to B in these two tori correspond to two circles which we denote by S_L and S_R corresponding to the left and right chambers. The first return maps on these circles are rotations which we denote by τ_θ. Reflection through the center barrier defines a symmetry of this system which interchanges the two chambers and interchanges S_L and S_R. Let ρ denote this involution. The maps ρ and τ_θ commute. Now let us consider the case when the size of the gap is positive. This has the effect of "coupling" the billiard flows in the two chambers. The points in the circles which correspond to the gap give us two intervals $I_L \subset S_L$ and $I_R \subset S_R$. These intervals are interchanged by the reflection. The first return map is described as follows. If p is not in I_R or I_L then p returns to $\tau_\theta(p)$. If p is in I_R or I_L then p returns to $\rho(\tau_\theta(p))$ and hence it jumps to the other circle. The resulting dynamical system can be viewed as a skew product built over a rotation of the circle with a two point fiber. A precise criterion for unique ergodicity of such maps was given by Veech in [V1]. In particular this criterion can be used to show that there are minimal non-ergodic directional flows for most values of the gap length.

§4. Billiard Techniques: Minimality and Closed Orbits

It might appear from the previous section that the study of polygonal billiards is another form of the study of interval exchange transformations. If we focus only on the flow in a fixed direction then this is largely true. The distinction between the areas appears when we ask not what behavior can occur for various directional flows (our question 1) but rather what behavior is typical (our question 2). Three of the significant problems are to understand the set of directions in which the flow is minimal, the set of directions in which we have periodic billiard orbits and the set of directions in which the flow is uniquely ergodic.

The first result concerns minimal directions.

Theorem 4.1. [ZK]. *The set of directions for which the directional flow is not minimal is countable.*

As we have seen in §3 this involves showing that there are only countably many vertex connections. This follows from the fact that there can be at most one vertex connection in each relative homotopy class of paths between vertices. This result is valid for all translation structures.

The second question we consider is whether closed billiard trajectories exist. We might ask whether (as in the result of [ZK]) we can use the association between curves and homotopy classes to construct closed trajectories. We might imagine picking a homotopically nontrivial curve on the surface M and shrinking it to obtain a curve of minimal length. This can be done and the resulting curve will be a geodesic in the sense of metric spaces, that is to say that locally it will minimize distance between points. When M is the torus the resulting curve will be a periodic trajectory. When M has genus greater than one the resulting curve will usually not be a periodic trajectory. Typically the geodesic will consist of a sequence of geodesic segments running between vertices so that no two successive segments travel in the same direction.

Despite the fact that this "variational" argument does not work the following theorem of Masur shows that periodic billiard trajectories do exist.

Theorem 4.2. [M2]. *There is a periodic trajectory on every translation surface.*

We will discuss the proof of this result in a moment but we first give a sophisticated extension of this result which is also due to Masur. The most important invariant of a closed orbit is its length. We can ask about the number of families of closed trajectories of length less than N. In the case of the square this question reduces to the question of counting points in the plane with pairs of relatively prime integers as coordinates. The number of such points is asymptotic to cN^2 where $c = \pi/\zeta(2) = 6/\pi$. There is a corresponding asymptotic expression for periodic billiard trajectories in the square or any integrable polygon.

Theorem 4.3. [M3]. *For any translation surface there are constants $0 < c_1 \leq c_2$ such that the number of closed geodesics of length less than N is bounded below by $c_1 N^2$ and bounded above by $c_2 N^2$.*

We will sketch a proof of Theorem 4.2 which shows the usefulness of certain techniques that play an important role in much of the theory of rational billiards and general translation structures. The argument we sketch here follows the logic of Masur's original proof but it replaces the Teichmüller space techniques with techniques based on the geometry of translation structures. A good reference for the techniques we use is [MS1].

We begin by observing that there is a geometric criterion for the existence of a closed geodesic. Let us call a a subset of M isometric to the product of a circle and an interval a *cylinder*. If we fix the genus of M, the number of vertices and the area then the only way for a translation structure on M to have large diameter is for M to contain a long cylinder (that is to say a cylinder where the interval factor is long). (See [MS1] for the simple proof.) If M contains a cylinder then it contains a family of periodic billiard trajectories.

Let M be a surface with a translation structure of area one with no closed billiard trajectories. If M has diameter larger than some constant D then it contains a periodic trajectory and we are done. The important observation is that we can vary the translation structure on M to produce a family of new metrics and apply the geometric argument above to any of these translation structures, not just the original translation structure. We think of a translation structure as being given by an atlas of charts $\{\phi_j\}$ where $\phi_j : U_j \to \mathbf{R}^2$. If we are given a linear map $\alpha : \mathbf{R}^2 \to \mathbf{R}^2$ we can use it to construct a new atlas of charts $\{\phi_j \circ \alpha\}$. Let us denote by $\alpha(M)$ the new translation structure on M. If we take $\alpha \in SL(2, \mathbf{R})$ then $\alpha(M)$ also has area 1. If $\alpha \in SO(2, \mathbf{R})$ then the new translation structure has the same metric as the original one. In general, however, the metric geometry of the new translation structure will be different from the original metric even though the underlying affine structure is the same. A given curve is a periodic trajectory if it does not contain vertices and if it maps to a straight line in any coordinate chart $\phi_i : U \to \mathbf{R}^2$. This is a property of the affine geometry of M which is independent of the metric geometry of M. So if a curve is a periodic trajectory for $\alpha(M)$ it is also a periodic trajectory for M. Of course the diameter of M does depend on the metric. So our hypothesis on M implies that D bounds the diameters of all the translation structures $\alpha(M)$.

Our next objective is to choose a linear transformation which will allow us to exploit this bound on diameter. Let e_1 be the shortest edge in M. We can change the translation structure (and hence the metric) to make e_1 as short as we wish at the expense of making perpendicular directions longer. Choose the translation structure so that e_1 has length less than some constant C_1 to be determined later. Let $\alpha_1(M)$ denote this translation structure. Choose a second edge e_2 disjoint from e_1. Since the diameter of the surface $\alpha_1(M)$ is bounded by hypothesis, we can assume that the length of e_2 is bounded above by some constant, D. We then change the translation structure to get a new translation structure $\alpha_2(M)$ with respect to which e_2 is shorter than some constant C_2. We can continue until we run out of disjoint edges. Since the number of disjoint edges in M is bounded above by some constant n we can construct disjoint edges $e_1, \ldots, e_n$. Let $\epsilon > 0$ be given. We can choose constants $C_1, \ldots, C_n$ so that each edge $e_1, \ldots, e_n$ has length less than ϵ with respect to the final translation structure, $\alpha_n(M)$. We will work backwards from C_n to C_1. Let $C_n = \epsilon$. The last change in the translation structure decreases the length of some edge by a factor of D/C_n. It can increase the length of any other curve by at most this factor. If we choose C_{n-1} to be sufficiently small we can insure that even after performing the last alteration of the translation structure the edge e_{n-1} still has length less than ϵ. Continuing in this way we determine constants $C_1 \ll C_2 \ll C_3 \ll \ldots C_n = \epsilon$.

A maximal collection of disjoint edges in M partition M into a collection of triangles and the number of triangles (say m) depends only on the topology of M. Choose ϵ so that a triangle for which all edges have length less than ϵ

has area less than $1/m$. Since all edges have length less than ϵ with respect to the translation structure $\alpha_n(M)$, the total area of $\alpha_n(M)$ is less than one. This contradicts our original assumption that M (and therefore $\alpha_n(M)$) has area one. We conclude that it is not possible for all metrics affinely equivalent to our initial metric M to have bounded diameter. This contradiction proves the existence of a closed trajectory.

An analysis of this argument gives something that Masur's original argument did not give. It produces an explicit upper bound on the length of shortest trajectory.

§5. Billiard Techniques: Unique Ergodicity

In the preceding proof we saw the utility of changing the translation structure and hence the Riemannian metric on a surface while preserving the affine structure. In this section we will show that this technique is also useful in analyzing the ergodic properties of the directional flows.

We will begin by describing a method of changing the translation structure on a surface that gives an analog of Rauzy induction. Let us fix a surface M with a translation structure and consider the directional flow in the vertical direction. To analyze this flow via the methods of interval exchanges we choose a transverse interval which we can take to be horizontal and of length one. The first return map to this interval is an interval exchange. The method of Rauzy induction involves considering a sequence of subintervals $I \supset I_1 \supset I_2 \ldots$ and the sequence of first return maps to these intervals. At each stage we rescale the interval I_n and its subintervals by multiplying its length by λ_n so that its has the same length as I. Now we can achieve the rescaling directly by changing the translation structure so that we multiply the lengths in the horizontal direction by λ_n. To preserve the area of the surface M we can rescale in the vertical direction by multiplying by $(\lambda_n)^{-1}$. We can think of the vertical rescaling as changing the speed of the flow so that the average return time remains constant. Let M_n denote this new translation surface.

Let us define

$$g_t = \begin{pmatrix} e^t & 0 \\ 0 & e^{-t} \end{pmatrix}$$

Then g_t is a one-parameter subgroup of $SL(2, \mathbf{R})$ and the surfaces M_n are just the surfaces $g_{t_n}(M)$ where $t_n = \log \lambda_n$. (In the literature g_t is often defined with a different normalization.)

As we will see there is a criterion for unique ergodicity of the vertical flow on M in terms of the geometry of the surfaces $g_t(M)$ for $t \geq 0$. We call the parametrized collection of translation structures, $\{g_t(M)\}$, a *ray* of translation structures. A ray of translation structures is *divergent* if for any $\epsilon > 0$ there is a T so that for $t > T$ the translation surface $g_t(M)$ has a vertex connection of length less than ϵ. If a ray is not divergent then it is *recurrent*.

We will give two examples of the behavior of rays. Assume that the vertical flow is not minimal. As we have seen this implies that M has a vertical vertex connection. If the length of this vertex connection in M is C then the length of this vertex connection with respect to the metric corresponding to the translation structure $g_t(M)$ is $e^{-t}C$. In particular this ray of translation structures is divergent. Let us consider a second example. Take M to be the torus. Let L be a linear map with integral entries which induces an automorphism of M. Assume that L is hyperbolic with eigenvalues λ and λ^{-1} satisfying $\lambda > 1 > \lambda^{-1} > 0$. Now assume that the translation structure on M is chosen so that the expanding eigenvector of L is horizontal and the contracting eigenvector of L is vertical. In this case the map L induces an isomorphism of translation structures between M and $g_{t_0}(M)$ where $t_0 = \log \lambda$. It follows that $g_t(M)$ is isometric to $g_{t+t_0}(M)$ for any t. We can summarize by saying that, unlike the first case, the geometric invariants of $g_t(M)$ are periodic functions of t. When M has genus greater than one then this periodic behavior occurs precisely when the horizontal and vertical foliations of M arise from a pseudo-Anosov diffeomorphism.

Theorem 5.1. [M4]. *If the vertical flow is not uniquely ergodic then the ray $g_t(M)$ is divergent.*

If the vertical flow is minimal but not uniquely ergodic then instead of having a single edge get short as $t \to \infty$, as in our first example, there will be a sequence of different edges so that the first edge gets short and then as it starts to lengthen a second edge gets short and so on.

To deal with flows in directions other than the vertical we can rotate the translation structure to make the direction vertical and then apply the above criterion. Let

$$r_\theta = \begin{pmatrix} \cos\theta & \sin\theta \\ -\sin\theta & \cos\theta \end{pmatrix}$$

If the flow in direction $\pi/2 - \theta$ is not uniquely ergodic then by Masur's criterion the ray $g_t r_\theta(M)$ is divergent. This test for non-unique ergodicity plays a key role in the following:

Theorem 5.2. [KMS]. *For each translation surface the set of directions for which the flow fails to be uniquely ergodic has measure zero.*

We will discuss the proof of this result after giving a corollary to the theorem and an improvement of the result. Recall that the billiard flow on $T(P)$ is never ergodic when P is irrational. Using the previous result and the technique of approximating non-rational tables by rational tables leads to:

Corollary 5.3. [KMS]. *There exist (non-rational) polygonal billiard tables for which the billiard flow is ergodic on $T(P)$.*

The earlier theorem was improved by Masur to show:

Theorem 5.4. [M4]. *The Hausdorff dimension of the set of directions for which the flow is not uniquely ergodic is at most* $1/2$.

An exposition of the proof of the Theorem 5.2 is contained in [A]. The published proofs rely on a criterion for unique ergodicity which is not as powerful as Masur's criterion (Theorem 5.1). Using theorem 5.1 allows the proof to be simplified somewhat.

We will not describe the proof of Theorem 5.2 here but we will explain how questions about the geometry of translation structures arise in the proof. Let M be a surface with a translation structure. For almost every θ we must show that the ray $g_t r_\theta(M)$ is not divergent as $t \to \infty$. The technique of the proof is to show that for t large and ϵ small the set of θ values for which $g_t r_\theta(M)$ has a segment of length less than ϵ has small measure. Fix t large and assume that for some θ the surface $g_t r_\theta(M)$ has a short edge. Call it e. Now as the translation structure changes we can keep track of the length of e with respect to the translation structure $g_t r_\theta(M)$ as a function of θ. The interval of θ values for which the length of e is greater than ϵ is much larger than the interval for which it has length less than ϵ. The problem is that making the shortest curve longer does not rule out the possibility that some other curve may have gotten shorter in the process. This problem does not arise when M is the torus. In the case of the torus when one curve is short all curves that cross it are long. When M has higher genus though it is possible to have many short curves simultaneously. The solution is to focus on a certain class of curves which behave like curves in the torus. These curves have the property that they are short but not crossed by other short curves.

§6. Dynamics on Moduli Spaces

We have seen how questions about billiard dynamics in rational polygons lead to the study of certain translation structures. In this section we will consider the collection of all translation structures and not restrict ourselves to those arising from polygons. If we identify "geometrically equivalent" translation structures on a given surface then the set of these equivalence classes of structures forms a "moduli space". These moduli spaces possess some interesting and useful geometric structures. The operation of changing the translation structure by elements of $SL(2, \mathbf{R})$ gives a group action of $SL(2, \mathbf{R})$ on each moduli space. We have seen in the previous section how the dynamical properties of the billiard flow on M translate into geometric properties of the various translation structures $\alpha(M)$ for $\alpha \in SL(2, \mathbf{R})$. We will see that the investigation of dynamical properties of the group action on moduli space leads to results about the dynamics of billiard flows for general translation structures as well as to some seemingly unrelated results.

Moduli spaces of translation structures also arise in the theory of quadratic differentials and Teichmüller space. The moduli spaces we will describe are called *strata* because they appear in a stratification of the unit tangent bundle to Teichmüller space. The flow g_t defined in §5 is called the *Teichmüller flow* in this context. For those familiar with Teichmüller space this connection is a source of insight and inspiration. Those not familiar with Teichmüller space methods should not be discouraged. The moduli spaces that we consider can be constructed with no reference to complex analysis (cf. [MS1], [V4] and [V6]) and furthermore none of the theorems discussed in this survey *require* complex methods for their proofs. Whether or not one chooses to use complex methods there is no reason not to use the terminology which arises from the study of Teichmüller space.

Let us say that two translation surfaces M and M' are *topologically equivalent* if the surfaces M and M' are homeomorphic and the number and cone angles of the vertices correspond. (In this survey we are not considering half-integral structures. If we were to do so then there would be an additional piece of data.) Let $f : M \to M'$ be a homeomorphism of translation structures that takes vertices to vertices. If f is a smooth map we can think of the derivative as a map from $\mathbf{R}^2$ to $\mathbf{R}^2$. We say that f is *affine* if the derivative is constant. We say that an affine diffeomorphism f is an *equivalence of translation structures* if the derivative is the identity map. When such an f exists we say that the translation surfaces M and M' are *geometrically equivalent*. For a fixed translation surface M let us denote by $\mathcal{M}(M)$ the moduli space of translation structures topologically equivalent to M with area one. We will call this moduli space a *stratum*. The question of precisely which strata are non-empty is answered in [MS2].

The construction of moduli spaces, that is to say the definition of a topology and other structure for the sets defined above, is rather involved. We will limit ourselves to a description of the construction in the simplest case, that of the surface of genus one. Let T be a torus with a translation structure. Let $\tilde{T}$ denote the universal cover of T. The translation structure gives a canonical way to identify $\tilde{T}$ with $\mathbf{R}^2$. The covering group acts by translation so we can identify it with a lattice $\Lambda \subset \mathbf{R}^2$. The moduli space we want to construct can be identified with the space of lattices in $\mathbf{R}^2$. To build this space we introduce some additional structure. Let us call a translation structure on T together with a choice of a basis of $\pi_1(T)$ a *marked translation structure*. A marked translation structure gives rise to a lattice in $\mathbf{R}^2$ together with a choice of a basis, v and w. Viewing the pair of vectors as a matrix $[vw]$ we can identify the set of marked translation structures with $GL(2, \mathbf{R})$. Now to construct the space of translation structures we analyze the effect of changing the marking on the space of marked translation structures. The mapping class group of the torus can be identified with the group $GL(2, \mathbf{Z})$. This group acts transitively on the sets of markings. Thus the space of translation structures can be identified with $GL(2, \mathbf{R})/GL(2, \mathbf{Z})$. If we restrict ourselves to translation structures that give the torus area 1 then appropriate moduli space is $SL(2, \mathbf{R})/SL(2, \mathbf{Z})$. The natural action of $SL(2, \mathbf{R})$ is the action by left multiplication. The moduli space for the torus has a smooth finite measure which is invariant under the action of $SL(2, \mathbf{R})$.

The construction of other moduli spaces is somewhat more complicated but it follows the outlines of this construction (see [V4] or [MS1]). As in the case of the torus there is an action of $SL(2, \mathbf{R})$ on the moduli space and a finite smooth measure μ defined on the moduli space invariant under this action. The orbits of the $SL(2, \mathbf{R})$ action are affinely equivalent translation surfaces. Unlike the case of the torus the action of $SL(2, \mathbf{R})$ is not transitive. A second distinction between the higher genus moduli spaces and that of the torus is that the higher genus moduli spaces are not connected in general, but have a finite number of components. This phenomenon was first discovered by Rauzy in the context of interval exchanges.

The moduli space for the torus can be identified with the unit tangent bundle of the modular surface. The Teichmüller flow g_t is just the geodesic flow on the modular surface (though with our normalization of g_t, geodesics travel at twice the usual speed). This geodesic flow is one of the classic examples in dynamical systems. (See Chapter 7 section 5 of this volume for a discussion.) A number of authors have described interesting connections between the geodesic flow on the modular surface and continued fraction expansions. One very elegant way of making this connection is described by Arnoux in [A3] where he uses the fact that each point in $SL(2, \mathbf{R})/SL(2, \mathbf{Z})$ can be interpreted as a flow on the torus. It is a classic result that the geodesic flow for the modular surface is ergodic. The corresponding ergodicity results for the higher genus moduli spaces were proved in general by Veech (certain important cases were established in [M1] and [Re]).

Theorem 6.1. [V4]. *The flow g_t is ergodic on each component of each stratum.*

The flow g_t is uniformly hyperbolic in the genus one case. In the higher genus case it is non-uniformly hyperbolic with respect to the natural measure μ (see [V4] and [KZ]). (For a discussion of non-uniform hyperbolicity see Ch. 7 section 2.8 of this volume) This is a key ingredient in a result of Veech which counts closed orbits for the flow g_t. As we have seen in §5, closed orbits of the Teichmüller flow on $\mathcal{M}(M)$ correspond to pseudo-Anosov diffeomorphisms with prescribed numbers and types of singular points. Let $N(t)$ denote the number of primitive conjugacy classes of pseudo-Anosov diffeomorphisms with the same singularities as M.

Theorem 6.2. [V4].

$$\liminf_{t\to\infty}\frac{\log N(t)}{t}\geq 6g-6+2n$$

where g is the genus of M and n is the number of singular points.

A corollary to this theorem is the existence of pseudo-Anosov diffeomorphisms of all possible topological types (that is to say all possible patterns of topological data) [see MS2].

Considering the $SL(2, \mathbf{R})$ action on moduli space also leads to information about "generic" translation structures.

Theorem 6.3. [MS1]. *For each component of each higher genus moduli space there is a $\delta > 0$ so that for almost every translation surface M in moduli space the set of directions for which the flow is not ergodic has positive Hausdorff dimension.*

The ergodicity of the flow g_t implies that the Hausdorff dimension of the set of non-ergodic directions is constant almost everywhere on each component of each stratum.

Theorem 6.3 shows that there are translation structures of all possible topological types with large sets of non-ergodic directions. The result does not show that there are polygonal billiard tables with this property because the set of translation structures arising from polygonal billiard tables has measure zero in the space translation structures. On the other hand it would certainly be interesting to know to what extent rational billiard tables behave like "generic" translation structures.

Every interval exchange transformation arises as the first return map for some translation structure. Thus the Theorem 6.3 has implications for interval exchange transformations.

Corollary 6.4. [MS1]. *Consider the simplex of interval exchange transformations with a given irreducible permutation not equivalent to a rotation. The subset of the simplex corresponding to not ergodic interval exchange transformations has codimension strictly less than one.*

§7. The Lattice Examples of Veech

While the results we have described in this survey give a great deal of information about general billiard flows they do not in general allow us to say, for a given polygon and a given direction, what the behavior of the directional flow is. In this section we will focus on polygons where such a precise description of the dynamics is possible.

Let us return to a discussion of the billiard flow for the square. In this case each directional flow has one of two types of behavior and we have a straightforward criterion for deciding which type of behavior occurs. If the slope of the direction is rational then all non-singular orbits are closed. If the slope of the direction is irrational then all non-singular orbits are uniformly distributed. Recall that polygonal tables which have only these two types of behavior are said to have the dichotomy property. In this section we will consider polygons where the "dichotomy property" holds and where there is an explicit description of which directions have which behavior.

The list of such "well behaved" tables starts with the integrable polygons. These are precisely the polygons P for which $\tilde{P}$ is the torus. The list of well behaved polygons was extended by Gutkin in his construction of "almost integrable" billiard tables [Gu1]. This class of polygons includes the regular hexagon and it includes polygons all of whose sides are horizontal or vertical and for which the coordinates of vertices are all rational. This list of well behaved polygons was extended by Veech. Veech showed that the list contains not just the regular polygons with 3, 4 and 6 sides but *every* regular polygon.

To describe Veech's result more precisely let us introduce some terminology. If M is a translation surface let $\Gamma(M)$ be the group of affine automorphisms of M. That is to say that Γ consists of homeomorphisms of M which take singularities to singularities and are differentiable with constant derivative away from the singularities. If we consider only orientation preserving maps then the derivative of such a map is an element of $SL(2, \mathbf{R})$. The derivative gives a homeomorphism D from Γ to $SL(2, \mathbf{R})$. We say that M has the *lattice property* if the image of Γ is a lattice in $SL(2, \mathbf{R})$. We say that a polygon P has the lattice property if the translation surface $\tilde{P}$ has the lattice property.

Any polygon that tiles the plane by reflection has the lattice property because $\tilde{P}$ is the torus, T, and $\Gamma(T) = SL(2, \mathbf{Z})$ which is a lattice in $SL(2, \mathbf{R})$. For the almost integrable polygons P the surface $\tilde{P}$ is not the torus but it is a branched cover of the torus where the branch points have rational coordinates. In this case $D\Gamma$ is commensurable to $SL(2, \mathbf{Z})$. Veech's examples also have the lattice property but they differ from the previous examples in that the group Γ is not commensurable to $SL(2, \mathbf{Z})$. In this sense Veech's examples represent a significant new phenomenon in the study of polygonal tables.

Theorem 7.1. [V5], [V7]. *The regular n-gon has the lattice property. A right triangle with one angle of the form π/n has the lattice property. An isosceles triangle with smallest angle of the form $2\pi/n$ has the lattice property.*

The statement that lattice polygons have easily described dynamics is contained in the following theorem.

Theorem 7.2. [V5]. *Lattice examples have the dichotomy property. Furthermore the directions in which all orbits are closed are just the directions fixed by parabolic elements of $D\Gamma$.*

When $D\Gamma$ is commensurable to $SL(2, \mathbf{Z})$ the directions fixed by parabolics are just the rational directions. When $D\Gamma$ is not commensurable to $SL(2, \mathbf{Z})$ these directions are mostly not rational [B]. In addition to identifying directions of closed trajectories it is also possible to analyze the growth rate of closed trajectories (cf. Theorem 4.3).

Theorem 7.3. [V5]. *For any lattice surface there is a constant c such that the number of closed geodesics of length less than N is asymptotic to cN^2.*

Furthermore it is possible, as in the case of the torus, to compute precisely the constants c that arise (see [V5], [V7] and [GJ2]).

We have seen that the properties of the billiard flow on M are captured in the behavior of the orbit of M in moduli space under the $SL(2, \mathbf{R})$ action. This orbit is parametrized by $SL(2, \mathbf{R})/\Gamma$. When Γ is a lattice this space has a special structure which is very much like the structure of $SL(2, \mathbf{R})/SL(2, \mathbf{Z})$. In particular we can identify it with the tangent space to a complete hyperbolic surface of finite area. Any such surface decomposes into a compact piece and a finite number of "cusps". According to Masur's criterion divergent rays correspond to geodesics which eventually remain in a single cusp. Geodesics that remain in a single cusp have a very special form and these correspond to translation structures all of whose vertical trajectories are periodic. It is in this way that the dichotomy property follows from the structure of the $SL(2, \mathbf{R})$ orbit.

Veech's discovery raises the question of whether it will be possible to describe explicitly the dynamics of the billiard flow for other rational tables. An initial question to ask is: Which polygon's have the lattice property? Ward [W] and Vorobets[Vo2] discovered some additional examples among rational triangles and proved that certain specific triangles do not have the lattice property. Kenyon and Smillie [KS] show that among right triangles the examples of Veech are the only ones that have the lattice property. They also analyze a large number, (10^{11}), of acute rational triangles and show that other than Veech's examples only three of these triangles have the lattice property.

There are many open questions related to the description of the billiard flow in explicit rational polygons. If the dynamical behavior of typical translation structures can be taken as a guide to the dynamical behavior of typical rational billiards we would expect that most tables possess non-ergodic directions. On the other hand it seems that no one has yet constructed a single example of a non-ergodic direction in an acute triangle.

It would be very interesting to know about the structure orbits of the $SL(2, \mathbf{R})$ action on moduli space. Is there some analog of Ratner's solution of Raghunathan's conjecture (cf. article by Dani in this volume) which would characterize orbit closures?

As we mentioned in the introduction, rational billiards have been considered as test cases for questions involving quantum chaos in part because their dynamics is close to the dynamics of integrable systems. As we have seen the dynamics of the lattice examples are particularly well behaved. This motivates the following question of Sinai: Do rational polygons with the lattice property have distinctive quantum mechanical behavior?

Bibliography

[A1] P. Arnoux, Échanges d'intervalles et flots sur les surfaces. (French) Ergodic theory (Sem., Les Plans-sur-Bex, 1980) (French), pp. 5–38, Monograph. Enseign. Math., 29, Univ. Genève, Geneva, 1981.

[A2] P. Arnoux, Ergodicité générique des billards polygonaux (d'après Kerckhoff, Masur, Smillie). Séminaire Bourbaki, Vol. 1987/88. Astérisque No. 161–162 (1988), Exp. No. 696, 5, 203–221 (1989).

[A3] P. Arnoux, Le codage du flot géodésique sur la surface modulaire. Enseign. Math. (2) 40 (1994), no. 1–2, 29–48.

[BKM] C. Boldrighini M. Keane, F. Marchetti, Billiards in polygons, Ann. of Prob. 6, 1978 532–540.

[B] M. Boshernitzan, Billiards and rational periodic directions in polygons, American Math. Monthly 99, 1992, 522–529.

[EG] C. J. Earle, F. P. Gardiner, Teichmüller disks and Veech's F-structures. *Extremal Riemann surfaces* 165–189 Contemp. Math., 201 AMS Providence, RI, 1997.

[FK] R. Fox and R. Kershner, Concerning the transitive properties of geodesics on a rational polyhedron, Duke Math J 2, 1936, 147–150.

[Ga] G. Galperin Non-periodic and not everywhere dense billiard trajectories in convex polygons and polyhedrons, Comm. Math. Phys. 91, 1983 187–211.

[GKT] G. Galperin, T. Kruger S. Troubetzkoy, Local instability of orbits in polygonal and polyhedral billiards, Comm. Math. Phys.

[Gu1] E. Gutkin, Billiards in almost integrable polyhedral surfaces. *Erg. Th. & Dyn. Syst.* **4** (1984), 569–584.

[Gu2] E. Gutkin Billiards in polygons, Physica D **19**, 1986 311–333.

[Gu3] E. Gutkin, Billiards in polygons: survey of recent results. J. Statist. Phys. 83 (1996), no. 1–2, 7–26.

[GT] E. Gutkin and S. Troubetzkoy Directional flows and strong recurrence for polygonal billiards. International Conference on Dynamical Systems (Montevideo, 1995), 21–45, Pitman Res. Notes Math. Ser., 362, Longman, Harlow, 1996.

[GJ1] E. Gutkin and C. Judge, The geometry and arithmetic of translation surfaces with applications to polygonal billiards, *Math. Res. Lett.* **3** (1996) 391–403.

[GJ2] E. Gutkin and C. Judge, Affine mappings of translation surfaces: geometry and arithmetic, preprint (1997).

[HW] Hardy, G. H.; Wright, E. M. An introduction to the theory of numbers. Fifth edition. The Clarendon Press, Oxford University Press, New York, 1979.

[Ka] Katok, A., In variant measures for the flows on oriented surfaces, Dokl. Akad. Nauk SSSR 211, 775–778 (1973) (Russian) English translation: Sov. Math. Dokl. 14, 1104–1108.

[K] M. Keane, Non-ergodic interval exchange transformations, Israel J. Math. 26 (1977), 188–196.

[KS] R. Kenyon, J. Smillie, Billiards on rational-angled triangles, to appear in Commentarii Math. Helv.

[Ke] Kerckhoff, S. P. Simplicial systems for interval exchange maps and measured foliations. Ergodic Theory Dynamical Systems 5 (1985), no. 2, 257–271.

[KMS] S. Kerckhoff, H. Masur, J. Smillie, Ergodicity of Billiard Flows and Quadratic Differentials, Ann. of Math. **124** 1986 p. 293–311.

[KN] H. Keynes and D. Newton, A minimal non-uniquely ergodic interval exchange transformation, Math. Z. **148** (1976), 101–105.

[KS] D. König and A. Szücs, Mouvement d'un point abandonné à l'intérieur d'un cube, Rendiconti del circulo matematico di Palermo, 36 (1913) 79–90.

[KZ] M. Kontsevich and A. Zorich, Lyapunov exponents and Hodge theory, IHES preprint (1997).

[M1] H. Masur, Interval exchange transformations and measured foliations, Ann. of Math. **115** (1982), 169–200.

[M2] H. Masur, Closed trajectories for quadratic differentials with an application to billiards, Duke Math. J. **53** (1986), 307–314.

[M3] H. Masur, The growth rate of trajectories of a quadratic differential. Ergodic Theory Dynamical Systems 10 (1990), no. 1, 151–176.

[M4] H. Masur, Hausdorff dimension of the set of nonergodic foliations of a quadratic differential. Duke Math. J. 66 (1992), no. 3, 387–442.

[MS1] H. Masur, J. Smillie. Hausdorff dimension of sets of nonergodic measured foliations *Annals of Math.* **134** (1991), 455–543.

[MS2] H. Masur. and J. Smillie. Quadratic differentials with prescribed singularities and pseudo-Anosov diffeomorphisms. Comment. Math. Helv. 68 (1993), no. 2, 289–307.

[Ra] G. Rauzy, Échanges d'intervalles et transformations induites, Acta Arith, 34 (1979), 315–328.

[Re] M. Rees, An alternative approach to the ergodic theory of measured foliations on surfaces, Ergodic Theory and Dynamical Systems, Vol. 1 (1981), 461–488.

[RB] P. J. Richens and M. V. Berry, Pseudointegrable systems in classical and quantum mechanics, *Physica* D **2:3** (1981) 495–512.

[Ta] S. Tabachnikoff, Billiards. Panoramas et Synthèses **1** Soc. Math. France (1995).

[V1] W. Veech, Strict ergodicity in zero dimensional dynamical systems and the Kronecker-Weyl theorem mod 2, Trans. A. M. S. **140** (1969), 1–34.

[V2] W. Veech, Interval exchange transformations, J. Analyse Math., 33 (1978), 222–272.

[V3] W. Veech, Gauss measures for transformations on the space of interval exchange maps. Ann. of Math. (2) 115 (1982), no. 1, 201–242.

[V4] W. Veech, The Teichmüller geodesic flow. Ann. of Math. (2) 124 (1986), no. 3, 441–530.

[V5] W. Veech, Teichmüller curves in moduli space, Eisenstein series and an application to triangular billiards. Invent. Math. 97 (1989), no. 3, 553–583.

[V6] W. Veech, Moduli spaces of quadratic differentials. J. Analyse Math. 55 (1990), 117–171.

[V7] W. Veech, The billiard in a regular polygon. Geom. Funct. Anal. 2 (1992), no. 3, 341–379.

[Vo1] Ya. B. Vorobets, Plane structures and billiards in rational polyhedra. (Russian) *Uspekhi Mat. Nauk* 51 (1996), no. 1(307), 145–146; translation in Russian Math. Surveys 51 (1996), no. 1, 177–178

[Vo2] Ya. B. Vorobets, Plane structures and billiards in rational polygons: the Veech alternative. (Russian) *Uspekhi Mat. Nauk* 51 (1996), no. 5(311), 3–42. translation in Russian Math. Surveys 51:5 (1996), 779–817.

[W] C. Ward, Fuchsian groups and polygonal billiards, Thesis, Rice University 1996.

[ZK] A. Zemlyakov, A. Katok. Topological transitivity of billiards in polygons, *Math. Notes of the USSR Acad. Sci.* **18:2** (1975), 291–300. (English translation in Math. Notes **18:2** (1976) 760–764.)

[Z1] A. Zorich, Asymptotic flag of an orientable measured foliation on a surface. Geometric study of foliations (Tokyo, 1993), 479–498, World Sci. Publishing, River Edge, NJ, 1994.

[Z2] A. Zorich, Finite Gauss measure on the space of interval exchange transformations. Lyapunov exponents. Ann. Inst. Fourier (Grenoble) 46 (1996), no. 2, 325–370.

[Z3] A. Zorich, Deviation for interval exchange transformations, Ergodic Theory Dynam. Systems 17 (1997), no. 6, 1477–1499.

V. Dynamical Systems of Statistical Mechanics and Kinetic Equations

Contents

As mentioned in the Preface to the English edition, two new contributions have been added for this second edition, i.e. Chapters 10 and 11 by S.G. Dani and J. Smillie. The numbering of the formulae in Chapters 12 and 13 of the present edition and any cross-references used in the body of the text refer to the structure of the previous edition, where these two chapters were entitled "Chapter 10" and "Chapter 11".

Chapter 12
Dynamical Systems of Statistical Mechanics

R.L. Dobrushin, Ya.G. Sinai, Yu.M. Sukhov

§ 1. Introduction

The motion of a system of N particles in d dimensions is described in Statistical Mechanics by means of a Hamiltonian system of $2Nd$ differential equations, which generates the group of transformations of the phase space. The object of the investigation is the time evolution of probability measures on the phase space determined by this group of transformations. The principal feature of problems in Statistical Mechanics is the fact that one deals with systems consisting of a large number of particles of the same type (a mole of a gas contains $6 \cdot 10^{23}$ particles). Therefore, only those results in which all estimates are uniform with respect to the number of degrees of freedom are of interest here. This restriction, which is unusual from the point of view of the standard theory of dynamical systems, specifies the mathematical feature of the problems of Statistical Mechanics.

Thus, the problem of studying asymptotically the properties of a system, as $N \to \infty$, arises naturally. In our view, the fundamental mathematical approach to this problem is the explicit consideration of infinitely-dimensional dynamical systems arising as the limit, as $N \to \infty$, of the system of the equations of motion of N particles. Unlike infinite systems of equations usually considered in mathematical physics which arise in other domains of application, one is interested here in systems for which all the degrees of freedom are completely "equal in importance".

Traditionally, statistical mechanics is divided into equilibrium and non-equilibrium mechanics. In the equilibrium statistical mechanics one studies properties of a special class of measures invariant with respect to the dynamics which are determined by the well-known Gibbs postulate. This wide theme which is the subject of numerous investigations (cf [S6], [D5], [R2]) remains mainly outside our exposition, although a part of the corresponding theory which is related to applications to dynamical systems with hyperbolic properties has been touched in Sect. 6 of Chap. 3 of Part I and in Chaps. 7, 8 of Part II. In the following section we shall discuss the main facts about Gibbs random fields used in further sections of this chapter.

The mathematical investigation of problems of non-equilibrium statistical mechanics is now in its initial stage. The results obtained so far are fairly odd; the remaining gaps are filled here by mathematically formulated conjectures and sometimes even by "physical" considerations. Among earlier review texts concerning this theme we mention the articles [GO], [AGL2], [DS1], [L3] (some parts of the papers [DS2], [L4] are a kind of review as well). A series of problems related to our theme is discussed in the book [CFS].

We have chosen for detailed discussion several crucial and most elaborated topics. First, there is the problem of the existence of the "infinite-particle" dynamics to which Sect. 3 and a part of Sect. 4 are devoted. Secondly, we discuss, in particular simple cases, ergodic properties of infinite-particle dynamical systems with invariant measures (see Sect. 5). The fundamental question about asymptotical properties of time evolution, as $t \to \pm\infty$, is closely connected with the problem of describing the set of invariant measures. This subject is considered in Sect. 4. The matter of Sects. 4, 5 is immediately related to the problem of mathematical foundation of the Gibbs postulate. In Sect. 6 we discuss some results which concern deriving kinetic equations, i.e., equations describing time evolution of expectation values of the principal physical variables.

The models of dynamics which are under investigation in statistical mechanics are fairly varied. In what follows we consider only one such model. This is the Newton dynamics of a system of point particles moving in Euclidean space subject to internal interaction forces—a model which is well-known from elementary courses in mechanics. As to other types, we only give references to some books and main papers where a more complete bibliography may be found. In the literature the classical "spin" dynamics is often investigated (cf [BPT], [LL], [LLL]). In models of spin dynamics one considers the evolution of coordinates which describe internal degrees of freedom of particles fastened at points of a regular lattice. Among other dynamical models usually studied are the gradient models in which, for the sake of simplicity, one deals with a system of differential equations of first order for the positions of particles (cf [FF], [F3], [L]).

Conceptually, papers on stochastic dynamics (cf [Du], [Gri], (Li]) which form now a wide chapter of probability theory, namely, the theory of Markov processes with local interaction, are close to the range of problems discussed here, as well as papers on dynamics of quantum systems with infinitely many degrees of freedom (cf [BM], [BR1], [BR2]). These themes require special reviews.

The English translation of Chapters 10 and 11 was prepared with the help of Prof. B. Hajek. The authors express to him the deep gratitude.

§2. Phase Space of Systems of Statistical Mechanics and Gibbs Measures

In this section we introduce basic notions of equilibrium statistical mechanics which will be used later: the configuration and phase spaces of dynamical systems of statistical mechanics and probability measures defined on such spaces.

2.1. The Configuration Space. According to tradition, we shall consider systems of particles in d dimensions, where d is an arbitrary integer. Although the case $d = 3$ is, of course, the most interesting, the low dimensions admit a physical interpretation as well ($d = 2$ is the case of a gossamer pellicle or face). The principal reason for such a generalization is that the properties of systems of statistical mechanics depend essentially on the dimensionality and it is interesting to investigate this dependence from both the mathematical and physical points of view.

A *particle configuration* in Euclidean space R^d is defined as a finite or countable subset $\mathbf{q} \subset R^d$. Let

$$\nu_{\mathcal{O}}(\mathbf{q}) = |\mathbf{q}_{\mathcal{O}}|, \qquad \mathcal{O} \subseteq R^d. \tag{10.1}$$

Here and below $\mathbf{q}_{\mathcal{O}} = \mathbf{q} \cap \mathcal{O}$, and the symbol $|\cdot|$ denotes the cardinality of a given set. One imposes the following condition:

$$\nu_{\mathcal{O}}(\mathbf{q}) < \infty \text{ for any bounded } \mathcal{O} \subset R^d \tag{10.2}$$

which means that infinite particle accumulations are inadmissible. The set of all such $\mathbf{q}$'s is denoted by Q^0.

In some models of dynamical systems one admits that sometimes several particles may be found at the same point $q \in R^d$. Thereby it is convenient to extend the above definition, and to treat a configuration as a pair $(\mathbf{q}, n_{\mathbf{q}})$ where $\mathbf{q} \in Q^0$ and $n_{\mathbf{q}}$ is an integer-valued non-negative function giving the number of particles at points $q \in \mathbf{q}$. The set of all such pairs $(\mathbf{q}, n_{\mathbf{q}})$ is denoted by Q. Given $\mathcal{O} \subset R^d$ and $(\mathbf{q}, n_{\mathbf{q}}) \in Q$, we denote by $\nu_{\mathcal{O}}(\mathbf{q}, n_{\mathbf{q}})$ the number of particles in the set $\mathcal{O}$:

$$\nu_{\mathcal{O}}(\mathbf{q}, n_{\mathbf{q}}) = \sum_{q \in \mathbf{q}_{\mathcal{O}}} n_{\mathbf{q}}(q) \tag{10.3}$$

By $Q_{\mathcal{O}}$ one denotes the set of configurations concentrated in $\mathcal{O}$, i.e., such that $\nu_{\mathcal{O}^c}(\mathbf{q}, n_{\mathbf{q}}) = 0$ (here and below $\mathcal{O}^c$ denotes the complement of $\mathcal{O}$). In the same way one introduces the set $Q^0_{\mathcal{O}}$.

The space Q is equipped with natural topology: the convergence of the sequence $(\mathbf{q}_s, n^{(s)}_{\mathbf{q}_s})$, $s = 1, 2, \ldots$, to $(\mathbf{q}, n_{\mathbf{q}})$, means that for any bounded open $\mathcal{O} \subset R^d$ with $\nu_{\partial\mathcal{O}}(\mathbf{q}, n_{\mathbf{q}}) = 0$, where $\partial\mathcal{O}$ is the boundary of the set $\mathcal{O}$, and for any sufficiently large s,

$$\nu_{\mathcal{O}}(\mathbf{q}_s, n^{(s)}_{\mathbf{q}_s}) = \nu_{\mathcal{O}}(\mathbf{q}, n_{\mathbf{q}}).$$

We define the σ-algebra $\mathcal{Q}$ of subsets of Q as the smallest one generated by the functions $\nu_{\mathcal{O}}$ where $\mathcal{O}$ is an arbitrary bounded Borel subset of R^d. It is not hard to show that $\mathcal{Q}$ is the Borel σ-algebra with respect to the topology introduced above. The configuration space of a particle system in R^d is defined as the measurable space $(Q, \mathcal{Q})$.

The configurations $(\mathbf{q}, n_{\mathbf{q}})$ with $n_{\mathbf{q}} = 1$ form an everywhere dense Borel subset of Q which will be identified with Q^0. The subsets of Q^0 which belong to $\mathcal{Q}$ form the σ-algebra which is denoted by $\mathcal{Q}^0$. Almost all probability distributions which appear in the sequel will be concentrated on Q^0, and we shall often determine probability measures on $(Q^0, \mathcal{Q}^0)$ to begin with, without specifying their possible interpretation as measures on $(Q, \mathcal{Q})$.

The action of the group of space translation $\{T_y, y \in R^d\}$ is defined on $(Q, \mathcal{Q})$:

$$T_y(\mathbf{q}, n_{\mathbf{q}}) = (\mathbf{q} + y, n_{\mathbf{q}+y})$$

where

$$\mathbf{q} + y = \{q \in R^d: q - y \in \mathbf{q}\}, n_{\mathbf{q}+y}(q) = n_{\mathbf{q}}(q - y), q \in \mathbf{q} + y.$$

A probability measure P on $(Q, \mathcal{Q})$ is called *translation invariant* if $P(T_y A) = P(A)$ for any $A \in \mathcal{Q}$ and $y \in R^d$.

When we speak about the convergence of probability measures on $(Q, \mathcal{Q})$ (and $(Q^0, \mathcal{Q}^0)$) we mean the weak convergence with respect to the above topology.

A probability measure on $(Q, \mathcal{Q})$ describes a "configuration state" of a particle system in R^d (cf [R2]). According to probability theory language, it is interpreted as the probability distribution of a random point field, and many results, useful for our aims may be found in monographs and papers related to the theory of random point processes and fields (see, for instance, [K1], [MKM]).

For a given Borel $\mathcal{O} \subset R^d$ it is possible to introduce the σ-subalgebra $\mathcal{Q}_{\mathcal{O}} \subset \mathcal{Q}$ generated by the functions $v_{\tilde{\mathcal{O}}}$ where $\tilde{\mathcal{O}}$ is an arbitrary bounded Borel subset of $\mathcal{O}$. In the same way one defines the σ-algebra $\mathcal{Q}^0_{\mathcal{O}}$. For any measure P on $(Q, \mathcal{Q})$ (or on $(Q^0, \mathcal{Q}^0)$) we denote by $P_{\mathcal{O}}$ the restriction of P to $\mathcal{Q}_{\mathcal{O}}$ (or to $\mathcal{Q}^0_{\mathcal{O}}$). It is possible to identify the measure $P_{\mathcal{O}}$ with a measure concentrated on $Q_{\mathcal{O}}$ (or $Q^0_{\mathcal{O}}$).

2.2. Poisson Measures. The simplest natural example of a probability measure on $(Q^0, \mathcal{Q}^0)$ is the *Poisson measure* P^0_z with the parameter $z > 0$. It is completely determined by the following two conditions:

1) for any bounded Borel set $\mathcal{O} \subset R^d$, the random variable $v_{\mathcal{O}}$ has the Poissonian distribution with the average value $zl(\mathcal{O})$ where $l(\mathcal{O})$ is the Lebesgue measure of $\mathcal{O}$, i.e.,

$$P^0_z(\{\mathbf{q} \in Q^0 \colon v_{\mathcal{O}}(\mathbf{q}) = k\}) = \frac{(zl(\mathcal{O}))^k}{k!} \exp(-zl(\mathcal{O})) \tag{10.4}$$

2) for any collection of pairwise disjoint Borel sets $\mathcal{O}_1, \ldots, \mathcal{O}_n \subset R^d$, the random variables $v_{\mathcal{O}_1}, \ldots, v_{\mathcal{O}_n}$ are mutually independent.

According to physical terminology, the Poisson measure P^0_z determines the equilibrium configuration state of the ideal gas. The parameter z determines the density of particles in this state.

We shall consider the "unnormalized" Poisson measure as well (or, as one says sometimes, the Lebesgue-Poisson measure). More precisely, given any bounded Borel $\mathcal{O} \subset R^d$, we introduce the measure $L_{\mathcal{O}}$ on the σ-algebra $\mathcal{Q}^0_{\mathcal{O}}$ which is determined by the condition: for any finite collection of pairwise disjoint Borel sets $\mathcal{O}_1, \ldots, \mathcal{O}_n \subseteq \mathcal{O}$ with $\bigcup_{j=1}^n \mathcal{O}_j = \mathcal{O}$ and for all nonnegative integers $k_1, \ldots, k_n$,

$$L_{\mathcal{O}}(\{\mathbf{q} \in Q^0 \colon v_{\mathcal{O}_j}(\mathbf{q}) = k_j, j = 1, \ldots, n\}) = \prod_{j=1}^n \frac{(l(\mathcal{O}_j))^{k_j}}{k_j!}. \tag{10.5}$$

It is possible to identify the measure $L_{\mathcal{O}}$ in the natural way with a measure concentrated on the set $Q^0_{\mathcal{O}}$. We shall often use such a mode in the sequel without specifying it every time again. The Poisson measure P^0_z is characterized by the fact that the restriction $(P^0_z)_{\mathcal{O}}$ is absolutely continuous with respect to the measure $L_{\mathcal{O}}$ and

$$\frac{d(P^0_z)_{\mathcal{O}}}{dL_{\mathcal{O}}}(\mathbf{q}) = z^{|\mathbf{q}|} \exp(-zl(\mathcal{O})), \mathbf{q} \in Q^0_{\mathcal{O}}. \tag{10.6}$$

2.3. The Gibbs Configuration Probability Distribution. In this section we shall introduce the definition of a configurational Gibbs distribution. This definition is a natural generalization of the well-known definition of the grand canonical ensemble in statistical mechanics to the case of an infinite particle system. The fundamental postulate of statistical mechanics is the possibility of using this ensemble for describing equilibrium states of a particle system (see Sect. 4 for a more detailed account).

First of all we introduce a Gibbs distribution in the simplest case of a pairwise interaction which is invariant with respect to the Euclidean group of motion, and then indicate some possible generalizations. Interaction between particles is described by the pair *interaction potential* which, in the case under consideration, is a fixed measurable function $U\colon [0, \infty) \to R^1 \cup \{\infty\}$. The value $U(r), 0 \leqslant r < \infty$, is interpreted as the potential energy of a pair of point particles at the distance r. We shall suppose as well that the numbers $z > 0$ and $\beta > 0$ are fixed. In statistical mechanics the parameter z is called the *activity* (*fugacity*) of the system and the parameter β is inversely proportional to the absolute *temperature*. The so-called chemical potential $\mu = \beta^{-1} \ln z$ is sometimes used instead of the activity.

For any bounded Borel $\mathcal{O} \subset R^d$ one defines the potential energy of a configuration $\mathbf{q} \in Q^0_{\mathcal{O}}$ by the equality[1]

$$V(\mathbf{q}) = \frac{1}{2} \sum_{\substack{q, q' \in \mathbf{q}:\\ q \neq q'}} U(\|q - q'\|). \tag{10.7}$$

The Gibbs configuration distribution in a volume $\mathcal{O}$ with the free boundary condition, interaction potential U and parameters (z, β), is the probability measure on $Q^0_{\mathcal{O}}$ which is determined by the density with respect to the measure $L_{\mathcal{O}}$ (regarded as a measure on $Q^0_{\mathcal{O}}$) given by

$$\Xi_{\mathcal{O}}^{-1} z^{|\mathbf{q}|} \exp(-\beta V(\mathbf{q})), \qquad \mathbf{q} \in Q^0_{\mathcal{O}}, \tag{10.8}$$

where $\Xi_{\mathcal{O}}$ is the normalization factor which is called the *partition function*

$$\Xi_{\mathcal{O}} = \int_{Q^0_{\mathcal{O}}} L_{\mathcal{O}}(d\mathbf{q}) z^{|\mathbf{q}|} \exp(-\beta V(\mathbf{q})). \tag{10.9}$$

For the given definition to be correct, one needs to assume, of course, that $\Xi_{\mathcal{O}} < \infty$.

A wider class of probability distributions on $Q^0_{\mathcal{O}}$ is obtained by introducing a *boundary condition* $\bar{\mathbf{q}} \in Q^0_{\mathcal{O}^c}$. We set

$$V(\mathbf{q} | \bar{\mathbf{q}}) = V(\mathbf{q}) + \sum_{q \in \mathbf{q}, q' \in \bar{\mathbf{q}}} U(\|q - q'\|), \mathbf{q} \in Q^0_{\mathcal{O}}. \tag{10.10}$$

The probability measure on $Q^0_{\mathcal{O}}$ determined by the density with respect to $L_{\mathcal{O}}$

$$\Xi_{\mathcal{O}}(\bar{\mathbf{q}})^{-1} z^{|\mathbf{q}|} \exp(-\beta V(\mathbf{q} | \bar{\mathbf{q}})), \mathbf{q} \in Q^0_{\mathcal{O}}, \tag{10.11}$$

where

$$\Xi_{\mathcal{O}}(\bar{\mathbf{q}}) = \int_{Q^0_{\mathcal{O}}} L_{\mathcal{O}}(d\mathbf{q}) z^{|\mathbf{q}|} \exp(-\beta V(\mathbf{q} | \bar{\mathbf{q}})), \tag{10.12}$$

[1] Here and below one assumes that for any $a \in R^1$ $a + \infty = \infty$, for any $a > 0$ $a\infty = \infty$, and $\exp(-\infty) = 0$.

will be called the Gibbs configuration distribution in the volume $\mathcal{O}$ with the boundary condition $\mathbf{q}$, interaction potential U and parameters (z, β).

For the existence of this distribution, it is sufficient to suppose that the series in the right-hand side of (10.10) is absolutely convergent for almost every $\mathbf{q} \in Q^0_{\mathcal{O}}$ and the integral $\Xi_{\mathcal{O}}(\bar{\mathbf{q}})$ is finite.

Definition 2.1. A probability measure P on $(\mathbf{Q}^0, \mathcal{Q}^0)$ is called a Gibbs configuration distribution in "infinite volume", or shortly: a Gibbs configuration distribution, with the interaction potential U and parameters (z, β) if for any bounded Borel $\mathcal{O} \subset R^d$ the following conditions are valid:

1) for P-almost all $\bar{\mathbf{q}} \in Q^0$ there exists the Gibbs configuration distribution in the volume $\mathcal{O}$ with the condition $\mathbf{q}_{\mathcal{O}^c}$, potential U and parameters (z, β),

2) for P-almost all $\bar{\mathbf{q}} \in Q^0$ the restriction onto $\mathcal{Q}_{\mathcal{O}}$ of the conditional probability distribution $P(\cdot | \mathcal{Q}_{\mathcal{O}^c})$ with respect to the σ-algebra $\mathcal{Q}_{\mathcal{O}^c}$ coincides with the Gibbs configuration distribution in the volume $\mathcal{O}$ with the boundary condition $\bar{\mathbf{q}}_{\mathcal{O}^c}$, potential U and parameters (z, β).

Let us briefly explain the meaning of this definition. As already mentioned, this is a generalization of the definition (10.8) to the case of the infinite particle system. It is impossible to extend the definition (10.8) to this case directly since the total energy of the infinite particle system is infinite. However, it is not difficult to calculate from (10.8) that, given $\tilde{\mathcal{O}} \subset \mathcal{O}$ and a configuration $\mathbf{q}_{\mathcal{O} \setminus \tilde{\mathcal{O}}}$ in the complement $\mathcal{O} \setminus \tilde{\mathcal{O}}$, the conditional probability density of the distribution of a configuration $\mathbf{q}_{\tilde{\mathcal{O}}} \in Q^0_{\tilde{\mathcal{O}}}$ in the volume $\tilde{\mathcal{O}}$ has the form (10.12). So, property 2) of the configurational Gibbs distribution is an analog of such a property of the usual Gibbs density (10.8). As to property 1), this is of technical character and is necessary for the correct formulation of 2).

One of the principal arguments for naturality of the given definition is that the measure P may be obtained as a limit of Gibbs configuration distributions in the volume $\mathcal{T}$ with (in general, random) boundary conditions $\mathbf{q}_{\mathcal{T}^c}$, as $\mathcal{T} \nearrow R^d$ [2]. Therefore, the term "a limit Gibbs configuration distribution" is sometimes used. The idea of such a construction arises in the work [BK] which has become classical (see also [BPK], [M1], [M2], [R1]).

Conditions 1) and 2) figuring in Definition 2.1 are often called, in literature, the DLR (Dobrushin, Lanford, Ruelle) *conditions* (cf [D2]–[D5], [LR], [R3]).

2.4. Potential of the Pair Interaction. Existence and Uniqueness of a Gibbs Configuration Probability Distribution. We shall now discuss the conditions on the interaction potential imposed in studying the configurational Gibbs distributions as well as for investigation of dynamical systems of statistical mechanics, which will be given in following sections. The simplest examples of the interaction potentials are the potential of ideal gas

[2] Here and below $\mathcal{T} \nearrow R^d$ denotes the inclusion-directed set of the d-dimensional cubes $\mathcal{T} = [-a, a]^d$, $d > 0$.

$$U(r) \equiv 0, \qquad r \geqslant 0, \tag{10.13}$$

and that of the gas of hard (or absolutely elastic) spheres (hard rods for $d = 1$)

$$U(r) = \begin{cases} \infty, & 0 \leqslant r < r_0, \\ 0, & r \geqslant r_0, \end{cases} \tag{10.14}$$

where r_0 is the diameter of a sphere (the length of a rod). For these cases the configurational Gibbs distribution depends on one parameter z only; for the case of ideal gas it coincides with the Poisson measure P_z^0.

A typical example of a particle interaction potential which is used in physical computations (for $d = 3$) is the Lenard-Jones potential

$$U(r) = \frac{a_1}{r^{12}} - \frac{a_2}{r^6} \tag{10.15}$$

where $a_1, a_2 > 0$. The interaction between atoms of noble gases is described well enough by means of this potential.

The interactions arising in various physical situations are of very different character and hence the results which hold under minimal assumptions on the interaction potential are of principal interest. However, different problems require different classes of interaction potentials and the corresponding formulations in a precise form are often cumbersome. Moreover, the assumptions which one introduces thereby, do not seem to be the final ones. Therefore, we shall not formulate the precise conditions under which a given result may be proved and shall restrict ourselves to giving qualitative descriptions and referring to the original papers [BPK], [BK], [D4]–[D6], [M1], [M2], [Su2], [GMS], [R1] [R3]. Let us list the main types of restrictions which are usually imposed in the literature:

I) a condition of boundedness from below for the interaction potential:

$$\inf_{r \geqslant 0} U(r) > -\infty.$$

This condition is physically natural and we shall suppose in the sequel that it is fulfilled.

II) conditions of increase as $r \to 0$. A typical example is

$$U(r) \geqslant cr^{-\gamma} \text{ for sufficiently small } r > 0 \tag{10.16}$$

where $c > 0$. Usually one supposes that $\gamma > d$. One considers often the potential with a hard core for which

$$U(r) \begin{cases} = \infty, & 0 \leqslant r < r_0, \\ < \infty, & r > r_0, \end{cases} \tag{10.17}$$

where $r_0 > 0$ is a diameter of the core (cf (10.14)). The advantage of the last condition is that it simplifies essentially several constructions. Sometimes totally bounded interaction potentials are considered, but in these cases one usually assumes that the values $U(r)$ for r near 0 are positive and the positive part of interaction dominates over the negative one.

The physical meaning of such conditions is that they correspond to a sufficiently strong repulsion of particles at short distances which is necessary for preventing the collapse of particles which should be excluded in an equilibrium situation.

III) conditions of decrease as $r \to \infty$. A typical example of such a condition is

$$|U(r)| \leqslant cr^{-\gamma} \text{ for sufficiently large } r \tag{10.18}$$

where $c > 0$. As above it is supposed that $\gamma > d$. For simplicity, one sometimes introduces the condition of *finite range* for the potential U:

$$U(r) = 0 \qquad \text{for } r > r_1. \tag{10.19}$$

The physical meaning of the conditions of III) is connected with an assumption about the locality of the interaction, i.e. its weakening at distances much longer than the average distance between particles.

IV) conditions of smoothness. It is supposed that the interaction potential has a certain number of continuous derivatives in the domain where it has finite values. One often introduces in addition some restrictions on the behaviour of these derivatives as $r \to 0$ ($r \to r_0$ in the case of potential with hard core) and $r \to \infty$.

These last conditions usually arise in the proof of existence of a time dynamics (cf Sections 3, 4).

We shall now discuss the question about the existence and uniqueness of a Gibbs configuration distribution. For simplicity assume that the potential U has a hard core of diameter $r_0 > 0$ and that the condition (10.19) with $\gamma > d$ is satisfied. For extending these results to other classes of interaction potentials see the papers cited above.

The theorem of existence of a configurational distribution is formulated as follows.

Theorem 2.1 (cf [D4]). *Given $z > 0$ and $\beta > 0$, there exists at least one translation invariant Gibbs configuration distribution with the interaction potential U and parameters (z, β). The set of Gibbs configuration distributions $\mathfrak{P}_{U,z,\beta}$ with the potential U and parameters (z, β) forms a convex compact set in the space of probability measures on $(Q, \mathcal{Q})$ (and hence coincides with the closure of the convex envelope of the set of its extreme points).*

The formulation of the uniqueness theorem is different for the multidimensional ($d \geqslant 2$) and one-dimensional ($d = 1$) case. Let $\mathcal{T}(y, b)$ denote the d-dimensional cube centered at a point $y \in R^d$ with the edges parallel to the coordinate axes and with the edge length $2b$.

Theorem 2.2 (cf [D4], [M1], [M2], [R1]). *Assume that $d \geqslant 1$. Then for any $\beta > 0$ one can find the value $z_0 = z_0(\beta) > 0$ such that for all $z \in (0, z_0)$ there exists* *only one Gibbs configuration distribution with the potential U and parameters* (z, β) *(i.e., the set* $\mathfrak{P}_{U,z,\beta}$ *consists of one point). This distribution P is translation invariant and has the following mixing property:*

$$\sup_{y \in R^d} \sup_{\substack{A_1 \in \mathcal{Q}_{\mathcal{T}(y,u)},\\ A_2 \in \mathcal{Q}_{\mathcal{T}(y,u+s)^c}}} |P(A_1 \cap A_2) - P(A_1)P(A_2)| \leqslant c_1 u^{d-1} s^{-\gamma'}, \; u, s > 0, \quad (10.20)$$

with constants $\gamma' \geqslant \gamma - d$ *and* $c_1 > 0$.

Theorem 2.3 (cf [D4], [GMS], [Su2]). *Assume that* $d = 1$ *and that the potential U has property* (10.18) *with* $\gamma > 2$. *Then for any* $z > 0$ *and* $\beta > 0$ *there exists only one Gibbs configuration distribution with the potential U and parameters* (z, β). *This distribution is translation invariant and has the Rosenblatt mixing property:*

$$\sup_{u \in R^1} \sup_{\substack{A_1 \in \mathcal{Q}_{(-\infty,u)},\\ A_2 \in \mathcal{Q}_{(u+s,\infty)}}} |P(A_1 \cap A_2) - P(A_1)P(A_2)| \leqslant c_1 s^{-(\gamma-2)}, \; s > 0, \quad (10.21)$$

where $c_1 > 0$ *is a constant.*

Let us discuss the condition $z \in (0, z_0)$ figuring in the formulation of Theorem 2.2. From the physical point of view, it means that we consider particle systems with low density. According to a wide-spread conjecture, for large values z, i.e. for high densities of particles, phase transitions may occur; this can be manifested, in particular, by a nonuniqueness of the Gibbs configuration distribution. On the other hand, in Theorem 2.3 there is no restriction on the value of the parameter z: this is explained by the fact that, according to physical pictures, there is no phase transitions in one-dimensional systems (under wide conditions on the interaction potential).

2.5. The Phase Space. The Gibbs Probability Distribution. To construct dynamical systems of statistical mechanics it is necessary to introduce a more detailed description of a particle system where one considers both particle positions q and momenta p (in the following we assume that the mass of every particle is equal to 1; this fact allows us to identify the momentum and the velocity of a particle). The situation where two particles have the same positions and momenta usually does not arise and we shall restrict ourselves to considering the phase space which is analogous to the configuration space Q^0.

Let M denote the collection of all finite or countable subsets $\mathbf{x} \subset R^d \times R^d$ satisfying the condition

$$\nu_{\mathcal{O} \times R^d}(\mathbf{x}) < \infty \text{ for any bounded } \mathcal{O} \subset R^d. \quad (10.22)$$

Here, as above,

$$\nu_{\mathcal{D}}(\mathbf{x}) = |\mathbf{x}_{\mathcal{D}}|, \qquad \mathcal{D} \subset R^d \times R^d, \quad (10.23)$$

and $\mathbf{x}_{\mathcal{D}} = \mathbf{x} \cap \mathcal{D}$. Let us denote by $M_{\mathcal{O}}$, $\mathcal{O} \subset R^d$, the collection of those $\mathbf{x} \in M$ for which $\nu_{\mathcal{O}^c \times R^d}(\mathbf{x}) = 0$.

The space M is endowed with natural topology: the convergence of the sequence $\mathbf{x}_s$, $s = 1, 2, \ldots$, to $\mathbf{x}$ means that for any bounded open $\mathcal{O} \subset R^d$, any open $C \subseteq R^d$ such that $\nu_{\partial(\mathcal{O} \times C)}(\mathbf{x}) = 0$ and for all sufficiently large s,

$$\nu_{\mathcal{O} \times C}(\mathbf{x}_s) = \nu_{\mathcal{O} \times C}(\mathbf{x}).$$

We shall define the σ-algebra $\mathcal{M}$ of subsets of M as the σ-algebra generated by the functions $\nu_{\mathcal{O} \times C}$ where $\mathcal{O}$ is an arbitrary bounded Borel subset and C is an arbitrary Borel subset of R^d. It is easy to show that $\mathcal{M}$ is the Borel σ-algebra with respect to the topology on $\mathcal{M}$ introduced above. The measurable space $(M, \mathcal{M})$ is called the phase space of a particle system in R^d.

On the space $(M, \mathcal{M})$, as well as on $(Q, \mathcal{Q})$, the action of the group of space translations $\{T_y, y \in R^d\}$ is given:

$$T_y \mathbf{x} = \{(q, p) \in R^d \times R^d : (q - y, p) \in \mathbf{x}\}.$$

A probability measure P on $(M, \mathcal{M})$ is called translation invariant if for any $A \in \mathcal{M}$ and $y \in R^d$ $P(T_y A) = P(A)$. Convergence of probability measures on $(M, \mathcal{M})$ will be weak convergence with respect to the mentioned topology.

According to probability theory language, the measure on the space $(M, \mathcal{M})$ is interpreted as the probability distribution of a random marked point field (cf [MKM]).

For a given Borel $\mathcal{O} \subset R^d$ it is possible to introduce as above the σ-subalgebra $\mathcal{M}_{\mathcal{O}} \subset \mathcal{M}$ generated by the functions $\nu_{\tilde{\mathcal{O}} \times C}$, where $\tilde{\mathcal{O}}$ is an arbitrary bounded Borel subset of $\mathcal{O}$ and C is an arbitrary Borel subset of R^d. For any measure P on $(M, \mathcal{M})$, denote by $P_{\mathcal{O}}$ the restriction of P to $\mathcal{M}_{\mathcal{O}}$. It is possible to identify the measure $P_{\mathcal{O}}$ in a natural way with a measure concentrated on $\mathcal{M}_{\mathcal{O}}$.

Consider a measurable map $\Pi: \mathbf{x} \in M \mapsto (\mathbf{q}, n_{\mathbf{q}}) \in Q$ obtained by omitting the particle momenta in $\mathbf{x}$:

$$\mathbf{q} = \{q \in R^d : \nu_{\{q\} \times R^d}(\mathbf{x}) \geqslant 1\}, \qquad n_{\mathbf{q}}(q) = |\{p \in R^d : (q, p) \in \mathbf{x}\}|. \tag{10.24}$$

This map allows us to associate with any probability distribution P on $(M, \mathcal{M})$ its "projection" ΠP which is a probability measure on $(Q, \mathcal{Q})$.

This mapping provides a natural method to construct a probability measure P on $(M, \mathcal{M})$ having a given projection ΠP. Namely, for constructing the measure P it is sufficient to determine, in addition to ΠP, the conditional probability distribution $P(\cdot | \mathcal{M}^Q)$ with respect to the σ-algebra $\mathcal{M}^Q \subset \mathcal{M}$ induced by the map Π. From the physical point of view, given a particle configuration, we consider the conditional momenta distribution. Assuming for simplicity that the measure ΠP is concentrated on Q^0, we may introduce this conditional probability distribution by means of the family of measures $P_{\mathbf{q}}$ on $(M, \mathcal{M})$ depending on $\mathbf{q} \in Q^0$. In addition, since the set of points $\mathbf{x} \in M$ with given $\mathbf{q} = \Pi\mathbf{x}$ is identified, in a natural way, with the space $R^{\mathbf{q}}$, we may interpret $P_{\mathbf{q}}$ as a probability measure on $R^{\mathbf{q}}$.

Fix a pair interaction potential U, numbers $z > 0$, $\beta > 0$ and a vector $p_0 \in R^d$.

Definition 2.2. A probability measure P on $(M, \mathcal{M})$ is called a Gibbs distribution with the interaction potential U and parameters (z, β, p_0) if

1) its projection ΠP is a configurational Gibbs distribution with the potential U and parameters z, β,

2) the conditional probability measures $P_{\mathbf{q}}$ correspond to the joint distribution of independent d-dimensional Gaussian vectors with the expectation value p_0 and covariance matrix $\beta^{-1}I$.

The conditional probability measure for particle momenta figuring in condition 2) is called the Maxwell distribution. In an analogous way, one introduces the notion of the Gibbs distribution in a bounded volume $\mathcal{O} \subset R^d$ with an extra condition $\mathbf{x} \in M_{\mathcal{O}}$.

Theorems of existence and uniqueness of a Gibbs distribution may be easily obtained from the corresponding theorems for the configurational Gibbs distribution contained in Sect. 2.3.

Sometimes it is useful to consider a more general class of measures which arises when the Gaussian probability distribution in the condition 2) is replaced by an arbitrary probability distribution σ on R^d. When $U \equiv 0$, i.e., the configurational Gibbs distribution is the Poisson measure on $(Q^0, \mathcal{Q}^0)$, the corresponding probability measure on $(M, \mathcal{M})$ is denoted by $P^0_{z,\sigma}$. We shall call the measure $P^0_{z,\sigma}$ the Poisson measure on $(M, \mathcal{M})$ with the parameters (z, σ). In the one-dimensional case we shall use this construction in Sect. 5 for the ideal hard rod potential (cf (10.14)). The corresponding measure on $(M, \mathcal{M})$ is denoted by $P^{r_0}_{z,\sigma}$.

2.6. Gibbs Measures with a General Potential. In this section we shall discuss the notion of a Gibbs measure corresponding to a general "potential" which depends on positions and momenta of arbitrary finite collections of particles. Such a generalization is meaningful, in particular, because the probability measures arising thereby may be distinguished in the class of all measures on $(M, \mathcal{M})$ by means of certain conditions which are connected with the decay of "correlations" in some natural sense (cf [Ko]). Hence, the limits of the mathematical applications of Gibbs distributions are much wider than those dictated by "physical traditions" and some of the results formulated below which hold true for a large class of such "generalized" Gibbs distributions may be considered as rather universal ones.

Now we shall call a potential a sequence $\Phi = (\Phi^{(1)}, \Phi^{(2)}, \ldots)$, where $\Phi^{(n)}$ is a symmetric function of variables $x_1, \ldots, x_n$ on the set

$$\{(x_1, \ldots, x_n) \in (R^d \times R^d)^n \colon x_{j_1} \neq x_{j_2}, 1 \leqslant j_1 < j_2 \leqslant n\} \tag{10.25}$$

with values in $R^1 \cup \{\infty\}$ where $x_j = (q_j, p_j) \in R^d \times R^d$, $j = 1, \ldots, n$. The function $\Phi^{(n)}$ is called the n-particle potential (we reserve the previous term "pair potential" for $n = 2$). Sometimes it will be convenient to consider, as the argument of the function $\Phi^{(n)}$, the point $\mathbf{x} \in M$ with $|\mathbf{x}| = n$.

For any bounded Borel $\mathcal{O} \subset R^d$ and any $\mathbf{x} \in M_{\mathcal{O}}$ and $\bar{\mathbf{x}} \in M_{\mathcal{O}^c}$ denote

$$h^{(\Phi)}(\mathbf{x}|\bar{\mathbf{x}}) = \sum_{n \geq 1} \sum_{\substack{\mathbf{x}' \subseteq \mathbf{x} \cup \bar{\mathbf{x}}: |\mathbf{x}'| = n, \\ \mathbf{x}' \cap \mathbf{x} \neq \varnothing}} \Phi^{(n)}(\mathbf{x}'). \tag{10.26}$$

The definition of a Gibbs probability distribution on $(M, \mathcal{M})$ corresponding to the potential Φ is analogous to the definition of the configurational Gibbs distribution given in Sect. 2.3. The main difference is that one considers σ-algebras $\mathcal{M}_{\mathcal{O}}$ and $\mathcal{M}_{\mathcal{O}^c}$ instead of $\mathcal{Q}_{\mathcal{O}}$ and $\mathcal{Q}_{\mathcal{O}^c}$, respectively, and the underlying measure $L_{\mathcal{O}}$ on $\mathcal{Q}_{\mathcal{O}}$ is replaced by the measure $\bar{L}_{\mathcal{O}}$ on the σ-algebra $\mathcal{M}_{\mathcal{O}}$. The measure $\bar{L}_{\mathcal{O}}$ is determined by the following conditions: 1) projection $\Pi \bar{L}_{\mathcal{O}}$ coincides with $L_{\mathcal{O}}$ 2) the conditional measure with respect to the σ-algebra $\mathcal{M}^Q$ generated by $\bar{L}_{\mathcal{O}}$ is the product of $|\mathbf{q}|$ copies of the Lebesgue measure l. The density (with respect to the measure $\bar{L}_{\mathcal{O}}$) of the restriction to $\mathcal{M}_{\mathcal{O}}$ of the conditional probability distribution $P(\cdot|\mathcal{M}_{\mathcal{O}^c})$ is

$$\Xi_{\mathcal{O}}(\bar{\mathbf{x}})^{-1} \exp(-h^{\Phi}(\mathbf{x}|\bar{\mathbf{x}})), \qquad \mathbf{x} \in M_{\mathcal{O}}, \tag{10.27}$$

where

$$\Xi_{\mathcal{O}}(\mathbf{x}) = \int_{M_{\mathcal{O}}} \bar{L}_{\mathcal{O}}(d\mathbf{x}) \exp(-h^{(\Phi)}(\mathbf{x}|\bar{\mathbf{x}})). \tag{10.28}$$

In order to prove the existence (and uniqueness) of a Gibbs distribution corresponding to a "general" potential $\Phi = (\Phi^{(1)}, \Phi^{(2)}, \ldots)$, one should impose on Φ some conditions. We shall not give here their explicit formulations since they are cumbersome; these conditions are discussed briefly in one of the possible cases in Sect. 4.2 below.

Let us comment on the connection between the general definition and Definition 2.2 given in Sect. 2.5. A Gibbs distribution with the potential U and parameters (z, β, p_0) in the context of the general definition, corresponds to the potential $\Phi = \Phi_{U;z,\beta,p_0}$ with $\Phi^{(n)} \equiv 0$ for $n \geq 3$ and

$$\Phi^{(1)}(x) = \gamma + \frac{\beta}{2} \|p - p_0\|^2, \qquad x = (q, p), \tag{10.29a}$$

$$\Phi^{(2)}(x_1, x_2) = \beta U(\|q_1 - q_2\|), \qquad x_j = (q_j, p_j),\ j = 1, 2, \tag{10.29b}$$

where $\gamma = -\ln z - d/2 \ln(2\pi\beta)$. The corresponding function h^{Φ} is of the form

$$h^{\Phi}(\mathbf{x}|\bar{\mathbf{x}}) = \gamma|\mathbf{x}| + \beta/2 \sum_{(q,p) \in \mathbf{x}} \|p - p_0\|^2 + \beta V(\Pi\mathbf{x}|\Pi\bar{\mathbf{x}}). \tag{10.30}$$

2.7. The Moment Measure and Moment Function. It is often convenient to determine the probability distributions on $(M, \mathcal{M})$ and $(Q, \mathcal{Q})$ by means of the

[3] In mathematical and physical literature one also frequently uses for these objects the terms "correlation measure", "correlation function" and "distribution function".

so-called *moment measure* or *moment function*[3] (for details, see [B1], [Le1], [Le2], [Z2]). This method is similar to the well-known method of describing random processes and fields by means of joint moments. For the sake of brevity we shall restrict ourselves by considering probability measures on $(M, \mathscr{M})$.

Let a probability distribution P on $(M, \mathscr{M})$ be given. For any $n = 1, 2, \ldots$, we define the n-moment measure $K_P^{(n)}$ on the σ-algebra of those Borel subsets of the set (10.25) which are invariant under the permutations of particles x_j, $j = 1, \ldots, n$, by the equality

$$K_P^{(n)}(A) = \int_M P(d\mathbf{x}) \sum_{\mathbf{x}' \subset \mathbf{x}: |\mathbf{x}'| = n} \chi_A(\mathbf{x}') \tag{10.31}$$

(as usual, χ_A denotes the indicator function of a set A). The density of the measure $K_P^{(n)}$ with respect to the Lebesgue measure $\prod_{j=1}^n dq_j \times dq_j$ (whenever it exists) is called the n-moment function and is denoted by $\mathscr{k}_P^{(n)}$. This is a symmetric function of the variables $x_j = (q_j, p_j) \in R^d \times R^d$, $j = 1, \ldots, n$. Intuitively, $\mathscr{k}_P^{(n)}((q_1, p_1), \ldots, (q_n, p_n)) \prod_{j=1}^n dq_j \times dp_j$ is interpreted as the probability that in the small volumes dq_j centered at the points q_j there are particles with momenta belonging to the small volumes dp_j centered at the points p_j, $j = 1, \ldots, n$. The sequence $K_P = (K_P^{(1)}, K_P^{(2)}, \ldots)$ is called the moment measure and the sequence $\mathscr{k}_P = (\mathscr{k}_P^{(1)}, \mathscr{k}_P^{(2)}, \ldots)$ the moment function of the probability distribution.

For the Poisson probability distribution $P^0_{z,\sigma}$ (cf Sect. 2.5) the n-moment measure is given by the formula

$$K^{(n)}_{P^0_{z,\sigma}}\left(\prod_{j=1}^n dq_j \times dp_j\right) = Z^n \prod_{j=1}^n dq_j \sigma(dp_j). \tag{10.32}$$

The necessary and sufficient condition for the one-to-one correspondence between a probability distribution P and the moment measure K_P is given by the divergence of the series

$$\sum_{n \geqslant 0} [K_P^{(n)}((\mathscr{O} \times R^d)^n) 1/n!]^{-1/n} \tag{10.33}$$

for any bounded Borel $\mathscr{O} \subset R^d$. This result is a consequence of well-known results connected with the so-called moment problem (cf [Le1], [Z2]). It is possible to show that condition (10.33) holds for a wide class of general Gibbs distributions.

A specific role is played in the following by the 1-moment measure $K_P^{(1)}$ and 1-moment function $\mathscr{k}_P^{(1)}$. Notice, in connection with this, that the measure $K_P^{(1)}$ gives the average values of random variables $\nu_{\mathscr{D}}$ with respect to the probability distribution P

$$K_P^{(1)}(\mathscr{D}) = \int_M P(d\mathbf{x}) \nu_{\mathscr{D}}(\mathbf{x}). \tag{10.34}$$

Thus, the function $\mathscr{k}_P^{(1)}$ may be considered as the density of the particle distribution in the one-particle phase space $R^d \times R^d$.

When speaking of the convergence of a sequence of measures $K_s^{(1)}$ on $R^d \times R^d$, $s = 1, 2, \ldots$, to a measure $K^{(1)}$ we shall have in mind the fact that for any bounded continuous function $f: R^d \times R^d \to R^1$ with support in $\mathcal{O} \times R^d$ where $\mathcal{O}$ is a bounded subset of R^d, the integral $\int K_s^{(1)}(dq \times dp) f(q, p)$ converges, as $s \to \infty$, to $\int K^{(1)}(dq \times dp) f(q, p)$.

§ 3. Dynamics of a System of Interacting Particles

3.1. Statement of the Problem. We shall suppose that the particle interaction is described by a pair potential U depending on the distance between the particles only (cf Sects. 2.3, 2.4). The classical equations of motion for a finite system of identical particles of mass one, may be written in the form

$$\dot{q}_i(t) = p_i(t), \dot{p}_i(t) = -\sum_{j: j \neq i} \operatorname{grad} U(\|q_j(t) - q_i(t)\|). \tag{10.35}$$

Here i and j run over a set of indices which label the particles, $q_i(t) \in R^d$ is the position vector, and $p_i(t) \in R^d$ the momentum vector of the i-th particle at time $t \in R^1$. The system of equations (10.35) may be written in the Hamiltonian form with the Hamiltonian

$$H = \sum_i \frac{1}{2} \|p_i\|^2 + \sum_{j \neq i} U(\|q_j - q_i\|). \tag{10.36}$$

We impose on U the conditions which were discussed in Sect. 2.4. Provided $U(\|q\|)$ is a smooth function of the variable $q \in R^d$, a solution of the Cauchy problem for system (10.35) exists and is unique on the whole time axis for all initial conditions $\{(q_i(0), p_i(0))\}$. If the potential U has a hard core of diameter r_0 and $\lim_{r \to r_0} U(r) = \infty$, then this assertion holds for the initial conditions with $\min_{i_1 \neq i_2} \|q_{i_1}(0) - q_{i_2}(0)\| > r_0$. If, otherwise, $\lim_{r \to r_0+} U(r) < \infty$, then one has to complete the system (10.35) with boundary conditions corresponding to collisions of particles when $\|q_{i_1}(t) - q_{i_2}(t)\| = r_0$ for some pair of indices $i_1 \neq i_2$. These boundary conditions usually correspond to the elastic collisions (compare with analogous conditions for the systems of the billiard type in Chap. 8, Sect. 1). The multiple collisions and other degenerations are neglected since they occur for subsets of initial data of Lebesgue measure zero.

Sometimes it is convenient to consider the motion of particles in a bounded domain $\mathcal{O} \subset R^d$ with a smooth or piecewise smooth boundary $\partial\mathcal{O}$. The motion equations are completed by boundary conditions for $q_i(t) \in \partial\mathcal{O}$. Usually one introduces the conditions of the elastic reflection of particles from the boundary $\partial\mathcal{O}$ (again compare with the billiard systems from Chap. 8, Sect. 1).

Formally, it is possible to write down equations (10.35) for an infinite particle system as well. However, the Hamiltonian (10.36) for this case will be infinite and therefore, some specific constructions are necessary. A family of smooth functions $\{(q_i(t), p_i(t)), t \in I\}$ for which the equations (10.35) are fulfilled for any i and $t \in I$

(and, in particular, the series in the right hand side of (10.35) converges absolutely) is called a solution of the infinite system (10.35) in the time interval $I \subseteq R^1$.

Provided the solution has the following property: for any bounded $\mathcal{O} \subset R^d$ and any $t \in I$ the number of indices i for which $q_i(t) \in \mathcal{O}$ is finite, it may be interpreted as a trajectory $\mathbf{x}(t)$ in the space M. These definitions are extended in a natural way to the case of motion with the elastic collisions of particles.

In the simplest case when $U \equiv 0$ (ideal gas) the trajectory $\mathbf{x}(t)$ with initial date $\mathbf{x}(0) = \mathbf{x}$ may be written in the explicit form

$$\mathbf{x}(t) = \{(q, p): (q - tp, p) \in \mathbf{x}\}. \tag{10.37}$$

However, already in this case some difficulties arise which become much more serious for a system with interaction. It is easy to indicate initial $\mathbf{x} \in M$ for which infinitely many particles collapse into a bounded domain $\mathcal{O} \subset R^d$ at a finite time t (because the velocities of distant particles may be large enough and "directed into the domain $\mathcal{O}$") and $\mathbf{x}(t)$ comes out of the space M. Therefore, the trajectory $\mathbf{x}(t)$ cannot be defined for arbitrary $\mathbf{x} \in M$.

For interacting particle systems one can construct, with the help of analogous arguments, examples of initial data $\mathbf{x} \in M$ for which solution (10.35) does not exist or, on the contrary, the trajectory belonging to M exists but it not unique. All these facts lead to the problem of finding a sufficiently "massive" measurable subset $\hat{M} \subset M$ such that for any point $\mathbf{x} \in \hat{M}$ there exists a unique trajectory $\mathbf{x}(t) \in \hat{M}$ on the whole axis R^1. Having such a subset, we define the one-parameter group of measurable transformations S_t: $\hat{M} \to \hat{M}$ by the formula $S_t\mathbf{x} = \mathbf{x}(t)$.

Definition 3.1. *The pair* $(\hat{M}, \{S_t, t \in R^1\})$ *is called the dynamics determined by the interaction potential* U *on the set* $\hat{M}$.

The construction of dynamics $(\hat{M}, \{S_t\})$ is the principal theme of this section. This construction will be meaningful, provided the set $\hat{M}$ is sufficiently massive, in the sense that the class of probability measures concentrated on $\hat{M}$ is large enough. If a measure P is of this class, we can introduce the family of measure

$$P_t(A) = P(S_{-t}(A \cap \hat{M})), \; A \in \mathcal{M}, \qquad t \in R^1. \tag{10.38}$$

Definition 3.2. The family of probability measures $\{P_t, t \in R^1\}$ given by equality (10.38) is called *time evolution of the initial measure* $P_0 = P$ generated by the dynamics $(\hat{M}, \{S_t\})$.

It is natural to claim that the class of measures for which the time evolution may be defined includes Gibbs distributions under certain restrictions on the potential (cf Sects. 2.3–2.6). In particular, if a Gibbs distribution with the interaction potential U and parameters (z, β, p_0) is concentrated on $\hat{M}$ and is not changed under the space translations in the direction of the vector p_0, then it is invariant with respect to transformations S_t (cf Sect. 4.1).

Observe that, unlike the finite-dimensional situation where there exists a distinguished class of measures (the measures which are absolutely continuous with respect to the Lebesgue measure), in the "infinite particle" situation the natural measure classes (for example, Gibbs measures with different pair potentials depending on the distance between particles) are mutually singular. Therefore, constructing trajectories for almost all initial data with respect to an "individual" probability measure does not give the complete solution of the problem under consideration.

3.2. Construction of the Dynamics and Time Evolution. A natural way of constructing the trajectory $\mathbf{x}(t)$ for $\mathbf{x} \in M$ is to pass to the limit $\lim_{\mathcal{T} \nearrow R^d} \mathbf{x}^{\mathcal{T}}(t)$ (in the topology of the space M) where $\mathbf{x}^{\mathcal{T}}(t)$ is a trajectory determining the motion of the finite particle system with the initial date $\mathbf{x}_{\mathcal{T} \times R^d}$ (i.e., the motion of the particles which are inside the cube $\mathcal{T}$ at the initial moment of time). There exist several natural possible ways to proceed here. First, it is possible "to ignore" the particles from $\mathbf{x}_{\mathcal{T}^c \times R^d}$ defining $\mathbf{x}^{\mathcal{T}}(t)$ by means of equations (10.35) with the initial data $\mathbf{x}_{\mathcal{T} \times R^d}$. On the other hand, it is possible to introduce the boundary condition of elastic reflection from the boundary $\partial\mathcal{T}$ and to consider the motion of the particles $(q, p) \in \mathbf{x}_{\mathcal{T} \times R^d}$ in the constant potential field generated by the "frozen" particles from $\mathbf{x}_{\mathcal{T}^c \times R^d}$. Although there are no general results about independence of the limit, as $\mathcal{T} \nearrow R^d$, on the choice of approximating motion, such an independence may be proved for all concrete situations which are discussed below.

Observe that for the motion of a finite system of perticles in R^d determined by equations (10.35), the well-known conservation laws are valid:

$$|S_t\mathbf{x}| = |\mathbf{x}| \tag{10.39}$$

(the law of conservation of the number of particles),

$$H(S_t\mathbf{x}) = H(\mathbf{x}) \tag{10.40}$$

(the law of conservation of the energy) and, finally,

$$\sum_{(q,p) \in S_t\mathbf{x}} p = \sum_{(q,p) \in \mathbf{x}} p \tag{10.41}$$

(the law of conservation of the momentum). For the particle motion in a domain $\mathcal{O}$ with the elastic reflections from the boundary $\partial\mathcal{O}$, the law of conservation of the momentum is not valid, and for motion in a potential field, one has to take into consideration, in the formulation of the law of conservation of the energy, the particle interaction with the field.

The conservation laws play a crucial role for constructing the limiting trajectory $\mathbf{x}(t)$. In particular, the proof of compactness of the family $\{\mathbf{x}^{\mathcal{T}}(t)\}$ for given $\mathbf{x}$ and t is based on these laws. The latter fact is the key point in the proof of existence of the limit. According to the definition of the topology in the space M, it is sufficient to control, as $\mathcal{T} \nearrow R^d$, the restriction $\mathbf{x}^{\mathcal{T}}(t)_{\mathcal{O} \times R^d}$ on the fixed bounded domain $\mathcal{O} \subset R^d$. Possible violations of compactness are explained by two (mutually connected) reasons: a) individual particles gain infinite speed in finite time:

$$\max[\|p\|: p \in \mathbf{x}^{\mathscr{T}}(t)_{\mathcal{O} \times R^d}] \to \infty$$

and b) accumulations of particles ("collapses") arise:

$$|\mathbf{x}^{\mathscr{T}}(t)_{\mathcal{O} \times R^d}| \to \infty.$$

The estimations guaranteeing compactness are deduced by means of uniform in $\mathscr{T}$ estimates for $|\mathbf{x}^{\mathscr{T}}(t)_{\mathcal{O} \times R^d}|$ and $H(\mathbf{x}^{\mathscr{T}}(t)_{\mathcal{O} \times R^d})$ which are obtained by using the conservation laws for the number of particles and the energy. The fact is, that an increase in the number of particles and the energy within the domain $\mathcal{O}$ can occur only because of an "influx" through the boundary of the domain, and one is sometimes able to estimate such as influx by means of the corresponding values for a larger domain. This idea was first realized in the papers of Lanford [L1], [L2], [L4], initiating the mathematical study of the infinite particle dynamics. In these papers the one-dimensional case ($d = 1$) was considered and $U(|q|)$ was supposed to be a smooth function of $q \in R^1$ with compact support. Under these conditions on the potential, the energy can be estimated by the number of particles. Hence, Lanford succeeded in producing necessary estimations by using the law of conservation of the number of particles only. He constructed the dynamics on the subset $\hat{M}^{(1)} \subset M$, characterized by the conditions

$$\sup_{(q,p) \in \mathbf{X}} \frac{|P|}{\ln_+ |q|} < \infty, \qquad \sup_{y \in R^1} \frac{|\mathbf{x}_{\mathscr{T}(y, \ln_+ |y|) \times R^1}|}{\ln_+ |y|} < \infty, \tag{10.42}$$

where $\ln_+ r = \max[1, \ln r]$, $r > 0$. It is not hard to show that this set is "massive" in the sense discussed in Section 3.1.

In the series of papers [DF], [F4], [FD], [MPP], the construction of dynamics was realized for a wide class of interaction potentials for the dimensions $d = 1$, 2. Here the law of conservation of the energy plays an essential role. The existence of the dynamics is proved for the subset $\hat{M}^{(2)} \subset M$, characterized by the conditions

$$\sup_{y \in R^d} \sup_{r < \ln + |y|} r^{-d}[H(\mathbf{x}_{\mathscr{T}(y,r) \times R^d}) + A|\mathbf{x}_{\mathscr{T}(y,r) \times R^d}|] < \infty \tag{10.43}$$

where A is a constant depending on the potential U. The set $\hat{M}^{(2)}$ is massive in the previous sense.

Notice also the paper [GS3] concerning the one-dimensional case where the dynamics is constructed for potentials U which admit elastic collisions of particles (this case is excluded by the conditions in the papers cited above).

The fact that the results of such a type have not yet been extended to the physically realistic dimension $d = 3$ seems to indicate an essential difficulty. Let us explain, on an intuitive level, why the estimates based on the law of conservation of the energy are not applicable for $d = 3$. Let a particle $(q, p) \in \mathbf{x}^{\mathscr{T}}(t)$ be fixed. Furthermore, let a number $r(t)$ be such that in the time interval from 0 to t this particle interacts directly with the particles which are at the moment 0 within the cube $\mathscr{T}(q, (\frac{1}{2})r(t))$ only. Assuming that the full energy of particles within $\mathscr{T}(q,(\frac{1}{2})r(t))$ at the time moment 0 is proportional to $r(t)^d$ (i.e., to the volume of this cube) and taking, as an estimate from above, the case where the full energy is transmitted to the fixed particle, we see that the module of its momentum is bounded from above by the value $r(t)^{d/2}$. Assuming that this estimate holds for other particles too, we conclude that the rate with which $r(t)$ increases does not exceed the order $r(t)^{d/2}$. Thus we come to the differential inequality

$$\dot{r}(t) \leqslant \text{const} \cdot r(t)^{d/2}.$$

The solutions of this inequality are bounded for $d \leqslant 2$ on any finite time interval, but may be unbounded for $d = 3$. This argument forms the basis of all proofs in the papers [DF], [F4], [FD], [MPP].

A somewhat different way of constructing the time evolution of a probability measure on the phase space was suggested recently by Siegmund-Schultze [SS]. He proved the following fact. Let the dimension d be arbitrary, U be a smooth nonnegative function with compact support, and P be an arbitrary translation invariant probability distribution on $(M, \mathscr{M})$ for which the mean energy of an individual particle is finite. Then, for initial conditions forming a subset of P-probability 1 in M, there exists a trajectory $\mathbf{x}(t) \in M$. It seems that the conditions on the potential listed above may be weakened. However, the methods of Siegmund-Schultze do not yet permit us to prove the uniqueness of a solution in some reasonable sense and moreover, to control either properties of the solution.

3.3. Hierarchy of the Bogolyubov Equations. As was remarked in Sect. 3.1, the construction of a dynamics permits us to define the time evolution $\{P_t, t \in R^1\}$ of an initial measure $P_0 = P$ (cf (10.38)). For certain purposes (for instance, for deducing kinetic equations; cf Sect. 6) it is convenient to use another, more direct method of describing the time evolution which is especially popular in physical literature. This method is based on studying a hierarchy of equations for moment function (see Sect. 2.7) $k_P^{(n)}$, $n = 1, 2, \ldots$, which are called the Bogolyubov equations [B1] (another frequently used term is the BBGKY hierarchy equations (Bogolyubov–Born–Green–Kirkwood–Yvon)). If one supposes that the interaction potential U "does not admit" collisions of particles, the *Bogolyubov hierarchy* takes the form

$$\begin{aligned}
&\frac{\partial}{\partial t} k^{(n)}(t;(q,p_1),\ldots,(q_n,p_n)) \\
&\quad = \{k^{(n)}(t;(q_1,p_1),\ldots,(q_n,p_n)), H((q_1,p_1),\ldots,(q_n,p_n))\} \\
&\qquad + \int dq_0 dp_0 \{k^{(n+1)}(t;(q_0,p_0),(q_1,p_1),\ldots,(q_n,p_n)), \sum_{j=1}^{n} U(\|q_0 - q_j\|)\} \qquad (10.44)
\end{aligned}$$

(for $n = 0$ only the integral part remains in the equation). Here $k^{(n)}(t;\cdot)$ is the n-moment function at the time moment t.

The connection between the two methods of description of time evolution has not yet been completely investigated. It is natural to expect that, for a large class of situations, the moment functions k_{P_t}, $t \in R^1$, give a solution of equations (10.44) (at least, in a weak sense). One can mention here the paper [GLL] and the note [Su3] where this fact is proved for the one-dimensional dynamics constructed, respectively, in [L1] and [GS3]. A series of papers is devoted to the proof of this fact for the case where the initial measure P is absolutely continuous or "almost" absolutely continuous with respect to a Gibbs distribution with the potential U and parameters (z, β, p_0) (cf [T1], [T2], [Z1] and also [Pu]).

Observe that another approach is simultaneously developed intensively and fruitfully. In this approach the problems of existence and uniqueness of a solution of the Bogolyubov hierarchy equations are treated using functional analysis. Under such an approach the Bogolyubov hierarchy equations are considered as an abstract evolution equation. In the first step, a solution is constructed in a Banach space of sequences of functions describing states of finite particle systems. Then one performs the thermodynamic limit passage to infinite systems. Here, as in Sect. 3.2, properties of the finite particle dynamics play an essential role. The functional analytic approach is developed in the papers of the Kiev specialists in mathematical physics [G], [GP1], [M], [Pet]. An explanatory review of this direction is contained in the recent paper of Petrina and Gerasimenko [GP2] to which we refer the reader for details. The consecutive exposition of all the circle of questions connected with the present state of the problem of solutions of the Bogolyubov hierarchy equations, is contained in the monograph of Petrina, Gerasimenko and Malyshev [GMP].

Notice also the paper [Sk] where the hierarchy of diffusion equations of the same type as the Bogolyubov diffusion equations is investigated from the functional-analytic positions.

§4. Equilibrium Dynamics

4.1. Definition and Construction of Equilibrium Dynamics. Let us consider the Hamiltonian system (10.35) in a bounded domain $\mathcal{O} \subset R^d$ with elastic reflection from the boundary $\partial\mathcal{O}$. The Liouville theorem and the laws of conservation of the number of particles (10.39) and energy (10.40) imply (under some weak additional assumptions) that the restriction of this dynamics to the hypersurface in the space $M_{\mathcal{O}}$, characterized by fixing the number of particles $|\mathbf{x}|$ and the energy $H(\mathbf{x})$, preserves the measure on this surface induced by the Lebesgue-Poisson measure $L_{\mathcal{O}}$. The corresponding normalized measure is called the microcanonical distribution. The Gibbs distribution in the volume $\mathcal{O}$ with the potential U and parameters z, β and $p_0 = 0$ is the result of averaging microcanonical distributions with some weight and hence is invariant, too. Analogously, the Gibbs distribution in the volume $\mathcal{O}$ with the boundary condition $\bar{\mathbf{x}}$, potential U and parameters

z, β and $p_0 = 0$ is invariant with respect to the motion in the potential field of the frozen particles from $\bar{\mathbf{x}}$.

Now let P be a probability measure on the phase space $(M, \mathscr{M})$. Let us consider a one-parameter automorphism group $\{S^t, t \in R^1\}$ of the measure space $(M, \mathscr{M}, P)$ which is generated by the infinite system of differential equations (10.35). Formally, this means that for any pair of smooth "local" functions f, g depending on positions and momenta of particles which are concentrated in a bounded domain of the space R^d, the following relation is valid

$$\frac{d}{dt}\int_M P(d\mathbf{x})f(S^t\mathbf{x})\bar{g}(\mathbf{x})\bigg|_{t=0} = \int_M P(d\mathbf{x})\mathscr{L}f(\mathbf{x})\bar{g}(\mathbf{x}) \tag{10.45}$$

where

$$\mathscr{L}f(\mathbf{x}) = \sum_{(q,p)\in\mathbf{x}}\Bigg((\mathrm{grad}_q f(\mathbf{x}), p) - \Bigg(\mathrm{grad}_p f(\mathbf{x}), \sum_{\substack{(q',p')\in\mathbf{x}:\\ q\neq q'}} \mathrm{grad}\, U(\|q'-q\|)\Bigg)\Bigg) \tag{10.46}$$

We say that the measure P is translation invariant in the direction of the vector p_0 if $P(T_{sp_0}A) = P(A)$ for any $A \in \mathscr{M}$ and $s \in R^1$. For Gibbs distributions, this property is known to be valid under the uniqueness conditions (cf Theorems 2.2 and 2.3).

Definition 4.1. *Let P be a Gibbs distribution with the interaction potential U and parameters (z, β, p_0) which is translation invariant in the direction of the vector p_0. Then $(M, \mathscr{M}, P, \{S_t\})$ is called the equilibrium dynamical system.*

The connection of Definition 4.1 with the constructions from Sect. 3 is the following. In the situation we discuss below where the equilibrium dynamics is constructed, for P-almost all $\mathbf{x}$, the trajectory $S^t\mathbf{x}$ gives a solution of the system of equations (10.35). On the other hand, if a dynamics is constructed, in the sense of Definition 3.1, on a set $\hat{M}$ which has full measure with respect to a Gibbs distribution P from the class we described just above, then P will be an invariant measure which determines the equilibrium dynamical system in the sense of Definition 4.1. The condition $P(\hat{M}) = 1$ may be easily verified in the situations discussed in Sect. 3.2. However, an equilibrium dynamical system may be constructed by other methods in a much more general situation (see papers [Z], [S3] treating the one-dimensional case and [S4], [PPT] devoted to the multi-dimensional case).

The main idea of these constructions which distinguishes them from the constructions of Sect. 3 is that one can get, for any fixed t, uniform estimates for trajectories $\mathbf{x}^{\mathscr{T}}(t)$. This holds due to the invariance of a Gibbs distribution with the parameter $p_0 = 0$ with respect to the approximating motion (cf Sect. 3.2). The passage to an arbitrary value p_0 is possible due to the obvious relation

$$S^t\Gamma_{p_0}\mathbf{x} = \Gamma_{p_0}T_{tp_0}S^t\mathbf{x} \tag{10.47}$$

where

$$\Gamma_{p_0}\mathbf{x} = \{(q,p): (q,p-p_0) \in \mathbf{x}\}.$$

In the case where the potential U has a finite range it is possible to verify a "cluster" character of an equilibrium dynamical system. More precisely, in dimension $d = 1$ [S3] and in dimension $d \geqslant 2$ for small values of z [S4] (cf Theorem 2.2) the typical (with respect to the corresponding Gibbs distribution) point $\mathbf{x}$ has the following property: one can divide the particles $(q,p) \in \mathbf{x}$ into finite pairwise disjoint groups (clusters) in such a way that the particles from different clusters will not interact in the course of motion within a given bounded interval of time. After that one introduces a new partition of particles into clusters which move again independently, etc.

The Gibbs distributions mentioned in Definition 4.1 are connected with the $(d+2)$-parameter family of invariants $\mathbb{N}(P)$, $\mathbb{H}(P)$ and $\mathbb{J}(P)$ (cf Sect. 3.2). It is not hard to check that for a Gibbs distribution P with the potential U and parameters (z, β, p_0) the specific average momentum coinsides with p_0. The parameter z is responsible for the specific average density $\mathbb{N}(P)$ and the parameter β for the specific average energy $\mathbb{H}(P)$ in the sense that, for fixed U, β and p_0 (U, z and p_0), $\mathbb{N}(P)$ changes monotonically with z (correspondingly, $\mathbb{H}(P)$ changes monotonically with β). The Gibbs distribution in the uniqueness region is completely determined by the values $\mathbb{N}(P)$, $\mathbb{H}(P)$ and $\mathbb{J}(P)$.

The question about existence of an equilibrium dynamical system is closely related, at least at the level of main ideas, to the question about existence of a solution of the Bogolyubov hierarchy equations (10.44) for an initial moment function k_P corresponding to a "local perturbation" of an invariant Gibbs distribution (this means that the measure P on $(M, \mathcal{M})$ is absolutely continuous with respect to one of the invariant Gibbs distributions). Such results are obtained, by using the functional-analytic method, in the series of papers [G], [GP1], [M], [Pet], [GP2] previously mentioned. On the other hand, in the cycle of papers [T1], [T2], [Z1] this question is investigated by means of the properties of the corresponding equilibrium dynamics.

4.2. The Gibbs Postulate. The fundamental postulate of statistical mechanics asserts that systems with a large number of particles in thermodynamical equilibrium are described by Gibbs distributions. The question of distinguishing the class of Gibbs distributions by means of some a priori physically natural conditions is extremely important for the mathematical foundations of statistical mechanics.

In the traditional "finite-particle" approach (cf [AA]) one refers usually to the ergodic theorem of Birkhoff-Khinchin and the theorem of equivalence of ensembles, which asserts that a microcanonical distribution, considered as a probability measure on $M_{\mathcal{T}}$, converges, as $\mathcal{T} \nearrow R^d$, to a Gibbs distribution with some values of parameters z, β and p_0. This approach is based on the well-known *ergodic hypothesis* for the dynamical system described by equations (10.35) with a microcanonical distribution. The well-known result here is the theorem about ergodicity of the system of two hard balls with elastic collisions (cf Chap. 8, Sect. 1).[4] On the other hand (although there are as yet no explicitly formulated results), the method of constructing invariant measures provided by the theory of Kolmogorov–Arnold–Moser gives us little hope that such an ergodicity takes place in a general situation.

Within the limits of the infinite particle approach, developed in this paper, the analog of the assertion about ergodicity of a microcanonical distribution is the hypothesis that the class of all "good enough" invariant measures is exhausted by Gibbs distributions which figure in Definition 4.1. Of course, it is not hard to construct trivial counter examples. For instance, for an interaction potential of the finite range r_1, any measure such that with probability 1 all the particles are at a distance more than r_1 from each other and have zero momenta is invariant. It seems that a priori natural restrictions on a class of measures which exclude such examples, may consist of assumptions of the following types:

1) the restrictions of these measures onto $\mathcal{M}_{\mathcal{O}}$ for bounded $\mathcal{O} \subset R^d$ should be determined by "nice enough" densities with respect to the measure $\bar{L}_{\mathcal{O}}$,

2) certain conditions of "space mixing" must be fulfilled (cf (10.20), (10.21)) which mean that events from the σ-algebras $\mathcal{M}_{\mathcal{O}_1}$ and $\mathcal{M}_{\mathcal{O}_2}$ for "distant" regions $\mathcal{O}_1, \mathcal{O}_2 \subset R^d$ are "almost independent".

The next reduction of the a priori class of measures is related to the assumption that all these measures are Gibbs distributions in the sense of the general definition of Sect. 2.6, under reasonable restrictions on their potentials Φ. This problem has been studied in such a context in the cycle of papers [GS1], [GS2].

In the papers [GS1], [GS2] the interaction potential U is supposed to have a finite range and a hard core of diameter $r_0 > 0$ and to obey $\lim_{r \to r_0+} U(r) = \infty$. The problem of describing invariant measures is solved here in an a priori introduced class of Gibbs distributions corresponding to potentials $\Phi = (\Phi^{(1)}, \Phi^{(2)}, \ldots)$ which satisfy some qualitative restrictions. Two restrictions among these are of principal character: (a) there exists $n_0 = n_0(\Phi) \geqslant 2$ such that $\Phi^{(n)} \equiv 0$ for $n > n_0$, (b) for all $n = 2, \ldots, n_0$ and $(q_1, p_1), \ldots, (q_n, p_n) \in R^d \times R^d$

$$|\Phi^{(n)}((q_1, p_1), \ldots, (q_n, p_n))| \leqslant \psi\left(\max_{1 \leqslant j_1 < j_2 \leqslant n} \|q_{j_1} - q_{j_2}\| \right) \tag{10.48}$$

where ψ is a monotone function which decreases rapidly enough.

The problem of finding invariant measures has been stated in [GS1], [GS2] as the problem of finding stationary-in-time solutions of the Bogolyubov hierarchy equations, i.e., as the problem of finding moment functions for which the left-hand side of the equations (10.44) is equal to zero. The main theorem of [GS1], [GS2] asserts that, inside the class of measures described above, every stationary solution is the moment function of one of the Gibbs distributions figuring in Definition 4.1.

[4] Recently, Kramli, Simany and Szasz have proved the same result for the system of three balls.

The proof of this theorem uses the following important notion. A function h^Φ of the form

$$h^\Phi(\mathbf{x}) = \sum_{n\geqslant 1} \sum_{\substack{\mathbf{X}'\subset\mathbf{X}:\\ |\mathbf{x}'|=n}} \Phi^{(n)}(\mathbf{x}'), \qquad \mathbf{x}\in M, |\mathbf{x}|<\infty, \tag{10.49}$$

(cf (10.26)) is called the first integral of the motion in the space R^d if

$$h^\Phi(S^t\mathbf{x}) = h^\Phi(\mathbf{x}), \qquad t\in R^1, \tag{10.50}$$

(cf (10.39)–(10.41)). In the first step one proves that h^Φ is a first integral of the motion whenever the moment function k_P of a Gibbs distribution P with a potential Φ from the indicated class gives a stationary solution of the Bogolyubov hierarchy equations.

In the second step, one investigates the summatory first integrals of the motion of the form (10.49). It turns out that under the imposed conditions they are reduced to linear combinations of "canonical" first integrals (10.39)–(10.41). This leads to equalities (10.29a, b) according to which P will be one of the Gibbs distributions with the potential U.

4.3. Degenerate Models. The connection between the summatory first integrals and the class of invariant measures is illustrated by the example of "degenerate" models, where, in addition to the "canonical" first integrals (10.39)–(10.41), there exist other summatory first integrals. For instance, in the case of an ideal gas where the momenta of particles do not change during the motion, there exist first integrals of the form

$$h(\mathbf{x}) = \sum_{(q,p)\in\mathbf{X}} \varphi(p), \qquad \mathbf{x}\in M, \qquad |\mathbf{x}|<\infty, \tag{10.51}$$

where φ is an arbitrary measurable function. Correspondingly, all Poisson measures $P^0_{z,\sigma}$ are invariant in this case (cf Sect. 2.5). An analogous situation takes place for the model of one-dimensional ($d=1$) hard rods of length $r_0>0$ where the particles change momenta under collisions. Here the measures $P^{r_0}_{z,\sigma}$ are invariant (see again Sect. 2.5). From the results of Section 5 (cf Sects. 5.2, 5.4) it follows that there are no other invariant measures for these two models inside a wide class of probability distributions on $(M, \mathcal{M})$.

The one-dimensional system (10.35) with the interaction potential

$$U(r) = (\operatorname{sh}(Ar))^{-2}, \qquad r>0, \tag{10.52}$$

with $A=\text{const}>0$, gives another interesting example of a degenerate system. Given any number of particles $N<\infty$, this system is integrated by the method of the inverse problem of scattering theory (cf [C]). Therefore, for this model there exists an infinite series of nontrivial summatory first integrals. In the paper [C] the simplest of these integrals is considered. It has the form (10.49) with some explicitly written Φ, and one proves the existence and invariance property of the Gibbs distribution with the potential $\beta\Phi, \beta>0$.

In the note [Gu] a result is given which extends, in the one-dimensional case, the constructions of Section 4.2 to a new class of interaction potentials including the potential (10.52). It turns out that, within this class of interaction potentials, the potential (10.52) (and multiples of it) exhausts the set of potentials for which there exist additional invariant Gibbs measures. Recently, Gurevich investigated the multidimensional situation, too. Here, under quite general conditions on the interaction potential U, the only potential for which "non-canonical" invariant Gibbs distributions may appear is that of ideal gas.

4.4. Asymptotic Properties of the Measures P_t. The question about the asymptotic behaviour of the measures P_t, $t \in R^1$, giving the time evolution of an initial measure $P_0 = P$ (cf Definition 3.2) seems to be very important. A natural conjecture is that, for non-degenerate models of motion and "good enough" (in the sense of Sect. 4.2) initial measures P, the measures P_t converge, as $t \to \pm\infty$, to an invariant Gibbs distribution with the potential U for which the values of the specific average particle number, energy and momentum are the same as for P. This conjecture may be considered as an analog of the classical ergodic Boltzmann hypothesis.

A similar role in the equilibrium dynamics concept is played by the hypothesis that, for non-degenerate models of motion, the equilibrium dynamical system has mixing properties. From this hypothesis one can derive the convergence of the measures P_t for initial measures P which are absolutely continuous with respect to an invariant Gibbs distribution with the potential U. Physically, this fact corresponds to an assertion about asymptotic "dispersing" local fluctuations in equilibrium dynamical systems.

Both these problems are very difficult and results at the mathematical level are obtained here only for degenerate models of ideal gas and of one-dimensional hard rods which will be considered in the next sections.

§ 5. Ideal Gas and Related Systems

5.1. The Poisson Superstructure. We begin by studying the dynamical system on $(M, \mathscr{M})$ corresponding to the potential $U \equiv 0$. It is convenient to give a general definition including the example of the ideal gas as a particular case.

Let $(N^1, \mathscr{N}^1, \pi)$ be a metric space with a Borel σ-finite measure π, $\{\tau_t^1, t \in R^1\}$ be a one parameter measurable group of bijective transformations of N^1 which preserve the measure π. By analogy with the spaces $(Q^0, \mathscr{Q}^0)$ and $(M, \mathscr{M})$ (cf Sects. 2.1, 2.5) we introduce the measurable space $(N, \mathscr{N})$ the points of which are "locally finite" subsets of N^1. We shall fix $z > 0$ and, by analogy with Poisson measures P_z^0 and $P_{z,\sigma}^0$, introduce a probability measure $\mathbb{P}_{z,\pi}^0$ on $(N, \mathscr{N})$ defined by the same conditions 1), 2) (cf Sect. 2.2), replacing R^d by the space N^1 and the Lebesgue measure by the measure π. We shall define the flow $\{\tau_t, t \in R^1\}$ in

$(N, \mathscr{N}, \mathbb{P}^0_{z,\pi})$ by setting

$$\tau_t X = \{x \in N^1 : \tau^1_{-t} x \in X\}. \tag{10.53}$$

It is not hard to check that the measure $\mathbb{P}^0_{z,\pi}$ is invariant with respect to the flow $\{\tau_t\}$.

Definition 5.1. The dynamical system $(N, \mathscr{N}, \mathbb{P}^0_{z,\pi}, \{\tau_t\})$ is called the *Poisson superstructure* over $(N^1, \mathscr{N}^1, \pi, \{\tau^1_t\})$. The quadruple $(N^1, \mathscr{N}^1, \pi, \{\tau^1_t\})$ is called the one-particle dynamical system.

In the particular case where (i) $N^1 = R^d \times R^d$, (ii) π is a measure $l \times \sigma$ where l is the Lebesgue measure and σ is a probability distribution on R^d, and, finally, (iii) $\tau^1_t(q, p) = (q + tp, p)$, $(q, p) \in R^d \times R^d$, we get the dynamical system of ideal gas. Here $\mathbb{P}^0_{z,\pi}$ coincides with the probability measure $\mathbb{P}^0_{z,\sigma}$ (cf Sect. 2.5).

Clearly, all ergodic properties of the Poisson superstructure are completely determined by the one-particle dynamical system. However, generally the problem of obtaining necessary and sufficient conditions of ergodicity and mixing is difficult enough. The situation is simplified if one assumes in addition that a so-called trend to infinity takes place for the one-particle dynamical system $(N^1, \mathscr{N}^1, \pi, \{\tau^1_t\})$, i.e., there exists a set $C \in \mathscr{N}^1$ with $\pi(C) < \infty$ and a number $t_0 > 0$ such that $\pi(N^1 \backslash (\bigcup_{t \in R^1} \tau^1_t C)) = 0$ and the intersection $C \cap \tau^1_t C = \varnothing$ for all t with $|t| > t_0$. For ideal gas this condition is fulfilled provided the probability distribution σ has no atom at 0.

Theorem 5.1 (cf [CFS]). *If a trend to infinity takes place in a one-particle dynamical system, then the corresponding Poisson superstructure is a B-flow.*

The proof of this theorem is a verification of the B-property of the partition given by the intersection $X \cap C$, $X \in N$.

The physical explanation of the fact that the Poisson superstructure has such strong ergodic properties is connected with specific features of infinite particle systems. Any probability measure absolutely continuous with respect to $\mathbb{P}^0_{z,\pi}$ is determined by the corresponding density $f(X)$, $X \in N$. The function f is "local" (i.e., depends on the positions and momenta of the particles concentrated in a bounded domain of the space R^d) or may be approximated by local functions in the sense of L_1-convergence. The "shifted" function depends essentially on the positions and momenta of the particles which are at a distance of order $|t|$ from the origin. It follows from the definition of the measure $\mathbb{P}^0_{z,\pi}$ that the functions $f(\tau_{-t}X)$ and $f(X)$ are "almost" independent for large $|t|$, and this fact leads to good ergodic properties.

Another popular example of Poisson superstructure is the so-called *Lorentz gas* (cf Chap. 8). Let a countable set of points ("immovable scatterers") in the space R^d be thrown about. Every scatterer generates a potential field which decreases rapidly enough at a large distance from the scatterer. The one-particle system will be the dynamical system corresponding to the motion of a particle in the potential field given by the total potential of the scatterers. The Lorentz gas, in a general form, is the Poisson superstructure over such a system.

The particular case where the potentials of the scatterers are those of hard spheres (not necessarily of a constant radius) may be investigated at the mathematical level. This case corresponds to the motion of a particle with a constant velocity and with the elastic reflection from the spheres. The Lorentz gas is studied in this case by methods of the theory of dispersing billiards (cf [BS1], [BS2] and Chap. 8). The case of a periodic configuration of scatterers is directly reduced to the dispersing billiard on the torus. In the paper [BS2] it is proved that the Lorentz gas is a K-system under wide assumptions about the configuration of scatterers.

Notice the close connection of the models regarded in this section with infinite systems of independently moving particles (see Chap. 7 of the book [MKM] and the references therein). In particular, the invariance of the Poisson measure $\mathbb{P}^0_{z,\sigma}$ with respect to the dynamics of ideal gas, as well as the mixing property, follow from results which were already established in the 1950's (cf [D1], [Do]).

5.2. Asymptotic Behaviour of the Probability Distribution P_t as $t \to \infty$. In this section we discuss the question about the asymptotic behaviour of the probability measures P_t, $t \in R^1$, which describe the time evolution of an initial measure P (cf (10.38)) induced by the dynamics of ideal gas (10.37). Since we do not assume that the measure P is absolutely continuous with respect to one of the invariant measure $\mathbb{P}^0_{z,\sigma}$, this question is beyond the constructions of the preceding section.

The investigation of asymptotic properties of the probability measures P_t was initiated in the early paper [D1] as has been previously mentioned. A detailed analysis of this question is given in the more recent paper [DS2], the results of which are discussed below. Let us introduce the following coefficient of asymptotic mixing (comp. (10.20)):

$$\alpha_P(r, s) = \sup_{y \in R^d} \sup_{\substack{A_1 \in \mathcal{M}_{\mathcal{T}(y,r)},\\ A_2 \in \mathcal{M}_{\mathcal{T}(y,r+s)^c}}} |P(A_1 \cap A_2) - P(A_1)P(A_2)|. \tag{10.54}$$

The following conditions are imposed onto the initial measure P:

(I) for any $b > 0$

$$\lim_{s \to \infty} \alpha_P(bs, s) = 0,$$

(II) the 1-moment measure $K_P^{(1)}(dq \times dp)$ is absolutely continuous with respect to a measure $\hat{K}_P^{(1)}(dq) \times dp$ on $R^d \times R^d$ with $\sup_{y \in R^d} \hat{K}_P^{(1)}(\mathcal{T}(y, b)) < \infty$ for any $b > 0$ and has a bounded density with respect to this measure,

(III) the 2-moment measure $K_P^{(2)}(\prod_{i=1,2} dq_j \times dp_j)$ is absolutely continuous with respect to a measure $\hat{K}_P^{(2)}(dq_1 \times dq_2)\prod_{j=1,2} dp_j$ on $(R^d \times R^d)^2$ with $\sup_{y_1, y_2 \in R^d} \hat{K}_P^{(2)}(\times_{j=1,2} \mathcal{T}(y_j, b_j)) < \infty$ for any b_1, $b_2 > 0$ and has a bounded density with respect to this measure.

Theorem 5.2 (cf [DS2]). *Let the probability distribution P satisfy conditions* (I)–(III). *Then the probability distributions P_t converge, as $t \to \pm\infty$, to the Poisson* *measure* $\mathbb{P}^0_{z,\sigma}$ *if and only if the* 1-*moment measures*

$$K^{(1)}_{P_t}(A) = K^{(1)}_P(\{q,p): (q - tp, p) \in A\})$$

converge to the measure $z(l \times \sigma)$.

As to the convergence of measures $K^{(1)}_{P_t}$, it may be established by wide assumptions about the 1-moment measure $K^{(1)}_P$. For instance, if $K^{(1)}_P$ has the periodicity property with respect to the space translations $(q,p) \mapsto (q + y, p)$, then the measures $K^{(1)}_{P_t}$ converge to a measure of the form $z(l \times \sigma)$, which is obtained by space averaging the measure $K^{(1)}_P$.

The conditions of Theorem 5.2 may be verified for Gibbs distributions under the uniqueness conditions (cf Theorems 2.2 and 2.3). The proof of Theorem 5.2 is based on a variant of the Poisson limit theorem for sums of weakly dependent random variables.

Recently the result of Theorem 5.2 was extended by Willms who suggested a more general variant of the condition (I) (cf [W]). Notice also the paper [Su5] where an "abstract" theorem of convergence to a Poisson measure $\mathbb{P}^0_{z,\pi}$ on the space $(M, \mathcal{M})$ (cf Sect. 5.1) is established. From the results proved in [Su5] it follows that one can require property (I) not for P but for the conditional distribution $P(\cdot \mid \mathcal{M}^Q)$ with respect to the σ-algebra $\mathcal{M}^Q$ (cf Sect. 2.5) induced by the projection map Π. This allows us to extend essentially the class of initial probability distributions for which the assumptions of the convergence theorem are valid.

A result similar to Theorem 5.2 was obtained in [KS] for the Lorentz gas.

5.3. The Dynamical System of One-Dimensional Hard Rods. The dynamics of one-dimensional hard rods corresponding to the potential (10.14) is closely connected with the dynamics of ideal gas. The connection between the two dynamics is established by using special transformations of "dilatation" and "contraction" in the phase space $(M, \mathcal{M})$. Let $\mathbf{x} \in M$ and a particle $(q,p) \in \mathbf{x}$ be given. We enumerate the particles $(\tilde{q}, \tilde{p}) \in \mathbf{x}$ by integers in order of increasing coordinates $\tilde{q} \in R^1$; with the number 0 given to the distinguished particle (q,p). Under the "dilatation" transformation, $\mathbb{D}_q$, the particle (q_j, p_j) passes into $(q_j + jr_0, p_j)$, $j \in Z^1$, where r_0 is the hard rod length. The transformation of "contraction", $\mathbb{C}_q$, is inverse to $\mathbb{D}_q$ (it is correctly defined only if $q_{j+1} - q_j > r_0$ for all j). If one denotes the time shift in the dynamics of ideal gas by S^0_t and the time shift in the dynamics of hard rods by $S^{r_0}_t$, then the following formula is valid:

$$S^{r_0}_t \mathbf{x} = T_{-r_0 n_t} \mathbb{D}_{q+tp} S^0_t \mathbb{C}_q \mathbf{x}, \qquad t \in R^1, \tag{10.55}$$

where $n_t = n_t((q,p), \mathbb{C}_q \mathbf{x})$ is the total "algebraic" number of the intersections of the trajectory of the particle (q,p) by the trajectories of other particles $(\tilde{q}, \tilde{p}) \in \mathbb{C}_q \mathbf{x}$ under the free motion in the time interval from 0 to t.

Ergodic properties of the hard rods system with the invariant measure $P^{r_0}_{z,\sigma}$ (cf Sect. 2.5) have been investigated on the basis of this connection in the papers

[S2], [AGL1], [P]. The most general result is proved in the paper of Aizenman, Goldstein and Lebowitz [AGL1].

Theorem 5.3. *Let σ be a probability measure on R^1 with a finite first moment ($\int_{R^1} \sigma(dp)|p| < \infty$) such that $\sigma(\{0\}) < 1$. Then for any $z > 0$, the dynamical system $(M, \mathcal{M}, P^{r_0}_{z,\sigma}, \{S^{r_0}_t\})$ is a K-flow. If z and σ satisfy the additional condition $\sigma((p^0_{z,\sigma} - \varepsilon, p^0_{z,\sigma} + \varepsilon)) = 0$ for some $\varepsilon > 0$ where*

$$p^0_{z,\sigma} = \frac{zr_0}{1 + zr_0} \int_{R^1} \sigma(dp)p, \tag{10.56}$$

then the dynamical system $(M, \mathcal{M}, \mathbb{P}^{r_0}_{z,\sigma}, \{S^{r_0}_t\})$ is a B-flow.

The possibility of omitting this additional condition remains an open problem.

Formula (10.55) provides the basis for a generalization of the results of Sect. 5.2 to the dynamics of the hard rods given in [DS2]. Let us suppose that the measure P is translation invariant and concentrated on the set of points $\mathbf{x} \in M$ having the property: $|q - q'| > r_0$ for any pair of different particles (q, p), $(q', p') \in \mathbf{x}$. By means of the transformations $\mathbb{D}_q$ and $\mathbb{C}_q$, one can associate with P a corresponding "contracted" translation invariant measure $P^{(0)}$. If the measure $P^{(0)}$ satisfies conditions (I)–(III) (cf Sect. 5.2), then the measure $P^{(0)}_t$ obtained from $P^{(0)}$ in the course of the ideal gas dynamics converges, as $t \to \pm\infty$, to a Poisson measure $\mathbb{P}^0_{z,\sigma}$. Using formula (10.55) one can prove that the measures P_t obtained from P in the course of the hard rod dynamics converge to the corresponding measure $\mathbb{P}^{r_0}_{z,\sigma}$. Notice that the conditions on the initial measure P are formulated here in terms of the measure $P^{(0)}$ whose connection with P is not so transparent. However, it is possible to verify such conditions for a wide class of Gibbs distributions (cf [DS2]).

§ 6. Kinetic Equations

6.1. Statement of the Problem. According to Definition 3.2, the time evolution of a probability measure P on the space $(M, \mathcal{M})$ is represented by the family of measures P_t, $t \in R^1$, determined by formula (10.38), or, equivalently, by the family of moment functions $\mathscr{k}(t, \cdot)$ satisfying the hierarchy of the Bogolyubov equations (10.44). Kinetic equations of statistical mechanics are used for an approximate description of the time evolution in more simple terms. The purpose of rigorous mathematical investigations initiated here in recent years is to provide the mathematical background for the approximations used, and to give us a deeper understanding in this difficult area of problems. Meanwhile, these investigations are at an initial level, and a large part of what is said below must be treated only as preliminary conjectures.

Notice the profound review of the papers in this subject by Spohn [Sp1] where the majority of the themes we touch in this paragraph are discussed in detail.

The problems of kinetic theory for the Lorentz gas are considered in the review of van Bejeren [B].

In this section we shall try to describe, without pretending to be mathematically precise, some general features in stating problems of derivation of the basic kinetic equations: Boltzmann equation, Vlasov equation, Landau equation, and finally, hydrodynamic Euler equation, taking, as a starting point, the ideas developed in Sect. 3. After that we shall proceed to a consequent discussion of various equations and formulate the (non-numerous) mathematical results which are known here.

Let $F = F_P$ be a functional defined on some subset of the space of probability measures on $(M, \mathcal{M})$ with values in the space $\mathcal{F}$ of (generally speaking, vector-valued) functions depending on q, $p \in R^d$. For Boltzmann, Vlasov and Landau equations, $F_P = \ell_P^{(1)}$ (the 1-moment function of a measure P). Suppose that a family of transformations $\mathbb{R}_t$, $t \in R^1$, of the space $\mathcal{F}$ is given, with the semi-group (or group) property

$$\mathbb{R}_{t_1+t_2} = \mathbb{R}_{t_1}\mathbb{R}_{t_2}, t_1, t_2 \geqslant 0 \qquad \text{or } t_1, t_2 \leqslant 0\,(t_1, t_2 \in R^1). \tag{10.57}$$

Suppose furthermore that for some classes of interaction potentials and initial measures P_0 on $(M, \mathcal{M})$ and for some interval $I \subset R^1$

1) an approximate equality takes place

$$F_{P_{t'}} \approx \mathbb{R}F_{P_0} \quad \text{where } t' = \kappa t, t \in I. \tag{10.58}$$

Here $\kappa > 0$ is a constant giving the "time-scaling";

2) the map $P \mapsto F_P$ is approximately invertible on some subset of the space of measures on $(M, \mathcal{M})$, including the measures $P_{t'}$, $t' \in R^1$, so that the measure $P_{\kappa t}$ may be "almost reconstructed" from the value of the functional $F_{P_{\kappa t}}$.

In addition, it is natural to assume that the transformations $\mathbb{R}_t$ may be described by means of a solution of a differential equation of the form

$$\frac{d}{dt}F(t) = AF(t) \tag{10.59}$$

where A is some operator in $\mathcal{F}$ which does not depend on t. Equations of this type are called kinetic equations.

In recent mathematical papers the approximate relations in conditions 1) and 2) are interpreted as asymptotic equalities with respect to an auxiliary parameter $\varepsilon \to 0_+$. One considers a family $\{P_0^\varepsilon, \varepsilon > 0\}$ of initial measures on $(M, \mathcal{M})$ onto which some restrictions are imposed. First, one assumes that the 1-moment function $\ell_{P_0^\varepsilon}^{(1)}$ satisfies the relation

$$\ell_{P_0^\varepsilon}^{(1)}(q, p) = \varepsilon^{\alpha_1} f_0(\varepsilon^{\alpha_2} q, p) \tag{10.60}$$

where f_0 is a fixed function $R^d \times R^d \to [0, \infty)$ and α_j are some constants (of course, one can consider a more general situation when equality (10.60) is fulfilled in the limit, as $\varepsilon \to 0_+$). Further, one supposes that $\kappa = \varepsilon^{-\alpha_3}$ (cf (10.58)) and finally, that the particle motion is determined by the interaction potential U^ε of the form

$$U^{\varepsilon}(r) = \varepsilon^{\alpha_4} U(\varepsilon^{\alpha_5} r), \qquad r \geqslant 0, \tag{10.61}$$

where U is a fixed function $[0, \infty) \to R^1 \cup \{\infty\}$ satisfying the conditions of Sect. 2.4.

Observe that the choice of constants α_j is not unique. This property is connected essentially with the fact that, under the simultaneous scale change $t = \varepsilon t'$, $q = \varepsilon q'$, $p = p'$, equations (10.35) are transformed into the equations of the same form with the new potential $U_{\varepsilon}(r) = U(\varepsilon^{-1} r)$ and the 1-moment function takes the form $\ell^{(1)'}(t'; (q', p')) = \varepsilon^{-d} \ell^{(1)}_{P^{\varepsilon}_{\varepsilon^{-1}t}}(\varepsilon^{-1} q, p)$.

For comparison, we present below a table indicating the choice of the parameters for the two most frequently used situations: $\alpha_2 = \alpha_3 = 1$ and $\alpha_2 = \alpha_3 = 0$. From the physical point of view, these situations correspond to the choice of the scale for measuring time and distance in "micro-" and "macro-world". We shall therefore speak about "micro"- and "macro-variables". The parameter ε gives the ratio of these scales and therefore may be considered as small in many applications.

The second part of the table is obtained from the first one by the change of variables $t = \varepsilon t'$, $q = \varepsilon q'$, $p = p'$.

		Microvariables: $t' = \varepsilon^{-1} t$		*Macrovariables:* $t' = t'$	
Type of the limit passage	Equation	1-moment function $\ell^{(1)}_{P^{\varepsilon}_0}(q, p)$	Potential	1-moment function $\ell^{(1)}_{P^{\varepsilon}_0}(q, p)$	Potential
Low density approximation	Boltzmann	$\varepsilon f_0(\varepsilon q, p)$	$U(r)$	$\varepsilon^{-(d-1)} f_0(q, p)$	$U(\varepsilon^{-1} r)$
Mean field approximation	Vlasov	$f_0(\varepsilon q, p)$	$\varepsilon^d U(r)$	$\varepsilon^{-d} f_0(q, p)$	$\varepsilon^d U(r)$
Weak coupling approximation	Landau	$f_0(\varepsilon q, p)$	$\varepsilon^{1/2} U(r)$	$\varepsilon^{-d} f_0(q, p)$	$\varepsilon^{1/2} U(\varepsilon^{-1} r)$
Hydrodynamic approximation	Euler for a compressible fluid	$f_0(\varepsilon q, p)$	$U(r)$	$\varepsilon^{-d} f_0(q, p)$	$U(\varepsilon^{-1} r)$

The conditions on the 1-moment function listed in the table are, of course, not sufficient for our purposes. Additional conditions on the whole measure P^{ε}_0 are formulated in a different form for different problems and it is too cumbersome to describe their form in detail at this stage of the development of the theory. It seems that, in the microvariables, these conditions must correspond, at a qualitative level, to the general conditions 1), 2) discussed in Sect. 4.2 and, in particular, they must be fulfilled for a large class of Gibbs measures.

In the macrovariables, the condition of the decay of correlations is sometimes formulated in terms of asymptotics of higher moment functions subject to the scaling transformations:

$$f_0^{(n)}((q_1, p_1), \ldots, (q_n, p_n)) = \varepsilon^{n\rho} k_{P_n^\varepsilon}^{(n)}((q_1, p_1), \ldots, (q_n, p_n)) \tag{10.62}$$

where $\rho = d - 1$, d, d and d for the corresponding lines of the table. This condition has the form

$$\lim_{\varepsilon \to 0} f_0^{(n)}((q_1, p_1), \ldots, (q_n, p_n)) = \prod_{j=1}^{n} f_0(q_j, p_j) \tag{10.63}$$

and is often called the "hypothesis of chaos" because it means (in macrovariables) the asymptotical "complete independence" of events which "occur" for different particles. For the case of the low density approximation, condition (10.63) leads to the well-known Boltzmann rule for calculating the number of collisions (Stoßzahlansatz). One can expect that in the situations we listed above, the fulfilment of conditions of the type (10.63) at the moment $t = 0$ involves the fulfilment of analogous conditions for all t.

The scheme described above leads to kinetic equations in the whole space R^d. Without special modifications of this construction one can get equations which describe the motion in a bounded domain with smooth or piecewise smooth boundary. In microvariables one can, for instance, consider a family $\{\mathcal{O}^\varepsilon, \varepsilon > 0\}$ of homotetic domains whose linear sizes increase like ε^{-1}. For any given ε one supposes that the probability measure P_0^ε is concentrated on the set $M_{\mathcal{O}^\varepsilon}$ and the time evolution P_t^ε is induced by the motion of particles in $\mathcal{O}^\varepsilon$ with boundary conditions on $\partial\mathcal{O}^\varepsilon$. For definiteness we consider below the boundary conditions of elastic reflection. Passing to microvariables, we get the motion in the fixed domain $\mathcal{O} \subset R^d$. The same is true for the limit kinetic equation as well.

6.2. The Boltzmann Equation. This equation is considered under the assumption that the dimension $d > 1$. It has the form

$$\begin{aligned} \frac{\partial}{\partial t} f_t(q, p) = -(p, grad_q f_t(q, p)) + \int_{(S^{d-1} \times R^d)_p^+} & |p - p_1| \\ \times B(p - p_1, e)[f_t(q, p') f_t(q, p_1') & \\ - f_t(q, p) f_t(q, p_1)]\, de\, dp_1, \; q, p \in R^d. & \end{aligned} \tag{10.64}$$

Here

$$(S^{d-1} \times R^d)_p^+ = \{(e, p_1) \in S^{d-1} \times R^d \colon (p - p_1, e) > 0\},$$

S^{d-1} is a unit sphere in R^d. Vectors p', p_1' are connected with p, p_1 by the relations

$$\begin{aligned} p' &= p - (p - p_1, e)e, \qquad e \in S^{d-1}, \\ p_1' &= p_1 - (p_1 - p, e)e, \qquad e \in S^{d-1}, \end{aligned}$$

and $B(\tilde{p}, e)$, $\tilde{p} \in R^d$, $e \in S^{d-1}$, is a function called differential cross section and is determined by the potential U. Physically, p, p_1 and p', p_1' correspond to the momenta of two colliding particles before their collision and after it, respectively.

The relation which connects them is equivalent to the natural conservation laws: $p + p_1 = p' + p_1'$, $\|p\|^2 + \|p_1\|^2 = \|p'\|^2 + \|p_1'\|^2$. For the function B one can write an explicit expression by integrating the equations of motion of two particles interacting via the potential U. In the case of the hard sphere potential (10.14) one has $B(\tilde{p}, e) = (\tilde{p}, e)$.

Problems of existence and uniqueness of a solution of the Boltzmann equation are extremely nontrivial. Chapter 11 of this volume is devoted to a review of results obtained here (see also the review papers in [No]).

As already mentioned above, the Boltzmann equation is obtained by the low density limit passage which is called also the *Boltzmann–Grad limit passage* (cf [Gr]). The physical meaning of this limit passage is most transparent under consideration in microvariables. The density of particles converges to 0 like ε and thereby (if we suppose for simplicity that the radius of interaction of the potential U is finite) particles move with constant momenta most of the time, changing their values now and then during relatively short intervals of collision interaction. The length of the time interval between subsequent collisions of a given particle with other particles (the time of free motion) has the order ε^{-1} and the time of each collision has the order of a constant. Most of these collisions are pairwise and asymptotically, as $\varepsilon \to 0$, one can neglect the multiple collisions. The normalization $t' = \varepsilon^{-1} t$ is natural because, during a time of the order ε^{-1}, each particle passes through a finite number of intervals of free motion and a finite number of collisions. The first term in the right hand side in (10.64) describes the free motion of particles while the second one corresponds to collisions.

In macrovariables, one considers particles of small diameter ε and their density, having the order $\varepsilon^{-(d-1)}$, diverges to ∞. The choice of macrovariables is more convenient for discussing details of the derivation of Boltzmann equation (see below).

An important result concerning the derivation of the Boltzmann equation was obtained by Lanford [L4]. He considered the motion generated by the potential of hard spheres of diameter $r_0 = \varepsilon$ (cf (10.14)) in a domain $\mathcal{O} \subset R^d$. It was assumed that for initial probability measures P_0^ε there exist moment functions $k_{P_0^\varepsilon}$ with the following properties:

1) for some positive constant z, β and all ε and n

$$\varepsilon^{(d-1)n} k_{P_0^\varepsilon}^{(n)}((q_1, p_1), \dots, (q_n, p_n)) \leqslant z^n \exp\{-\beta(p_1^2 + \cdots + p_n^2)\}, \qquad (10.65)$$

2) there exists $s > 0$ such that for any n, uniformly on every compact subset of $\Gamma_{\mathcal{O}}^{(n)}(s)$, the following convergence takes place

$$\lim_{\varepsilon \to 0} \varepsilon^{(d-1)n} k_{P_0^\varepsilon}^{(n)}((q_1, p_1), \dots, (q_n, p_n)) = \prod_{j=1}^{n} f_0(q_j, p_j) \qquad (10.66)$$

where f_0 is a smooth function $\mathcal{O} \times R^d \to [0, \infty)$.

Here and below $\Gamma_{\mathcal{O}}^{(n)}(s)$ denotes the set

$$\{(x_1, \dots, x_n) \in (\mathcal{O} \times R^d)^n : q_{\mathcal{O}}^0(\tilde{s}, x_{j_1}) \neq q_{\mathcal{O}}^0(\tilde{s}, x_{j_2}) \text{ for any } \tilde{s} \in [-s, 0] \text{ and } j_1 \neq j_2\} \qquad (10.67)$$

where $q^0_{\mathcal{O}}(\tilde{s}, (q, p))$ is the position, at the time moment $\tilde{s}$ of the particle having, at time zero, the position $q \in \mathcal{O}$ and the momentum vector $p \in R^d$ under the free motion in the domain $\mathcal{O}$ (with the condition of the elastic reflection from the boundary).

Theorem 6.1 (cf [L4]). *Let condition* 1) *and* 2) *be fulfilled. Then there exists a constant* $t_0(z, \beta) > 0$ *such that for any* $t \in [0, t_0(z, \beta))$ *the moment functions* $k^{(n)}_{P^\varepsilon_t}((q_1, p_1), \ldots, (q_n, p_n))$, $n \geq 1$, *of the measure* P^ε_t *satisfy the relation*

$$\lim_{\varepsilon \to 0} \varepsilon^{(d-1)n} k^{(n)}_{P^\varepsilon_t}((q_1, p_1), \ldots, (q_n, p_n)) = \prod_{j=1}^{n} f_t(q_j, p_j)$$

for any n uniformly on every compact subset of $\Gamma^{(n)}_{\mathcal{O}}(t + s)$. *Here* $f_t(q, p)$ *is the solution of the Boltzmann equation* (10.64) *in the domain* $\mathcal{O}$ *with the initial date* $f_0(q, p)$.

An analogous assertion may be formulated as well for $t \in (-t_0(z, \beta), 0]$ (one assumes here that $s < 0$ in condition 2) and the interval $[-s, 0]$ in definition (10.67) is replaced by the interval $[0, -s]$ where $s < 0$). Instead of (10.64), the equation with the opposite sign in front of the integral term arises. This fact illustrates the well-known property of "irreversibility" of the Boltzmann equation.

Notice that, before passing to the limit, the system (10.35) has the reversibility property: if one replaces, at the moment zero, the values of momenta of all particles by the opposite values, then the motion in the "positive" direction will be the same as the motion in the "negative" direction for particles with the original values of momenta. The paradox of irreversibility which we encounter is explained by the fact that the sets $\Gamma^{(n)}_{\mathcal{O}}(s)$, $s > 0$, are not invariant under the change of time. Hence, condition 2) for the moment functions obtained by "reversing" the values of the particle momenta at time $t \in (0, t_0(z, \beta))$ does not follow from the assertion of Theorem 6.1 at all.

So, Theorem 6.1 guarantees a convergence to the solution of the Boltzmann equation on a finite interval of time. This restriction is closely connected with the fact that the existence of a solution of the Boltzmann equation itself, is proved in a general case only locally in time (cf Chap. 11). Conditions 1), 2) (reformulated in terms of microvariables) may be verified for a wide class of Gibbs measures with the value of activity $z(\varepsilon)$ depending on ε.

In the (unpublished) dissertation of King, the results of Lanford are extended onto a wider class of interaction potentials.

Lanford's method is based on the fact that the moment function $k_\varepsilon(t) = k_{P^\varepsilon_t}$ which gives the solution of Bogolyubov hierarchy equations (10.44) is expanded into the series of perturbation theory which converges for small enough t. The main term of the expansion is the term corresponding to the free motion of particles, and the terms corresponding to interaction between particles play the role of a "small" perturbation. After that, one passes separately to the limit for all terms, and thereby an analogous limiting series of perturbation theory arises which gives a solution of the Boltzmann equation.

In the recent paper [IP] the Boltzmann-Grad limit is performed for all times $t \in R^1$ for the two-dimensional case of a rarified gas.[5]

A problem which seems to be interesting is the study of the trajectory of a given particle under the conditions of the Boltzmann-Grad limit passage. It is natural to expect that this behaviour is described by some nonlinear Markov process whose infinitesimal generator at the time moment t is determined by the unconditional probability distribution for the state of the process at this moment. This generator is connected with an operator which describes the linearized Boltzmann equation. The trajectory of the process consists of intervals of deterministic motion with a constant speed along a straight line, interrupted by moments of random jump change of the velocity. Papers [Ta1], [Ta2], [U] are devoted to the study of such processes from a probabilistic point of view. As Spohn showed in [Sp1], under the conditions of Theorem 6.1, the probability distribution of the trajectory of a given particle converges weakly, for $0 \leqslant t \leqslant t_0(z, \beta)$, to the corresponding nonlinear Markov process. The problem of studying time fluctuations in the Boltzmann-Grad limit is of great interest as well. In terms of microvariables this problem in stated in the following way. Let $\mathbf{x}^\varepsilon(t)$, $t \in [0, \infty)$, be a trajectory of the motion in a domain $\mathcal{O} \subset R^d$ subjected to an interaction potential U_ε. An initial probability distribution P_0^ε for $\mathbf{x}^\varepsilon(0)$ determines the probability measure on the space of trajectories and hence defines the joint distribution of random variables

$$\xi_g^\varepsilon(t) = \sum_{(q,p) \in \mathbf{X}^\varepsilon(t)} g(\varepsilon q, p) \tag{10.68}$$

where g is a smooth function with a compact support. Change of the average value

$$E^\varepsilon(t, g) = \varepsilon^{d-1} E\xi_g^\varepsilon(t) = \varepsilon^{d-1} \int_{R^d \times R^d} g(q, p) \mathscr{k}_{P^\varepsilon_{\varepsilon^{-1}t}}(\varepsilon^{-1} q, p)\, dq\, dp \tag{10.69}$$

is described in the limit, as $\varepsilon \to 0$, by the Boltzmann equation. The standard deviation of $\xi_g^\varepsilon(t)$ is of the order $\varepsilon^{-(d-1)/2}$ and therefore it is natural to consider the normalized value

$$\eta_g^\varepsilon(t) = \varepsilon^{(d-1)/2}(\xi_g^\varepsilon(t) - E^\varepsilon(t, g)). \tag{10.70}$$

In papers [Sp1], [Sp2] it is proved that, under the conditions of Theorem 6.1 and the assumption that the initial probability measure P_0^ε is a Gibbs distribution in a volume $\mathcal{O}^\varepsilon$ with a potential U_ε, inverse temperature β and an appropriate value of the activity $z(\varepsilon)$, there exists a limit, as $\varepsilon \to 0$, of the covariance $\mathrm{Cov}(\eta_{g_1}^\varepsilon(t)\eta_{g_2}^\varepsilon(t))$. Evolution of limiting covariances may be described by the linearized Boltzmann equation. The natural conjecture that the values $\eta_g^\varepsilon(t)$ have in the limit a Gaussian distribution remains open.

In the review [Sp1] one noticed the possibility of extending Lanford's construction to the model of Lorentz gas (cf Sect. 5.1) where, in the low density limit,

---

[5] This result is now extended by Illner and Pulvirenti to the three-dimensional case, too.

a linear equation arises of the Kolmogorov type equation in the theory of Markov processes.

6.3. The Vlasov Equation. Under the assumption that $U(\|q\|)$ is a smooth function of the variable $q \in R^d$, $d \geqslant 1$, this equation takes the form

$$\frac{\partial}{\partial t} f_t(q,p) = -(p, \operatorname{grad}_q f_t(q,p)) - (\operatorname{grad}_p f_t(q,p), \int_{R^d \times R^d} dq_1\, dp_1 \operatorname{grad}_q U(\|q - q_1\|) f_t(q_1, p_1)). \tag{10.71}$$

If one considers a motion in a bounded domain $\mathcal{O} \subset R^d$, then the integration is on the set $\mathcal{O} \times R^d$ and one introduces the condition of elastic reflection.

The physical meaning of the Vlasov equation is illustrated by the corresponding line of the table. In microvariables, in the course of the mean field limit passage, the radius of interaction of the potential increases, but the value of the interaction potential between fixed particles tends to 0 in such a way that the force of interaction between a given particle and all other particles in the system has the order ε. Therefore, a finite change of the particle position occurs during the time of the order ε. After passing to macrovariables the trajectory of every particle becomes deterministic in the mean field limit and is given by the equations

$$\begin{aligned} \dot{q}(t) &= p(t), \\ \dot{p}(t) &= -\int_{R^d \times R^d} f_t(q,p) \operatorname{grad} U(\|q - q(t)\|)\, dq\, dp. \end{aligned} \tag{10.72}$$

The Vlasov equation describes the change of the density $f_t(q,p)$ induced by the dynamics of particles of form (10.72).

For the case of a smooth potential U considered in this section, the mathematically rigorous investigation of the limit is not difficult under the assumption that the total mass of the system is finite: one requires only weak additional conditions on the initial distribution (typically, the validity of a law of large numbers). For accuracy we shall consider the motion in a bounded domain $\mathcal{O} \subset R^d$ (analogous statements are valid for the motion in the space R^d).

Theorem 6.2 (cf [Ma], [BH]). *Let a smooth nonnegative integrable function f_0 on $\mathcal{O} \times R^d$ be fixed. Suppose that the initial probability distribution P_0^ε is concentrated on the set $M_{\mathcal{O}}$ and that for any smooth function $g\colon \mathcal{O} \times R^d \to R^1$ with compact support, the random variable $\varepsilon^d \xi_g^\varepsilon(0)$* (cf (10.68)) *converges in the distribution, as $\varepsilon \to 0$,* to the integral

$$\int_{\mathcal{O} \times R^d} dq\, dp\, g(q,p) f_0(q,p).$$

Then for any $t \in [0, \infty)$ the random variable $\varepsilon^d \xi_g^\varepsilon(t)$ converges in distribution, as $\varepsilon \to 0$, to the integral

$$\int_{\mathcal{O}\times R^d} dq\,dp\,g(q,p)f_t(q,p)$$

where f_t is a solution of the Vlasov equation in the domain $\mathcal{O}$ with the initial function f_0.

Notice that the relations (10.60), (10.62) and (10.63) in the weak sense follow from the condition of Theorem 6.2.

Under the same conditions, one proves the convergence of the trajectory of a single particle to a solution of the equation (10.72).

Existence of the solution of the Vlasov equation (10.71) for all $t \in [0, \infty)$ follows from Theorem 6.2. Uniqueness of the solution was proved in papers [D7], [N] on the basis of the fact that equations (10.72) determine characteristics of the Vlasov equation and some probabilistic constructions.

The problem of the investigation of time fluctuations in the mean field limit is stated in the same way as for the low density limit, was described in Sect. 6.2. The difference is that in (10.70) and (10.71) one must replace $d-1$ by d and the linearized Vlasov equation arises instead of the linearized Boltzmann equation. The complete investigation of this problem including the proof of asymptotic normality of fluctuations was done in the paper of Braun and Hepp [BH]. Results connected with the derivation of the Vlasov equation may be extended to the Lorentz gas as well, where a linearized variant of this equation arises.

Notice however that the integrability of the initial data f_0 (finite total mass) as well as the smoothness of $U(\|q\|)$ are very essential in all the proofs. One of the difficulties arising when f_0 is assumed only to be bounded, is that for slowly decreasing potentials, the integral in the right hand side of (10.71) may diverge. However, only such potentials (for instance, the Coulomb potential $U(r) = 1/r$ in the dimension $d = 3$) are needed in most applications of the Vlasov equation. Another difficulty appears when one takes a potential which is singular near the origin.

6.4. The Landau Equation. This equation has the form

$$\begin{aligned}\frac{\partial}{\partial t} f_t(q,p) = {} & -(p, \operatorname{grad}_q f_t(q,p)) \\ & + \left[\left[\int_{R^d} dp_1\, f_t(q,p_1)\left(\sum_{j=1}^{d} \frac{\partial}{\partial p^j} a^j(p-p_1)\right)\right] f_t(q,p)\right. \\ & + \left[\left[\int_{R^d} dp_1\, f_t(q,p_1)\left(\sum_{j_1,j_2=1}^{d} \frac{\partial}{\partial p^{j_1}} D_{j_1,j_2}(p-p_1)\frac{\partial}{\partial p^{j_2}}\right) f_t(q,p).\right.\right.\end{aligned} \tag{10.73}$$

Here $a(p) = (a^1(p), \ldots, a^d(p))$ is the drift vector and $D(p) = (D_{j_1,j_2}(p), j_1, j_2 = 1, \ldots, d)$ is the diffusion matrix at a point $p \in R^d$ determined by the potential U.

The physical meaning of the weak coupling limit which leads to the Landau equation may be understood by considering the trajectory of a given particle under conditions which are written in the corresponding line of the table. In microvariables, the particle density has the order of a constant and the interaction radius does not depend on ε. Hence, it is natural to expect that, in time of the order ε^{-1}, approximately ε^{-1} "interaction acts" (collisions) of a given particle with other particles occur. The potential is multiplied by a small factor $\varepsilon^{1/2}$ and hence, during every collision act, the momentum of a particles changes a little and the dispersion of this change is proportional to ε. So, the dispersion of a general change of the momentum during time of the order ε^{-1} is of the order of a constant. It is natural to expect that the momentum of a particle will be determined in the limit by a non-linear Markov process of diffusion type. The Landau equation describes the change of the density of particle distribution whose motion is given by this non-linear Markov process.

At present, there are no mathematical results connected with the derivation of the Landau equation. As Spohn observed in [Sp1], some corollaries of results of Kesten and Papanicolau [KP] may be interpreted as a derivation of the linearized Landau equation for a Lorentz gas.

6.5. Hydrodynamic Equations. The special feature of the hydrodynamic limit passage is that, in microvariables, the radius of interaction of the potential and the density of particles (in the order of value) do not depend on ε. This fact is an imposing obstacle for rigorous investigation of the hydrodynamic limit passage. A large amount of literature is devoted to a discussion of physical and mathematical problems arising here, from which we mention the fundamental papers of Bogolyubov [B1], [B2] (see also the development of these ideas in [Zu1], [Zu2]).

We restrict ourselves to a brief discussion of the statement of the problem. As to the initial probability distribution P_0^ε it is natural to suppose, in addition to conditions of a general character (see conditions 1), 2) in Sect. 4.2), that it is locally translation invariant, i.e., "changes only a little" under space shifts of the order $o(\varepsilon^{-1})$; this agrees with the consideration of the 1-moment function of the form $f(\varepsilon q, p)$ (see the table drawn above).

One can expect that the property of local translation invariance is preserved in the course of the time evolution and moreover, in accordance with the conjectures discussed in Sect. 4.2, the probability distribution $P^\varepsilon_{\varepsilon^{-1}t}$ is approximated, as $\varepsilon \to 0$, in a neighborhood of the point $\varepsilon^{-1}q$ of the order $o(\varepsilon^{-1})$ by a Gibbs distribution with the potential U and parameters (z, β, p_0) depending on q and t. Therefore (at least in the domain of absence of phase transitions), the distribution $P^\varepsilon_{\varepsilon^{-1}t}$ may be approximately described by indicating "local" values of parameters (z, β, p_0) or corresponding "local" values of the invariants of the motion: the specific average density, specific average energy and specific average momentum (cf Sect. 4.1). Unlike the cases considered in the preceding sections, we now write kinetic equations for the set of the above listed parameters of the Gibbs distribution, not for the 1-moment function. More precisely, we connect with the probability measure P_t the following family of functions

$$G_P(q) = (\bar{n}_P(q), \bar{E}_P(q), \bar{p}_P(q)) \tag{10.74}$$

where

$$\bar{n}_P(q) = \int_{R^d} k_P^{(1)}(q,p)\,dp, \tag{10.75a}$$

$$\bar{p}_P(q) = \int_{R^d} p k_P^{(1)}(q,p)\,dp, \tag{10.75b}$$

$$\begin{aligned}\bar{E}_P(q) = &\int_{R^d} \frac{\|p\|^2}{2} k_P^{(1)}(q,p)\,dp \\ &+ \frac{1}{2}\int_{(R^d)^3} U(\|\tilde{q}\|) k_P^{(2)}((q,p_1),(q+\tilde{q},p_2))\,d\tilde{q}\,dp_1\,dp_2)\end{aligned} \tag{10.75c}$$

where $k_P^{(1)}$, $k_P^{(2)}$ are the 1- and 2-moment functions of measure P, respectively. Under the assumption that the limit for the initial family of probability distributions P_0^ε, $\varepsilon > 0$,

$$G_0(q) = \lim_{\varepsilon\to 0} G_{P_0^\varepsilon}(q), \tag{10.76}$$

exists, it is natural to expect the existence for any $t \in [0, \infty)$ of the limit

$$G_t(q) = \lim_{\varepsilon\to 0} G_{P^\varepsilon_{\varepsilon^{-1}t}}(q) \tag{10.77}$$

which gives a solution of the Euler equation for a compressible fluid with the initial date G_0.

Notice the interesting paper of Morrey [M], where an attempt to investigate the hydrodynamic limit was done in an approach closely connected with that described above. Unfortunately, Morrey introduced additional complicated conditions not only on the initial distribution P_0^ε, but on the distributions P_t^ε obtained in the course of the time evolution and the question of compatibility of these conditions remains open. One expects that, in the next asymptotic (in ε) approximation, the hydrodynamic equation contains additional terms of the order ε and describes the evolution of the system until the time of the order $\varepsilon^{-2}t$ (equations of *Navier-Stokes type*), but this circle of problems is even less clear.

We shall not discuss in detail the problems of existence and uniqueness of a solution of hydrodynamic equations, but we refer the reader to the wide literature existing here (see [Te] and the references therein).

It is worth mentioning an important direction of investigation where one takes the Boltzmann equation as the initial point for the derivation of hydrodynamic equations. These results are described in Sect. 5 of Chap. 11. Notice only that, although these results may be considered as an important stage in the foundation of the hydrodynamic limit passage for the case of a rarefied gas, there are some principal difficulties here from the point of view of the concepts developed in this section. Firstly, as was observed in Sect. 6.2, the derivation of the Boltzmann equation is performed now only for small t while, in the hydrodynamic limit passage, $t \to \infty$. Secondly, the possibility of changing the two limit passages: the low density and hydrodynamic is not clear.

Notice also an interesting attempt to derive hydrodynamic equations from Hamiltonian equations of the so-called motion of vortices [MP].

The situation is essentially simplified for the degenerate models of motion considered in Sect. 5. The simplest model where a non-trivial hydrodynamic equation arises is the model of one-dimensional hard rods. Since the invariant state $P^{r_0}_{z,\sigma}$ is determined here by its 1-moment function $k^{(1)}_{P^{r_0}_{z,\sigma}}$, the analog of the Euler equation is the following equation for the function $f_t(q,p) = \lim_{\varepsilon\to 0} k_{P^{\varepsilon}_{\varepsilon^{-1}t}}(\varepsilon^{-1}q,p)$:

$$\begin{aligned}\frac{\partial}{\partial t} f_t(q,p) = & -p\frac{\partial}{\partial q} f_t(q,p) + r_0\frac{\partial}{\partial q}\Big[f_t(q,p) \\ & \times \int_{R^1} dp_1 (p-p_1) f_t(q,p_1) \\ & \times \left(1 - r_0 \int_{R^1} dp_2 f_t(q,p_2)\right)^{-1}\Big].\end{aligned} \tag{10.78}$$

This equation was first written in the paper [Pe]. In [BDS2] it was proved that, under some sufficiently general assumptions on the initial state, the function $f_t(q,p)$ satisfies equation (10.78). In [BDS2] one proved as well the uniqueness of the solution of equation (10.78). The method used in this paper is based on the fact that the transformation described in Sect. 5.3, which reduces the one-dimensional hard rod system to the ideal gas, may also be extended to the limit equation (10.78).

Bibliography

In the monograph [B1] the hierarchy of equations is written and investigated which describes the change in the time of the moment functions of a probability measure in the course of the motion of interacting particles. A new method of derivation of kinetic equations (Boltzmann, Vlasov and Landau) from the hierarchy equations for moment functions is developed, on the basis of profound general considerations. The series of fundamental facts characterizing the process of convergence to an equilibrium state was formulated for the first time in this literature. In the paper [B2] the first derivation of hydrodynamic equations (Euler equations for a compressible ideal fluid) from the hierarchy equations for moment functions is given. Ideas of the book [B1] and paper [B2] compose the ground of modern conceptions about the connection between kinetic equations and equations which describe the motion of a large system of particles.

The paper [GO] contains a review of mathematical papers concerning the themes of the present chapter which are published mainly from 1968 to 1975. In the monograph [Zu1] the results developing the approach proposed in [B1], [B2] are presented. A review of further results in this direction is contained in [Zu2]. Papers [Ma], [MT] are devoted to results which connect kinetic equations with different types of hierarchy equations (similar to the Bogolyubov hierarchy equations) arising in the description of motion of various physical systems.

The paper [GP2] contains a systematized analysis of results on existence of a solution of the Bogolyubov hierarchy equations obtained by means of functional-analytic approach.

Papers [AGL2], [L3], [L4], which are close to this chapter in methods and style, are devoted to detailed and consequent exposition of results which are discussed in Sections 3, 4, 5.

In the paper [Gr] a limiting procedure was formulated explicitely (the low density, or Boltzmann-Grad limit) by means of which one must get the Boltzmann equation from equations describing the motion of particles (in particular, from the Bogolyubov hierarchy equations). A rigorous derivation of the Boltzmann equation from the Bogolyubov hierarchy equations in the course of the low density limit is given in the paper [L4].

Different aspects of the material presented in Chapter 10 are also discussed in reviews and papers by the authors of this chapter which partly have the character of a review ([DS], [DS1], [DS2], [S1]).

Finally, in connection with the last paragraph of this chapter, we mention papers [BDS2], [H] and the important review [Sp1] which contains the first systematized analysis of the problem of derivation of kinetic equations on the basis of a unified and precisely formulated approach.

For the convenience of the reader, references to reviews in Zentralblatt für Mathematik (Zbl.), compiled using the MATH database, have, as far as possible, been included in this bibliography.

[AGL1] Aizenman, M., Goldstein, S., Lebowitz, J.L.: Ergodic properties of an infinite one-dimensional hard rod system. Commun. Math. Phys. *39*, 289–301 (1975). Zbl. 352.60073

[AGL2] Aizenman, M., Goldstein, S., Lebowitz, J.L.: Ergodic properties of infinite systems. Lect. Notes Phys. *38*, 112–143 (1975). Zbl. 316.28008

[A] Alexander, R.: Time evolution for infinitely many hard spheres. Commun. Math. Phys. *49*, 217–232 (1976)

[AA] Arnold, V.I., Avez, A.: Ergodic problems of classical mechanics. New York—Amsterdam: Benjamin 1968, 286 p. Zbl. 167.229

[B] Beieren, H. van: Transport properties of stochastic Lorentz models. Rev. Mod. Phys. *54*, No. 1, 195–234 (1982)

[B1] Bogolyubov, N.N.: Problems of Dynamical Theory in Statistical Physics [Russian]. Moscow—Leningrad: OGIZ, Gostechnizdat 1946; see also Bogolyubov, N.N., Selected Papers, Vol. 2, 99–196. Kiev: Naukova Dumka 1970 [Russian]. Zbl. 197.536

[B2] Bogolyubov, N.N.: Equations of hydrodynamics in statistical mechanics [Ukrainian]. Sb. Tr. Inst. Mat. Akad. Nauk USSR *10*, 41–59 (1948); see also Bogolyubov, N.N., Selected Papers, Vol. 2, 258–276. Kiev: Naukova Dumka 1970 [Russian]. Zbl. 197.536

[BK] Bogolyubov, N.N., Khatset, B.I.: On some mathematical questions in the theory of statistical equilibrium. Dokl. Akad Nauk SSSR *66*, 321–324 (1949) [Russian] Zbl. 39.217, see also Bogolyubov, N.N., Selected Papers, Vol. 2, 494–498. Kiev: Naukova Dumka 1970 [Russian]. Zbl. 197.536

[BPK] Bogolyubov, N.N., Petrina, D.Ya., Khazet, B.I.: Mathematical description of an equilibrium state of classical systems on the basis of the canonical ensemble formalism. Teor. Mat. Fiz. *1*, No. 2, 251–274 (1969) [Russian]

[BDS1] Boldrighini, C., Dobrushin, R.L., Sukhov, Yu.M.: Hydrodynamics of one-dimensional hard rods. Usp. Mat. Nauk. *35*, No. 5, 252–253 (1980) [Russian]

[BDS2] Boldrighini, C., Dobrushin, R.L., Sukhov, Yu.M.: One dimensional hard rod caricature of hydrodynamics. J. Stat. Phys. *31*, 577–615 (1983)

[BPT] Boldrighini, C., Pellgrinotti, A., Triolo, L.: Convergence to stationary states for infinite harmonic systems. J. Stat. Phys. *30*, 123–155 (1983)

[BM] Botvich, D.A., Malyshev, V.A.: Unitary equivalence of temperature dynamics for ideal and locally perturbed Fermi-gas. Commun. Math. Phys. *91*, 301–312 (1983). Zbl. 547.46053

[BR1] Bratteli, O., Robinson, D.: Operator Algebras and Quantum Statistical Mechanics. I. New York–Heidelberg–Berlin, Springer-Verlag 1979. Zbl. 421.46048

[BR2] Bratteli, O., Robinson, D.: Operator Algebras and Quantum Statistical Mechanics. II. New York–Heidelberg–Berlin, Springer-Verlag 1981. Zbl. 463.46052

[BH] Braun, W., Hepp, K.: The Vlasov dynamics and its fluctuations in the 1/N limit of interacting classical particles. Commun. Math. Phys. *56*, 101–113 (1977)

[Bu] Bunimovich, L.A.: Dynamical systems with elastic reflections. Usp. Mat. Nauk. *39*, No. 1, 184–185 (1984) [Russian]

[BS1] Bunimovich, L.A., Sinai, Ya.G.: Markov partitions for dispersed billiards. Commun. Math. Phys. *78*, 247–280 (1980). Zbl. 453.60098

[BS2] Bunimovich, L.A., Sinai, Ya.G.: Statistical properties of Lorentz gas with a periodic configuration of scatterers. Commun. Math. Phys. *78*, 479–497 (1981). Zbl. 459.60099

[C] Chulaevsky, V.A.: The method of the inverse problem of scattering theory in statistical physics. Funkts. Anal. Prilozh. *17*, No. 1, 53–62 (1983) [Russian]

[CFS] Cornfeld, I.P., Fomin, S.V., Sinai, Ya.G.: Ergodic theory. New York–Heidelberg–Berlin, Springer-Verlag 1982. 486 p. Zbl. 493.28007

[DIPP] De Masi, A., Ianiro, N., Pellegrinotti, A., Presutti, E.: A survey of the hydrodynamical behaviour of many-particle systems. In: Nonequilibrium phenomena. II, pp. 123–294. Stud. Stat. Mech. 11. Amsterdam–Oxford–New York, North-Holland 1984. Zbl. 567.76006

[D1] Dobrushin, R.L.: On the Poisson law for the distribution of particles in space. Ukr. Mat. Z. *8*, No. 2, 127–134 (1956) [Russian]. Zbl. 73.352

[D2] Dobrushin, R.L.: Gibbsian random fields for lattice system with pair interaction. Funkts. Anal. Prilozh. *2*, No. 4, 31–43 (1968) [Russian]

[D3] Dobrushin, R.L.: The uniqueness problem for a Gibbsian random field and the problem of phase transitions. Funkts. Anal. Prilozh. *2*, No. 4, 44–57 (1968) [Russian]. Zbl. 192.617. English transl.: Funct. Anal. Appl. *2*, 302–312 (1968)

[D4] Dobrushin, R.L.: Gibbsian random fields. General case. Funkts. Anal. Prilozh. *3*, No. 1, 27–35 (1969) [Russian]. Zbl. 192.618. English transl.: Funct. Anal. Appl. *3*, 22–28 (1969)

[D5] Dobrushin, R.L.: Gibbsian random fields for particles without hard core. Teor. Mat. Fiz. *4*, No. 1, 101–118 (1970) [Russian]

[D6] Dobrushin, R.L.: Conditions for the absence of phase transitions in one-dimensional classical systems. Mat. Sb., Nov. Ser. *93*, No. 1, 29–49 (1974) [Russian]. Zbl. 307.60081. English transl.. Math. USSR, Sb. *22* (1974) 28–48 (1975)

[D7] Dobrushin, R.L.: Vlasov equations. Funkts. Anal. Prilozh. *13*, No. 2, 48–58 (1979) [Russian]. Zbl. 405.35069. English. transl.: Funct. Anal. Appl. *13*, 115–123 (1979)

[DF] Dobrushin, R.L., Fritz, J.: Nonequilibrium dynamics of one-dimensional infinite particle systems with a singular interaction. Commun. Math. Phys. *55*, No. 3, 275–292 (1977)

[DSS] Dobrushin, R.L., Siegmund-Schultze, R.: The hydrodynamic limit for systems of particles with independent evolution. Math. Nachr. *105*, No. 1, 199–224 (1982)

[DS] Dobrushin, R.L., Sinai, Ya.G.: Mathematical problems in statistical mechanics. Math. Phys. Rev. *1*, 55–106 (1980). Zbl. 538.60094

[DS1] Dobrushin, R.L., Sukhov, Yu.M.: On the problem of the mathematical foundation of the Gibbs postulate in classical statistical mechanics. Lect. Notes Phys. *80*, 325–340 (1978). Zbl. 439.70016

[DS2] Dobrushin, R.L., Sukhov, Yu.M.: Time asymptotics for some degenerated models of evolution of systems with infinitely many particles. In: Itogi Nauki Tekh., Ser. Sovrem. Probl. Mat. *14*, 147–254. (1979) [Russian]. Zbl. 424.60096. English transl.: J. Sov. Math. *16*, 1277–1340 (1981)

[Do] Doob, J.: Stochastic processes. Chichester–New York–Brisbane–Toronto, Wiley 1953. Zbl. 53.268

[Du] Durret, R.: An introduction to infinite particle systems. Stochast. Process Appl. *11*, No. 2, 109–150 (1981)

[FF] Fichtner, K.H., Freudenberg, W.: Asymptotic behaviour of time evolutions of infinite particle systems. Z. Wahrscheinlichkeitstheorie Verw. Gebiete. *54*, 141–159 (1980). Zbl. 438.60088

[F1] Fritz, J.: An ergodic theorem for the stochastic dynamics of quasi-harmonic crystals. In: Random fields. Rigorous results in statistical mechanics and quantum field theory. Coll. Math. Soc. J; Bolyai, 27 vol. 373–386. Amsterdam–Oxford–New York, North-Holland 1981. Zbl. 489.60066

[F2] Fritz, J.: Infinite lattice systems of interacting diffusion processes. Existence and regularity properties. Z. Wahrscheinlichkeitstheor. Verw. Gebiete. *59*, 291–309 (1982). Zbl. 477.60097

[F3] Fritz, J.: Local stability and hydrodynamical limit of Spitzer's one-dimensional lattice model. Commun. Math. Phys. *86*, 363–373 (1982). Zbl. 526.60092

[F4] Fritz, J.: Some remarks on nonequilibrium dynamics of infinite particle systems. J. Stat. Phys. *34*, 539–556 (1984). Zbl. 591.60098

[FD] Fritz, J., Dobrushin, R.L.: Nonequilibrium dynamics of two-dimensional infinite particle systems with a singular interaction. Commun. Math. Phys. *57*, 67–81 (1977)

[GLL] Gallavotti, G., Lanford, O.E., Lebovitz, J.L.: Thermodynamic limit of time dependent correlation functions for one dimensional systems. J. Math. Phys. *13*, No. 11, 2898–2905 (1972)

[GMS] Gallovotti, G., Miracle-Sole, S.: Absence of phase transition in hard-core one-dimensional systems with long-range interaction. J. Math. Phys. *11*, No. 1, 147–155 (1970)

[G] Gerasimenko, V.I.: The evolution of infinite system of particles interacting with nearest neighbours. Dokl. Akad. Nauk Ukr. SSR, Ser. A. No. 5, 10–13 (1982) [Russian]. Zbl. 486.45009

[GMP] Gerasimenko, V.I., Malyshev, P.V., Petrina, D.Ya.: Mathematical foundations of classical statistical mechanics. Kiev. Naukova Dumka 1985 [Russian]

[GP1] Gerasimenko, V.I., Petrina, D.Ya.: Statistical mechanics of quantum-classical systems. Nonequilibrium systems. Teor. Mat. Fiz. *42*, No. 1, 88–100 (1980) [Russian]

[GP2] Gerasimenko, V.I., Petrina, D.Ya.: Mathematical description of the evolution of the state of infinite systems in classical statistical mechanics. Usp. Mat. Nauk. *38*, No. 5, 3–58 (1983) [Russian]. Zbl. 552.35069. English transl.: Russ. Math. Surv. *38*, No. 5, 1–61 (1983)

[Gr] Grad, H.: Principles of the kinetic theory of gases. In: Handbuch der Physik, B. XII, Sekt. 26, pp. 205–294. Berlin, Springer-Verlag, 1958

[Gri] Griffeath, D.: Additive and cancellative interacting particle systems. Lect. Notes Math. *724* (1979). Zbl. 412.60095

[Gu] Gurevich, B.M.: Additive integrals of motion of point particles in one dimension. Math. Res. *12*, 59–63 (1982). Zbl. 515.70017

[GO] Gurevich, B.M., Oseledets, V.I.: Some mathematical problems connected with non-equilibrium statistical mechanics of an infinite number of particles. In: Probab. Teor. Mat. Stat., Teor. Kybern. *14*, 5–40. 1977 [Russian]. Zbl. 405.60093

[GSS] Gurevich, B.M., Sinai, Ya.G., Sukhov, Yu.M.: On invariant measures of dynamical systems of one dimensional statistical mechanics. Usp. Mat. Nauk. *28*, No. 5, 45–82 (1973) [Russian]. Zbl. 321.28011. English transl.: Russ. Math. Surv. *28*, No. 5, 49–86 (1973)

[GS1] Gurevich, B.M., Sukhov, Yu.M.: Stationary solutions of the Bogolyubov hierarchy equations in classical statistical mechanics. 1–4. Commun. Math. Phys. *49*, No. 1, 69–96 (1976); *54*, No. 1, 81–96 (1977); *56*, No. 3, 225–236 (1977); *84*, No. 4, 333–376 (1982)

[GS2] Gurevich, B.M., Sukhov, Yu.M.: Stationary solutions of the Bogolyubov hierarchy in classical statistical mechanics. Dokl. Akad. Nauk SSSR *223*, No. 2, 276–279 (1975) [Russian]. Zbl. 361.60090. English transl.: Sov. Math., Dokl. *16*, 857–861 (1976)

[GS3] Gurevich, B.M., Sukhov, Yu.M.: Time evolution of the Gibbs states in one-dimensional classical statistical mechanics. Dokl. Akad. Nauk SSSR. *242*, No. 2, 276–279 (1978) [Russian]. Zbl. 431.46054. English transl.: Sov. Math., Dokl. *19*, 1088–1091 (1978)

[H] Hauge, E.G.: What can one learn from Lorentz models? Lect. Notes Phys. *31*, 337–353 (1974)

[IP] Illner, R., Pulvirenti, M.: Global validity of the Boltzmann equation for a two-dimensional rare gas in vacuum. Commun. Math. Phys. *105*, No. 2, 189–203 (1986). Zbl. 609.76083

[K1] Kallenberg, O.: Random measures: London–New York–San Francisco, Academic Press., 104 p. 1976. Zbl. 345.60032; Akademie-Verlag 1975. Zbl. 345.60031

[K2] Kallenberg, O.: On the asymptotic behaviour of line processes and systems of non-interacting particles. Z. Wahrscheinlichkeitstheorie Verw. Gebiete *43*, N 1, 65–95 (1978). Zbl. 366.60086 (Zbl. 376.60059)

[KP] Kesten, H., Papanicolau, G.: A limit theorem for turbulent diffusion. Commun. Math. Phys. *65*, No. 2, 97–128 (1979). Zbl. 399.60049

[K] Kinetic theories and the Boltzmann equation. ed. C. Cercignani Lect. Notes Math. *1048*, 1984. Zbl. 536.00019

[Ko] Kozlov, O.K.: Gibbsian description of point random fields. Teor. Veroyatn. Primen. *21*, No. 2, 348–365 (1976) [Russian]. Zbl. 364.60086. English transl.: Theory Probab. Appl. *21* (1976), 339–356 (1977)

[KS] Kramli, A., Szász, D.: On the convergence of equilibrium of the Lorentz gas In: Functions, series, operators. Colloq. Math. Soc. János Bolyai *35*, vol. II. 757–766. Amsterdam–Oxford–New York, North-Holland 1983. Zbl. 538.60099

[L1] Lanford, O.E.: The classical mechanics of one-dimensional systems of infinitely many particles. I. An existence theorem. Commun. Math. Phys. *9*, 176–191 (1968). Zbl. 164.254

[L2] Lanford, O.E.: The classical mechanics of one-dimensional systems of infinitely many particles. II. Kinetic theory. Commun. Math. Phys. *11*, 257–292 (1969). Zbl. 175.214

[L3] Lanford, O.E.: Ergodic theory and approach to equilibrium for finite and infinite systems. Acta Phys. Aust., Suppl. *10*, 619–639 (1973)

[L4] Lanford, O.E.: Time evolution of large classical systems. Lect. Notes Phys. *38*, 1–111 (1975). Zbl. 329.70011

[LL] Lanford, O.E., Lebowitz, J.L.: Time evolution and ergodic properties of harmonic systems. Lect. Notes Phys. *38*, 144–177 (1975). Zbl. 338.28011

[LLL] Lanford, O.E., Lebowitz, J.L., Lieb, E.: Time evolution of infinite anharmonic systems. J. Stat. Phys. *16*, No. 6, 453–461 (1977)

[LR] Lanford, O.E., Ruelle, D.: Observable at infinitely and states with short range correlations in statistical mechanics. Commun. Math. Phys. *13*, 194–215 (1969)

[L] Lang, R.: On the asymptotic behaviour of infinite gradient systems. Commun. Math. Phys. *65*, 129–149 (1979). Zbl. 394.60098

[LS] Lebowitz, J.L., Spohn, H.: Steady self-diffusion at low density. J. Stat. Phys. *29*, 39–55 (1982)

[Le1] Lenard, A.: Correlation functions and the uniqueness of the state in classical statistical mechanics. Commun. Math. Phys. *30*, 35–44 (1973)

[Le2] Lenard, A.: States of classical statistical mechanical systems of infinitely many particles. I, II. Arch. Rat. Mech. Anal. *50*, 219–239; 241–256 (1975)

[Li] Liggett, T.M.: The stochastic evolution of infinite systems of interacting particles. Lect. Notes Math. *598*, 187–248 (1977). Zbl. 363.60109

[M] Malyshev, P.V.: Mathematical description of evolution of classical infinite system. Teor. Mat. Fiz. *44*, No. 1, 63–74 (1980) [Russian]

[MPP] Marchioro, C., Pellegrinotti, A., Pulvirenti, M.: Remarks on the existence of non-equilibrium dynamics. In: Random fields. Rigorous results in statistical mechanics and quantum field theory, Colloq. Math. Soc. Janos Bolyai *27*, vol. II, 733–746. Amsterdam–Oxford–New York, North-Holland 1981. Zbl. 496.60100

[MP] Marchioro, C., Pulvirenti, M.: Vortex methods in two-dimensional fluid dynamics. (Roma, Edizione Klim), 1984. Zbl. 545.76027. Lect. Notes Phys. *203*. Berlin etc.: Springer-Verlag III, 137 p. (1984)

[Ma] Maslov, V.P.: Equations of the self-consisted field. In: Itogi Nauki Tekh., Ser. Sovrem. Probl. Mat. *11*, 153–234. (1978) [Russian]. Zbl. 404.45008. English transl.: J. Sov. Math. *11*, 123–195 (1979)

[MT] Maslov, V.P., Tariverdiev, C.E.: The asymptotics of the Kolmogorov-Feller equation for a system of a large number of particles. In: Itogi Nauki Tekh., Ser. Teor. Veroyatn. Mat. Stat. Teor. Kibern. *19*, 85–125 (1982) [Russian]. Zbl. 517.60100. English transl.: J. Sov. Math. *23*, 2553–2579 (1983)

[MKM] Matthes, K., Kerstan, J., Mecke, J.: Infinitely divisible point processes. Chichester–New York—Brisbane–Toronto, Wiley 1978. 532 p. Zbl. 383.60001

[M1] Minlos, R.A.: The limit Gibbs distribution. Funkts. Anal. Prilozh. *1*, No. 2, 60–73 (1967) [Russian]. Zbl. 245.60080. English transl.: Funct. Anal. Appl. *1*, 140–150 (1968)

[M2] Minlos, R.A.: Regularity of the limit Gibbs distribution. Funkts. Anal. Prilozh. *1*, No. 3, 40–53 (1967) [Russian]. Zbl. 245.60081. English transl.: Funct. Anal. Appl. *1*, 206–217 (1968)

[M] Morrey, C.B.: On the derivation of the equations of hydrodynamics from statistical mechanics. Commun. Pure Appl. Math. *8*, 279–326 (1955)

[N] Neunzert, H.: The Vlasov equation as a limit of Hamiltonian classical mechanical systems of interacting particles. Trans. Fluid Dynamics *18*, 663–678 (1977)

[No] Nonequilibrium phenomena I. The Boltzmann equation. Studies in Statistical Mechanics, Z (J.L. Lebowitz, E.W. Montroll, Eds.) Amsterdam–New York–Oxford, North-Holland 1983. Zbl. 583.76004

[P] de Pazzis, O.: Ergodic properties of a semi-infinite hard rods systems. Commun. Math. Phys. *22*, 121–132 (1971). Zbl. 236.60071

[Pe] Perkus, J.K.: Exact solution of kinetics of a model of classical fluid. Phys. Fluids. *12*, 1560–1563 (1969)

[Pet] Petrina, D.Ya.: Mathematical description for the evolution of infinite systems in classical statistical physics. Locally perturbed one dimensional systems. Teor. Mat. Fiz. *38*, No. 2, 230–250 (1979) [Russian]

[PPT] Presutti, E., Pulvirenti, M., Tirozzi, B.: Time evolution of infinite classical systems with singular, long range, two body interactions. Commun. Math. Phys. *47*, 85–91 (1976)

[Pu] Pulvirenti, M.: On the time evolution of the states of infinitely extended particles systems. J. Stat. Phys. *27*, 693–709 (1982). Zbl. 511.60096

[R1] Ruelle, D.: Correlation functions of classical gases. Ann. Phys. *25*, 109–120 (1963)

[R2] Ruelle, D.: Statistical Mechanics. Rigorous results. New York–Amsterdam, Benjamin 1969. Zbl. 177.573

[R3] Ruelle, D.: Superstable interactions in classical statistical mechanics. Commun. Math. Phys. *18*, 127–159 (1970)

[SS] Siegmund-Schultze, R.: On non-equilibrium dynamics of multidimensional infinite particle systems in the translation invariant case. Commun. Math. Phys. *100*, 245–265 (1985). Zbl. 607.60097

[S1] Sinai, Ya.G.: Ergodic theory. Acta Phys. Aust., Suppl. *10*, 575–606 (1973)

[S2] Sinai, Ya.G.: Ergodic properties of a gas of one dimensional hard rods with an infinite number of degrees of freedom. Funkts. Anal. Prilozh. *6*, No. 1, 41–50 (1972) [Russian]. Zbl. 257.60036. English. transl.: Funct. Anal. Appl. *6*, 35–43 (1972)

[S3] Sinai, Ya.G.: Construction of dynamics in one dimensional systems of statistical mechanics. Teor. Mat. Fiz. *11*, No. 2, 248–258 (1972) [Russian]

[S4] Sinai, Ya.G.: Construction of cluster dynamics for dynamical systems of statistical mechanics. Vestn. Mosk. Gos. Univ. Mat. Mekh. *29*, No. 3, 152–158 (1974) [Russian]

[S5] Sinai, Ya.G.: Ergodic properties of the Lorentz gas. Funkt. Anal. Prilozh. *13*, No. 3, 46–59 (1979) [Russian]. Zbl. 414.28015. English transl.: Funct. Anal. Appl. *13*, 192–202 (1980)

[S6] Sinai, Ya.G.: The theory of phase transitions. M.: Nauka 1980 [Russian]. Zbl. 508.60084

[Sk] Skripnik, V.I.: On generalized solutions of Gibbs type for the Bogolyubov-Streltsova diffusion hierarchy. Teor. Mat. Fiz. *58*, No. 3, 398–420 (1984) [Russian]

[Sp1] Spohn, H.: Kinetic equations from Hamiltonian dynamics: Markovian limits. Rev. Mod. Phys. *52*, No. 3, 569–615. (1980)

[Sp2] Spohn, H.: Fluctuation theory for the Boltzmann equation. In: Nonequilibrium phenomena I. The Boltzmann equation. Studies in statistical mechanics; pp. 418–439. (J.L. Lebowitz, E.W. Montroll, Eds.). Amsterdam–Oxford–New York, North-Holland 1983

[Sp3] Spohn, H.: Hydrodynamical theory for equilibrium time correlation functions of hard rods. Ann. Phys. *141*, No. 2, 353–364 (1982)

[Su1] Sukhov, Yu.M.: Random point processes and DLR equations. Commun. Math. Phys. *50*, No. 2, 113–132 (1976). Zbl. 357.60056

[Su2] Sukhov, Yu.M.: Matrix method for continuous systems of classical statistical mechanics. Tr. Mosk. Mat. O.-va. *24*, 175–200 (1971) [Russian]. Zbl. 367.60117. English transl.: Trans. Mosc. Math. Soc. *24*, 185–212 (1974)

[Su3] Sukhov, Yu.M.: The strong solution of the Bogolyubov hierarchy in one dimensional classical statistical mechanics. Dokl. Akad. Nauk SSSR *244*, No. 5, 1081–1084 (1979) [Russian]. Zbl. 422.45007. English transl.: Sov. Math., Dokl. *20*, 179–182 (1979)

[Su4] Sukhov, Yu.M.: Stationary solutions of the Bogolyubov hierarchy and first integrals of the motion for the system of classical particles. Teor. Mat. Fiz. *55*, No. 1, 78–87 (1983) [Russian]

[Su5] Sukhov, Yu.M.: Convergence to a Poissonian distribution for certain models of particle motion. Izv. Akad. Nauk SSSR, Ser. Mat. *46*, No. 1, 135–154 (1982) [Russian]. Zbl. 521.60063. English transl.: Math. USSR, Izv. *20*, 137–155 (1983)

[T1] Takahashi, Y.: A class of solutions of the Bogolyubov system of equations for classical statistical mechanics of hard core particles. Sci. Papers Coll. Gen. Educ., Univ. Tokyo. *26*, No. 1, 15–26 (1976). Zbl. 402.70013

[T2] Takahashi, Y.: On a class of Bogolyubov equations and the time evolution in classical statistical mechanics. In: Random fields. Rigorous results in statistical mechanics and quantum field theory, vol. II. Amsterdam–Oxford–New York. Colloq. Math. Soc. János Bolyai, *27*, 1033–1056 (1981). Zbl. 495.28017

[Ta1] Tanaka, H.: On Markov processes corresponding to Boltzmann's equation of Maxwellian gas. In: Proc. Second Japan–USSR Sympos. Prob. Theory, Kyoto. 1972, Lect. Notes Math. *330*, 478–489 (1973). Zbl. 265.60095

[Ta2] Tanaka, H.: Stochastic differential equation associated with the Boltzmann equation of Maxwellian molecules in several dimensions. In: Stochastic analysis. London–New York San Francisco, Acad. Press 1978. 301–314. Zbl. 445.76056. Proc. int. Conf., Evanston. Ill.

[Te] Temam, R.: Navier-Stokes equations. Theory and numerical analysis. Stud. Math. Appl. *2* Amsterdam–New York–Oxford, North-Holland P.C., 1977. 500 p. Zbl. 383.35057

[U] Ueno, T.: A stochastic model associated with the Boltzmann equation. Second Japan—USSR Symp. Prob. Theory. v. 2, 183–195, Kyoto 1972

[VS] Volkovysskij, K.L., Sinai, Ya.G.: Ergodic properties of an ideal gas with an infinite number of degrees of freedom. Funkt. Anal. Prilozh. *5*, No. 3, 19–21 (1971) [Russian]. Zbl. 307.76005. English transl.: Funct. Anal. Appl. 5 (1971), 185–187 (1972)

[W] Willms, J.: Convergence of infinite particle systems to the Poisson process under the action of the free dynamics. Z. Wahrscheinlichkeitstheor. Verw. Gebiete *60*, 69–74 (1982). Zbl. 583.60096. Infinite dimensional analysis and stochastic processes, Semin. Meet., Bielefeld 1983, Res. Notes Math. *124*, 161–170 (1985)

[Z] Zemlyakov, A.N.: The construction of the dynamics in one dimensional systems of statistical physics in the case of infinite range potentials. Usp. Mat. Nauk. *28*, No. 1, 239–240 (1973) [Russian]

[Z1] Zessin, H.: Stability of equilibria of infinite particle systems. III. International Vilnius Conf. Prob. Theory Math. Stat. vol. III, 371–372, Vilnius 1981

[Z2] Zessin, H.: The method of moments for random measures. Z. Wahrscheinlichkeitstheor. Verw. Gebiete *62*, 395–409 (1983). Zbl. 503.60060

[Z3] Zessin, H.: BBGKY hierarchy equations for Newtonian and gradient systems. In: Infinite dimensional analysis and stochastic processes Semin. Meet., Bielefeld 1983, Res. Notes Math. *124* (S. Albeverio, Ed.), pp. 171–196. Boston–London–Melbourne: Pitman Adv. Publ. Program, 1985. Zbl. 583.60097

[Zu1] Zubarev, D.N.: Nonequilibrium statistical thermodynamics. Moscow, Nauka 1973 [Russian]

[Zu2] Zubarev, D.: Modern methods of statistical theory of nonequilibrium processes. Itogi Nauki Tekh. Ser. Sovrem. Probl. Mat. *15*, 131–226 (1980) [Russian]. Zbl. 441.60099. English transl.: J. Sov. Math. *16*, 1509–1572 (1981)

Chapter 13
Existence and Uniqueness Theorems for the Boltzmann Equation

N.B. Maslova

§1. Formulation of Boundary Problems. Properties of Integral Operators

1.1. The Boltzmann Equation. In general, the situation which arises when mathematically studying the Boltzmann equation may be outlined as follows. There are two well-studied limiting regimes. The first of them is the so-called free-particle flow, where the particles do not interact. The second one corresponds to thermodynamic equilibrium which is described by the Maxwell distribution. Almost all theorems which are known at present guarantee existence of a solution of boundary problems in situations which are close to one of the mentioned regimes. The only problem for which one succeeds to prove global eixtence without serious restrictions on initial deta, is the Cauchy problem for a homogeneous gas.

Flows which are close to free-particle flows are described by the non-linear Boltzmann equation with a small parameters ε in front of the collision integral (the value ε^{-1} is called the Knudsen number). Local existence theorems for non-stationary problems (such theorems are now proven under very general conditions) may be regarded as statements about solutions of the Boltzmann equation for large Knudsen numbers.

More difficult and deeper results are related to the study of those flows which are close to equilibrium. The first approximation for describing these flows is given by the linearized Boltzmann equation. Boundary problems for this equation turn out to be hard enough. However, these problems have now been studied quite well so that one can effectively use them for investigating solutions of non-linear equations. In this way one has obtained theorems which guarantee existence for stationary problems and global existence for non-stationary problems.

Below, only the simplest variant of the Boltzmann equation is considered. This is a kinetic equation which describes the motion in R^3 of a system of identical particles which interact via a pair potential depending on the distance between particles (cf Chap. 10, Sect. 2). One may include into consideration, without difficult mixtures of particles of different types which are moving, subject to extra forces.

As noted in Sect. 6 of Chap. 10, the main characteristic of a gas in kinetic theory is the 1-moment function (*Boltzmann distribution density*) which is denoted in the sequel[1] as $F(t, \xi, x)$ and describes the distribution of particles in coordinates $x\ (x \in \Omega, \Omega \subseteq R^m, m = 1, 2, 3)$ and momenta $\xi (\xi \in R^3)$. More precisely, this function describes the density of a distribution of particles in the phase space $R^3 \times \Omega$ at the time moment t. In the limit of low density, the function $F(t, \xi, x)$ must satisfy the Boltzmann equation

$$\frac{\partial}{\partial t} F + DF = J(F, F).$$

Here $D = \sum_{\alpha=1}^{m} \xi_\alpha \dfrac{\partial}{\partial x_\alpha}$, J is the *collision integral* given, at the point t, ξ, x, by the formulae:

$$\begin{aligned} J(F, F) = \int_{R^3 \times \Sigma} \Big[& F(t, \xi', x) F(t, \xi'_1, x) \\ & - F(t, \xi, x) F(t, \xi_1, x) \Big] B(|\xi - \xi_1|, |\xi - \xi_1|^{-1} |\langle \xi - \xi_1, \alpha \rangle|)\, d\alpha\, d\xi, \end{aligned}$$

$$\Sigma = \{\alpha \in R^3 \,|\, |\alpha| = 1\},$$

$$\xi' = \xi - \alpha \langle \xi - \xi_1, \alpha \rangle, \quad \xi'_1 = \xi_1 + \alpha \langle \xi - \xi_1, \alpha \rangle.$$

The non-negative function B, continuous on $(0, \infty) \times (0, 1)$, is uniquely determined by the interaction potential between particles. The main mathematical results are obtained for the class of "*hard*" *potentials*, introduced by Grad [G2]. This class is determined by the following conditions on the function B:

$$B(v, z) \leqslant b^1 z (v + v^{\beta - 1}), \tag{11.1}$$

$$\int_0^1 B(v, z)\, dz \geqslant b^0 v (v + 1)^{-1}, \tag{11.2}$$

where $\beta \in (0, 1)$ and $b^i (i = 0, 1)$ are positive constants. The potential of hard spheres (balls) (cf (10.14)) for which $B = b^1 vz$ is, of course, in this class. For the case of power potentials $U(r) = cr^{-s}$, classical mechanics gives the relation

[1] The system of notations adopted in this chapter differs somewhat from that of Chap. 10. This fact is explained by traditions established in literature in this area.

$$B(v,z) = v^p z b(z) b_1(z),\ p = 1 - 4/s, \tag{11.3}$$

where the function b is continuous on $[0,1]$ and $b_1(z) = z^{p-3}$. From the physical point of view, the singularity of the function b_1 at the origin is connected with the interaction at large distances. If this interaction is essential, then the Boltzmann equation gives a bad approximation. Replacing the function b_1 by a constant (which is called the "angular cut-off") guarantees the fulfilment of Grad's conditions for $s \geqslant 4$.

For slowly decreasing potentials ($s < 4$), the second Grad condition which guarantees boundedness from below of the frequency of collisions is violated. Such potentials are included in a class of "*soft*" *potentials*, which was introduced by Grad, and is determined by the relation (1.1) and condition

$$\int_0^1 B(v,z)\,dz \leqslant b^1(1 + v\beta^{-1}),\ \beta \in (0,1) \tag{11.4}$$

which guarantees boundedness of the frequency of collisions from above. A specific position in kinetic theory is occupied by the "Maxwell molecules" determined by the relation (11.3) for $s = 4$. This class was introduced by Maxwell because of the simplicity of the structure of corresponding moment equations. Within this class one has found nontrivial exact solutions of the Boltzmann equation (cf [B1]).

For most interaction potentials for particles in a neutral gas (cf Chap. 10, Sect. 2.4), conditions (11.1), (11.2) are fulfilled. In particular, the functions B, obtained in the course of quantum mechanical computations of collisions, satisfy these conditions. Below we assume that conditions (11.1), (11.2) are fulfilled, whenever a contrary assumption is not explicity specified.

The *Maxwell distribution*

$$\omega(\xi) \equiv \omega(\rho, h, V, \xi) = \rho(h/2\pi)^{3/2} \exp(-h/2|\xi - V|^2)$$
$$(\rho > 0, h > 0, V \in R^3)$$

for any constant values of parameters β, h, V gives an exact solution of the Boltzmann equation. A large area of physical applications and mathematical investigations is connected with the study of perturbations of the state $\omega = \omega(1,1,0,\xi)$. The perturbation f determined by the equality $F = \omega(1 + f)$ should be found from the equation

$$\frac{\partial}{\partial t} f + Df = Lf + \Gamma(f,f). \tag{11.5}$$

Here, the *linearized collision operator*, L, and the operator Γ are given by the formulae

$$\begin{aligned} \omega Lf &= J(\omega(1+f), \omega(1+f)) - J(\omega f, \omega f), \\ \Gamma(f,f) &= J(\omega f, \omega f). \end{aligned} \tag{11.6}$$

Omitting from (11.5) the nonlinear part, Γ leads to the linearized Boltzmann equation.

Let S be the boundary of a domain Ω in R^m, $n(x)$ be the internal normal vector to S at a point x. Let us denote by F^- and F^+ the densities of distributions of falling and reflected molecules on S, respectively:

$$F^+ = F\chi(\langle\xi, n(x)\rangle), F^- = F - F^+, \qquad x \in S;$$

χ is the indicator function of the interval $(0, \infty)$.

Boundary conditions on the surface S lead to a connection between the densities of distributions of falling and reflected molecules:

$$F^+ = \mathscr{R}F^- + \Phi^+, \qquad x \in S,$$

$$(\mathscr{R}F^-)(t, \xi, x) = \int_{R^3} r(t, x, \xi, \eta)F^-(t, \eta, x)\, d\eta,$$

with some given functions Φ^+, r.

In the investigation of linear problems, one usually assumes that the operator $\mathscr{R}$ admits a representation $\mathscr{R} = \mathscr{R}_0 + \mathscr{R}_1$, such that $\omega^+ = \mathscr{R}_0\omega^-$ and the function $\mathscr{R}_1\omega^- F^-$ is negligibly small. Then the boundary condition takes the form

$$F^+ = RF^- + \Phi^+$$

with $R = (\omega^+)^{-1}\mathscr{R}_0\omega^-$ and some given function Φ^+. If the domain Ω is unbounded, one adds to the boundary conditions on S the condition at infinity

$$F \to F_\infty \qquad \text{as } |x| \to \infty. \tag{11.7}$$

Hence, the following list of boundary problems may be connected with the Boltzmann equation:

1) the *internal stationary problem*

$$DF = J(F, F), x \in \Omega;\ F^+ = \mathscr{R}F^- + \Phi, x \in S; \tag{11.8}$$

considered for a bounded domain Ω;

2) the *external stationary problem*: to find functions satisfying the relations (11.8), (11.7) where Ω is the complement of the closure of a bounded domain in R^m;

3) the *nonstationary boundary problem*

$$\frac{\partial}{\partial t}F + DF = J(F, F), x \in \Omega, t \in (0, T), \tag{11.9}$$

$$F^+ = \mathscr{R}F^- + \Phi^+, x \in S, t \in (0, T), \tag{11.10}$$

$$F = F_0, x \in \Omega, t = 0; \tag{11.11}$$

4) the *Cauchy problem* (11.9), (11.11) with $\Omega = R^m$.

One formulates in an analogous way the corresponding problems for the linearized equation.

A specific role is played by problems on initial and boundary Knudsen strata (layers) for applications of the above. The first of them consists in finding a solution of the Cauchy problem independent of the coordinates x. Solutions of this problem describe a "fast" process of approaching local thermodynamical equilibrium (i.e., Maxwell distribution with variable parameters ρ, h, V) and give the possibility to find asymptotical initial data for hydrodynamic equations. The second problem is a generalization of the *Kramers problem* and consists of the description of a gas flow over a plane under the action of a given momentum flow $B_\alpha (\alpha = 1, 2, 3)$ and energy flow B_4. In the linearized stationary version, this problem is reduced to finding a function $f(\xi, x)$ satisfying the conditions:

$$\xi_1 \frac{\partial}{\partial x} f = Lf + g, x > 0, f^+ = Rf^- + \Phi^+, x = 0, \tag{11.12}$$

$$\lim_{x \to \infty} \int_{R^3} (\psi_j - \delta_{j4}(2/3)^{1/2}\psi_0) f\omega \, d\xi = B_j, j = 1, \ldots, 4; \tag{11.13}$$

$$\psi_0 = 1, \psi_\alpha = \xi_\alpha, \alpha = 1, 2, 3, \psi_4 = (|\xi|^2 - 3)6^{-1/2}. \tag{11.14}$$

In the asymptotical approach, a solution of the problem (11.12), (11.13) determines the velocity of sliding and temperature jump on the boundary of a streamlined body and thereby, the correct boundary conditions for hydrodynamic equations.

1.2. Formulation of Boundary Problems. A solution of boundary problems and properties of the collision integral are described below in terms of spaces $L_p(X)$ and $L_p(X, \varphi) = \{f | \varphi f \in L_p(X)\}$ where φ is a nonnegative weight function. The spaces $H = L_2(R^3, \omega^{1/2})$, $\mathscr{H} = L_2(R^3 \times \Omega, \omega^{1/2})$ and

$$\mathscr{H}(S) = L_2(R^3 \times S, \omega^{1/2}\varphi), \qquad \varphi = |\langle \xi, n(x) \rangle|^{1/2},$$

play the main role in the investigation of linear problems.

Notations $\langle f, g | X \rangle$ and $\| f | X \|$ are used below for the scalar product and norm in a space X.

Let us write the definition of a solution of the above-formulated boundary problems in a more precise form. For fixed ξ, x we set

$$\hat{F}(\tau) = F(\xi, x + \xi^{(m)}\tau), \tau \in R^1, \xi^{(m)} = \{\xi_1, \ldots, \xi_m\}.$$

Associated with the space $X = L_p(R^3 \times \Omega, \varphi)$ is the functional class $\mathscr{B}(X)$ consisting of the functions F which satisfy the following conditions: 1) $F \in X$, 2) $F^+ \in L_p(R^3 \times S, \varphi_1)$, $\varphi_1 = \varphi |\langle \xi, n \rangle|^{1/p}$, 3) for almost all ξ, x the function $\hat{F}$ has the generalized first order derivative and $(1 + |\xi|)^{-1} \frac{\partial}{\partial \tau} \hat{F} \in X$. Functions F from $\mathscr{B}(X)$ do not need to have all the derivatives figuring in the Boltzmann equation. However, for these functions, the value of the operator $D\left(DF = \frac{\partial}{\partial \tau} \hat{F}\right)$ and boundary value F^- are correctly defined. A solution of the stationary problem (11.8) in $L_p(R^3 \times \Omega, \varphi)$ is a function F from a corresponding class $\mathscr{B}$ satisfying relations (11.8) almost everywhere. In an analogous way, one defines a solution of the non-stationary problems in $L_p([0, T] \times R^3 \times \Omega, \varphi)$. In general, such solutions have no derivatives in t and x_α, but the action of the operator $\dfrac{\partial}{\partial \tau} + D$ is correctly defined for them.

In constructing solutions of boundary problems, one uses essentially integral forms of kinetic equations. For their description, consider an auxiliary problem

$$Df = g_1 - fg_2, \qquad x \in \Omega, \qquad f^+ = \Phi^+, \qquad x \in S, \tag{11.15}$$

where $g_i (i = 1, 2)$, Φ^+ are given functions and $g_2 \geqslant 0$. If the ray $\{y \in \Omega | y = x - \xi^{(m)}\tau, \tau > 0\}$ does not intersect the boundary S, then one adds to relation (11.15) the following condition

$$f(\xi, x - \xi^{(m)}\tau) \to 0, \qquad \tau \to \infty.$$

A solution, $f = U(g_1, g_2) + E(\Phi^+, g_2)$, of the problem (11.15) leads to the following integral form of the problem (11.6):

$$\begin{aligned} f &= VF, \qquad x \in \Omega, \qquad F^+ = \mathscr{R}F^- + \Phi^+, \qquad x \in S, \\ VF &= U(\Phi(F, F), Q(F)) + E(F^+, Q(F)), \\ \Phi(F, F) &= J(F, F) + FQ(F), \\ Q(F) &= 2\pi \int_{R^3} F(\eta, x) \left[\int_0^1 B(|\xi - \eta|, z)\, dz \right] d\eta. \end{aligned} \tag{11.16}$$

By setting

$$\nu = Q(\omega), \qquad Kf = Lf + \nu f,$$

we obtain the integral form of the stationary linear problem

$$f = UKf + Ef^+, x \in \Omega, \qquad f^+ = Rf^- + \Phi^+, \qquad x \in S. \tag{11.17}$$

One constructs analogously integral kinetic equations connected with non-stationary problems.

The problem of constructing a solution of an integro-differential equation in $\mathscr{B}(X)$ for an appropriate space X is equivalent to the problem of constructing a solution of the corresponding integral equation in X.

1.3. Properties of the Collision Integral. Investigation of properties of the collision integral J for the gas of elastic balls was done in the papers of Carleman. Below we give some bounds generalizing Carleman's results.

We set

$$\varphi = \exp\{s|\xi|^2\}(1 + |\xi|^2)^{r/2} \tag{11.18}$$

If $r = 0$, $s > 0$, then the bound for $\Phi(F, F)$ is obtained in quite a simple way: due to the equality $\Phi(\varphi^{-1}, \varphi^{-1}) = \varphi^{-1}Q(\varphi^{-1})$, one gets

$$|\varphi\Phi(F,F)| \leqslant Q(\varphi^{-1})\|F|L_\infty(R^3,\varphi)\|^2. \tag{11.19}$$

For soft potentials, one deduces from this estimate the boundedness of Φ in $L_\infty(R^3,\varphi)$. For hard potentials this estimate is not too useful because of unboundedness of $Q(\varphi^{-1})$ for large $|\xi|$.

Lemma 1.1. *If $s > 0$, $r > 2$, then the operator Φ is bounded in $L_\infty(R^3,\varphi)$* (cf [M1]).

Lemma 1.1 permits to prove the existence of a local solution for initial-boundary values problems. However, one does not succeed to prove the existence of a global solution in the class of exponentially decreasing (in ξ) functions. Therefore, following Carleman, we consider distribution functions with inverse-power decreasing.

Lemma 1.2 (cf [MC1], [MC2]). *Let φ be defined by equality* (11.18) *with $s = 0$, $r > 5$ and $B(v,z) \leqslant z(b_0 v^{1-\gamma} + b_2 v^{\varepsilon-1})$ with $\gamma \in [0,1]$, $\varepsilon \in (0,1)$. Then there exists a positive constant C such that, for all F from $L_\infty(R^3,\varphi)$, the following inequality is valid*:

$$|\varphi\Phi(F,F)| \leqslant C\|F|L_\infty(R^3,\varphi)\|^2$$
$$+ 4\pi b_0(r-2)^{-1}(1+|\xi|)^{1-\gamma}\|F|L_\infty(R^3,\varphi)\|\,\|F|L_1(R^3)\|.$$

The bound stated in lemma cannot be improved, and therefore, the operator Φ is unbounded in $L_\infty(R^3,\varphi)$. However, if $\int_0^1 B(v,z)\,dz \geqslant b_1(v^{1-\gamma}+1)$, then the operator $\Phi[Q+\alpha]^{-1}$, for large enough α and r, is bounded and moreover, has a small norm. This fact provides a basis for the proof of the existence of a global solution of the Cauchy problem for a space-homogeneous gas and local solution of some initial-boundary value problems.

Functions ψ_j given by equalities (11.14) are called *invariants of collisions*. The conservation laws for the mass, momentum and energy guarantee the fulfilment of the relations

$$\int_{R^3} \psi_j J(F,F)\,d\xi = 0 \, (j = 0,1,\ldots,4) \tag{11.20}$$

for all good enough functions F. We set

$$P_0 f = \sum_{j=0}^{4} \psi_j \langle \psi_j, f|H\rangle, \qquad Pf = f - P_0 f, \tag{11.21}$$

and denote by $P_0 H$ and PH the corresponding subspaces of the space H. Let be $\varphi_r = \omega^{1/2}(1+|\xi|)^r$. The following lemma contains a description of properties of the collision integral which guarantee the "extinguishing" of the small perturbations.

Lemma 1.3. 1) *For any real r, the operator $K = L + vI$ is continuous as an operator from $L_2(R^3,\varphi_r)$ to $L_2(R^3,\varphi_{r+1/2})$ and is completely continuous in H.*

2) *There exists a constant* $l > 0$ *such that*

$$\langle f, Lf|H\rangle \leqslant -l\|v^{1/2}Pf|H\|^2$$

for all f from $L_2(R^3, (\omega v)^{1/2})$. 3) $\langle f, Lg|H\rangle = \langle g, Lf|H\rangle$, $P_0L = LP_0 = 0$.

§ 2. Linear Stationary Problems

2.1. Asymptotics. In the investigation of boundary problems one essentially uses information about solutions of the non-homogeneous linearized equation

$$Df = \varepsilon^{-1}Lf + \varepsilon g,\ x \in R^m,\ \varepsilon \in (0, 1]. \tag{11.22}$$

The investigation of asymptotics of f, as $|x| \to \infty$ and $\varepsilon \to 0$, is closely connected with the investigation of the connection of the Boltzmann equation with hydrodynamic equations. The pressure, velocity and temperature of a gas are determined by functions $A_j = \langle \psi_j, f|H\rangle$ and so, the subspace P_0H contains, in a certain sense, the whole hydrodynamic information about the gas. A traditional method of obtaining such information from kinetic equations is to cut (in quite an arbitrary way) the infinite hierarchy equations for moments. The first equations of this hierarchy (the conservation laws) are obtained by projecting (11.22) onto the subspace P_0H. Together with the functions A_j, the higher moments figure in the system of the conservation laws: the strain (tension) tensor $\tau_{\alpha\beta} = \langle P\psi_\alpha\psi_\beta, f|H\rangle$ and the vector of the thermal flow $q_\alpha = \langle P\psi_\alpha\psi_4, fH\rangle$. We define functions φ_{α_j} by the relations

$$\varphi_{\alpha_j} = L^{-1}(P\psi_\alpha\psi_j), \qquad P_0\varphi_{\alpha_j} = 0. \tag{11.23}$$

By scalar multiplying (11.22) in H by the functions φ_{α_j} we get

$$\begin{aligned}\varepsilon^{-1}\tau_{\alpha\beta} &= -\mu\left(\frac{\partial}{\partial x_\alpha}A_\beta + \frac{\partial}{\partial x_\beta}A_\alpha\right) + 2/3\mu\delta_{\alpha\beta}\sum_{\gamma=1}^{m}\frac{\partial}{\partial x_\gamma}A_\gamma \\ &\quad + \sum_{\gamma=1}^{m}\frac{\partial}{\partial x_\gamma}\langle\varphi_{\alpha\beta}\xi_\gamma, Pf|H\rangle - \varepsilon\langle\varphi_{\alpha\beta}, g|H\rangle,\end{aligned} \tag{11.24}$$

$$\begin{aligned}\varepsilon^{-1}q_\alpha &= -\lambda\frac{\partial}{\partial x_\alpha}A_4 + \sum_{\gamma=1}^{m}\frac{\partial}{\partial x_\gamma}\langle\varphi_{\alpha 4}\xi_\gamma, Pf|H\rangle \\ &\quad -\varepsilon\langle\varphi_{\alpha 4}, g|H\rangle,\end{aligned} \tag{11.25}$$

where μ, λ are positive coefficients which have the sense of coefficients of viscosity and thermal conductivity.

If one throws away the two last terms in the right hand side of (11.25), (11.24), these formulae give Newton and Fourier laws, correspondingly. The system of conservation laws completed by formulae (11.25), (11.24) makes it possible to get an exact estimate for P_0f in terms of initial and boundary data of the problem under consideration and to describe asymptotic properties of f. (Notice that the functions $\varphi_{\alpha j}$ are polynomials in ξ only for Maxwell molecules. In general, the system of moment equations does not include the relations (11.25), (11.24).)

It is convenient to formulate the main result in terms of the Fourier transform $\tilde{f}$ (in the variables x) of the function $\tilde{f}$. The function $\tilde{f}$ has to satisfy the equation

$$\tilde{D}\tilde{f} = \varepsilon^{-1}L\tilde{f} + \varepsilon\tilde{g}, \qquad \tilde{D} = i\sum_{\alpha=1}^{m} k_\alpha\xi_\alpha, \varepsilon \in (0,1]. \tag{11.26}$$

The structure of the power decomposition in ε

$$\tilde{f} = \sum_{n\geqslant 0} \varepsilon^n \tilde{f}_n \tag{11.27}$$

was investigated, for the nonlinear nonstationary problem, by Hilbert. Formal substitution of the Hilbert series (11.27) into (11.22) gives the equations

$$Lf_0 = 0, \qquad Lf_{n+1} = Df_n - g\delta_{n1}. \tag{11.28}$$

The condition $P_0(Df_n - g\delta_{n1}) = 0$ of existence of a solution of equation (11.28) is equivalent to the equations of linear hydrodynamics. It is not difficult to check that $\tilde{f}_n = 0(|k|^{n-2})$ for fixed ξ and $k \to 0$.

Theorem 2.1 (cf [M9], [M10]). *If $\tilde{g} \in H$, then the equation* (11.26) *has, for $k \neq 0$, a unique solution in H. If $(1 + |\xi|)^{n-1}g \in H$, then for $n \geqslant 1$ this solution admits the representation*

$$\tilde{f} = \sum_{l=0}^{n} \varepsilon^l \tilde{f}_l + \varepsilon^{n+1}\tilde{F}_n,$$

$$\|\tilde{F}_n|H\| \leqslant C(1 + |k|)^{n-1}\|(1 + |\xi|)^{n-1}\tilde{g}|H\|$$

where C is a positive constant independent of k and ε.

2.2. Internal Problems. Consider the problem of finding a function f satisfying the conditions

$$Df = Lf, \qquad x \in \Omega, \qquad f^+ = Rf^- + \Phi^+, \qquad x \in S, \tag{11.29}$$

in a bounded domain Ω. The first result concerning a solution of problem (11.29) was obtained by Grad [G1]. Grad proved the existence and uniqueness of a solution of problem (11.29) for $m = 1$, $R = 0$, $\Omega = (0, d)$ and small enough d. Guiraud investigated conditions of existence of a solution of problem (11.29) for the gas of elastic balls (cf [Gu1]).

Lemma 1.3 and Theorem 2.1 give a possibility of obtaining the following a priori estimates for a solution of problem (11.29) in $\mathscr{H}$:

$$\tfrac{1}{2}\|f^-|\mathscr{H}(S)\| + l\|v^{1/2}Pf|\mathscr{H}\|^2 \leqslant \tfrac{1}{2}\|f^+|\mathscr{H}(S)\|^2,$$

$$\|P_0 f|\mathscr{H}\| \leqslant C[\|Pf|\mathscr{H}\| + \|f|\mathscr{H}(S)\|].$$

These estimates guarantee the uniqueness of a solution of the problem in $\mathscr{H}$ provided $\|R|\mathscr{H}(S)\| < 1$. To prove its existence, we consider the system (11.17).

Lemma 2.1. *The operator UK is completely continuous in $\mathscr{H}$.*

From Lemma 2.1, one easily deduces the Grad result and even the following, much more general theorem:

Theorem 2.2. *If $\Phi^+ \in \mathscr{H}(S)$, $\|R|\mathscr{H}(S)\| < 1$, then problem* (11.29) *has a unique solution in $\mathscr{H}$.*

The lack of conservation laws for the interaction of molecules with the boundary is guaranteed by the condition $\|R|\mathscr{H}(S)\| < 1$. However, in applications, one often encounters situations where this condition is not fulfilled. Let us set for $x \in S$, $\varphi = |(\xi, n(x))|$,

$$\Pi_0 f^{\pm} = (\varphi, f^{\pm}|H)(\varphi, 1^{\pm}|H)^{-1}, \qquad \Pi f^{\pm} = f^{\pm} - 1^{\pm}\,\Pi_0 f^{\pm}.$$

Suppose the operator R satisfies the following conditions (cf [M1], [Gu1], [Gu3]):

$$R1^- = 1^+, \qquad R\Pi = \Pi R, \qquad \|R\Pi|\mathscr{H}(S)\| < 1. \tag{11.30}$$

From the physical point of view, these conditions mean that, for the interaction with the boundary, the law of conservation of mass is fulfilled but the others are not valid.

Theorem 2.3 (cf [M1]). *Let $\Phi^+ \in \mathscr{H}(S)$ and conditions* (11.30) *be fulfilled. Then the fulfilment of the condition $\langle \Phi^+, 1|\mathscr{H}(S)\rangle = 0$ is necessary and sufficient for the existence of a solution of problem* (11.29) *in $\mathscr{H}$. Moreover, the difference of any two solution is constant.*

2.3. External Problems. Usually, the external problem is formulated as being the problem of finding a function f satisfying the conditions (11.29) and the relation

$$f \to f_\infty, \qquad |x| \to \infty. \tag{11.31}$$

One supposes that the domain $R^m \backslash \bar{\Omega}$ is bounded, $m \geqslant 2$. Experts in classical mechanics are usually unanimous about the questions of existence and uniqueness of a solution of problems they formulated. This is no longer the situation for external stationary problems for the Boltzmann equation. The review [Gu3] shows how contradictory opinions were announced in connection with these problems, on the basis of "physical" considerations.

To start off, it is convenient to leave the traditional statement of the problem and to consider the problem of constructing a function f satisfying the conditions (11.29) and the relation $f \in \mathscr{H}_k$, where

$$\mathscr{H}_k = \{f|(1+|x|)^{-k} f \in \mathscr{H}\}.$$

The condition $f \in \mathscr{H}_k$ gives a restriction on the rate of increase of the function f, as $|x| \to \infty$. In particular, for $k \leqslant m/2$, this condition means that f converges, in some sense, to zero, as $|x| \to \infty$. We set $\hat{\psi}_j = \psi_j - \delta_{j4}(2/3)^{1/2}$ and

$$Q_j = \langle \hat{\psi}_j \operatorname{sgn}\langle \xi, n(x) \rangle, f | \mathscr{H}(S) \rangle,$$

$$p = \langle \psi_0 + \psi_4 (2/3)^{1/2}, f | H \rangle,$$

$$f_0 = \sum_{j=0}^{4} a_j(x) \hat{\psi}_j, \qquad a_0 = C_0,$$

$$a_\alpha = C_\alpha + \sum_{\beta=1}^{3} Q_\beta H_{\alpha\beta}, \qquad a_4 = C_4 + Q_4 H,$$

where C_j are constants, $\{H_{\alpha\beta}\}$ is a matrix of fundamental solutions of the Stokes system, H is a fundamental solution of the Laplace equation. Constants Q_j determine the force and thermal flow acting on the "body" $R^m \backslash \Omega$, function p determines the pressure corresponding to the distribution f.

Under sufficiently general assumptions about the operator R, the following theorem holds.

Theorem 2.4 (cf [M5], [M10]). *If $\Phi^+ \in \mathscr{H}(S)$, then for any given constants B_j, the problem* (11.29) *has a unique solution satisfying the condition*

$$p - B_0 \in \mathscr{H}_1, \qquad Q_j = B_j, \qquad j \geqslant 1.$$

This solution admits the representation $f = f_0 + f_1$ where $f_1 \in \mathscr{H}_\varepsilon$ for any positive ε.

Theorem 2.4 gives a possibility of examining conditions of existence of a solution of the external problems in the traditional statement. If $m = 3$, then, due to the theorem, one has $f - f_0 \in \mathscr{H}_1$ for any Q_j. Hence, one gets the unique possible condition at infinity:

$$f_\infty = \sum_{j=0}^{4} C_j \hat{\psi}_j. \tag{11.32}$$

The arbitrariness in the selecting of constants B_j prescribed by Theorem 2.4 provides the possibility of determining in an arbitrary way constants C_j (i.e. the pressure, velocity and temperature of the gas at infinity). This fact allows us to prove the following theorem.

Theorem 2.5 (cf.[M5], [M9]). *If $m = 3$, $\Phi^+ \in \mathscr{H}(S)$, then for any constants C_j the problem* (11.29) *has a unique solution in $\mathscr{H}_2$.*

The situation is changed for the plane case. If the force and thermal flow differ from zero, then the solution of the problem (11.29) increases like $\ln|x|$, as $|x| \to \infty$. In order to fulfil condition (11.32) or even only a condition of boundedness of the solution, one needs to put $Q_j = 0$. Hence, due to Theorem 2.4, one obtains the result that a phenomenon which is completely analogous to Stokes' paradox in hydrodynamics takes place for the plane problem: it is impossible to define in an arbitrary way the velocity and temperature of a gas at infinity, because these parameters are uniquely determined by the requirement of boundedness of the solution.

2.4. Kramers' Problem. The problem (11.12)–(11.13) was widely discussed in connection with the applications we mentioned above (cf [G3], [Gu3]). In particular, Grad [G3] formulated conjectures about conditions of existence and uniqueness of a solution of this problem and asymptotic properties of solutions. Set

$$f_0 = \sum_{j=0}^{4} C_j \hat{\psi}_j + \sum_{j=2}^{4} B_j(\varphi_{1j} + x\hat{\psi}_j) < \varphi_{1j}, L\varphi_{1j} | H >^{-1}, \tag{11.33}$$

where φ_{1j} are the functions determined by the relations (11.23).

Theorem 2.6 (cf [M4], [M8]). *If* $g = 0$, $R = 0$, $|\xi_1|^{1/2}\Phi^+ \in II$, *then for any* ℓ *from* $[2, +\infty)$ *the problem* (11.12)–(11.13) *has a unique solution in* $\mathscr{H}_\ell$. *This solution admits the representation* $f = f_0 + f_1$ *where* $f_1 \in \mathscr{H}$ *and* f_0 *is determined by formula* (11.33) *with* $B_0 = C_0$.

Constants $C_j (j \geqslant 2)$ determine the velocity of sliding and temperature jump. Theorem 2.6 remains true under sufficiently general restrictions onto the function g and operator R (cf [M4], [M8]). It is natural to consider this theorem as a confirmation and precision of the Grad conjecture.

§ 3. Nonlinear Stationary Problems

In situations near equilibrium, it is possible to construct a solution of the stationary problem (11.8) in a bounded domain Ω by methods of iterations taking, as first approximation, the corresponding solution of the linearized equation. Guiraud [Gu1] proved this way the existence and uniqueness of a solution of the problem (11.8) for the gas of elastic balls. Theorems 2.2 and 2.3 give a possibility to obtain an analogous result for all hard potentials under conditions about the form of the boundary and the law of interaction between molecules and the boundary which are more general than in the Guiraud paper (cf [G]).

The external stationary problem, in particular the problem of describing a flow around a body, is more difficult. Let us consider the problem (11.7), (11.8) with $F_\infty = \omega(1, 1, V, \xi)$, $V = \{|V|, 0, 0\}$. The Mach number M is connected with V by the equality $M = (3/5)^{1/2}|V|$. By setting $f = \omega^{-1}(F - \omega)$, $\omega = F_\infty$, $v = \xi - V$ and assuming that the function F^+ is given, we obtain the following reformulation of the problem (11.7), (11.8):

$$\sum_{\alpha=1}^{3} (|V|\delta_{\alpha 1} + v_\alpha)\frac{\partial}{\partial x_\alpha} f = Lf + \Gamma(f, f), \qquad x \in \Omega, \tag{11.34}$$

$$f^+ = \Phi^+, \qquad x \in S, \qquad f \to 0, \qquad |x| \to \infty. \tag{11.35}$$

In situations near equilibrium the velocity V and the nonlinear term in (11.34) are small. By omitting these terms one gets the problem, whose solution has Stokes' asymptotics, as $|x| \to \infty$, due to Theorem 2.4. However, as in hydrodynamic problems, the corresponding method of iterations leads to the appearance of secular terms for large $|x|$.

Set

$$\|f|B_p(\Omega)\| = \sup(1 + |\xi|)^3 \omega^{1/2} \|f|L_p(\Omega)\|,$$

$$\|f|B_p(S)\| = \sup(1 + |\xi|)^4 \omega^{1/2} \|f|L_p(S)\|$$

and denote by $B_p(\Omega)$, $B_p(S)$ the corresponding Banach function spaces.

Let be $\Phi^+ \in B_\infty(S)$, $B_{p,\infty} = B_p(\Omega) \cap B_\infty(\Omega)$. Since functions f from $B_{p,\infty}$ are bounded, the index p measures the velocity of decrease of f, as $|x| \to \infty$ (the smaller p is, the faster the decrease).

Theorem 3.1 (cf [M6]). *Let f be a solution of the problem* (11.34), (11.35) *in $B_{p,\infty}$, $p < 12/5$. Then f admits the representation*

$$f(\xi, x) = \sum_{j=0}^{4} a_j(x)\hat{\psi}_j(v) + \hat{f}(\xi, x), \qquad (v = \xi - V),$$

$$\hat{f} \in B_2(\Omega), \qquad a_j \in L_\beta(\Omega), \qquad \beta > 2.$$

Functions a_j give the main terms for the asymptotics of the hydrodynamic parameters (pressure, velocity, temperature) at large distances from the boundary. Exact formulae for the Fourier transform of a_j show that these functions decrease, as $|x| \to \infty$, more slowly, of course, than functions from L_2. For computing the functions a_j, one needs only information about the flows Q_j. In particular, for $M < 1$,

$$a_\alpha(x) = \sum_{\beta=1}^{3} Q_\beta \hat{H}_{\alpha\beta}(x), \qquad \alpha = 1, 2, 3,$$

where $\{\hat{H}_{\alpha\beta}\}$ is the matrix of fundamental solutions of Oseen's system of equations. For subsonic flows (below the chock barrier) the functions a_j decrease like $|x|^{-2}$ outside the trace behind the body. For $M > 1$, along with the trace some new domains of slow decreasing of the functions a_j appear which are connected with Mach's cone.

Theorem 3.1 has a conditional character: one supposes there the existence of a solution of the problem (11.34), (11.35) from $B_{p,\infty}$. The following theorem shows that the problem (11.34), (11.35) may in fact be solved for small Φ^+.

Theorem 3.2 (cf [M6]). *There exists a positive number δ with the following property: if $\|\Phi^+|B_\infty(S)\| < \delta$, then the problem* (11.34), (11.35) *has a unique solution in $B_{p,\infty}$, $p \in (2, 12/5)$ satisfying the condition $\|f|B_{p,\infty}\| \leqslant C\delta$.*

From Theorem 3.2 it follows that the problem (11.34), (11.35) has a unique solution for the diffusive reflection if Mach's number is small.

The uniqueness of the solution of stationary problems may be proven also for another limiting case, for the case of a strongly rarefied gas (cf [M2], [M3]). In this case the integral operator determined by the equality (11.16) is a contraction in a suitable functional space. The rate of convergence of the iterations $F^{(n)} = VF^{(n-1)}$ depends on the dimension of the space. In particular, for $m = 3$, $\|F - F^{(n)}\| = 0(\varepsilon^n)$ and for $m = 1$, $\|F - F^{(n)}\| = O\left(\left(\varepsilon \ln \frac{1}{\varepsilon}\right)^n\right)$.

§ 4. Non-Stationary Problems

4.1. Relaxation in a Homogeneous Gas. The problem of describing the relaxation is to find a function $F(t, \xi)$ satisfying the following conditions

$$\frac{\partial}{\partial t} F = J(F, F), \qquad t \in (0, T), \qquad F = F_0, \qquad t = 0. \tag{11.36}$$

The first results concerning a solution of the problem (11.36) were obtained by Carleman for the gas of elastic balls.

The theorems we formulate below (cf [MC1], [MC3]) are the generalizations of Carleman's theorems to the class of potentials satisfying the following conditions

$$B(v, z) \leqslant b^1 z(v^{1-\gamma} + 1), \int_0^1 B(v, z)\, dz \geqslant b^0 v^{1-\gamma}, \gamma \in [0, 1). \tag{11.37}$$

Let

$$\varphi = (1 + |\xi|)^r, \qquad r \geqslant r_0, \qquad r_0 = 2 + \max\{3, 8b^1(b^0)^{-1}\}. \tag{11.38}$$

Theorem 4.1. *If $F_0 \in L_\infty(R^3, \varphi)$, then for all T the problem* (11.36) *has a unique solution in $L_\infty([0, T] \times R^3, \varphi)$. This solution satisfies the condition*

$$\sup_t \|F|L_\infty(R^3, \varphi)\| < \infty.$$

Denote by ω the Maxwell distribution with the parameters satisfying the condition $\langle \psi_j, F_0 - \omega | L_2(R^3) \rangle = 0$.

Theorem 4.2. *Let the condition of Theorem* 4.1 *be fulfilled and $F_0 \in C(R^3)$. The solution of the problem* (11.36) *is continuous in ξ uniformly with respect to t and satisfies the relation $\sup_\xi |F(t, \xi) - \omega| \to 0$, $t \to \infty$.*

For applications, it is important to have an estimate of the rate of approach to the equilibrium state ω. For sufficiently general initial conditions, so far one has such estimates only for the case of the Maxwell molecules (cf [B2], [V]). By reducing drastically the class of initial distributions, Grad proved the exponentially rapid convergence of solutions of the problem (11.36) to the equilibrium. Let $F = \omega(1 + f)$, and

$$\varphi = (1 + |\xi|)^3 \omega^{1/2}, \qquad N(f) = \|f | L_\infty(R^3, \varphi)\|.$$

Grad's theorem asserts that there exists a positive constant β for which $\sup_t N(f)\exp(\beta t) < \infty$ provided the norm $N(f)$ is sufficiently small for $t = 0$. However, there is no reason to hope to drop the condition on $N(f)$ for $t = 0$. Moreover, Bobylev's results show that for the Maxwell molecules and compact support initial data the perturbation may leave the space H in finite time (cf [B2], [B3]).

Notice also that from Povzner's results one can deduce that the global uniqueness for the problem (11.36) is preserved for a much wider class of initial distributions (cf [P], [M1]). However, for conditions of Povzner's theorem one does not even succeed to show the boundedness of F for large t.

4.2. The Cauchy Problem. Now let us consider the Cauchy problem in a general case:

$$\frac{\partial}{\partial t} F + DF = J(F, F), \qquad t \in (0, T), \qquad F = F_0, \qquad t = 0. \tag{11.39}$$

Under weak restrictions on the initial distribution it is possible to prove the existence of a unique solution of the problem (11.39) on the time interval $(0, T)$ the length of which depends on initial data (cf [M1]). In particular, the following theorem holds.

Theorem 4.3 (cf [MC4]). *If* $F_0 \in L_\infty(R^6, \varphi)$, $F_0 \geqslant 0$, *and conditions* (11.38), (11.37) *are fulfilled, then there exists a positive number* T_1 *such that for* $T \leqslant T_1$ *the problem* (11.39) *has a unique solution in* $L_\infty([0, T] \times R^6, \varphi)$.

If φ is determined by the formula (11.18), then for $s > 0$, $r > 2$ the local uniqueness of a solution of the problem (11.39) in $L_\infty(R^6, \varphi)$ takes place for all potentials satisfying the first Grad condition (11.1) (cf [M1]). For the potentials satisfying the condition (11.4) the problem (11.39) has a locally unique solution in

$$L_\infty([0, T] \times R^6, \varphi) \quad \text{for } s > 0, r = 0.$$

For each case mentioned the solution may be obtained as the limit of a sequence of iterations in an integral kinetic equation. The proof of convergence of the iterations is based on the estimates contained in Lemmas 1.1, 1.2. It is not hard to prove that the solution is nonnegative for nonnegative F_0 and has the same smoothness in x, ξ as the initial distribution.

Now consider the problem (11.39), supposing that the initial distribution is close to the Maxwell distribution $\omega = \omega(1, 1, 0, \xi)$. In this situation it is possible to prove the existence of a unique solution of the problem on the infinite time interval. Let us set

$$\|f | B_{p,\infty}\| = \|f | B_p(R^3)\| + \|f | B_\infty(R^3)\|,$$

$$\|f | \hat{B}_p\| = \sup_{t \in [0, T]} \|f | B_{p,\infty}\|.$$

Theorem 4.4 (cf [M7], [MF1], [MF2]). *There exists a positive number δ such that, if $\|f_0|B_{2,\infty}\| < \delta$, then for all T the problem* (11.39) *has a unique solution satisfying the condition $\omega^{-1}(F - \omega) \in \hat{B}_2$. Moreover, the following estimate is valid*:

$$\sup_{t \geqslant 0} \|\omega^{-1}(F - \omega)|B_{2,\infty}\| < \infty.$$

The function F has the same smoothness in x as the initial distribution. Under additional assumptions about smoothness of F_0 and velocity of decrease of f_0, as $|x| \to \infty$, it is possible to obtain estimates of the rate of approach to the equilibrium state ω. In particular, if f_0 and its derivatives in x up to the second order are contained in $B_{1,\infty}$, then

$$\|f|B_{2,\infty}\| \leqslant C(1 + t)^{-3/4}, \qquad \|f|H\| \leqslant C(1 + t)^{-3/2} \text{ (cf [M1])}.$$

The difference h between the solutions of the nonlinear and linearized Boltzmann equations satisfies the condition $\|h|B_{2,\infty}\| \leqslant C(1 + t)^{-5/4}$. In the paper [KMN] similar estimates for the difference between the moments (ψ_j, f) and solutions of Navier-Stokes equations were obtained. An analogous theorem for soft potentials was obtained in the paper [UA1].

4.3. Boundary Problems. The existence of unique solutions of nonstationary boundary problems may be proven under sufficiently general assumptions about initial and boundary conditions (cf [M1], [MC4]).

As well as for the Cauchy problem, it is possible to obtain global theorems only in situations which are close to equilibrium ones. The first results in this direction were obtained in the papers [F], [U] where a domain Ω is a cube with the normally reflecting boundary. For this case the boundary problem is reduced to the Cauchy problem with periodic initial conditions.

In a general case such a reduction is, of course, impossible but for homogeneous boundary conditions one reduces the solution of the nonstationary problem, as well as the solution of the Cauchy problem, to the investigation of the equation

$$f(t) = T(t)f_0 + \int_0^t T(t - \tau)\Gamma(f(\tau), f(\tau))\, d\tau,$$

where $T(t)f_0$ is the solution of the nonstationary boundary problem for the linearized Boltzmann equation.

A solution of this equation requires information about properties of the semi-group $T(t)$. Guiraud [Gu2] investigated its properties for the gas of elastic balls contained in a bounded domain with a smooth boundary. Using some additional assumptions of a technical character, he proved exponential bounds for the decrease of $T(t)$ for growing t and the existence of a unique solution of the initial-boundary value problem (11.9)–(11.11). In Guiraud's proof, his unpublished result is essentially used for the complete continuity of the tenth degree of the operator UK figuring in the integral Boltzmann equation (11.17). From Lemma 2.1 the compactness of the first degree of this operator follows for all hard potentials. This fact and the above formulated theorems about solutions of linear stationary problems allow us to obtain estimates for the semi-group $T(t)$ and furthermore to use them for the investigation of nonlinear problems. As an example of such an application, we consider the problem (11.9)–(11.11) with $\Phi^+ = \omega(1,1,0,\xi)$, $\mathscr{R} = 0$. Assuming $f = \omega^{-1}(F - \omega)$ we obtain the following formulation of this problem:

$$\frac{\partial}{\partial t} f = -Df + Lf + \Gamma(f,f), \qquad t \in (0,T), \qquad x \in \Omega, \tag{11.40}$$

$$f^+ = 0, \qquad t \in (0,T), \qquad x \in S, \qquad f = f_0, \qquad t = 0, \qquad x \in \Omega. \tag{11.41}$$

If Ω is a bounded domain with a smooth boundary S, then the following theorem is valid for all potentials satisfying the conditions (11.1), (11.2).

Theorem 4.5 (cf [M7]). *There exists a positive constant δ such that if $\|f_0|B_\infty(\Omega)\| < \delta$, then for all T the problem* (11.40), (11.41) *has a unique solution in $\hat{B}_\infty$. This solution satisfies the condition*

$$\sup_t \|f|B_\infty(\Omega)\| \exp\{\sigma t\} \leqslant C\|f_0|B_\infty(\Omega)\|$$

with some positive constant σ.

§ 5. On a Connection of the Boltzmann Equation with Hydrodynamic Equations

5.1. Statement of the Problem. Investigation of the connection of the Boltzmann equation with the equations of hydrodynamics is one of the classical problems of statistical physics. In papers of Maxwell, for the first time in the literature, an infinite chain of equations for the moments of the Boltzmann distribution function appeared and the problem of the foundation of hydrodynamics was formulated as a problem of "closing" this chain. From the Boltzmann equation and the equality (11.20) it follows that the moments $M_j = \int_{R^3} \psi_j F\, d\xi$ $(j = 0, 1, \dots 4)$ must satisfy the system of equations

$$\frac{\partial}{\partial t} M_j = -\sum_{\alpha=1}^{3} \frac{\partial}{\partial x_\alpha} T_{\alpha_j}, \qquad T_{\alpha_j} = \int_{R^3} \xi_\alpha \psi_j F\, d\xi. \tag{11.42}$$

Equations (11.42) are the first five equations of the above-mentioned chain. In hydrodynamics this chain is closed by means of the Newton and Fourier laws which express the moments T_{α_j} in terms of the functions M_j and their space derivatives. One should expect that, in situations close to equilibrium, a "synchronisation" of higher moments is rapidly taking the place which leads at a possibility of a simplified description. However, the physical and mathematical nature of this process is not clear so far.

One of the directions of investigations is related to the study of the asymptotics of a solution of the Cauchy problem for the Boltzmann equation

$$\frac{\partial}{\partial t} F^{\varepsilon} + DF^{\varepsilon} = \varepsilon^{-1} J(F^{\varepsilon}, F^{\varepsilon}), \qquad F^{\varepsilon}|_{t=0} = F^{\varepsilon}(0) \qquad (x \in \Omega,\, t \in [0, T]) \quad (11.43)$$

with the large parameter ε^{-1} in front of the collision integral, which corresponds to the hydrodynamic limit passage discussed in Chap. 10, Sect. 6. This direction has its own meaningful history which is connected with the names of Hilbert and Carleman. Hilbert described a structure of formal power expansions $F^{\varepsilon} = \sum_{n \geqslant 0} \varepsilon^n F_n$. The main term of the Hilbert series, F_0, is a Maxwell distribution whose moments satisfy the Euler equations. The Hilbert paradox discussed in 20-th is that, to construct all terms of the Hilbert series, one needs only information on hydrodynamic moments $M_j (j = 0, \ldots, 4)$ of an initial distribution. The resolution of the paradox is that the Hilbert series (whenever it converges) guarantees the validity of initial conditions only for initial data of a very special form.

The existence theorems formulated in Sect. 4 do not give any background for discussing the hydrodynamic limit passage since local theorems (in particular Theorem 4.3) guarantee the existence of a solution on a time interval of order ε only, and global theorems allow us to consider only small perturbations (of order ε).

In the papers of Japanese mathematicians [N], [UA2], theorems are obtained which give a foundation for the main term of the Hilbert series for initial data which are close to a Maxwell distribution with constant parameters and are analytic in x. The time of existence of a solution of the Cauchy problem guaranteed by these theorems depends on a norm of the initial perturbation in the corresponding Banach space of analytic functions and decreases when this norm is growing.

In fact, solutions of the Cauchy problem (11.43) exist at least on an interval where Kato's theorem guarantees the existence of solutions of Euler equations in Sobolev spaces. The precise formulation of this statement is given below, in Theorem 5.1. The only essential condition for an application of this theorem is a closeness of $F^{\varepsilon}(0)$ to a local Maxwell distribution. However, it is sufficient to suppose a closeness in L_{∞} without introducing any restriction on gradients of the hydrodynamic momenta M_j.

One of the corollaries of Theorem 5.1 is the possibility to give the complete *foundation* for the Hilbert method: for smooth initial data the partial sum of the Hilbert series differs from the exact solution by a quantity of order $\varepsilon^n + e^{-t/\varepsilon}$ (cf Theorem 5.2).

As to global solutions of the problem (11.43), one succeeds in constructing them only for small initial perturbations (of order ε in W_2^2) of hydrodynamic moments. The principal term of the asymptotics is determined by a solution of Navier-Stokes equations.

5.2. Local Solutions. Reduction to Euler Equations. Before passing to the discussion of the problem (11.43), we consider the Cauchy problem for the Euler equation. Let $M^{(F)} = (M_0^{(F)}, \dots, M_4^{(F)})$ be the vector of the hydrodynamic momenta of a function F. The density ρ, velocity V and temperature θ are connected with $M^{(F)}$ by the relations

$$M_0^{(F)} = \rho, \quad M_j^{(F)} = \rho V_j (j = 1, 2, 3), \sqrt{6} M_4^{(F)} = \rho V^2 + 3\rho(\theta - 1). \quad (11.44)$$

Euler equations admit the following representation

$$\frac{\partial}{\partial t} M(t) + \sum_{j=1}^{3} A_j(M(t)) \frac{\partial}{\partial x_j} M(t) = 0 \quad (11.45)$$

where A_j are smooth matrix functions. If, for $t = 0$, the condition

$$M_j(0) - \delta_{j0} \in W_2^l(\Omega),\ l \geqslant 2,\ \inf \rho > 0,\ \inf \theta > 0. \quad (11.46)$$

holds, then there exists a time interval $[0, T]$ such that the problem (11.45) has a unique solution in $L_\infty([0, T], W_2^l(\Omega))$ satisfying the condition: $\inf \rho > 0$, $\inf \theta > 0$.

By $\omega = \omega(M, \xi)$ we denote the Maxwell distribution with the momenta M. If the condition (11.46) is valid, then there exists a Maxwell distribution $\bar{\omega}$ with constant parameters such that

$$\omega(M(t), \xi) \leqslant \bar{\omega}. \quad (11.47)$$

Given a distribution function F for which the condition (11.46) holds, we denote by $M^{(F)}(t)$, $t \in [0, T]$, the local solution of (11.45) with the initial data $M^{(F)}$.

Now let us turn to the problem (11.43), assuming for simplicity that $\Omega = R^3$ and the initial distribution does not depend on ε. Principal statements which are given below remain true under the assumption that the initial function F^ε is a polynomial as well as for the case of where Ω is a torus. Denote by $H_m(\omega)$ the Hilbert space of functions F with the norm

$$\|F | H_m(\omega)\| = \| \|F | W_2^m(\Omega)\| | L_2(R^3, \omega^{-1/2})\|. \quad (11.48)$$

We shall consider initial data which belong to the set

$$\begin{aligned} \mathbb{B}(m, \alpha, c) = \{F | M_j^{(F)} - \delta_{j0} \in W_2^m(\Omega),\ \inf \rho > 0,\ \inf \theta > 0, \\ (1 + |\xi|)^\alpha (F - \omega) \in H_m(\omega),\ \|F - \omega | L_2(R^3, \omega^{-1/2})\| \leqslant C\} \quad (11.49) \end{aligned}$$

Assume that the function B which characterizes the interaction potential satisfies, in addition to (11.1), (11.2), the following condition uniformly in v:

$$b_2 |v|^p \leqslant \left(\int_0^1 B^2(v, z)\, dz \right)^{1/2} \leqslant b_3(1 + |v|^p), \qquad p \geqslant 0, \qquad b > 0. \quad (11.50)$$

This inequality holds, in particular, for the hard spheres as well as for power potentials $U(r) = cr^{-s}$, $s \geqslant 4$, with an "angular cut-off".

It follows from the inequality (11.50) that, in a gas with the distribution ω, for the collision frequency ν (cf Sect. 1) and the operator Γ, the following estimates hold uniformly in x, t

$$\nu_1 \leqslant \nu(1 + |\xi|)^{-p} \leqslant \nu_2 \qquad (0 < \nu_1 < \nu_2),$$

$$\|h_\beta \nu^{-1}\Gamma(f_1, f_2)|H\| \leqslant \gamma(\|h_\beta f_1|H\|\,\|f_2|H\| + \|f_1|H\|\,\|h_\beta f_2|H\|), \quad H = L_2(R^3, \omega^{1/2}),$$

$$h_\beta = (1 + |\xi|)^\beta, \quad h_\beta f_j \in H, \quad \beta \geqslant 0, \quad \gamma > 0.$$

The bounds from Lemma 1.3 are valid, too, uniformly in x, t.

Theorem 5.1. *Let $F \in \mathbb{B}(m, \alpha, c)$, $m \geqslant 5$, $\alpha \geqslant 4$. Then there exists a positive constant $c^* = c^*(l, \gamma)$ such that for $c \leqslant c^*$ the problem* (11.43) *has for all ε a unique solution on an interval $[0, T]$ satisfying the conditions: $h_\beta(F - \omega) \in L_\infty([0, T], H_{m-3}(\bar{\omega}))$, and*

$$\sup_{x,t} \|F - \omega|L_2(R^3, \bar{\omega}^{-1/2})\| \leqslant (2\gamma)^{-1}.$$

A sketch of the proof is as follows. Let $S_0 = S_0(n, \varepsilon)$ be a partial sum of the Hilbert series. For smooth initial data the method of Hilbert allows us to reduce equation (11.43) to the non-homogeneous Boltzmann equation

$$\left(\frac{\partial}{\partial t} + D\right)\Phi = 2\varepsilon^{-1}J(S_0, \Phi) + \varepsilon^{-1}J(\Phi, \Phi) + R_0 \tag{11.51}$$

with the function R_0 of order ε^n. The initial value of Φ under assumptions of the theorem does not need to be small. We introduce into equation (11.51) a "rapid" time $\tau = \varepsilon^{-1}t$ and consider a formal power expansion $\Phi(\tau) = \sum_{n \geqslant 0} \varepsilon^n \Pi_n(\tau)$. The function $\Pi_0 + F_0(0)$ must be a solution of the Carleman problem (11.36). For $n > 0$, the functions Π_n are solutions of linear equations corresponding to the problem (11.36). The Hilbert series terms and functions Π_n are uniquely determined by the following conditions

$$\Pi_n(0) + F_n(0) = F(0)\delta_{n0}, \quad \Pi_n(\tau) \to 0, \quad \tau \to \infty. \tag{11.52}$$

By setting $S = \sum_{j=0}^{n} \varepsilon^j(\Pi_j + F_j)$, $G = F^\varepsilon - S$, we obtain the following problem for the function G

$$\left(\frac{\partial}{\partial t} + D\right)G = 2\varepsilon^{-1}J(S, G) + \varepsilon^{-1}J(G, G) + R, \quad G\Big|_{t=0} = 0,$$

with the function R satisfying the condition $\|R|L_2([0, T], H_s(\omega))\| \leqslant C\varepsilon^n$, $s \leqslant m - 2n - 1$, uniformly in ε.

To prove a possibility of applying a perturbation theory to the problem (11.51), one necessarily needs uniform in ε estimates for solutions of the linear problem

$$\left(\frac{\partial}{\partial t} + D\right)G = 2\varepsilon^{-1}J(S, G) + X, \quad G\Big|_{t=0} = 0, \tag{11.53}$$

where X is a given function from $L_2([0, T], H_s(\bar{\omega}))$. Therefore, a key role in the proof of Theorem 5.1 is played by the following lemma.

Lemma 5.1. *The problem* (11.52) *has a unique solution in* $L_\infty([0,T], H_s(\varpi))$. *This solution satisfies the inequality*

$$\|G\|_\infty \leqslant C(T)\|v^{-1/2}X\|_2$$

uniformly in ε *where* $C(T)$ *is a positive constant and*

$$\|G\|_p = \|G|L_p([0,T], H_0(\varpi)\|, \qquad p \in [1,\infty].$$

The main role in the proof of this lemma is played by the bounds for the operator L which are given by Lemma 1.3, together with a rapid convergence of solutions of the Carleman problem (11.36) to a Maxwell distribution. In the course of the proof one uses as well some ideas of the paper [C].

Lemma 5.1 allows to obtain the bound

$$\|F^\varepsilon - S\|_\infty \leqslant C(T)\varepsilon^{n+1/2}, \alpha \geqslant \max\{n,4\}, \quad m \geqslant 2n+3. \tag{11.54}$$

From the bound (11.54) and Theorem 5.1 the following theorem is easily deduced.

Theorem 5.2. *Let the assumptions of Theorem* 5.1 *be valid with* $\alpha \geqslant n$ *and* $m \geqslant 2n+3$. *Then the solution of the problem* (11.43) *admits the representation*

$$F^\varepsilon = \sum_{j=0}^{n-1} \varepsilon^j F_j + R(n,\varepsilon).$$

The function $R(n,\varepsilon)$ *obeys*

$$\|R(n,\varepsilon)|H_0(\varpi)\| \leqslant C(\varepsilon^h + e^{-t/\varepsilon})$$

uniformly in ε, t.

5.3. A Global Theorem. Reduction to Navier-Stokes Equations. Let us return to the problem of evolution of small perturbations of the state $\omega = \omega(1,1,0,\xi)$ which are described by equation (11.5). The hydrodynamic parameters are uniquely determined by the function $P_0 f$: $M_j = \delta_{j0} + (\psi_j, P_0 f)$. Hence, the problem of closing the moment equations (11.42) may be formulated as a problem of describing the dependence of Pf on $P_0 f$. Equation (11.5) is equivalent to the system of equations

$$\frac{\partial}{\partial t} P_0 f = -P_0 Df, \qquad f = P_0 f + Pf, \tag{11.55}$$

$$\frac{\partial}{\partial t} PS = -PDf + \varepsilon^{-1} LPf + \varepsilon^{-1}\Gamma(f,f). \tag{11.56}$$

Let us denote by Y Banach space of functions φ with the norm

$$\|\varphi\|_Y = \|\varphi|L_\infty([0,T], H_m)\| + \|v^{1/2}\varphi|L_2([0,T], H_m)\|$$

where $H_m = H_m(\omega)$ is the space introduced in Sect. 5.2, and $m \geqslant 2$. We shall seek a solution of the Cauchy problem for equation (11.56) in the ball $S_1 = \{\varphi$:

$\|\varphi\|_Y \leqslant a_1\}$ for given values $Pf(0)$ and $P_0 f$ from the ball $S_0 = \{(Pf(0), P_0 f) \in H_m \times Y \colon \|Pf(0)|H_m\|^2 + \|P_0 f|L_\infty([0,T], H_m)\|^2 \leqslant a_0^2, m \geqslant 2\}$.

We denote by V the operator which associates with the point $(Pf(0), P_0 f)$ from S_0 a solution of equation (11.56).

Theorem 5.3. *There exist positive constants* a_0, a_1 such that for any ε *and* T *the Cauchy problem for equation* (11.56) *has a unique solution in the ball* S_1. *The operator* V *satisfies uniformly in* x *and* T *the condition*

$$\|V(P\varphi_1, P_0 f) - V(P\varphi_2, P_0 f)|H_m\| \\ \leqslant C\exp\{-ct/\varepsilon\}\|P(\varphi_1 - \varphi_2)|H_m\|, \; C, c > 0, (P\varphi_j, P_0 f) \in S_0.$$

From Theorem 5.3 it follows that the Boltzmann equation is reduced to a (non-local in x, t) system of equations for hydrodynamic moments. When $t\varepsilon^{-1} \to \infty$, the operator V is well-approximated by a local one. More precisely, for smooth and bounded uniformly in ε initial data and the moments M, it follows from Theorem 5.3 that the solution of the problem (11.43) satisfies the condition

$$F^\varepsilon = \omega(M, \xi) + O(\varepsilon) + O(e^{-ct/\varepsilon})$$

where, as above, $\omega(M, \xi)$ is the Maxwell distribution with the momenta M. If perturbations of the moments $(M_j - \delta_{jo})$ have an order ε, then the function Pf admits the representation

$$Pf = \varepsilon L^{-1} PDP_0 f + \Phi, \tag{11.57}$$

$$\Phi = P\omega^{-1}(\omega(M, \xi) - \omega) + \varphi_0 e^{-ct/\varepsilon} + O(\varepsilon^3) + O(\varepsilon^2 e^{-t/\varepsilon}), \tag{11.58}$$

where φ_0 is a function bounded in H_m uniformly in ε, t and depending on the initial date f only. This is the only situation where one succeeds to construct a global solution for the moments.

Denote by $E(t)$ the group in H generated by the operator $P_0 DP_0$ and by $N_\varepsilon(t)$ the semi-group in H generated by the operator $P_0 D(P_0 + \varepsilon L^{-1} PDP_0)$. The operators $E(t)$ and $N_\varepsilon(t)$ determine solutions of linear Euler and Navier-Stokes equations. The problem of constructing the moments is reduced, due to Theorem 5.3 and the relations (11.57), (11.58), to solving the equation

$$E(-t)P_0 f = U(P_0 f) + \bar{N}_\varepsilon(t) P_0 f(0) \tag{11.59}$$

where $\bar{N}_\varepsilon(t) = E(-t)N_\varepsilon(t)$ is the semigroup introduced in [EP] and the operator U is defined by

$$Uf = \int_0^t \bar{N}_\varepsilon(t - \tau)\Phi(\varphi(\tau), \tau)\,d\tau.$$

Let

$$\|\varphi\|_{Y_0}^2 = \|\varphi|L_\infty([0,T], H_m)\|^2 + \varepsilon\|\nabla_x \varphi|L_2([0,T], H_m)\|^2.$$

We shall consider initial data satisfying the condition

$$\varepsilon^{-1}\|P_0 f(0)|H_m\| + \|(1+|\xi|)^2 Pf(0)|H_{m+2}\| \leqslant C_0 \qquad (m \geqslant 2).$$

If $P_0\varphi \in Y_0$, then, uniformly in T, ε,

$$\|U(P_0\varphi)|Y_0\| \leqslant C(\varepsilon + \varepsilon^{-1}\|E(-t)P_0\varphi|Y_0\|^2)$$

where C is a constant depending on $\|P_0\varphi|Y_0\|$.

Theorem 5.4. *There exist constants* c_0, c_1 such that the Cauchy problem for the system (11.55), (11.56) *has, for any* ε *and* T, *a unique solution satisfying the condition*

$$\varepsilon^{-1}\|P_0\varphi|Y_0\| + \|Pf|Y\| \leqslant C_1.$$

Replacing the function Φ by the sum of the first two terms from the right hand side of (11.58), corresponds to passing from the Boltzmann equation to non-linear Navier-Stokes equations. An error generated by such a replacement is of order ε. More precisely, let $\varphi = \varepsilon^{-1}E(-t)P_0 f$ and denote by φ_N the described above Navier-Stokes approximation of this function. Under assumptions of Theorem 5.4 the following relations holds true

$$\sup_T \|\varphi - \varphi_N|Y_0\| = O(\varepsilon).$$

Bibliography

Results on existence, uniqueness and properties of solutions of various problems for the Boltzmann equation which were obtained before 1978 are presented in the paper [M1]. The new results are contained in the papers of Soviet [B2], [G], [M4]–[M10], [MC4], [V] and foreign [A1], [AEP], [U], [UA1] authors. See also the review papers in the issue [No] and references therein. We particularly mention the classical papers of Grad [G1]–[G3] which have been the starting points for many contemporary investigations.

The papers [C], [EP], [KMN], [N], [UA2] contain results of foreign authors about the connection between the Boltzmann equation and equations of hydrodynamics (see also [No]). In [L] bounds for the Boltzmann collision integral are obtained. The papers [BMT], [MT] concern the derivation of the Boltzmann equation from stochastic dynamical models.

For the convenience of the reader, references to reviews in Zentralblatt für Mathematik (Zbl.), compiled using the MATH database, have, as far as possible, been included in this bibliography.

[A1] Arkeryd, L.: On the Boltzmann equation in unbounded space far from equilibrium, and the limit of zero-mean free path. Commun. Math. Phys. *105*, 205–219 (1986). Zbl. 606.76094

[AEP] Arkeryd, L., Esposito, R., Pulvirenti, M.: The Boltzmann equation for weakly inhomogeneous data. Preprint 1986

[BMT] Belavkin, V.P., Maslov, V.P., Tariverdiev, S.E.: Asymptotic dynamics of a system with a large number of particles described by the Kolmogorov—Feller equations. Teor. Mat. Fiz. *49*, No. 3, 298–306 (1981) [Russian]

[B1] Bobylev, A.V.: Exact solutions of the Boltzmann equation. Dokl. Akad. Nauk SSSR *225*,

No. 6, 1296–1299 (1975) [Russian]. English transl.: Sov. Phys., Dokl. *20* (1975), 822–824 (1976). Zbl. 361.76082

[B2] Bobylev, A.V.: Asymptotic properties of solutions of the Boltzmann equation. Dokl. Akad. Nauk SSSR. *261*, No. 5, 1099–1104 (1981) [Russian]

[B3] Bobylev, A.V.: Exact solutions of the nonlinear Boltzmann equation and the theory of relaxation of a Maxwell gas. Teor. Mat. Fiz. *60*, No. 2, 280–310 (1984) [Russian]. Zbl. 565.76074. English transl.: Theor. Math. Phys. *60*, 820–841 (1984)

[C] Caflish, R.: The fluid dynamical limit of the nonlinear Boltzmann equation. Commun. Pure Appl. Math. *33*, 651–666 (1980). Zbl. 424.76060

[EP] Ellis, R., Pinsky, M.: The first and second fluid approximations to the linearized Boltzmann equation. J. Math. Pures Appl., IX. Ser. *54*, 125–156 (1975). Zbl. 297.35066

[F] Firsov, A.N.: On a Cauchy problem for the nonlinear Boltzmann equation. Aehrodinamika Razr. Gasov. *8*, 22–36 (1976) [Russian]

[G] Gejnts, A.G.: On the solvability of the boundary value problem for the nonlinear Boltzmann equation. Aehrodin. Razr. Gazov. *10*, 16 24 (1980) [Russian]. Zbl. 493.76073

[G1] Grad, H.: High frequency sound according to the Boltzmann equation. SIAM J. Appl. Math. *14*, 935–955 (1966). Zbl. 163.232

[G2] Grad, H.: Asymptotic theory of the Boltzmann equation. II. Rarefield Gas Dynamics *1*, 25–59 (1969)

[G3] Grad, H.: Singular and non-uniform limits of solutions of the Boltzmann equation. Transport Theory, New York 1967. SIAM AMS Proc. *1*, 269–308 (1969). Zbl. 181.285

[Gu1] Guiraud, J.P.: Problème aux limites intérieur pour l'équation de Boltzmann en régime stationnaire, faiblement non-lineaire. J. Méc, Paris *11*, No. 2, 183–231 (1972). Zbl. 245.76061

[Gu2] Guiraud, J.P.: An H-theorem for a gas of rigid spheres in a bounded domain. Theor. Cinet, class. relativ., Colloq. int. CNRS 236, Paris 1974, 29–58 (1975). Zbl. 364.76067

[Gu3] Guiraud, J.P.: The Boltzmann equation in kinetic theory. A survey of mathematical results. Fluid Dynamics Trans. 7, Part II, 37–84 (1976)

[KMN] Kawashima, S., Matsumura, A., Nishida, T.: On the fluid dynamical approximation to the Boltzmann equation at the level of the Navier-Stokes equation. Commun. Math. Phys. *70*, 97–124 (1979). Zbl. 449.76053

[L] Lukshin, A.V.: On a property of the collision integral. Zh. Vychisl. Mat. Mat. Fiz. *25*, No. 1, 151–153 (1985) [Russian]. Zbl. 586.76140. English transl.: USSR Comput. Math. Math. Phys. *25*, No. 1, 102–104 (1985)

[MT] Maslov, V.P., Tariverdiev, S.E.: The asymptotics of the Kolmogorov-Feller equation for a system of a large number of particles. In: Itogi Nauki Tekhn., Ser. Teor. Veroyatn. Mat. Stat. Teor. Kibern. *19*, 85–125. (1982) [Russian]. Zbl. 517.60100. English transl.: J. Sov. Math. *23*, 2553–2579 (1983)

[M1] Maslova, N.B.: Theorems on the solvability of the nonlinear Boltzmann equation. Complement II to the Russian translation of the book: Cercignani C.: Theory and applications of the Boltzmann equation. Edinburgh-London, Scottish Academic Press 1975. Zbl. 403.76065

[M2] Maslova, N.B.: Stationary problems for the Boltzmann equation for large Knudsen numbers. Dokl. Akad. Nauk SSSR *229*, No. 3, 593–596 (1976) [Russian]. Zbl. 355.45012. English translation: Sov. Phys., Dokl. *21*, 378–380 (1976)

[M3] Maslova, N.B.: The solvability of stationary problems for the Boltzmann equation for large Knudsen numbers. Zh. Vychisl. Mat. Mat. Fiz. *17*, 1020–1030 (1977) [Russian]. Zbl. 358.35067. English transl.: USSR Comput. Math. Math. Phys. *17* (1977), No. 4, 194–204 (1978)

[M4] Maslova, N.B.: Stationary solutions of the Boltzmann equation and the Knudsen boundary layer. [Aehrodinamika Razr. gazov *10*, 5–15 (1980)] [Russian]. Zbl. 493.76074. Molecular gas dynamics, Interuniv. Collect., Aerodyn. Rarefied Gases *10*, Leningrad 1980, 5–15 (1980)

[M5] Maslova, N.B.: Stationary solutions of the Boltzmann equations in unbounded domains. Dokl. Akad. Nauk SSSR *260*, (M 4), 844–848 (1981) [Russian]. Zbl. 522.76073. English transl.: Sov. Phys., Dokl. *26*, 948–950 (1981)

[M6] Maslova, N.B.: Stationary boundary value problems for the nonlinear Boltzmann equation. Zap. Nauchn. Semin. Leningr. Otd. Mat. Inst. Steklova 110. *115*, 100–104 (1981) [Russian]. Zbl. 482.76073. English transl.: J. Sov. Math. *25*, 869–872 (1984)

[M7] Maslova, N.B.: Global solutions of nonstationary kinetic equations. Zap. Nauchn. Semin. Leningr. Otd. Mat. Inst. Steklova *115*, 169–177 (1982) [Russian]. Zbl. 493.35027. English transl.: J. Sov. Math. *28*, 735–741 (1985)

[M8] Maslova, N.B.: The Kramers problem in kinetic theory of gases. Zh. Vychisl. Mat. Mat. Fiz. *22*, No. 3, 700–704 (1982) [Russian]. Zbl. 502.76089. English transl.: USSR Comput. Math. Math. Phys. *22*, No. 3, 208–209 (1982)

[M9] Maslova, N.B.: Stationary solutions of the linearized Boltzmann equation. Tr. Mat. Inst. Steklova *159*, 41–60 (1983) [Russian]. Zbl. 538.76069. English transl.: Proc. Steklov Inst. Math. *159*, 41–60 (1984)

[M10] Maslova, N.B.: Exterior stationary problems for the linearized Boltzmann equation. Aehrodinamika Razr. Gasov. *11*, 143–165 (1983) [Russian]

[MC1] Maslova, N.B., Chubenko, R.P.: Limiting properties of solutions of the Boltzmann equation. Dokl. Akad. Nauk SSSR *202*, No. 4, 800–803 (1972) [Russian]

[MC2] Maslova, N.B., Chubenko, R.P.: Estimations of the integral of collisions. Vestn. Leningr. Univ. No. *13*, 130–137 (1973) [Russian]

[MC3] Maslova, N.B., Chubenko, R.P.: Relaxation in a mono-atomic space-homogeneous gas. Vestn. Leningr. Univ. No. *13*, 90–97 (1976) [Russian]. Zbl. 373.45008

[MC4] Maslova, N.B., Chubenko, R.P.: On solutions of the nonstationary Boltzmann equation. Vestn. Leningr. Univ. No. *19*, 100–105 (1973) [Russian]. Zbl. 278.35075

[MF1] Maslova, N.B., Firsov, A.I.: The solution of the Cauchy problem for the Boltzman equation. I. The theorem of existence and uniqueness. Vestn. Leningr. Univ. No. *19*, 83–88 (1975) [Russian]. Zbl. 325.76105

[MF2] Maslova, N.B., Firsov, A.I.: The solution of the Cauchy problem for the Boltzmann equation. II. Estimations of solutions of the inhomogeneous linearized equation. Vestn. Leningr. Univ. No. *1*, 97–103 (1976) [Russian]. Zbl. 371.76062

[N] Nishida, T.: Fluid dynamic limit of the nonlinear Boltzmann equation to the level of the compressible Euler equation. Commun. Math. Phys. *61*, 119–148 (1978). Zbl. 381.76060

[No] Nonequilibrium phenomena I. The Boltzmann equation. Studies in Statistical Mechanics, X (J.L. Lebowitz, E.W. Montroll, Eds.). Amsterdam–New York–Oxford, North-Holland 1983. Zbl. 583.76004

[P] Povzner, A.Ya.: On the Boltzmann equation in kinetic theory of gases. Mat. Sb., Nov. Ser. *58*, No. 1, 65–86 (1962) [Russian]. Zbl. 128.225

[U] Ukai, S.: On the existence of global solutions of mixed problem for non-linear Boltzmann equation. Proc. Japan Acad., Ser. A *50*, 179–184 (1974). Zbl. 312.35061

[UA1] Ukai, S., Asano, K.: On the Cauchy problem of the Boltzmann equation with a soft potential. Publ. Res. Inst. Math. Sci., Kyoto Univ. *18*, 477–519 (1982). Zbl. 538.45011

[UA2] Ukai, S., Asano, K.: The Euler limit and initial layer of the non-linear Boltzmann equation. Hokkaido Math. J. *12*, 311–332 (1983). Zbl. 525.76062

[V] Vedenyapin, V.V.: Anisotropic solutions of the nonlinear Boltzmann equation for the Maxwell molecules. Dokl. Akad. Nauk SSSR *256*, No. 2, 338–342 (1981) [Russian]

Subject Index

Printing: Weihert-Druck GmbH, Darmstadt
Binding: Buchbinderei Schäffer, Grünstadt